FDA and USDA Nutrition Labeling Guide

Decision Diagrams, Checklists, and Regulations

TRACY A. ALTMAN, Ph.D.
Jetra, Inc.

TECHNOMIC
PUBLISHING CO., INC.
LANCASTER · BASEL

Ref
KF
1620
.F66
A95
1998

FDA and USDA Nutrition Labeling Guide
a**TECHNOMIC** publication

Technomic Publishing Company, Inc.
851 New Holland Avenue, Box 3535
Lancaster, Pennsylvania 17604 U.S.A.

Copyright ©1998 by Technomic Publishing Company, Inc.
All rights reserved

No part of this publication may be reproduced, stored in a
retrieval system, or transmitted, in any form or by any means,
electronic, mechanical, photocopying, recording, or otherwise,
without the prior written permission of the publisher.

Printed in the United States of America
10 9 8 7 6 5 4 3 2

Main entry under title:
 FDA and USDA Nutrition Labeling Guide: Decision Diagrams, Checklists, and Regulations

A Technomic Publishing Company book
Bibliography: p.

Library of Congress Catalog Card No. 98-85935
ISBN No. 1-56676-706-7

HOW TO ORDER THIS BOOK
BY PHONE: 800-233-9936 or 717-291-5609, 8AM–5PM Eastern Time
BY FAX: 717-295-4538
BY MAIL: Order Department
Technomic Publishing Company, Inc.
851 New Holland Avenue, Box 3535
Lancaster, PA 17604, U.S.A.
BY CREDIT CARD: American Express, VISA, MasterCard
BY WWW SITE: http://www.techpub.com

PERMISSION TO PHOTOCOPY–POLICY STATEMENT
Authorization to photocopy items for internal or personal use, or the internal or personal use of specific clients, is granted by Technomic Publishing Co., Inc. provided that the base fee of US $3.00 per copy, plus US $ .25 per page is paid directly to Copyright Clearance Center, 222 Rosewood Drive, Danvers, MA 01923, USA. For those organizations that have been granted a photocopy license by CCC, a separate system of payment has been arranged. The fee code for users of the Transactional Reporting Service is 1-56676/98 $5.00 + $ .25.

To my husband, Jeff ·
Special thanks also to my production assistant,
Ray Bohn, for his steadfast support

Table of Contents

Part A: Overview of Labeling Rules + Compliance Checklists

Chapter 1	**Agency Rulemaking Processes**	1.1
	Basic Labeling Concepts	1.3
	History of U.S. Nutrition Labeling Requirements and Current Status	1.4
Chapter 2	Glossary of Acronyms + Compliance Checklists	2.1

Part B: Nutrition Labeling Fundamentals

Chapter 3	**Applicability and Labeling Basics**	3.1
	Exemptions — FDA	3.3
	Exemptions — FSIS	3.6
	Formatting Requirements: Principal Display and Information Panels	3.7
	Alternative Formats	3.11
	Serving Size and Reference Amounts	3.16
	Voluntary Programs for Raw, Single-Ingredient Products	3.18
	USDA Prior Approval	3.19
	Restaurant Foods	3.20
Chapter 4	**Mandatory Nutrient Declarations**	4.1
	Voluntary Nutrients	4.3
	Declaring by Weight	4.4
	Daily Reference Value (DRV)	4.7
	Reference Daily Intake (RDI)	4.8
	Demonstrating Compliance	4.10

Part C: Claims Made on Labels

Chapter 5	**Nutrient Content Claims**	5.1
	Currently Authorized Claims	5.2
	Decision Diagram for Making a Claim	5.4
	"Relative" Claims	5.8
	Displaying a Nutrient Content Statement	5.9
	Requirements for Claiming a Specific Nutrient	5.10
	Claiming "Good Source," "High," "More," or "Lean"	5.12
	Conflicts with Standards of Identity	5.13
	FSIS Labeling Applications	5.14
	Requests for More Regulatory Flexibility	5.16
	FDA Modernization Act of 1997	5.20
Chapter 6	**Health Claims**	6.1
	FDA Modernization Act of 1997	6.2
	Claims Currently Authorized by FDA	6.3
	General Requirements for Making Claims	6.5
	The "Jelly Bean" Rule	6.6
	Petitioning for a Claim Authorization	6.7
	Medical Foods	6.9
	Recently Authorized Claims	6.11
	Prospects for Future Rule Changes	6.15
	Requirements for All Currently Approved Claims	6.18

Table of Contents

Part C: Claims Made on Labels – continued

Chapter 7	**Implied and Descriptive Claims**	7.1
	Using "Healthy" on Product Labels	7.3
	Recent Changes to the "Healthy" Requirements	7.4

Part D: Dietary Supplements

Chapter 8	**Nutrition Labeling Under the DSHEA**	8.1
	FDA's 1998 Proposal Concerning Structure/Function Claims	8.16
	Comparison of Allowed Structure/Function Statements to Prohibited Drug/Disease Claims	8.19

Part E: Regulatory Text

Chapter 9	FDA Nutrition Labeling Regulations	9.1
Chapter 10	FSIS Rules for Labeling Meat Products	10.1
Chapter 11	FSIS Rules for Labeling Poultry Products	11.1

Appendix:

- References for Key *Federal Register* Notices — A.1
- Forthcoming Actions Planned by FDA & FSIS — A.4
- Bibliography — A.5

Index:

- FDA Requirements — I.1
- FSIS Meat Product Requirements — I.21
- FSIS Poultry Product Requirements — I.25

List of Tables

1.1	Location of Key FDA and USDA/FSIS Regulations
2.1	Definitions of Common Acronyms
3.1	Labeling Exemptions Provided in the FDA Rules
3.2	Labeling Exemptions Provided in the USDA FSIS Rules
3.3	Footnotes Required by FDA and FSIS
3.4	Special Labeling Provisions Under FDA Program
3.5	FDA & FSIS Rules for Determining Serving Size
3.6	Use of Household Measures and Metric Equivalents
4.1	Required Nutrients (with Allowed Synonyms)
4.2	Nutrients that May Be Declared Voluntarily
4.3	Rules for Declaring Nutrients by Weight
4.4	Daily Reference Values
4.5	Reference Daily Intakes for Vitamins and Minerals
4.6	Rounding Rules for Declaring Vitamins and Minerals
4.7	Criteria for Demonstrating Compliance with FDA & FSIS Rules
5.1	Nutrient Content Claims Authorized by FDA & FSIS
5.2	Rules for Displaying Nutrient Content Claims
5.3	Requirements for Claiming Specific Nutrients
5.4	Requirements for Claiming Good Source, High, More, Light, Lean
5.5	FDA Actions on Dairy Product Standards
6.1	Health Claims Authorized or Prohibited by FDA
6.2	Disqualifying Nutrient Levels Under the Health Claim Rules
6.3	Criteria for Specific Health Claims
7.1	Rules for Making Implied Nutrient Content Claims
7.2	Criteria for "Healthy" Claims Under FDA & FSIS Rules
8.1	Primary Provisions of FDA's Dietary Supplement Rules
8.2	(b)(2)-Dietary Ingredients Required on Dietary Supplement Labels
8.3	Allowed Structure/Function Claims and Prohibited Disease References
A.1	Key FDA & FSIS Nutrition Labeling Rules Published Since NLEA
A.2	Forthcoming Actions Planned by FDA & FSIS

List of Figures

2.1	Sample "Nutrition Facts" Label
2.2	Nutrition Labeling Development Checklist
2.3	Nutrition Facts Labeling Revision History
3.1	Applicability of FDA & FSIS Nutrition Labeling Programs
3.2	Sample Nutrition Facts Label with Required Type Sizes
3.3	Basic Formatting Rules for Nutrition Labeling
3.4	FSIS and FDA Rules for Alternative Labeling Formats
3.5	Model Exemption Notice for Low-Volume Products
5.1	General Requirements for Making Nutrient Content Claims
5.2	Requirements for Making "Relative" Nutrient Content Claims
5.3	Synonyms for Listed Terms Under Proposal to Revise 21 *CFR* §101.13
5.4	Proposed Criteria for Declaring "Available" Fat
6.1	General Health Claim Requirements
6.2	Criteria for Making a Claim Relating Folate to NTD Risk
8.1–8.4	Dietary Supplement Label Examples
8.5	Nutrient Content Claim Requirements for Dietary Supplements

INTRODUCTION

Keeping Up With the Food Labeling Process

Nutrition labels on many food items must comply with numerous, ever-changing requirements. Products such as **meat and poultry**, food packages, and **dietary supplements** are subject to stringent federal regulations — the costs of compliance can be significant.

The Nutrition Labeling and Education Act of 1990 (**NLEA**) imposed new mandates for labeling of many packaged food products and restricted how various nutrition claims may be made; some products became subject to a voluntary labeling program. Following that lead, USDA has imposed parallel labeling requirements. In 1994, the Dietary Supplement Health and Education Act (DSHEA) added even more **complexity** to the nutrition labeling program.

Available products, and **legislation** targeting them, have become so numerous and complex that regulations now impose restrictions along what can be described as a **food/dietary supplement/drug spectrum** rather than setting simple requirements for basic food items.

In response, this book provides hands-on information about fundamental regulatory concepts, while also identifying the very latest changes in federal requirements. The guide consists of three components:

- **Discussions** of key statutory and regulatory labeling requirements and tips on complying,

- The **full text** of the FDA and USDA/FSIS **labeling** regulations current as of **May 1, 1998**, and

- An **index** of each chapter, including hundreds of regulatory citations.

How to Use This Guide

- Chapters 1–8 contain **plain-English analysis** of the FDA and FSIS labeling rules and their underlying legal requirements. Included are numerous decision diagrams and tables, along with the text of new statutory language. Also, citations point to the actual regulatory requirements.

- **Checklists** are provided in Chapter 2 to help in developing a system for tracking regulatory changes and then documenting whether they affect a given product.

- Additionally, the text of the following agency **regulations** is included in Chapters 9–11:

    — FDA food labeling rules in 21 *CFR* Part 101
    — FSIS meat labeling rules in 9 *CFR* Part 317, Subparts A and B
    — FSIS poultry labeling rules in 9 *CFR* Part 381, Subparts N and Y

- To quickly locate information on a particular item, turn to the **index** at the back of the guide. It references the discussion in the first eight chapters and also provides hundreds of regulatory citations, listed by topic.

Part A
OVERVIEW OF LABELING RULES + COMPLIANCE CHECKLISTS

- *Chapter 1* — AGENCY RULEMAKING PROCESSES & HISTORY OF U.S. NUTRITION LABELING REQUIREMENTS

- *Chapter 2* — GLOSSARY OF COMMON ACRONYMS + COMPLIANCE CHECKLISTS

Chapter 1—RULEMAKING PROCESSES & THE HISTORY OF NUTRITION LABELING REQUIREMENTS

> ### In This Chapter...
>
> - How FDA and FSIS go about establishing **new regulations**
>
> - **History** of labeling requirements and changes initiated by the Nutrition Labeling and Education Act of 1990 (NLEA)—differences between FSIS and FDA
>
> - FDA and USDA/FSIS nutrition labeling requirements for food, red meat, and poultry processors, including locations of **key regulations**

First, this opening chapter provides an overview of the administrative rulemaking process — the statutory and regulatory history of FDA and FSIS nutrition labeling requirements are then outlined. Finally, the locations of key federal regulations are identified for quick reference.

Agency Rulemaking Processes

How do the federal regulations come about? In the United States, various agencies "promulgate" rules through a process guided by the Administrative Procedure Act (APA), as described below.

Reminder: In Chapter 2, all acronyms used in this *Guide* are defined.

Legal Authority

Before a regulatory agency can issue or enforce any rules, it must have legal authority to do so. This legal authority emerges when Congress passes a statute (i.e., a law) 1) establishing certain requirements to be complied with by industry, and 2) authorizing a particular agency to administer those requirements. With respect to food, meat, and poultry processors, the following statutes guide many agency activities:

- *FDCA*— The Federal Food, Drug, and Cosmetic Act (first passed in 1938, and amended several times since) establishes requirements administered and enforced by the Food and Drug Administration (FDA) through its regulatory process. FDCA's provisions include requirements for appropriate labeling. Provisions specific to dietary supplement products were adopted in the Dietary Supplement Health and Education Act of 1994 (DSHEA); the DSHEA revised certain labeling requirements, such as provisions for listing the quantity of each dietary ingredient and the optional listing of the sources of various ingredients.

- *FMIA*— Among other things, authority to require proper handling and labeling of meat products is given to the U.S. Department of Agriculture (USDA) by the Federal Meat Inspection Act.

- *PPIA*— The Poultry Products Inspection Act is modeled largely after the FMIA. Again, various labeling (and other) requirements are spelled out in the statute and administered by USDA.

As will be discussed later, recent revisions to the FDCA set the nutrition labeling process in motion — USDA has since modeled many of its requirements after FDA rules.

Federal Agencies

One of FDA's mandates is to ensure the safety of the general food supply; this agency is a part of the U.S. Department of Health and Human Services (HHS). With respect to meat and poultry products, however, jurisdiction is given to USDA; within that department, the Food Safety and Inspection Service (FSIS) administers regulations governing inspection, processing, and labeling of meat and poultry products. Typically, foods containing less than 2% cooked meat or poultry, or less than 3% raw meat, are regulated by FDA, not FSIS.

These agencies carry out their statutory mandates by issuing and enforcing numerous regulations. They also establish informal policies by publishing guidance manuals and policy memoranda—these documents often provide answers to important technical questions related to regulatory compliance. When a particular issue is a source of confusion or concern to industry, an agency may issue a memo or guidance manual to clarify a regulatory requirement.

Rulemaking

As noted earlier, the FMIA and PPIA spell out conditions for the safe processing of red meat and poultry products; the FDCA establishes numerous requirements for other food processors. While these requirements are often significant, the language of the statutes is typically not specific enough to establish clear courses of action for the regulated industries. Consequently, after a piece of legislation becomes law, FDA and FSIS initiate a rulemaking process to implement the statutory requirements.

The CFR and FR

Regulations issued by federal agencies are much more specific (and lengthy) than the statutes they are designed to implement — these rules often establish required procedures for the regulated industry as well as for state or local governments. All the federal regulations are codified in a set of volumes known as the *Code of Federal Regulations*, or *CFR*. Updated *CFR* volumes are published once a year by the Government Printing Office (GPO). The *CFR* is organized first by title, and then grouped by part, subpart, section, and paragraph within each title. USDA/FSIS rules are in Title 9, while the FDA regulations are published in Title 21.

Example:
A reference to an FSIS rule in part 318 would be written as "9 *CFR* **Part 318**", while a citation of an FDA regulation in paragraph (a), section 9, part 101 is written "21 *CFR* §101.9(a)".

Throughout the year, agencies continually issue new regulations and revoke obsolete ones. Proposed and final changes are published in the daily *Federal Register* (*FR*). These items are identified according to the page number and volume (i.e., year) of publication — for example, a regulation published on page 2743 in the 1998 *Federal Register* would be referenced as "63 *FR* 2743". After 12 months have elapsed, all the final regulatory changes published in the *FR* during that year are incorporated into the *CFR*, and a new annual volume is published. However, these volumes are typically several months out of date when they first become available from GPO.

The APA

How does the rulemaking process work? How do the targets of various regulatory programs stay informed and/or participate in this process?

Federal agencies are required to follow rulemaking procedures outlined in the Administrative Procedure Act (APA). The APA ensures that the public 1) is notified of pending regulatory revisions, and 2) has an opportunity to comment on a proposal. Thus, the process is sometimes referred to as "notice-and-comment rulemaking." Under the APA framework, rules published in the *Federal Register* typically fall into one of three categories:

- *Advance notices of proposed rulemaking* describe regulations at the "prerule" stage, in which an agency discusses its plans to propose a particular requirement. Comments are solicited and incorporated into the formal proposal.

- A *proposed rule* outlines a set of planned regulations. The public is invited to formally comment on the forthcoming requirements for a specified period of time. Hearings are sometimes held in conjunction with this "notice-and-comment" period.

- *Final rules* represent formal adoption of regulatory requirements; these are incorporated into the *CFR* and have the effect of law. Frequently, a final rule will specify some later date on which the requirements actually take effect — this is intended to give regulated entities adequate time to revise their processes or labeling schemes in order to comply.

In addition, twice each year, the federal government publishes a summary of its plans to issue new rules affecting industry. This "semiannual regulatory agenda" or "unified agenda" provides a snapshot of the various programs being implemented by FDA or USDA.

Federal, Not State, Rules Take Precedence

While state governments do have substantial responsibilities with respect to ensuring a safe food supply, the individual states' regulations are not within the scope of this book. Congress has guaranteed national uniformity for its nutrition labeling requirements by specifying that federal provisions take precedence over state rules where requirements differ; this ensures that consumer products are labeled consistently, regardless of their state of origin.

Key Labeling Terms and Concepts

First, some fundamental terms are defined:

- *Label*—Generally, the term "label" refers to any printed material placed upon a food container. For example, the Federal Food, Drug, and Cosmetic Act (FDCA), which guides most FDA activities, defines this term as "a display of written, printed, or graphic matter upon the immediate container of any article [Section 201(k)]." FSIS regulations clarify that stickers are considered part of a label.

- *Labeling*—Likewise, labeling is defined to mean "all labels and other written, printed, or graphic matter (1) upon any article or any of its containers or wrappers, or (2) accompanying such article [Section 201(m)]."

With respect to nutrition labeling, product information is provided in two key places: the principal display panel (PDP) and the information panel. Generally, if an outside container or wrapper is provided, any information required to appear on a container label also must 1) appear on the outside wrapper, or 2) be easily legible through the outside container or wrapper. Precisely what the FSIS and FDA rules require on each of these panels is outlined in Chapter 3.

Differences Between the FDA and FSIS Nutrition Labeling Rules

Many nutrition labeling concepts are based on language in the FDCA as revised by the Nutrition Labeling and Education Act of 1990 (the NLEA); this statute specifies how FDA must ensure that food and nutrition labels are accurate and consistent. However, because USDA elected to follow suit when FDA began establishing its nutrition labeling program, most requirements discussed in this book are equally applicable to products sold by the food, red meat, and poultry industries. Where important differences do occur, they are noted. Both the NLEA and the subsequent rulemaking processes are explained below—the relationship between the FDA and FSIS regulations is also reviewed.

Pre-1990 Labeling Requirements

Labels have long been an important source of food product information for consumers. Although the NLEA imposed significant new labeling requirements for food processors, labeling was already an important component of the federal regulatory programs.

USDA

Among other things, USDA requires meat and poultry product labels to inform consumers as to the quality and content of items offered for sale. FSIS administers a prior label approval program under which labeling to be used on meat and poultry products must be approved by the agency prior to its use. In many cases, FSIS prescribes the content and design of product labels.

Prior to 1990, FSIS typically disseminated its nutrition labeling guidelines through various policy memoranda issued by the agency's food labeling division. Documents outlining nutrition labeling requirements (or agency preferences) included policy memos numbered 7, 16A, 46, 49C, 70B, 71A, 74A, 78, 85B, and 86. Most of these memos were issued during the 1980's; topics included fat and lean claims, sodium labeling, and information panels.

FDA

As noted previously, many of FDA's activities are governed by the FDCA. Since 1938, this statute has required FDA to ensure that products are labeled with:

- Name of the food,

- Name and address of the manufacturer,

- A statement of ingredients, and

- Net quantity of contents.

Furthermore, Congress in 1966 passed the Fair Packaging and Labeling Act (FPLA). This legislation also required FDA to administer rules concerning information provided on food labels. However, since many of the FPLA requirements had already been imposed under the FDCA, the most significant change inspired by the FPLA was a requirement that information concerning a product's net quantity of contents be provided in a standardized fashion.

During the 1970s, FDA adopted regulations in 21 *CFR* §101.9 prescribing a specific labeling format to be used when voluntary or mandatory nutrition information was provided. That section of the regulations has been rewritten in response to the NLEA, as discussed below.

NLEA Ushers in a New Era

The NLEA was signed into law by President Bush on November 8, 1990, revising several sections of the FDCA. This legislation represented the first major overhaul of food labeling requirements since the FDCA was passed in 1938.

Keep in mind that the FMIA and PPIA, which govern meat and poultry respectively, were not amended in 1990. However, FSIS has elected to establish, to the maximum extent possible, nutrition labeling rules that parallel FDA's regulations. Consequently, since 1990, both agencies have actively been issuing proposals and finalizing new rules. To clear the way for the nutrition labeling program, several FDA regulations have been revoked or revised, and numerous FSIS policy memos have been revoked in whole or in part.

Status of Labeling Rules

The locations of key labeling regulations administered by FDA or FSIS are listed in **Table 1.1**; the full text of these labeling rules is provided in Chapters 9–11. Note that some of these regulations encompass pre-1990 requirements, such as rules concerning net quantity of contents. In addition to the new nutrition labeling regulations, other key provisions are included so that this guide can provide a comprehensive reference to the FDA and FSIS labeling requirements.

Although the mandatory nutrition labeling program took shape relatively quickly, the agencies are still issuing rules at regular intervals. The current status of this program implementation is reviewed below.

TABLE 1.1
Key FDA and FSIS Labeling Regulations

Topic	CFR reference
FDA Rules—	
Principal display panels	21 *CFR* §101.1
Nutrition labeling: formats and required or mandatory nutrients, reference daily intakes (RDIs), and daily reference values (DRVs)	21 *CFR* §101.9
Restaurant foods: nutrition labeling	21 *CFR* §101.10
Reference amounts customarily consumed per eating occasion (RACC)	21 *CFR* §101.12
General principles for nutrient content claims and health claims	21 *CFR* §§101.13-101.14
Dietary supplement rules	21 *CFR* §101.36
Voluntary labeling of raw fruits, vegetables and fish	21 *CFR* §101.45
Specific nutrient content, health, or descriptive claims	21 *CFR* §§101.54-101.95
FSIS Rules—	
Labeling; general provisions, including approval	9 *CFR* Part 317, Subpart A and Part 381, Subpart N
Nutrition labeling: label content	9 *CFR* §§317.309, 381.409

TABLE 1.1
Key FDA and FSIS Labeling Regulations

Topic	CFR reference
Reference amounts customarily consumed per eating occasion (RACC)	9 CFR §§317.312, 381.412
General principles for nutrient content claims	9 CFR §§317.313, 381.413
Voluntary labeling of single-ingredient, raw products	9 CFR §§317.345, 381.445
Specific nutrient content, health, or descriptive claims	9 CFR §§317.354-317.380, §§381.454-381.480
Exemptions from nutrition labeling	9 CFR §§317.400, 381.500

FDA

After passage of the NLEA, FDA was faced with the prospect of proposing and issuing numerous complex regulations — certain deadlines were spelled out in the FDCA, requiring aggressive action on the part of FDA. The bulk of the nutrition labeling rules were issued on January 6, 1993, only slightly behind the statutory deadline. FDA published several different regulations on that day (at the same time, FSIS incorporated most of the FDA rules into its regulations by reference—more about that later).

At this point, FDA continues to publish nutrition labeling rules, both to complete the implementation process and to revise requirements where industry's attempts to comply have raised important issues. In addition, some new rulemaking actions are taken in response to newly available scientific information. The dates and *Federal Register* references for the key nutrition labeling regulations published by FDA and FSIS since 1990 are listed in the **Appendix**. Note that, to simplify compliance efforts, FDA periodically sets a so-called "uniform compliance date" as the uniform date on which food labeling regulations take effect. The most recent date set is January 1, 2000 — this will be the compliance date for labeling rules issued by the agency between January 1, 1997 and December 31, 1998.

FSIS Responds With Its Own Regulations

As a general rule, meat and poultry products are subject to nutrition labeling requirements that closely resemble those administered by FDA. By electing to issue rules that paralleled the FDA nutrition labeling provisions, FSIS encountered many of the same issues as FDA. However, although FSIS was required to engage in significant rulemaking activity in order to establish a formalized nutrition labeling program, it did not have the pressure of satisfying statutory deadlines.

Incorporated by Reference

FSIS published its new nutrition labeling regulations on January 6, 1993. At that time, the agency simply cross-referenced many of the FDA regulations, incorporating the FDA rules by reference where requirements were identical. For the most part, FSIS published its own codified language only where differences in the products being labeled required changes to the regulatory provisions. (One important component of the FDA rules has been omitted from the FSIS regulations thus far, however. FDA authorizes certain "health claims", which are statements that a particular dietary component may assist in preventing a particular disease (e.g., a claim about the relationship between calcium and osteoporosis). Although FSIS has not yet authorized any health claims, it plans to do so, as discussed in Chapter 6.)

Subsequently, FSIS received comments from the regulated community indicating that it was very difficult to refer back and forth between the FDA and FSIS rules; those comments suggested that FSIS codify all its nutrition labeling requirements together under 9 *CFR*. The agency agreed, and on January 3, 1995 published the full text of its nutrition labeling rules for incorporation into the FSIS regulations. The locations of the various sets of nutrition labeling regulations are listed in **Table 1.1** (page 1.5).

One Set of Rules?

Now that FSIS has established its own complete set of nutrition labeling regulations, how will it respond to regulatory changes made by FDA? While FSIS' decision to publish its own rules has been helpful to the regulated community, it has created some inconsistencies between the FDA and FSIS requirements.

FDA continues to issue new or revised rules to correct oversights, respond to new information, or streamline the regulations based on industry's experience with various labeling provisions. Since the FSIS rules often are "clones" of the FDA regulations, any FDA revisions ideally should be incorporated by FSIS — if and when they are relevant to the labeling of meat and poultry products. FSIS staff in the Product Assessment Division indicate that the agency intends to adopt FDA nutrition labeling changes in most cases. First, FSIS will make adjustments as necessary to establish rules that are appropriate for USDA-regulated products.

Because of ongoing revisions being made by both agencies, the FSIS rules are unlikely to ever again match the FDA regulations word-for-word. It might be said that nutrition labeling requirements are being developed by FSIS through a "trickle-down" process. This lag will continue to create discrepancies between the two agency's regulations.

Reminder:
Key FSIS and FDA nutrition labeling proposals and rules are listed in the Appendix.

Clearly, not every FDA nutrition labeling regulation will be adopted by FSIS. The dietary supplement rules being developed in response to the DSHEA, for example, do not impact producers of meat or poultry products and thus would not be incorporated into the FSIS regulations.

Chapter 2—GLOSSARY OF COMMON ACRONYMS + COMPLIANCE CHECKLISTS

> ### In This Chapter...
>
> - **Glossary** of acronyms commonly used in the food regulations
>
> - **Guidelines** for establishing a nutrition label records management system — checklists to:
>
>   – Ensure that new labels satisfy regulatory requirements
>
>   – Track nutrition labeling rule changes
>
>   – Document whether a given nutrition label is affected by regulatory changes

FIGURE 2.1
Sample Nutrition Facts Label

Nutrition Facts

Serving Size 1 cup (228g)
Servings Per Container 2

Amount Per Serving

Calories 260   Calories from Fat 120

	% Daily Value*
Total Fat 13g	**20%**
Saturated Fat 5g	**25%**
Cholesterol 30mg	**10%**
Sodium 660mg	**28%**
Total Carbohydrate 31g	**10%**
Dietary Fiber 0g	**0%**
Sugars 5g	
Protein 5g	

Vitamin A 4%   •   Vitamin C 2%
Calcium 15%   •   Iron 4%

* Percent Daily Values are based on a 2,000 calorie diet. Your daily values may be higher or lower depending on your calorie needs:

		Calories:	2,000	2,500
Total Fat	Less than		65g	80g
Sat Fat	Less than		20g	25g
Cholesterol	Less than		300mg	300mg
Sodium	Less than		2,400mg	2,400mg
Total Carbohydrate			300g	375g
Dietary Fiber			25g	30g

Calories per gram:
Fat 9 • Carbohydrate 4 • Protein 4

TABLE 2.1
Glossary of Acronyms Commonly Used in U.S. Food Labeling Regulations

Acronym	Definition
AOAC	Association of Official Analytical Chemists International
APA	Administrative Procedure Act
CFR	*Code of Federal Regulations*
CGMP	Current good manufacturing practice
CHD	Coronary heart disease
DRV	Daily reference value
DSHEA	Dietary Supplement Health and Education Act of 1994
FDA	Food and Drug Administration
FDCA	Federal Food, Drug, and Cosmetic Act
FMIA	Federal Meat Inspection Act
FPLA	Fair Packaging and Labeling Act
FR	*Federal Register*
FSIS	Food Safety and Inspection Service
GPO	Government Printing Office
HDL	High density lipoprotein (i.e., cholesterol)
HHS	Department of Health and Human Services
LDL	Low density lipoprotein (i.e., cholesterol)
NLEA	Nutrition Labeling and Education Act of 1990
NTD	Neural tube defect
PDP	Principal display panel
PDV	Percent daily value (or %DV)
PPIA	Poultry products Inspection Act
RACC	Reference amount customarily consumed
RDI	Reference daily intake
USDA	United States Department of Agriculture

Compliance Checklists

Determining If Products Are in Compliance

First and foremost, nutrition labels must satisfy all applicable FDA and FSIS requirements. In the event of either an internal or agency audit, how does one go about documenting that a product label has been adequately reviewed? A nutrition label recordkeeping system can save valuable time down the road — during an audit, or when the manufacturer launches another product with similar characteristics.

The form on **pages 2.5–2.6** provides a checklist to use in developing a new nutrition label — this ensures documentation of how regulatory compliance decisions were made and records the references that were used. Note that this checklist also provides a means of documenting cases where nutrition labels are not required (i.e., where an exemption applies).

Keeping Up With Regulatory Revisions: Management of Change

Once a nutrition label has been developed for a given product, ongoing regulatory changes that might affect the labeling information must be tracked. Documenting this process can avoid "reinventing the wheel" during a compliance audit.

The form on **pages 2.7-2.8** is an example of how the history of a product label can be chronicled; more importantly, it provides a means of documenting whether new labeling regulations require revisions to nutrition labels.

Developing a Management System

The following checklists might not fully suit the needs of a particular facility or regulatory affairs staff. However, they can serve as useful prototypes for developing a more customized recordkeeping system. In any case, cost-effective regulatory compliance requires systematic review and documentation of the impacts of agency actions.

Besides tracking regulatory revisions, another important way to confirm the compliance status of a product is to perform an internal audit. Table 4.7 summarizes the criteria and procedures used by FDA and FSIS to determine whether a particular product complies with the nutrition labeling rules. These may serve as guidelines for conducting a facility-specific evaluation.

FIGURE 2.2
NUTRITION LABELING DEVELOPMENT CHECKLIST

Attach copy of **Nutrition Facts** *label here*

Product name:
_____
_____

Product code:
_____

Manufacturing location:
_____
_____

Label required?

Yes ____ No ____ (*if No, provide regulatory citation for exemption*)
_____
_____

Date label created: _____ **By:** _____

Label generated by:

____ Software (*name* _____)

____ In-house staff (*list* _____)

____ Other (*list* _____)

Alternative format used for this label?

Yes ____ No ____ (*if Yes, provide regulatory citation and explanation*)

Type of alternative format: _____

NUTRITION LABELING DEVELOPMENT CHECKLIST

Compliance check:

Label component:	List declared value (*or attach software output*):	Remarks or regulatory references:	Approved by *(initial)*:
Serving size			
# servings			
Calories—			
Per serving			
From fat			
"Core" nutrients—			
Fat, total			
Saturated fat			
Cholesterol			
Sodium			
Carbohydrate, total			
Dietary fiber			
Sugars			
Protein			
Voluntary nutrients:			
Vitamins and minerals—			
Vitamin A			
Vitamin C			
Calcium			
Iron			
Voluntary vitamins and minerals:			

FIGURE 2.3
NUTRITION FACTS LABELING REVISION HISTORY

Attach copy of **Nutrition Facts** *label here*

Product:

_____

_____

Product code:

_____

Manufacturing location:

_____

_____

Date label created: _____

By: _____

Label revision history:

Date revised (attach copy of revised label):	Reason (regulatory revision, design change, etc.):	Revised by (initial):	Approved by (initial):

Nutrition Facts Labeling Revision History

Regulatory tracking:

Federal Register date:	Topic of FDA or FSIS rule:	Affect label? *(Yes or No)*	If *Yes*, has label been revised?	Checked by *(initial)*:

Part B
NUTRITION LABELING FUNDAMENTALS

- *Chapter 3* — LABELING BASICS: WHO IS EXEMPT? WHAT MUST BE DISPLAYED ON THE LABEL PANELS? HOW DOES THE VOLUNTARY PROGRAM WORK? WHAT ABOUT RESTAURANT FOODS?

- *Chapter 4* — REQUIRED INFORMATION: MANDATORY AND VOLUNTARY NUTRIENTS, REFERENCE AMOUNTS, PERCENT DAILY VALUES, AND DEMONSTRATING COMPLIANCE

Chapter 3—LABELING BASICS

> ### In This Chapter...
>
> - Who is **exempt** from the nutrition labeling rules? Is an exemption notice needed?
> - What information must be provided on the **principal display** and **information** panels, and what **formats** are acceptable?
> - How are **serving size** and number of servings per container determined?
> - Does USDA require prior label **approval**?
> - How does the **voluntary** program work for single-ingredient, raw products?
> - Are special rules provided for restaurant foods?

As noted in Chapter 1, "labeling" is usually taken to mean all written, printed, or graphic material placed on food in package form. Federal laws and corresponding agency rules require that certain specific information, such as net quantity of contents, be displayed on packages offered for sale.

What exactly are the requirements imposed by the NLEA and by subsequent FDA and FSIS regulations? For the most part, the NLEA-based rules prescribe what *nutrition* information must appear on food labels. While the more generic term "food labeling" often refers to activities such as naming a product, listing its ingredients, etc., "nutrition labeling" refers specifically to the information provided on the Nutrition Facts labels inspired by the NLEA.

During the past five years, both industry and government have expended significant energy in coming to grips with nutrition labeling requirements. Since then, passage of the Dietary Supplement Health and Education Act (the DSHEA) has led to even more label-oriented rulemaking activity (FDA's regulations implementing the DSHEA are discussed specifically in Chapter 8). Rules covering exemptions, label formats, serving sizes, and voluntary labeling for food, meat, and poultry processors are discussed below.

Who Is Covered . . .

Generally, all food products sold to consumers must provide nutrition information on their labels, although special provisions may apply to foods sold in bulk or to single-ingredient, raw products such as chicken breasts or fresh fruits. **Figure 3.1** (page 3.2) illustrates which products are subject to nutrition labeling requirements.

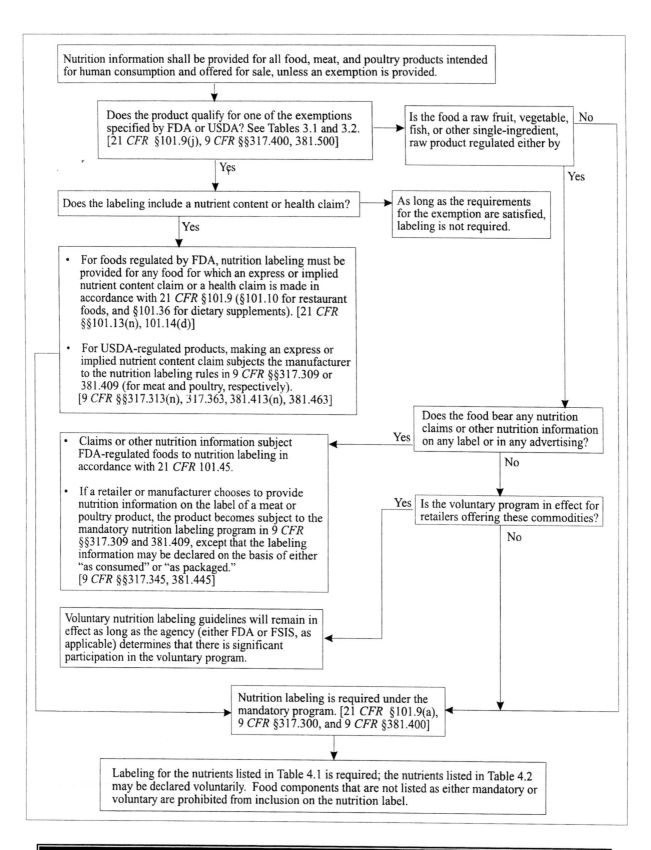

FIGURE 3.1
Applicability of FDA & FSIS Nutrition Labeling Programs

... and Who is Exempted?

A number of exemptions are available under the FDA and FSIS rules. Some of these are designed to assist small businesses — others exempt products where compliance with nutrition labeling rules would be impractical, such as those prepared and served in retail establishments. However, any product for which a nutrient content claim, health claim, or other nutrition information is provided automatically becomes subject to the nutrition labeling rules.

FDA

The regulations exempting certain products from nutrition labeling can be difficult to follow—in some cases, full exemptions are provided, while in others, exceptions are made to formatting or nutrient declaration requirements. The exemptions provided in §101.9 of the current FDA regulations are summarized in **Table 3.1**. Of particular interest has been the low-volume-product exemption, discussed further below.

TABLE 3.1
Labeling Exemptions Provided in the FDA Rules

Exemption provided	21 CFR §101.9
Small Businesses and Low-Volume Products	
Food offered for sale by a person who makes direct sales to customers (e.g., a retailer) who has:[1] • annual gross sales made (or business done) in sales to consumers that is not more than $500,000, *or* • annual gross sales made (or business done) in sales of *food* to consumers of not more than $50,000	(j)(1)(i)

Low-volume food products meeting the following requirements:[1,2]					
First introduced into interstate commerce	*If during this time period . . .*	*U.S. product sales were below*	*and on average, FTEs were fewer than*	*Product is eligible for exemption for the time period . . .*	
before May 8, 1994	May 8, 1994 to May 7, 1995	400,000 units	300	May 8, 1995, to May 7, 1996	(j)(18)
before May 8, 1994	May 8, 1995 to May 7, 1996	200,000 units	200	May 8, 1996 to May 7, 1997	
after May 8, 1994	any 12-month period	100,000 units [3]	100	the subsequent 12 months	

Products Not Distributed to Consumers	
Food products shipped in bulk form, not for distribution to consumers, that are 1) intended for use in the manufacture of other foods, or 2) to be processed, labeled, or repacked at a site other than where originally processed or packed	(j)(9)
Products Served for Immediate Consumption or Processed at Retail Establishments	
Food products which are:[1] • served in restaurants, • served in other establishments in which food is served for immediate human consumption (e.g., institutional food service establishments, such as schools) or sold only in such facilities, • used only in such facilities, and not served to the consumer in the package in which they are received (e.g., foods that are not packaged in individual serving containers), *or* • sold by a distributor who principally sells food to such facilities, except: 1) if there is a reasonable possibility that the product will be purchased directly by consumers, the manufacturer is responsible for providing nutrition information, and 2) this exemption does not apply to foods manufactured, processed, or repackaged by the distributor for sale to establishments other than those that serve food for immediate consumption	(j)(2)(i)–(v)

TABLE 3.1
Labeling Exemptions Provided in the FDA Rules

Exemption provided	*21 CFR §101.9*
Foods that are: [1] • ready for human consumption, • of the type described in §101.9(j)(2)(i) and (ii) (see above), • offered for sale to consumers, but not for immediate consumption, *and* • processed primarily in a retail establishment, and not offered for sale outside of that establishment (e.g., ready-to-eat foods prepared onsite and sold by independent bakeries or delis)	(j)(3) (i)–(v)
Foods with Insignificant Nutrient Content	
Foods containing insignificant amounts of all the nutrients and food components required to be declared on nutrition labels under §101.9(c), such as coffee beans (whole or ground), tea leaves, plain unsweetened instant coffee and tea, condiment-type dehydrated vegetables, flavor extracts, and food colors [1,4]	(j)(4)
Products for Infants and Children	
Infant formula subject to FDCA Section 412, and labeled in compliance with 21 *CFR* Part 107	(j)(7)
A medical food as defined in Section 5(b) of the Orphan Drug Act., i.e., a food consumed or administered enterally under the supervision of a physician and intended for the specific dietary management of a disease or condition for which distinctive nutritional requirements, based on recognized scientific principles, are established by medical evaluation. To be exempt, the food also must be: 1) specially formulated (as opposed to a naturally occurring foodstuff used in its natural state), 2) intended for a patient who, because of therapeutic or chronic medical needs, has limited or impaired capacity to ingest, digest, absorb, or metabolize ordinary foodstuffs or certain nutrients, 3) specifically modified for the management of the unique nutrient needs that result from the specific condition, 4) intended for use under medical supervision; and 5) intended only for a patient receiving ongoing medical supervision.	(j)(8) (i)–(v)
Single-Ingredient, Raw Products	
Raw fruits, vegetables, and fish [1] (labeling of such foods should adhere to guidelines in §101.45)	(j)(10)
Small Packages	
Foods in packages that have a total surface area available to bear labeling of less than 12 square inches. [1] To use this exemption, the manufacturer, packer, or distributor must provide an address or telephone number where a consumer can obtain nutrition information (e.g., "For nutrition information, call (800) 123-4567"). (See Figure 3.4 for alternative formatting for these packages.)	(j)(13) (i)

Note: Special FDA labeling exceptions and requirements are listed in Table 3.4. Rules for alternative label formats are illustrated in Figure 3.4.

[1] Provided the food bears no nutrition claims or other nutrition information in any context on labeling or in advertising. Claims or other nutrition information subject the food to the provisions of §101.9
[2] For these products to be exempt, a notice must be filed with FDA before the time period for which this exemption is claimed (see Figure 3.5 for a reproduction of the model exemption notice)
[3] In the case of a product that was not sold in the 12-month period preceding the period for which an exemption is claimed, FDA requires that fewer than 100,000 units are reasonably anticipated to be sold in the United States during the period for which this exemption is claimed
[4] An insignificant amount of a nutrient is that amount that allows a declaration of zero in nutrition labeling, except that for total carbohydrate, dietary fiber, and protein, it is an amount permitting a declaration of "less than 1 gram"

Low-Volume Products

In the wake of NLEA, some constituents expressed concern that the new law provided only a very narrow exemption for small businesses—NLEA exempted foods sold by firms with gross sales of 1) less than $500,000/year, or 2) annual gross sales from food of less than $50,000. Many small businesses indicated that they would have difficulty satisfying the NLEA requirements and/or qualifying for this exemption. Thus, Congress provided another exemption when it passed the Nutrition Labeling and Education Act Amendments of 1993; the added exemption applies to "low-volume food products".

Consequently, in addition to the original exemption based on gross annual sales, firms also may escape nutrition labeling requirements by qualifying for the low-volume food product exemption. For a food product to qualify as low-volume:

1. A "person" (i.e., a manufacturer, packer, or distributor) who meets agency requirements must file a timely exemption claim with FDA (these claims cover 12-month periods and thus must be renewed annually);

2. Units sold must fall below the thresholds listed in Table 3.1; *and*

3. Full-time employees (FTEs) must not exceed the limits specified in the table.

Once filed, each exemption notice covers the following 12-month period; information on product sales and FTEs for the previous 12-month time period is used to establish that a product qualifies for an exemption.

The language in Congress' 1993 NLEA amendments made the low-volume food product exemption "self-effectuating"; that is, the exemption took effect immediately even though there were no FDA regulations to codify this new exemption. Numerous small businesses immediately filed exemption notices with the agency indicating that one or more of their products was exempt from nutrition labeling rules under the low-volume exemption. As shown in the table, in §101.9(j)(18), FDA did eventually issue regulations formally detailing how firms should go about qualifying for the low-volume exemption established in the 1993 NLEA amendments. Note that these rules were not published until August 7, 1996.

FDA has also published a model exemption notice for low-volume food products, and instructions for submitting such a notice under §101.9(j)(18); these are reproduced in **Figure 3.5** at the end of this chapter (page 3.21). Among other things, FDA notes that firms have struggled with the meaning of "food product." The agency explains that a specific food product is defined by three parameters:

- Its being from a single manufacturer or bearing a single brand name,

- Bearing the same statement of identity, and

- Having a similar method of preparation.

This means that, in counting the number of units of a food product for purposes of claiming a low-volume exemption, firms must consider 1) the total number of units produced by the manu-facturer for sale to consumers in the United States, regardless of the brand name under which it is packaged; *and* 2) the total number of units labeled under one brand name, regardless of the number of manufacturers that produced it. If either number exceeds the low-volume criteria shown in Table 3.1, the product is not eligible for this exemption. Finally, note that the 1993 amendments provide a "very small business" exemption for a person selling fewer than 10,000 units of a given product annually.

FSIS

Table 3.2 identifies the meat and poultry product nutrition labeling exemptions provided by FSIS. The agency provides a small-business exemption based on number of employees and on production volume. A product qualifying for a nutrition labeling exemption must be 1) produced in annual volumes of less than 100,000 pounds, and 2) manufactured by a firm employing fewer than 500 people. Other exemptions apply to products that are not sold directly to consumers or that are prepared and served in retail establishments. Note that both FDA and FSIS exempt single-ingredient, raw products — which fall under a voluntary program.

TABLE 3.2
Labeling Exemptions Provided in the USDA FSIS Rules

Exemptions provided	9 CFR §317.400, §381.500
Small Businesses	

Meat, meat food products, and poultry products produced by small businesses are exempt from nutrition labeling as follows: [1]
• a product, for the purposes of this small-business exemption, is defined as a formulation, not including distinct flavors which do not significantly alter the nutritional profile, sold in any size package in commerce
• a small business is any single-plant facility or multi-plant company/firm that employs 500 or fewer people and produces no more than the following amounts of the product qualifying the firm for exemption: [2]
   — 250,000 pounds or less during the first year of nutrition labeling (July 1994 to July 1995)
   — 175,000 pounds or less during the second year of nutrition labeling (July 1995 to July 1996)
   — 100,000 pounds or less during subsequent years
	(a)(1)–(7)

Products Not Intended for Sale to the General Consumer	
Products intended for further processing [1]	(a)(2)
Products not intended for sale to consumers [1]	(a)(3)
Custom slaughtered or prepared products	(a)(5)
Products intended for export	(a)(6)
Products Prepared & Served at Retail Establishments	
Products prepared and served or sold at retail: 1) ready-to-eat products that are packaged or portioned at a retail store or similar retail-type establishment; or 2) multi-ingredient products (e.g., sausage) processed at a retail store or similar retail-type establishment	(a)(7)
Restaurant menus generally do not constitute labeling or fall within the scope of these regulations	(b)
Products for Infants and Children	
Foods represented to be specifically for infants and children less than 4 years of age must bear nutrition labeling, except that: • the labeling should not declare percent of daily value for total fat, saturated fat, cholesterol, sodium, potassium, total carbohydrate, or dietary fiber • nutrient names and quantitative amounts should be presented in two separate columns • the "Percent Daily Value" heading required in §§317.309(d)(6) and 381.409(d)(6) must be placed immediately below the quantitative information by weight for protein • the percent daily value for protein, vitamins, and minerals shall be listed immediately below the heading "Percent Daily Value" • labeling should not include the footnote specified in §§317.309(d)(9) and 381.409(d)(9)	(c)(2)
Foods represented to be specifically for infants and children less than 2 years of age must be labeled in accordance with §§317.400(c)(2) and 381.500(c)(2) (see above), except the labeling should not include calories from fat, calories from saturated fat, saturated fat, stearic acid, polyunsaturated or monounsaturated fat, or cholesterol	(c)(1)
Small Packages	
Products in small packages that are individually wrapped and of less than ½ ounce net weight [1]	(a)(4)

TABLE 3.2
Labeling Exemptions Provided in the USDA FSIS Rules

Exemptions provided	9 CFR §317.400, §381.500
Packages that have a total surface area available to bear labeling of less than 12 square inches are exempt from nutrition labeling:[1] • the manufacturer, packer, or distributor must provide an address or telephone number where consumers can obtain nutrition information (e.g., "For nutrition information call 1-800-123-4567") • when nutrition labeling *is* included on these products, either voluntarily or because nutrition claims or other nutrition information is provided, the required information generally must be in a type size no smaller than 6 point	(d)(1)–(2)
Single-Ingredient, Raw Products	
Nutrition labeling is not required for single-ingredient, raw products, although it may be provided voluntarily [1] — this exemption will continue as long as retailers comply substantially with FSIS' voluntary labeling program outlined in §§317.345, 381.445	§317.300, §381.400

Note: FSIS and FDA rules for alternative label formats are illustrated in Figure 3.4. Special FDA labeling exceptions and requirements are listed in Table 3.4.

[1] Provided the food bears no nutrition claims or other nutrition information in any context on labeling or in advertising. Claims or other nutrition information subject the food to the provisions of §101.9

[2] Calculation of the amount (in pounds) shall be based on the most recent 2-year average of business activity. Where firms have been in business less than 2 years or where products have been produced for less than 2 years, reasonable estimates must indicate that the annual pounds produced will not exceed the amounts specified

Source: 9 CFR §§317.300-317.400 & 381.400-381.500

Nutrition Facts Labels

Both FDA and FSIS specify formats for nutrition labels. **Figure 3.2** (page 3.8) presents a sample label and shows agency requirements for type faces and sizes. To promote consistency and ease of use for the consumer, all labels begin with a "Nutrition Facts" heading, and provide information in the same order. Information on serving size and number of servings is followed by a summary of the calories per serving. Details on specific nutrients are then provided.

Principal Display and Information Panels

Under the regulations, certain key information must be displayed on either the *principal display panel* (PDP) or the *information panel* of packages offered for sale. These include, for example, FDA and FSIS requirements that the name of the food product be displayed in bold-face type on the PDP. Specific markings are required on meat and poultry products, such as official inspection legends, or marks signifying that a product is approved for export.

The regulations establishing labeling requirements other than for Nutrition Facts labels and related nutrient claims are beyond the scope of this manual. However, for the purpose of complying with the nutrition labeling rules, some basic definitions are important [note that these rules appear in 21 *CFR* §§101.1-101.2, 101.9(j); and 9 *CFR* §§317.2, 381.116]:

• The PDP of a food package is the part that is most likely to be displayed or examined under customary conditions of retail sale, i.e., the portion of the package presented to the consumer. If a package has another surface that could serve as the PDP, it is an alternate PDP; information required on the PDP must be duplicated on all alternate PDPs.

- The information panel is the panel immediately to the right of the PDP (or an alternate PDP), as observed by the consumer. If this panel is too small to display the required information or is otherwise unusable, the next label panel immediately to the right may serve as the information panel. Also, if the top of a container serves as the PDP and there is no alternate PDP, the information panel is any panel adjacent to the PDP.

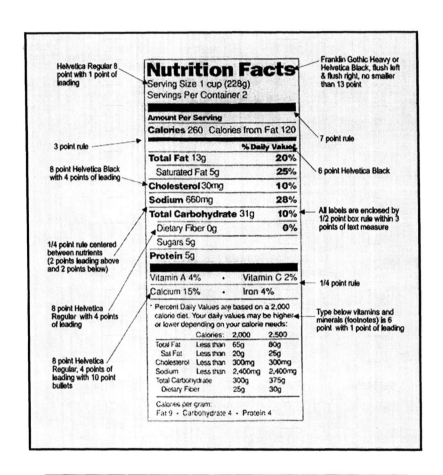

FIGURE 3.2
Sample Label with Required Type Sizes

Placement of "Nutrition Facts"

Nutrition information is to be placed on either the PDP or the information panel. However, if a package has total surface area available to bear labeling of more than 40 square inches, but the PDP and information panel do not provide sufficient space to accommodate all required information, any alternate panel may be used for the nutrition information. Additional exceptions to labeling requirements are identified in Figure 3.4 and Tables 3.1, 3.2, and 3.4.

Note: Federal agencies have actively been reviewing their regulations under a "reinventing government" initiative; one of the objectives is to remove obsolete or unnecessary rules from the *CFR*. Both FDA and FSIS have proposed and/or finalized some "housekeeping" revisions to their regulations.

For example, with respect to PDPs and information panels, on August 12, 1997 FDA deleted rules in §101.2(c)(1)-(3) and (c)(5)(iii) — these sections had provided type-size exemptions preceding the NLEA-based program (62 FR 43071).

Label Formats

The general requirements for nutrition label formats are illustrated in **Figure 3.3** (page 3.10). Note that a footnote is required to explain that nutrient information is based on a 2,000-calorie diet — this provision is illustrated in **Table 3.3** below.

TABLE 3.3
Footnotes Required by FDA & FSIS

Beneath the list of vitamins and minerals on the nutrition label, and separated from that list by a hairline, the following statement/table is required: [1]

Percent Daily Values are based on a 2,000 calorie diet. Your daily values may be higher or lower depending on your calorie needs.

	Calories:	2,000	2,500
Total fat	Less than	65 g	80 g
Saturated fat ..	Less than	20 g	25 g
Cholesterol	Less than	300 mg	300 mg
Sodium	Less than	2,400 mg	2,400 mg
Potassium[2]	Less than	3,500 mg	3,500 mg
Total carbohydrate...	....	300 g	375 g
Dietary fiber ...	....	25 g	30 g
Protein[3]	....	50 g	65 g

Note: FSIS and FDA rules for alternative label formats are illustrated in Figure 3.4. Special FDA labeling exceptions and requirements are listed in Table 3.4.

[1] Below the footnote, caloric conversion information on a per-gram basis is required, separated by a hairline. Conversion values for fat, carbohydrate, and protein are required; the information may be presented horizontally (i.e., "Calories per gram: fat 9, carbohydrate 4, protein 4") or vertically in columns
[2] Include this entry only if the % of daily value is given for potassium [as provided for in §101.9(c)(5) or 9 CFR §317.309(c)(5) and §381.409(c)(5)]
[3] Include this entry only if the % of daily value is given for protein [as provided for in §101.9(d)(7)(ii) or 9 CFR §§317.309(d)(7)(ii) and 381.409(d)(7)(ii)]

Source: 21 CFR §101.9(c)(9), (d)(9)-(10); 9 CFR §317.309(c)(9), (d)(9)-(10) and §381.409(c)(9), (d)(9)-(10)

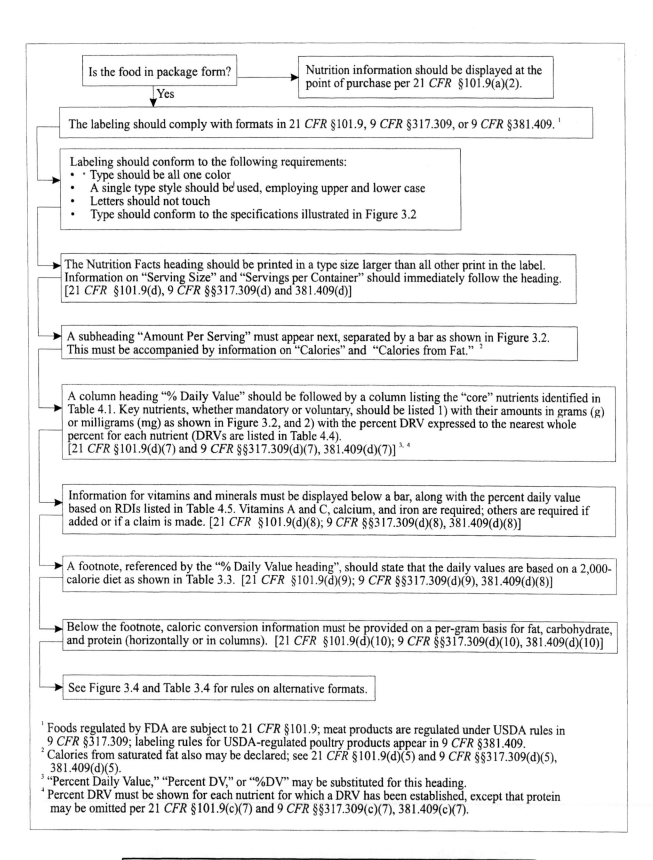

**FIGURE 3.3
Basic Formatting Rules for Nutrition Labeling**

Alternative Formats Available

Because food products come in all shapes and sizes, the nutrition label format does not always lend itself to placement on a given package. Additionally, the format sometimes is overly prescriptive when a given product contains insignificant amounts of several of the nutrients required for declaration. Thus, FDA and FSIS allow alternative formats in many cases. These include a simplified format, a linear display, and a tabular format.

If a package includes two or more individually packaged foods, the label may provide information for multiple products using an aggregate display. Also, where a manufacturer wants to provide information on a product both as packaged and as prepared (e.g., for a cake mix), a dual display may be used to present information for more than one form of the food.

Figure 3.4 details the rules for alternative labeling formats. For the most part, the FSIS regulations mirror the FDA rules. However, to make these requirements applicable to meat and poultry products, FSIS made certain changes, such as to the list of nutrients for which an insignificant amount qualifies the product for a simplified-format label. For each special provision or alternative format allowed, the regulatory citation is provided for quick reference to the actual regulatory text.

FDA Special Provisions

Besides the formatting alternatives described above, FDA has included other special provisions for items such as dietary supplements, products sold in bulk, or products packaged in individual containers. These are listed in **Table 3.4** (pages 3.14–3.15).

FIGURE 3.4
FSIS and FDA Rules for Alternative Labeling Formats*

If . . .	then . . .
Tabular Display	
• the continuous vertical space on the label cannot accommodate the required label components, up to and including the mandatory declaration for iron	. . . the tabular display may be used, as illustrated in 21 *CFR* §101.9(d)(11) [21 *CFR* 101.9(d)(11), 9 *CFR* §§317.309(d)(11), 381.409(d)(11)]
Aggregate Display	
• the package contains two or more individually packaged foods, intended to be eaten separately, *or* • the packaging is intended to be used interchangeably for similar types of food	. . . the aggregate display may be used as illustrated in 21 *CFR* §101.9(d)(13) [21 *CFR* 101.9(d)(13), 9 *CFR* §§317.309(d)(13), 381.409(d)(13)]
Dual Display	
• the labeling presents information for: – two or more forms of the food (e.g., as purchased and as prepared), – common combinations of food (e.g., dry cereal and cereal with milk), – two or more groups for which separate RDI's are established, *or* – different sets of units (e.g., slices of bread and per 100 grams)	. . . dual labeling is allowed per 21 *CFR* §101.9(e), and 9 CFR §§317.309(e) and 381.409(e), with the declarations sharing equal prominence [21 *CFR* §101.9(b), (e), and (h)(4); 9 *CFR* §§317.309(b), (c)(8), and (e), 381.409(b), (c)(8), and (e)]

FIGURE 3.4
FSIS and FDA Rules for Alternative Labeling Formats*

If . . .	then . . .
Simplified Format	
<ul><li>For FDA-regulated products, the food contains insignificant amounts of any seven of the following:<ul><li>calories,</li><li>total fat,</li><li>saturated fat,</li><li>cholesterol,</li><li>sodium,</li><li>total carbohydrate,</li><li>dietary fiber,</li><li>sugars,</li><li>protein,</li><li>vitamins A or C,</li><li>calcium, *or*</li><li>iron</li></ul></li><li>For products regulated by USDA, the simplified format is allowed when any required nutrients are present in insignificant amounts *other than* these core nutrients: calories, total fat, sodium, total carbohydrate, and protein</li></ul>	. . . information may be declared using the simplified format described in 21 *CFR* §101.9(f) (for FDA-regulated products) and 9 *CFR* §§317.309(f) or 381.409(f) (for meat and poultry products, respectively).[1] . . . the simplified format must include information on 1) total calories, total fat, total carbohydrate, protein, and sodium; 2) other required nutrients present in more than insignificant amounts, and 3) added vitamins or minerals either allowed or required under the FDA rules (other nutrients naturally present in "significant" amounts may be declared voluntarily). . . . the footnote shown in Table 3.3 above may be omitted and the statement "Percent daily values are based on a 2,000 calorie diet" substituted; if "DV" is used in the heading, it must be defined here. . . . under the simplified format, required nutrients present in insignificant amounts must be listed after the statement "Not a significant source of _____".[2] *Note:* The regulations define "insignificant amount" as that amount of a nutrient that allows a declaration of zero on the label (see Table 4.3). For total carbohydrate, dietary fiber, and protein, an amount < than 1 gram is "insignificant" (FSIS also includes sugars in this 1-gram exception).
Small and Intermediate Packages	
<ul><li>the food is contained in a package with a total surface area available to bear labeling of less than 12 square inches</li></ul>	. . . the product is exempt from nutrition labeling requirements 1) as long as no claim or other nutrition information is declared, and 2) if an address or telephone number is provided for consumers to obtain the required nutrition information. . . . *and* a tabular or linear display may be used as illustrated in 21 *CFR* §101.9(j)(13)(ii)(A). The linear format may be used only if the label will not accommodate a tabular display. If labeling *is* provided on these small packages, a type size of at least 6 point is generally required — see 21 *CFR* §101.9(j)(13)(i) and 9 *CFR* §§317.400(d), 381.500(d).
<ul><li>the food is in a package with a total surface area available to bear labeling of less than 40 square inches</li></ul>	. . . the labeling may be modified by one or more of the following means: 1) if the product has available space of less than 40 in^2, and the package shape or size cannot accommodate a standard vertical column or tabular display on any label panel, the tabular or linear formats shown in 21 *CFR* §101.9(j) (13)(ii)(A) may be employed. The linear format may be used only if the label will not accommodate a tabular display. 2) abbreviations may be used per 21 *CFR* §101.9(j)(13)(ii)(B) or 9 *CFR* §§317.309(g)(2), 381.409(g)(2). 3) the footnote concerning the 2,000-calorie diet may be revised per the regulations. 4) the required nutrition information may be presented on any label panel. [21 *CFR* §101.9(j)(13)(ii)(A)-(D); 9 *CFR* §§317.309(g)(1)-(4) and 381.409(g)(1)-(4)]

FIGURE 3.4
FSIS and FDA Rules for Alternative Labeling Formats*

If . . .	then . . .
Insufficient Space	
• the space beneath the vitamin and mineral listings is inadequate for the footnote, *or*
• the space below the mandatory declaration for iron is inadequate for the footnote and additional vitamins and minerals	. . . the information may be moved to the right of the nutrient/percent daily value declarations and set off by a line to set it apart. The caloric conversion information may be presented beneath either side or along the full length of the nutrition label. [21 *CFR* 101.9(d)(11), 9 *CFR* §§317.309(d)(11), 381.409(d)(11)]
• the food is in a package with a total surface area available to bear labeling greater than 40 square inches, but the principal display panel and information panel do not provide sufficient space to accommodate all required information	. . . any alternate panel that can be readily seen by consumers may be used for the nutrition label:
• the space needed for vignettes, designs, and other nonmandatory information on the principal display panel may be considered in determining the sufficiency of available space on the principal display panel for the placement of the nutrition label
• nonmandatory information on the information panel shall not be considered in determining the sufficiency of available space for the placement of the nutrition label
[21 *CFR* §101.9(j)(17)]

* *Note:* Special labeling rules for FDA-regulated products are outlined in Table 3.4.

[1] The rules vary slightly for foods intended for children less than 2 years of age. See 21 *CFR* §101.9(f) and (j)(5); 9 *CFR* §§317.309(c)(8)(i) and (e), 381.409(c)(8)(i) and (e).

[2] FDA and FSIS rules differ on the use of this statement; see 21 *CFR* §101.9(f)(4) and 9 *CFR* §§317.309(f)(4), 381.409(f)(4). Also, FSIS allows "Contains less than 2% of the daily value of [this nutrient]" as an alternative.

Q: Can the nutrition facts panel be oriented perpendicularly to the base of a package, instead of parallel?

A: Yes. There is no requirement that this information be printed parallel to the base of a package.

TABLE 3.4
Special Labeling Provisions in the FDA Rules

Requirement or exception	21 CFR §101.9
Foods for Infants and Children	
Foods, other than infant formula, represented to be specifically for infants and children less than 2 years of age must bear nutrition labeling, except 1) as provided in §101.9 (j)(5)(ii) (see below), and 2) labeling information should not include declarations of: calories from fat, calories from saturated fat, saturated fat, polyunsaturated or monounsaturated fat, and cholesterol as described in 21 *CFR* §101.9(c)(1)(ii)-(iii), (c)(2)(i)-(iii), and (c)(3)	(j)(5)(i)
Foods, other than infant formula, represented to be specifically for infants and children less than 4 years of age must bear nutrition labeling, except that: • such labeling does not include declarations of percent daily value for total fat, saturated fat, cholesterol, sodium, potassium, total carbohydrate, or dietary fiber, • nutrient names and quantitative amounts are to be presented in two separate columns, • the heading "Percent Daily Value" required in §101.9(d)(6) is to be placed immediately below the quantitative information by weight for protein; • percent of daily value for protein, vitamins, and minerals shall be listed immediately below the heading "Percent Daily Value"; and • The labeling should not include the footnote specified in §101.9(d)(9) (see Table 3.3)	(j)(5)(ii)
Dietary Supplements	
Dietary supplements of vitamins and minerals that have a DRV or an RDI established (see Tables 4.4 and 4.5) must be labeled in compliance with §101.36, except that vitamin or mineral supplements in conventional food form (e.g., breakfast cereals), herbs, and other similar nutritional substances should be labeled per §101.9	(j)(6)
Fish and Game	
Packaged, single-ingredient products that consist of fish or game meat (i.e., animal products not covered under the FMIA or PPIA, such as flesh products from deer, quail, or ostrich) may provide required nutrition information for a 3-ounce, cooked portion except that: • if claims are made based on nutrient values as packaged, the product label must provide information on an as-packaged basis, and • Nutrition information is not required for custom processed fish or game meats	(j)(11)
Game meats (i.e., animal products not covered under the FMIA or PPIA, such as flesh products from deer, quail, or ostrich) may provide required nutrition information in accordance with §101.9(a)(2)	(j)(12)
Eggs	
Shell eggs packaged in a carton with a top lid conforming to the shape of the eggs are exempt from outer carton label requirements if the required nutrition information is presented 1) immediately beneath the carton lid, or 2) in an insert that can be clearly seen when the carton is opened	(j)(14)

TABLE 3.4 Special Labeling Provisions in the FDA Rules ||
Requirement or exception	21 CFR §101.9
Individual or Bulk Containers	
Individual containers in a multiunit retail food package do not require labeling if: • the multiunit package label displays the required nutrition information, • the individual units are enclosed within the outer package and not intended to be separated under conditions of retail sale, and • each unit is labeled with the statement "This Unit Not Labeled For Retail Sale" in type size not less than 1/16-inch in height (this statement is not required when the inner units bear no labeling at all—the word "individual" may be used in lieu of or immediately preceding the word "Retail" in the statement)	(j)(15)
For foods sold from bulk containers, nutrition information required under §101.9 may be displayed 1) on the labeling of the bulk container (plainly in view), or 2) in accordance with §101.9(a)(2)	(j)(16)
Insufficient Space	
Foods in packages that have a total surface area available to bear labeling greater than 40 square inches, but whose principal display panel and information panel do not provide sufficient space to accommodate all required information, may use any alternate panel that can be readily seen by consumers for the nutrition label. • The space needed for vignettes, designs, and other nonmandatory information on the principal display panel may be considered in determining the sufficiency of available space on the principal display panel for the placement of the nutrition label. • Nonmandatory information on the information panel shall not be considered in determining the sufficiency of available space for the placement of the nutrition label.	(j)(17)
Note: Exemptions from the FDA and FSIS nutrition labeling requirements are listed in Tables 3.1 and 3.2, respectively. Rules for alternative label formats are illustrated in Figure 3.4.	

Label Content

Generally, nutrition labels must provide the following information:

- Serving size,

- Number of servings in the container or package,

- Total calories per serving and calories from fat (per serving),

- Amounts of certain "core" nutrients expressed in grams or milligrams, and

- "Percent daily values" for certain nutrients, vitamins, and minerals.

This information is provided on the basis of the product as packaged, although certain single-ingredient, raw products may be labeled on the basis of the product as consumed (more about that later in this chapter). Also, as noted earlier, nutrition labels may include "dual" declarations for a product on both an as-packaged and as-prepared basis, as long as preparation or cooking instructions are provided. Furthermore, although serving size and other nutrition information is typically based on a product as packaged, in some cases the liquid packing medium is not typically consumed (e.g., pickles in juice). In this type of situation, the label declarations should be based on drained solids.

Rules for determining serving size and number of servings are outlined below. Requirements for declaring specific nutrients using DRVs and RDIs are discussed in Chapter 4.

Serving Size

The labeling rules require that nutrition information be provided in relation to one serving of the product being labeled, as follows:

- The term "serving" or "serving size" is defined as an amount of food customarily consumed per eating occasion by persons four years of age or older. (If a product is specially formulated for infants or toddlers, a serving is the amount customarily consumed by an infant up to 12 months of age, or by a child 1 to 3 years of age, respectively.)

- Beside the per-serving information, the label may include a second column of figures declaring nutrient information per 100 grams, 100 milliliters, 1 ounce, or 1 fluid ounce of the product as packaged or purchased.

- Serving sizes for meal-type products should be declared as the entire edible content of the package — see 21 *CFR* §101.13(l) and 9 *CFR* §§317.313(l) and 381.413(l) for definitions of these products.

How is the serving size determined for a given food product? For the purposes of label declarations, manufacturers (and others responsible for nutrition labeling) must use reference values established by FDA and FSIS — these values are referred to as reference amounts customarily consumed (RACC). The RACC values are listed according to product category in lengthy tables published at 21 CFR §101.12(b) and 9 CFR §§317.312(b), 381.412(b).

Note that the agencies periodically revise these, often upon request from petitioners. For example, on November 18, 1997, FDA proposed to revise the reference amount for baking powder, baking soda, and pectin from 1 g to 0.6 g — the agency believes this value more accurately reflects the amount of these products customarily consumed (62 *FR* 61476). Similarly, FDA revised the RACC for salt, salt substitutes, and seasoning salts on December 2, 1997 (62 *FR* 63647). Also, reference amounts for three types of candies were proposed on January 8, 1998 (63 *FR* 1078).

Serving sizes must be expressed in common household measures, along with metric equivalents — for example, "1 Cup (228 g)". However, this poses a problem for products sold in discrete units (such as muffins or frozen pizza), in bulk (such as pancake mix), or in variety packs (such as a cheese/sausage tray). The regulations specify how serving size should be determined and declared in these situations, as shown in **Table 3.5** (page 3.17). Thus, for a pizza kit, the rules for large, discrete units allow declaration of the fractional portion that most closely approximates the RACC (e.g., "1/4 pizza").

Q: Can type sizes larger than 6-point and 8-point be used on my "Nutrition Facts" label?

A: Yes. The 6- and 8-point type sizes specified for various portions of the label are minimum requirements.

Q: Can the nutrition facts be printed on a sticker and affixed to a package?

A: Yes, as long as the sticker adheres to the product under the intended storage conditions. Some companies use generic cartons or bags and then affix product-specific labeling.

TABLE 3.5
FDA & FSIS Rules for Determining Serving Size in Special Cases

If the unit weighs:	The serving size is:*
Discrete units [1] —	
50% of reference amount* or less	# whole units most closely approximating reference amount
> 50% but < 67% of reference amount	1 or 2 units
≥ 67% but < 200% of reference amount	1 unit
> 150% but < 200% of reference amount [2]	unit is 1 or 2 servings
< 200% of reference amount and the product is packaged and sold individually (except see above for products with large reference amounts)	1 unit (i.e., it is treated as a single-serving container)
≥ 200% of reference amount	1 unit *if* it can reasonably be consumed on a single occasion
Large, discrete units [3, 4] —	Fractional slice most closely approximating the reference amount for that product category — use 1/2, 1/3, 1/4, 1/5, or 1/6 [5]
Nondiscrete bulk products [4, 6] —	Amount in household measures most closely approximating the reference amount
Variety packs [7] —	Based on reference amount for each variety or food — serving sizes are provided separately for each

*Use RACC values (reference amounts customarily consumed per eating occasion) to determine serving size. If a promoted use of a product differs in quantity by twofold or more from the use upon which the RACC was based, the label must provide a second column of nutrition information based on the amount consumed in the promoted use (e.g., coffee cream substitutes promoted for use on breakfast cereal). Ingredients such as flour or butter are exempt from this requirement.

[1] For example, muffins or individually packaged products within a multiserving package
[2] For products that have reference amounts of 100 grams (or milliliters) or larger and are individual units within a multiserving package
[3] Products usually divided for consumption, such as cake mix or pizza
[4] If two foods are packaged together (e.g., cake mix and can of frosting), with the main ingredient a large discrete unit or a nondiscrete bulk product, the serving size may consist of a portion of the main ingredient plus proportioned minor ingredients to make the reference amount
[5] Or smaller fractions generated by dividing by 2 or 3
[6] For example, pancake mix
[7] A package containing two or more varieties of single-serving units or two or more types of food

Source: 21 CFR §101.9(b); 9 CFR §§317.309(b), §381.409(b)

Units of Measure

As mentioned previously, serving sizes must be declared in common household measures, followed by metric equivalents in parentheses. Regulatory requirements for use of household measures and metric equivalents in declaring serving size are summarized in **Table 3.6**.

TABLE 3.6
Use of Household Measures and Metric Equivalents

For *household measures*:
— where other units are used to declare serving size, it may also be expressed in ounce/fluid ounce; these values should be declared in parentheses following the metric quantity, and separated by a slash—e.g., for sliced bread this might read as "Serving Size: 1 slice (28 g/1 oz)"
— these abbreviations may be used: tbsp, tsp, g, mL, oz, and fl oz

When declaring *metric equivalents*:
— fluid amounts are expressed in milliliters (mL); all other quantities are in grams (g)
— the gram or milliliter equivalent should be rounded according to the following rules:
  - for quantities less than 2 g (or mL), round to the nearest 0.1 g (mL)
  - between 2 and 5 g (mL), express in 0.5-g (mL) increments
  - for 5 g (mL) or more, round to nearest whole number

For *single-serving containers*:
— the metric quantity (which also is provided on the principal display panel as part of the net-weight statement), is not required (except where nutrition information is required on a drained-weight basis)

Source: 21 *CFR* §101.9(b), 9 *CFR* §§317.309(b) and 381.409(b)

Number of Servings per Container

Once serving size has been determined, the number of servings per container must be declared. The regulations specify that the number of servings be rounded to the nearest whole number, with some exceptions:

- If the number of servings is between 2 and 5 (and for random-weight products), the number should be rounded to the nearest half serving and declared using the term "about". For example, "about 3.5 servings".

- When the serving size is expressed on a drained-solids basis, and the number of servings varies because of natural product variations, the typical number of servings per container may be stated (e.g., usually 5 servings of pickled pigs feet).

In addition, manufacturers of random-weight products may choose to declare "varied" for the number of servings per container and provide nutrition information based on the RACC (e.g., "varied: approximately 6 servings per pound").

"Voluntary" Labeling of Single-Ingredient, Raw Products

Producers and retailers simply cannot provide a customized nutrition panel for every raw chicken breast or radish bunch purchased at the grocery store. Fortunately, nutrition labeling policies have been designed to accommodate these single-ingredient, raw products, while at the same time ensuring that consumers are able to access nutrition information on these products with relative ease.

In the opening sentences of the FSIS nutrition labeling regulations (in 9 *CFR* §§317.300 and 381.400), the agency explains that single-ingredient, raw products are not required to bear nutrition labeling, although distributors of these products may label them voluntarily. At the same time, the agency has established "voluntary" guidelines for labeling these products — as long as FSIS finds significant participation in this program, single-ingredient, raw products will be exempted from mandatory nutrition labeling requirements.

Similarly, the NLEA requires FDA to establish voluntary labeling guidelines for raw fruits, vegetables, and fish—these products also are exempt from mandatory nutrition labeling rules as long as retailers participate in the agency's voluntary program.

The Voluntary Programs

Regulations in 21 *CFR* §101.45 and 9 *CFR* §§317.345, 381.445 outline voluntary labeling guidelines for raw fruits, vegetables, and fish; single-ingredient, raw meat products, and single-ingredient, raw poultry products, respectively. The gist of the voluntary program is that nutrition information may be provided at the point-of-purchase, rather than on packages, by posting a sign, providing brochures, or through some other means (posters are commonly being used under this program). Labeling information should be provided in close proximity to where the products are displayed for sale.

Generally, nutrition information provided voluntarily must comply with the formats and other requirements of the mandatory labeling program—this promotes consistency among all food products providing nutrition information. However, in its voluntary guidelines, FDA requires that information on fish be provided on an as-cooked basis (the cooking method should not include breading or seasonings); nutrition information for fruits and vegetables should be based on the raw, edible portion. Specific nutrient values for use in labeling these products are listed in the FDA regulations at Part 101, Appendices C and D.

For products regulated by FSIS, the nutrition information may be given on either an as-consumed or an as-packaged basis.

What is "Significant Participation?"

Both FDA and FSIS plan to continue their voluntary nutrition labeling programs for single-ingredient, raw products as long as they find significant participation. If surveys of retailers show "substantial compliance" with the voluntary guidelines, the agencies will not impose mandatory nutrition labeling requirements for these products. FDA and FSIS each plan to conduct surveys of voluntary participation every two years.

Substantial compliance is defined to mean that at least 60% of food retailers sampled in a representative survey provide nutrition labeling information in compliance with agency guidelines.

- For FSIS-regulated products, a retailer is in compliance if it provides appropriate information for at least 90% of the single-ingredient, raw products it stocks. According to agency rules, only the "major cuts" of meat and poultry products are considered during FSIS' evaluation of significant participation; major cuts of single-ingredient, raw meat and poultry products are listed in the regulations at 9 *CFR* §§317.344 and 381.444, respectively.

- Under the FDA guidelines, a retailer is participating appropriately if it provides accurate nutrition labeling information for at least 90% of the foods it sells that are among the 20 most frequently consumed raw fruits, vegetables, and fish in the United States. So that retailers may provide this nutrition information, NLEA requires FDA to identify and formally list these 20 most frequently consumed raw fruits, vegetables, and fish; the agency updated its list on August 16, 1996 (see 21 *CFR* §101.44).

So far, FDA's investigations indicate that retailers are, in fact, participating significantly with the voluntary program. The agency's most recent biannual report, which presented the results of a December 1996 study, concluded that more than 70% of stores surveyed provided information for raw fruits, vegetables, and fish.

A parallel study sponsored by FSIS found that 92% of meat and poultry labels surveyed provided nutrition information within regulatory specifications. The agency analyzed 300 products for protein, fat, and other constituents (63 *FR* 200).

FSIS Prior Label Approval

A key difference between the FSIS and FDA rules is that FSIS often requires prior agency approval before labels may be placed on products offered for sale. Frequently, minor label revisions qualify for "generically approved labeling," which greatly speeds and streamlines the approval process. For example, 9 CFR §317.5(b)(2) provides that labeling for single-ingredient products, such as lamb chops, qualifies for generic approval— as long as the label makes no special claims pertaining to features such as quality, nutrient content, or geographical origin.

The rules governing poultry product labeling are found in §§381.132-381.133; regulations for meat products appear in §§317.4-317.5.

FSIS has numerous rules governing the labeling of meat and poultry products ; only the requirements related to nutrition labeling are detailed in this discussion..However, all these labeling provisions must be satisfied. For example, FSIS has issued a series of rules establishing when poultry may be labeled as "fresh" — the agency's final action, dated December 17, 1996, prohibits use of this term on raw product whose internal temperature has ever fallen below 26 °F (61 *FR* 66198).

Restaurants and Retail

Products prepared and/or served in restaurants and other retail establishments, or in institutions, often qualify for nutrition labeling exemptions or special formatting requirements, as outlined earlier in Figure 3.4 and Tables 3.1, 3.2, and 3.4. Moreover, FDA has published regulations specifically concerning restaurant foods in §101.10.

When it first issued its nutrition labeling rules, FDA gave restaurants extra leniency in demonstrating compliance with the strict nutrient content and health claim labeling requirements that apply to most foods (claims are discussed in Chapters 5–7). In §§101.10, 101.13, and 101.14, the agency exempted restaurant menus from the nutrient content and health claim requirements. However, in response to public comments and to a lawsuit filed by public interest groups, FDA proposed to eliminate the restaurant menu exemption on June 15, 1993. The lawsuit had claimed that the exemption violated FDCA as revised by the NLEA. A June 28, 1996 ruling by the U.S. District Court for the District of Columbia confirmed that claim, declaring that the menu exemption was in conflict with the FDCA. FDA recently closed the gap by publishing a final rule on August 2, 1996, eliminating the restaurant menu exemption.

Note, however, that restaurant foods are still exempt from many of the nutrition labeling provisions (as long as no claims are made). Furthermore, when nutrient content or health claims are made, compliance may be based on a "reasonable basis" other than analytical testing.

Q: Are cellophane windows on bags or boxes considered "space available to bear labeling"?

A: Yes, if the window is used for any labeling, including promotion stickers, However, if no labeling is present in the "window", it is not considered available space.

Q: When the statement "not a significant source of _____" is used for more than one nutrient, is a certain order required?

A: The footnote must identify the nutrients in the order in which they are usually listed on a label (i.e., in the order for the regular format). This statement may be used with any labeling format as long as the "insignificant" nutrient levels meet the required per-serving limits (e.g., < 5 calories from fat, < 1 g dietary fiber, etc.).

FIGURE 3.5
FDA's Model Small Business Food Labeling Exemption Notice*

1. Name of firm: _____

2. Address of firm:
   Street address _____
   City, State _____
   Zip or postal code _____ Country _____
   Telephone _____ FAX _____

3. Type of firm *(check all that apply)*:

   Manufacturer _____   Packer/Repacker _____
   Distributor _____   Importer _____
   Retailer _____

4. Twelve-month time period for which you are claiming exemption:
   FROM: __ / __ / __ *(mm/dd/yy)*
   TO: __ / __ / __ *(mm/dd/yy)*

5. Average number of full-time equivalent employees
   for 12-month period: _____

6. Report of units sold *(use continuation sheets if necessary)*:

   Product: _____
   No. of units: _____
   Manufacturer: _____

7. Name and address of manufacturer(s) or distributor(s) of product(s) in Item 6, if different from firm claiming exemption *(use continuation sheets if necessary)*:

   B. Name of manufacturer or distributor _____
      Address _____
      _____

   C. Name of manufacturer or distributor _____
      Address _____
      _____

8. Contact person _____

9. The undersigned certifies that the above information is a true and accurate representation of the operations of _____ *(name of firm)*. The undersigned will notify the Office of Food Labeling of the date on which the average number of full-time equivalent employees or the number of units of food products sold in the United States exceeds the applicable number for exemption which is being claimed herein.

Signature _____
Name *(type or clearly print)* _____
Title _____
Date _____

Note: This document is intended to facilitate notification to FDA that a low-volume food product qualifies for exemption from nutrition labeling under the 1993 NLEA amendments. See §101.9(j)(18).

Instructions for Completing FDA's
Model Small Business Food Labeling Exemption Notice

1. **Name of firm:** Enter the recognized legal name of your firm.

2. **Firm address:** Enter the mailing address for the principal location of your firm. Also, provide the telephone and FAX numbers.

3. **Type of firm:** Place a check mark or "x" in each block that is applicable to your firm. For example, if your firm manufactures all products that it sells place a check mark after "Manufacturer". If your firm also distributes a product that is manufactured by another firm, also place a check mark after "Distributor."

4. **Twelve-month time period for which you are claiming exemption:** Enter the specific time period for which you are requesting exemption for your products. For products initially introduced into interstate commerce before May 8, 1994, this time period will be from May 8 of the current year to May 7 of next year: e.g., "FROM 05/08/95 TO 05/07/96." For new products, the time period should start with the date on which sales in the United States are expected to begin: e.g., "FROM 11/01//95 TO 10/31/96."

5. **Average number of full-time equivalent employees for 12-month period:** Enter the average number of full-time equivalent employees of your firm and of all of its affiliates for the year preceding the year for which an exemption is claimed under Item 4. The average number should include all employees of your firm and of its affiliates (e.g., owners; officers; and all other personnel such as secretarial, production, and distribution employees). Firms are affiliates of each other when, either directly or indirectly:
   (1) One firm has the power to control the other,
   (2) a third party controls or has the power to control both, or
   (3) an identity of interest exists such that affiliation may be found.

The average number of full-time equivalent employees is to be determined by using the following formula:
   Total number of employee/hours paid divided by 2,080 hours =
            average number of full-time equivalent employees.

For example, 254,998 paid employee/hours ÷ 2,080 = 122. If the total number of actual employees for your firm and its affiliates is less than 100, you may enter the total number of actual employees instead of calculating the average number of full-time employees; e.g., if your firm has 24 employees that work full-time and 12 employees that work part-time, you may report 36 total actual employees instead of calculating the average number of full-time equivalent employees.

6. **Report of units sold** *(continuation sheets using the same format for Item 6 may be used if necessary)*:

Product: Under the column for product, enter the name, including the brand name, for each food product for which your firm is claiming an exemption. A food product is a food in any sized package which is manufactured by a single manufacturer or which bears the same brand name; which bears the same statement of identity; and which has a similar preparation method. In considering whether food products have similar preparation methods, consider all steps that go into the preparation of the food products, from the initial formulation steps to any finishing steps; for example, products having differing ingredients would be considered different food products and counted separately in determining the number of units.

No. of Units: Provide the approximate sales of your firm, in terms of units, for the product for the year immediately preceding the time period for the exemption entered under Item 4. For example, if the time period for which you are claiming exemption for a food product is May 8, 1996, to May 7, 1997, provide an approximation of your sales of that product from May 8, 1995, to May 7, 1996. If the product was not sold for the entire 12 months preceding the time period for the exemption entered under Item 4, provide an approximation of the sales expected to be made during the time period in Item 4. For example, if the time period being claimed in Item 4 is

November 1, 1995, to October 31, 1996, for a product that is going to be sold beginning November 5, 1995, provide an approximation of sales for the period from November 1, 1995, to October 31, 1996. The approximate total number of units is the summation of the number of units of the various package sizes of the food product in the form in which it is sold to consumers; for example, the total of all 2-pound bags of flour plus all 5-pound bags of flour plus all 10-pound bags of flour should be provided as the number of units sold by your firm in the United States. There may also be occasions where a food is sold in bulk or by individual pieces rather than in packaging; e.g., flour may be sold in bulk displays at grocery stores. In such a case, the number of units should be determined on the basis of the typical sales practice for the specific food product; e.g., if 2,000 pounds of flour are sold from bulk displays at grocery stores, and the typical practice for sales to consumers is to price the flour on a per pound basis, then the bulk sales would represent 2,000 units. If the firm sells the same product in package form, then the bulk sales, 2,000 units in the above example, should be added to the sum of the number of packages of the flour sold to determine the total number of units of flour sold by the firm in the United States.

Manufacturer: Under the column designated "Manufacturer" enter the letter that corresponds with the name of the manufacturer of the product. The letter "A" is used to designate the firm submitting the notice if it is the manufacturer of the product. If the firm submitting the notice is not the manufacturer of the product, use the letter from Item 7 (B or C), or from the continuation sheets for Item 7, that corresponds to the name and address of the manufacturer of the product.

7. **Name and address of manufacturer(s) or distributor(s) of product(s) in Item 6 if different from firm claiming exemption** *(continuation sheets may be used if necessary)*: Provide the name and addresses of the manufacturers of the food products for which exemption is being claimed if they are different from the firm claiming the exemption. If the name of the manufacturer is unknown, provide the name of the firm from which the product is purchased. Insert the letter corresponding to the name of the manufacturer ("A" corresponds to the firm submitting the notice) or distributor in the appropriate block for the name of the product under Item 6.

8. **Contact person:** Enter the name of a person that can act as a contact for your firm if any questions arise concerning the information included in the notice.

9. **Certification:** The form is to be signed by a responsible individual for the firm that can certify to the authenticity of the information presented on the form. The individual signing the form will commit to notify the Office of Food Labeling when the numbers of full-time equivalent employees or total numbers of units of products sold in the United States exceed the applicable number for an exemption.

The **completed form should be mailed** to: Office of Food Labeling (HFS-150), Food and Drug Administration, 200 C St., SW, Washington, DC 20204. **Questions** concerning a claim may be directed to the Office of Food Labeling at the above address or to (202) 205-4561.

Source: 61 *FR* 40963, August 7, 1996

Chapter 4—REQUIRED INFORMATION: MANDATORY AND VOLUNTARY NUTRIENTS, % DAILY VALUES; DETERMINING COMPLIANCE STATUS

In This Chapter...

- Describing **individual nutrients** on the "Nutrition Facts" labels

- Daily reference values (**DRVs**) and reference daily intakes (**RDIs**)

- Declaring **percent daily values** and following **rounding** rules

- Demonstrating **compliance** through **sampling** and **testing**

Which nutrients may be declared on nutrition labels? Which ones are required? How are percent daily values determined?

As discussed in Chapter 3, the nutrition label for a particular product must identify the:

- Serving size,

- Number of servings in the container or package,

- Total calories per serving and calories from fat (per serving),

- Amounts of certain nutrients expressed in grams or milligrams, and

- "Percent daily values" for certain nutrients, vitamins, and minerals.

The rules governing label formats and serving size declarations were reviewed in Chapter 3. Specific nutrients, and the requirements for declaring per-serving levels on nutrition labels, are discussed in this chapter.

What Nutrients Must Be (or May Be) Declared?

Both FDA and FSIS specify which nutrients may be declared on nutrition labeling. Some substances are mandatory, and some are voluntary. However, nutrients or other food components not specifically identified in the regulations as either mandatory or voluntary may *not* be included in the nutrition label; any information on nonlisted nutrients must be provided outside the nutrition label.

Nutrients Allowed on Labeling

Not only do the regulations identify which nutrients may or must be included on the "Nutrition Facts" labels; they also specify the order in which they must appear.

The nutrients that are required to be included on nutrition labels are listed in **Table 4.1** (see page 4.2). Note that, for vitamin A, the percent that is present as beta-carotene may be declared voluntarily. Declaration of the nutrients listed in **Table 4.2** is voluntary (turn to page 4.3).

TABLE 4.1
Required Nutrients*

Calories (or Total calories or Calories) Calories from fat
"Core" Nutrients— **Fat, total** (or Total fat) Saturated fat (or Saturated) **Cholesterol** **Sodium** **Carbohydrate, total** (or Total carbohydrate) Dietary fiber Sugars **Protein**
Vitamins and Minerals— Vitamin A Vitamin C Calcium Iron
* Allowed synonyms are in (). Items shown in bold should appear in bold on the label.
Source: 21 *CFR* §101.9(c), 9 *CFR* §§317.309(c) and 381.409(c)

Q: Why is declaration of the DRV for protein not mandatory?

A: The percent daily value for protein is required if 1) a protein claim is made for the product, or 2) the product is represented to be for use by infants or children under 4 years old. However, FDA does not require the %DV declaration in other cases because of the associated testing costs and because protein intake is not generally considered a concern.

TABLE 4.2
Nutrients that May Be Declared Voluntarily

"Core" Nutrients—	Vitamins and Minerals*—
Calories from saturated fat (or Calories from saturated)	Vitamin D
Stearic acid	Vitamin E
Polyunsaturated fat (or Polyunsaturated)	Vitamin K
Monounsaturated fat (or Monounsaturated)	Thiamin
Potassium	Riboflavin
Soluble fiber	Niacin
Insoluble fiber	Vitamin B_6
Sugar alcohol	Folate
Other carbohydrate	Vitamin B_{12}
	Biotin
	Pantothenic acid
	Phosphorus
	Iodine
	Magnesium
	Zinc
	Selenium
	Copper
	Manganese
	Chromium
	Molybdenum
	Chloride

* If any of these are added as a nutrient supplement, or if a claim is made about a voluntary vitamin or mineral, a nutrition label declaration is required.

Source: 21 CFR §101.9(c), 9 CFR §§317.309(c) and 381.409(c)

Making Declarations

Specific requirements for declaring nutrient weights and daily values are explained below.

Calories

After serving size and servings per container, the next information required is "calories" and "calories from fat." These values are provided on a per-serving basis. As shown in **Table 4.3** below, calories and calories-from-fat are expressed in 5- or 10-calorie increments, depending upon the declared value. A zero declaration is allowed when the per-serving level is below 5 calories. In addition, the synonym "energy" may be used in conjunction with calories by including it in parentheses immediately after "calories".

"Core" Nutrients

Information on key nutrients is listed after the calorie declarations. To simplify this discussion, this guide uses the term "core" nutrients to refer to those nutrients for which weights *and* percent daily values are declared — as opposed to calories (discussed above) or vitamins and minerals, for which only percent daily values are provided. Some of these core nutrients are mandatory (e.g., total fat) and others are voluntary (such as polyunsaturated fat). In either case, when declaring these key nutrients, the label must indicate the amount present on a weight-per-serving basis.

For the "core" mandatory and voluntary nutrients, the weight is given in grams or milligrams, as shown in the sample label in **Figure 3.2**. FDA and FSIS rules for declaring amounts of required or voluntary core nutrients are summarized in Table 4.3. Note that some of the voluntary nutrients must be declared if certain claims are made, or if certain nutrients are included on the label (e.g., stearic acid). The table also summarizes the rules specifying when zero declarations are allowed.

TABLE 4.3
Rules for Declaring Nutrients by Weight

Nutrient	Units to use:	Mandatory or voluntary?	Zero declaration allowed if level is:	For this level:	The required increment* is:
Calories, total (or Total calories or Calories)	cal	Mandatory	< 5	≤ 50 > 50	5 10
Calories from fat	cal	Mandatory	< 5	≤ 50 > 50	5 10
Calories from saturated fat (or Calories from saturated)	cal	Voluntary	< 5	≤ 50 > 50	5 10
Fat, total (or Total fat)	gram	Mandatory [2]	< 0.5	≤ 5 > 5	0.5 1
Saturated fat (or Saturated)	gram	Mandatory [3]	< 0.5	≤ 5 > 5	0.5 1
Stearic acid	gram	Voluntary [4]	< 0.5	≤ 5 > 5	0.5 1
Polyunsaturated fat (or Polyunsaturated)	gram	Voluntary [5]	< 0.5	≤ 5 > 5	0.5 1

TABLE 4.3
Rules for Declaring Nutrients by Weight

Nutrient	Units to use:	Mandatory or voluntary?	Zero declaration allowed if level is:	For this level:	The required increment* is:
Monounsaturated fat (or Monounsaturated)	gram	Voluntary [6]	< 0.5	≤ 5 > 5	0.5 1
Cholesterol	mg	Mandatory	< 2 [7]	> 5 2 to 5	5 (Less than 5 mg)
Sodium	mg	Mandatory	< 5	5 to 140 > 140	5 10
Potassium	mg	Voluntary	< 5	5 to 140 > 140	5 10
Carbohydrate, total (or Total carbohydrate)	gram	Mandatory	< 0.5	≥ 1 < 1	1 (Contains less than 1 gram) or (Less than 1 gram)
Dietary fiber	gram	Mandatory	< 0.5	≥ 1 < 1	1 declaration not required; (Contains less than 1 gram) or (Less than 1 gram)
Soluble fiber	gram	Voluntary [4]	< 0.5	≥ 1 < 1	1 (Contains less than 1 gram) or (Less than 1 gram)
Insoluble fiber	gram	Voluntary [4]	< 0.5	≥ 1 < 1	1 (Contains less than 1 gram) or (Less than 1 gram)
Sugars	gram	Mandatory	< 0.5	≥ 1 < 1	1 declaration not required; [8] (Contains less than 1 gram) or (Less than 1 gram)
Sugar alcohol	gram	Voluntary [9]	< 0.5	≥ 1 < 1	1 (Contains less than 1 gram) or (Less than 1 gram)
Other carbohydrate	gram	Voluntary	< 0.5	≥ 1 < 1	1 (Contains less than 1 gram) or (Less than 1 gram)
Protein	gram	Mandatory [10]	< 0.5	≥ 1 < 1	1 (Contains less than 1 gram) or (Less than 1 gram)

* For low nutrient levels, allowed alternative declarations are given in parentheses. In some cases, declaration is not required when levels are below a specified minimum.

TABLE 4.3
Rules for Declaring Nutrients by Weight

Nutrient	Units to use:	Mandatory or voluntary?	Zero declaration allowed if level is:	For this level:	The required increment* is:

1. Declaration not required if < 0.5 gram of fat per serving.
2. Defined as total lipid fatty acids and expressed as triglycerides.
3. Not required if 1) < 0.5 gram of total fat per serving, 2) no fat or cholesterol claims are made, and 3) "calories from saturated fat" is not declared. Under FDA rules, if saturated fat content is not declared, the statement "Not a significant source of saturated fat" is placed at the bottom of the label except as provided in the rules for simplified formats in 21 *CFR* §101.9(f). [21 *CFR* §101.9(c)(2)(i)]
4. Declaration required if claim is made. (Stearic acid declaration is allowed under FSIS rules.)
5. Required if 1) monounsaturated is declared, or 2) a cholesterol or fatty acid claim is made (other than "fat-free").
6. Required if 1) polyunsaturated fat is declared, or 2) a cholesterol or fatty acid claim is made (other than "fat-free").
7. Declaration not required if no fat, fatty acid, or cholesterol claim is made. Under FDA rules, if cholesterol is not declared, the statement "Not a significant source of cholesterol" is placed at the bottom of the label except as provided in the rules for simplified formats in 21 *CFR* §101.9(f). [21 *CFR* §101.9(c)(3)]
8. Sugars are the sum of all mono- and disaccharides. Declaration not required if no claim is made concerning sugars, sweeteners, or sugar alcohol. Under FDA rules, if not declared, the statement "Not a significant source of sugars" is placed at the bottom of the label except as provided in the rules for simplified formats in 21 *CFR* §101.9(f). [21 *CFR* §101.9(c)(6)(ii)]
9. Required when claim is made about sugars or sugar alcohol and sugar alcohol is present.
10. Additional requirements apply, depending on whether the product is for adults or children.

Source: 21 *CFR* §101.9(c), 9 *CFR* §§317.309(c) and 381.409(c)

Vitamins and Minerals

As noted earlier, for a vitamin or mineral, only the percent daily value, and not the nutrient amount, is included on the nutrition label. Procedures for determining %DV are outlined below.

Percent Daily Values

"Core" nutrients are declared by weight in grams or milligrams on the nutrition label. A "percent daily value" (or %DV) is also provided, both for the core nutrients *and* for vitamins and minerals.

Daily Reference Values

For most of the so-called core nutrients, daily reference values (DRVs) have been established based on a reference caloric intake of 2,000 calories. These are published for the purpose of declaring %DV (listed in a column to the right of the column identifying the core nutrients by weight). For a nutrient with an established DRV, the amount of the nutrient in the product is simply divided by the established DRV and multiplied by 100, and the result (i.e., the %DV) is expressed to the nearest whole percent.

The percent daily value column is left blank when no DRV is established, such as when declaring sugars. Furthermore, %DV is not required for protein unless 1) a protein claim is made on the product, or 2) the product is represented as being intended for use by infants or children less than 4 years of age [21 *CFR* §101.9(c)(7)(i); 9 *CFR* §§317.309(c)(7)(i), 381.409(c)(7)(i)]. More information on products intended for infants, children, and mothers is provided later in this chapter. The current DRVs are listed in **Table 4.4**.

TABLE 4.4
Daily Reference Values (DRVs)*

"Core" Nutrient	DRV (grams, unless specified otherwise)
Fat	65
Saturated fatty acids	20
Cholesterol	300 mg
Total carbohydrate	300
Fiber	25
Sodium	2400 mg
Potassium	3500 mg
Protein [1]	50 [2]

* Use these DRVs to determine and declare percent daily values for "core" nutrients.

[1] Declaration of percent daily value for protein is not always required.
[2] RDIs have also been established for protein; see discussion of products intended for infants and children.

Source: 21 CFR §101.9(c)(9) and 9 CFR §§317.309(c)(9), 381.409(c)(9)

Reference Daily Intakes (RDIs)

For vitamins and minerals, FDA and FSIS have adopted reference daily intakes (RDIs) for the purposes of declaring percent daily values. (Recall that only the daily values are declared for these substances, except in the case of dietary supplements, which are discussed in Chapter 8.) Current RDI values are listed in **Table 4.5** on the following page.

Note that FSIS has proposed amendments to its RDI list (December 13, 1996; 61 FR 65490). The agency intends to complete that action during October 1998.

Use these RDIs to declare percent daily values. This list indicates the order in which vitamins and minerals are to be declared on nutrition labeling, except on dietary supplements (as discussed in Chapter 8).

TABLE 4.5
Reference Daily Intakes for Vitamins and Minerals

Required vitamins and minerals—	
Vitamin A	5,000 International Units
Vitamin C	60 milligrams
Calcium	1,000 milligrams
Iron	18 milligrams
Required if added, or if a claim is made—	
Vitamin D	400 International Units
Vitamin E	30 International Units
Vitamin K	80 micrograms
Thiamin	1.5 milligrams
Riboflavin	1.7 milligrams
Niacin	20 milligrams
Vitamin B_6	2.0 milligrams
Folate	400 micrograms
Vitamin B_{12}	6 micrograms
Biotin	300 micrograms
Pantothenic acid	10 milligrams
Phosphorus	1,000 milligrams
Iodine	150 micrograms
Magnesium	400 milligrams
Zinc	15 milligrams
Selenium	70 micrograms
Copper	2.0 milligrams
Manganese	2.0 milligrams
Chromium	120 micrograms
Molybdenum	75 micrograms
Chloride	3,400 milligrams

Source: 21 *CFR* §101.9(c)(8); 9 *CFR* §§317.309(c)(8), 381.409(c)(8)

In declaring %DV for vitamins and minerals, the following synonyms may be used:

- *Vitamin C*—Ascorbic acid,
- *Thiamin*—Vitamin B_1,
- *Riboflavin*—Vitamin B_2, and
- *Folate*—either Folic acid or Folacin.

Synonyms are placed in parentheses immediately following the name of the nutrient, except folic acid or folacin may be listed without parentheses in place of folate under 21 *CFR* 101.9§(j)(8)(v).

Rounding

Unlike the core nutrients, the percent daily values for vitamins and minerals is not simply expressed to the nearest whole percent. The regulations specify how rounding should be handled for vitamins and minerals, as shown in **Table 4.6**. Special provisions are made for declaring vitamins and minerals present at less than 2% of the RDI level. In addition, when several nutrients are present in insignificant amounts, a simplified label format may be allowed — see Figure 3.4 for a summary of alternative label formats.

Foods for Infants, Children, and Mothers

Frequently, the nutrition labeling requirements differ for products intended for infants, children, and pregnant or lactating women. For example, the "adult" DRV for protein is 50 grams—separate RDIs are established for children less than 4 years of age, infants, pregnant women, and lactating women. These values are 16, 14, 60, and 65 grams of protein, respectively. Percent daily values are not required for protein unless 1) a protein claim is made on the product, or 2) the product is represented as being intended for use by infants or children less than 4 years of age [21 *CFR* §101.9(c)(7)(i); 9 *CFR* §§317.309(c)(7)(i), 381.409(c)(7)(i)].

In addition, special provisions apply when declaring percent daily values for these foods. Separate declarations may be required, and separate RDIs may be specified for an intended group. See 21 *CFR* §101.9(c)(8), (e) and 9 *CFR* §§317.309(c)(8), (e) and 381.409(c)(8) and (e) for specific requirements.

TABLE 4.6
Rounding Rules for Declaring Vitamins and Minerals

Level at which nutrient is present (% of RDI)	Declare % daily value to the nearest % increment
< 2%	0%*
≥ 2 and ≤ 10%	2%
≤ 50%	5%
> 50%	10%

*Declaration of amounts present at less than 2% of the RDI is not required. However, these vitamins and minerals may be declared:

1) as zero, or

2) by using an asterisk and footnote stating "Contains less than 2 percent of the Daily Value of this (these) nutrient(s)."

If vitamin A, vitamin C, calcium, or iron is present at a level below 2% of the RDI, label declaration is not required if the following statement is placed at the bottom of the table of nutrient values: "Not a significant source of _____ (listing the omitted nutrients)."

Source: 21 *CFR* §101.9(c)(8)(iii), 9 *CFR* §§317.309(c)(8)(iii) and 381.409(c)(8)(iii)

Establishing Compliance
(Or, How Close to the Bull's Eye Do I Need to Be?)

The natural variation of nutrient levels in various foods prevents manufacturers from offering consumer products with identical nutrient profiles in each production lot. At the same time, manufacturers typically print substantial quantities of nutrition labels for a given product.

How much "leeway" is there in the FDA and USDA requirements that a nutrition label represent the nutrient levels in a particular package? Fortunately, the regulations provide a range of acceptable levels for the purposes of demonstrating regulatory compliance. In establishing these ranges, the agencies take into account the variability of analytical methods as well as the possibility that indigenous nutrient levels often fluctuate. In addition, they allow compliance to be determined based on a composite sample from a production lot, rather than on a single item, recognizing the realities of manufacturing and production processes.

The criteria used by FDA and FSIS in establishing nutrition labeling compliance are summarized in **Table 4.7**. As a general rule, analytical results for a composite sample must be within 20% of the values stated on the nutrition label. However, where nutrients are added to fortified or fabricated foods (i.e., for so-called Class I nutrients), sample results must demonstrate nutrient levels at least equal to the declared values.

FDA may provide even more flexibility in extreme circumstances, such as where compliance is not "technologically feasible". Under 21 *CFR* §101.9(g)(9), the agency may permit an alternative means of compliance; however, the manufacturer must first request an exemption in writing.

TABLE 4.7
Establishing Regulatory Compliance Based on Sample Analysis

Criterion	*FDA Rules*	*FSIS Rules*
Sample	a composite of 12 subsamples (consumer units), taken 1 from each of 12 randomly selected shipping cases representing a lot	a composite of a minimum of 6 consumer units, each from a production lot (or randomly chosen to represent a production lot)
Lot	a collection of primary containers or units of the same size, type, and style produced under uniform conditions, designated by a common container code (or a day's production if no code is available)	a set of consumer units from one production shift *or* a collection of consumer units of the same size, type, and style produced under uniform conditions and designated by a common container code
Analysis	unless otherwise specified in §101.9(c), samples are analyzed using methods given in "Official Methods of Analysis of the AOAC International" (if no AOAC method is appropriate, another reliable procedure may be used)	samples analyzed in accordance with the "Chemistry Laboratory Guidebook" or, if no USDA method is appropriate, using methods given in "Official Methods of Analysis of the AOAC International" (or another reliable procedure) [1]
Class I nutrients — added nutrients in fortified or fabricated foods	the nutritional content of the composite must be at least equal to the value declared for that nutrient on the label [2]	the nutritional content of the composite must be at least equal to the value declared for that nutrient on the label [2]

TABLE 4.7
Establishing Regulatory Compliance Based on Sample Analysis

Criterion	FDA Rules	FSIS Rules
Class II nutrients — naturally occurring (i.e., indigenous) nutrients: – vitamin, mineral, protein, total carbohydrate, dietary fiber, other carbohydrate, polyunsaturated or monounsaturated fat, and potassium	the nutrient content of the composite must be $\geq$ 80% of value declared on label [2]	the nutrient content of the composite must be $\geq$ 80% of value declared on label [2]
– calories, sugars, total fat, saturated fat, cholesterol, or sodium	the nutrient content of the composite must not be > 20% over the value declared on the label [2]	the nutrient content of the composite must not be > 20% over the value declared on the label [2]
Serving size	compliance determinations are based on the metric measure for serving size stated on the label	compliance determinations are based on the metric measure for serving size stated on the label
Database values	compliance may be established using an FDA-approved database and where samples are handled to prevent nutrition loss	compliance provisions do not apply to single-ingredient, raw meat products (including those that have been previously frozen), when labeling is based on representative values in the USDA National Nutrient Data Bank or the Agriculture Handbook No. 8 series

[1] Individual subsample results may be averaged, or a composite sample may be analyzed. In either case, the analytical result will be treated as a composite for compliance determination.

[2] Reasonable differences are allowed due to the variability of analytical methods and within current good manufacturing practice.

Source: 21 CFR §101.9(g), 9 CFR §§317.309(h) and 381.409(h)

Part C
CLAIMS MADE ON PRODUCT LABELS

- *Chapter 5* — NUTRIENT CONTENT CLAIMS

- *Chapter 6* — HEALTH CLAIMS

- *Chapter 7* — IMPLIED & DESCRIPTIVE CLAIMS

Chapter 5—NUTRIENT CONTENT CLAIMS

In This Chapter...

- What is a **nutrient content claim**?

- Which claims are **allowed** by FDA and FSIS, and what are the criteria? How may various **terms** be used in making these claims?

- Is it possible to **petition** for approval of a new claim? Does FSIS have special approval requirements?

Recognizing some inconsistencies and weaknesses in the types of claims being made on product labels, the federal government has created a framework for making statements about the nutritional content of a product. For example, claims such as "low in cholesterol and saturated fat" and "sodium free" may be made only in accordance with established criteria. Under the NLEA, FDA has been required by law to develop regulations governing these types of statements. FSIS has adopted rules that parallel the FDA regulations.

Because they describe the nutrients in a given food product, these claims are sometimes referred to as "nutrient descriptors"; however, the more common phrase is *nutrient content claim*. In the NLEA, Congress defined this term as a statement that "expressly or by implication characterizes the level of any nutrient" [FDCA Section 403(r)(1)(A)]. This definition has guided the development of the FDA and FSIS regulations.

The FDA Modernization Act

Congress recently passed a bill amending the FDCA with respect to how manufacturers may make new nutrient content and health claims, among other things. The FDA Modernization Act of 1997 allows new claims if they are based on an authoritative statement of a U.S. government scientific body with official responsibility for public health protection or research. However, manufacturers first must submit premarket notification to FDA and allow 120 days for agency review. See **page 5.20** for more discussion and the full text of these provisions.

What Is a Nutrient Content Claim?

Most nutrient content claims are one of two distinct types of statements — express or implied:

- An *expressed* nutrient content claim is any direct statement about the level or range of a nutrient in a food, such as "low sodium" or "contains 100 calories."

- *Implied* nutrient content claims are statements suggesting that a particular nutrient is absent or is present at a certain level (e.g., claiming "high in oat bran" to indicate high fiber content).

Implied nutrient content claims (e.g., "healthy") are discussed in Chapter 7. Additionally, keep in mind that nutrient content claims are statements only about nutrient levels. Statements concerning how a particular food or nutrient relates to a disease or health-related condition are known as "health claims"— these are discussed in Chapter 6. Also, where FDA makes special provisions for dietary supplements, they are noted in Chapter 8.

How Are These Statements Regulated?

First, the agencies developed a set of general provisions, which set out requirements for any statement that is considered a nutrient content claim. In addition, certain claims have been authorized — these are listed in **Table 5.1**. Use of terms not specifically defined by FDA or FSIS is prohibited. However, as discussed later, new nutrient content claims are being added to the regulations, and manufacturers may petition for specific provisions.

- Under the FDA rules, general principles for making nutrient content claims are outlined in 21 CFR §101.13, while specific allowed claims are described in Part 101, Subpart D (§§101.54-101.69).

- FSIS has comparable requirements for meat and poultry products — general principles appear in 9 CFR §§317.313 and 381.413, respectively, while specific claims are described in §§317.354-317.363 and §§381.454-381.463. Labeling applications for nutrient content claims under the FSIS rules are addressed in §317.369 and §381.469.

TABLE 5.1
Nutrient Content Claims Authorized by FDA & FSIS

Claim allowed	FDA Rules (in 21 CFR Part 101, Subpart D)	FSIS Rules (in 9 CFR)
"good source," "high," and "more"	§101.54	§§317.354, 381.454
"light" or "lite"	§101.56	§§317.356, 381.456
calorie content	§101.60	§§317.360, 381.460
sodium content	§101.61	§§317.361, 381.461
fat, saturated fat, fatty acid, and cholesterol content	§101.62	§§317.362, 381.462
implied nutrient content claims and related label statements (e.g., "healthy")	§101.65	§§317.363, 381.463
butter	§101.67	-----
"high potency"*	§101.54(f) and (g)	-----
antioxidants*		-----

*Discussed in Chapter 8 because they pertain primarily to dietary supplements.

Which Statements Are *Not* Nutrient Content Claims?

Of course, not every label statement is considered a nutrient content claim. Statements about the nature of a product are not nutrient content claims, unless they are made in a context that would make them an implied claim. The following types of label statements are generally not considered implied claims and thus are typically not subject to the regulatory requirements for nutrient content claims:

- Claims that a specific ingredient or food component is absent from a product, when the purpose is to facilitate avoidance of the substance because of allergies, intolerance, religious beliefs, or dietary practices such as vegetarianism (e.g., "100% milk free")

- Statements about a substance that does not have

a nutritive function, (e.g., "contains no preservatives" or "no artificial colors")

- Claims about the presence of an ingredient that is perceived to add value to the product (e.g., "made with real butter," or "contains honey")

- Statements of identity in which an ingredient constitutes essentially 100% of the food (e.g, "corn oil")

- A statement of identity that names, as a characterizing ingredient, a substance associated with a nutrient benefit (e.g., "corn oil margarine") unless the claim is made in a context in which labeling statements, symbols, vignettes, or other forms of communication suggest that a nutrient is absent or present in a certain amount

- A label statement on a food for special dietary use, made in compliance with 21 *CFR* Part 105, where the claim identifies the special diet of which the food is intended to be a part.

Making Claims

The NLEA confirmed FDA's authority to regulate nutrient content claims—statements such as "light" or "good source"— on food labels. FSIS has the authority under the FMIA and PPIA to regulate labeling. Products must bear accurate labels, and they may not make claims about nutrient levels unless the claims use terms that are defined and designated in the regulations. Also, these claims must be stated in such a way that the public will be able to comprehend the claim and understand its significance in the context of total daily diet.

General Concepts

As noted earlier, FDA and FSIS outline general principles for nutrient content claims. FDA rules appear in 21 *CFR* §101.13, and parallel FSIS requirements are found in 9 *CFR* §§317.313 and 381.413. *All* products with labels making nutrient content claims must 1) comply with the general principles, 2) use terms as defined in one or more of the specifically authorized nutrient content claims, and 3) provide "Nutrition Facts" labeling, even if the product would otherwise qualify for an exemption.

Essentially, nutrient content claims fall into one of two categories (other than implied claims, which are discussed in Chapter 7):

1. Some claims compare nutrient levels in the labeled product to levels in a comparable food (e.g., "less fat", "more vitamin C") — these are called "relative" claims.

2. Other claims simply express the level of a given nutrient in the labeled product, or the absence thereof (e.g., "fat free" or "low cholesterol").

Simple, explicit statements about nutrient levels are allowed without any disclaimers. For example, a label may state "100 calories" or "3 grams of fat" because nothing is implied about whether this is a high or low level in the context of daily diet, or how the labeled product compares to similar foods. Statements implicitly characterizing nutrient levels are subject to a variety of restrictions, however.

The basic provisions for making nutrient content claims are illustrated in **Figure 5.1** (pages 5.4–5.6). Note that the figure documents the requirements as published by FDA. The FSIS rules have some minor differences, such as exemptions for products already named at the time that the nutrition labeling program was put into place [see §§317.313(q) and 381.413(q)].

"Low" and "Free"

Statements indicating that a food is "__ free" (e.g., sodium free) or "low" in a nutrient must comply with specific regulations for making the claim — the regulations prohibit a label from simply including these claims when the food inherently meets the required criteria. Thus, if a food has not been specially processed, altered, or formulated to make the nutrient qualify for the claim, a disclaimer must be included (e.g., "celery, a fat-free food"). (Note also that, as shown in the figure, additional requirements apply to "___ fat free" claims.)

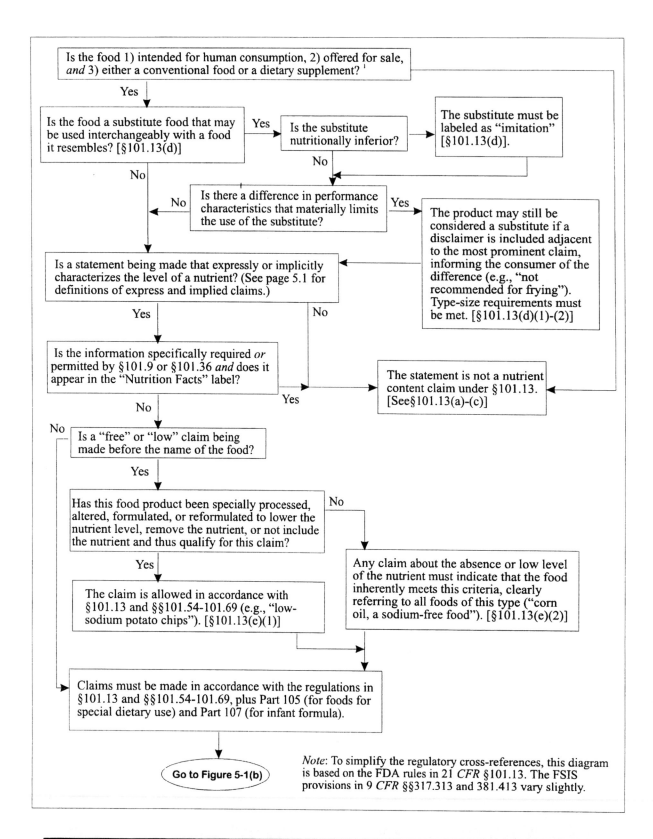

FIGURE 5.1(a)
General Requirements for Nutrient Content Claims

5.4 NUTRIENT CONTENT CLAIMS

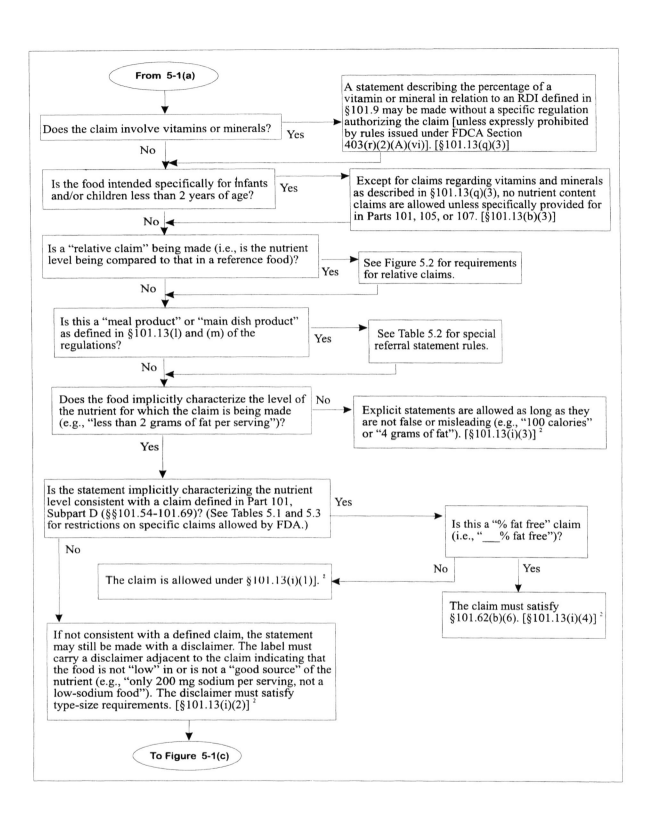

FIGURE 5.1(b)
General Requirements for Nutrient Content Claims

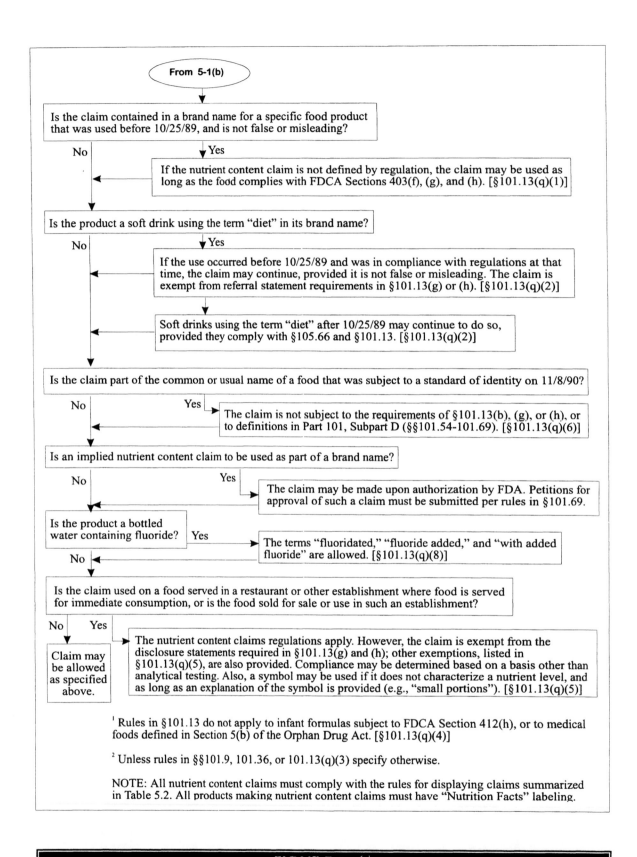

FIGURE 5.1(c)
General Requirements for Nutrient Content Claims

Relative Claims and Reference Foods

The general principles outlined by FDA and FSIS explain how relative claims must be made. As shown in **Figure 5.2** (page 5.8), the labeled food must be compared to an appropriate reference food; these relative claims are made with respect to a particular nutrient, such as fat or cholesterol.

In the claim, the identity of the reference food must be stated, plus the percentage or fraction of the nutrient amount in the reference food by which the labeled food differs. For example, a claim could state "1/3 fewer calories than [reference food]" or "50% fewer calories than [reference food]", as long as the reference food is identified. Selection of an appropriate reference food is a key part of regulatory compliance, and is based partly on which type of claim is being made. As shown in the figure, the rules for "less", "fewer", or "more" claims allow the use of a dissimilar product in the same food category—potato chips can be compared to pretzels, for example. On the other hand, for "light" claims, the nutrient value for the reference food must be based on a broad base of foods of that type, such as an average of the three top brands in the category.

Finally, note that if the reference food qualifies for a "low" claim for the nutrient in question, the labeled product is prohibited from making a claim about decreased levels for that nutrient.

Miscellaneous Items

Before discussing the individual claims allowed under the regulations, some miscellaneous regulatory provisions are reviewed.

- *Meals and main dishes*— For the purposes of making claims, FDA specifies definitions for meal-type and main-dish products; the FSIS regulations contain a definition only for meal-type products. These products are defined in terms of content and weight per serving. Rules for nutrient content claims often specify different "trigger" levels and other requirements for meal-type or main-dish products.

- *Spelling*— Reasonable variations in the spelling of the terms defined in the nutrient content claim regulations are allowed (e.g., "lo" or "hi").

- *Prominent location*— Certain information required by the nutrition labeling rules must be placed next to the "most prominent claim." Determining which use of a claim is the most prominent location is specified in the regulations; in descending order, the prominence is: 1) a claim immediately next to the statement of identity on the principal display panel (PDP), 2) another claim on the PDP, 3) a claim on the information panel, and 4) a claim on other parts of the labeling.

- *Criteria for claims*— For the purpose of making statements about nutrient levels, the reference amount customarily consumed (RACC) for the product or category is used (unless specified otherwise in a particular regulation). Recall that the RACC values are published in §101.12(b) of the FDA regulations and §§317.312(b) and 381.412(b) of the FSIS rules. If the labeled serving size differs from the RACC, the basis for the nutrient content claim must be explained on the label; the claim must be followed by the criteria for the claim (e.g., "very low sodium, 35 mg or less per 55 grams").

- *"Nutrition Facts" labeling*— Information placed within the nutrition label (either required or voluntary) is not considered a claim.

- *Determining compliance*— Generally, analytical testing is required for determining compliance with these FDA rules. However, restaurant-type foods may be evaluated based on other "reasonable" means. Additionally, FSIS allows the use of nutrient data bank values in many cases.

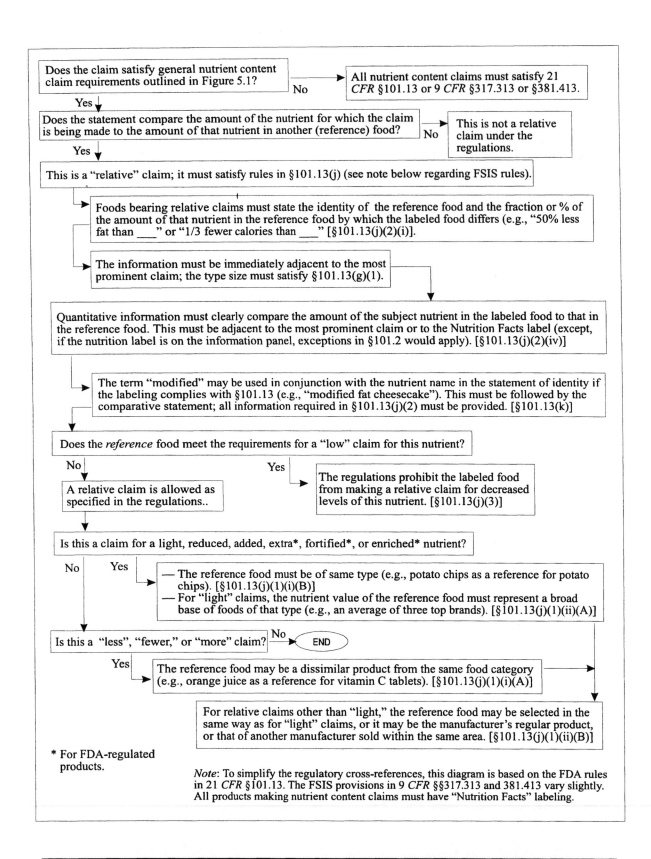

FIGURE 5.2
Requirements for "Relative" Nutrient Content Claims

Referral Statements

Nutrient content claims must be displayed according to certain rules; type-size requirements, among other things, apply. FDA requires referral statements adjacent to these claims, as outlined in **Table 5.2** below.

TABLE 5.2
Rules for Displaying Nutrient Content Claims*

Type size—

- A nutrient content claim must be in type size no larger than two times the statement of identity and must not be unduly prominent in style compared to the statement of identity

Referral statement without disqualifying nutrient levels—

- Food for which a nutrient content claim is made must include a prominent referral statement in immediate proximity to the claim:
  "See [*appropriate panel*] for nutrition information"
- If the claim appears on the same panel as the nutrition information, the referral statement may be omitted
- Use easily legible boldface print or type, in distinct contrast to other printed or graphic matter, and in a size no less than that required by §101.105(i) for the net quantity of contents statement, (if the size of the claim is less than two times the required size of the net quantity of contents statement, the referral statement should be no less than ½ the size of the claim but no smaller than 1/16 of an inch, unless the package complies with §101.2(c)(5), in which case the referral statement may be in type of not less than 1/32 thirty-second of an inch)
- The referral statement must be immediately adjacent to the nutrient content claim and may have no intervening material other than, if applicable, 1) information in the statement of identity or 2) other information required to be presented with the claim (e.g., relative claims), or information required under Subpart D (e.g., §§101.54)
- If the claim appears on more than one panel, the referral statement must be adjacent to the claim on each panel (except for the panel with the nutrition label)
- If a single panel contains multiple nutrient content claims or a single claim repeated several times, a single referral statement may be made

Referral statement with disqualifying nutrient levels—

- In addition to the referral statement described above, certain foods must disclose that a nutrient exceeding specified levels is present as follows: "See [*appropriate panel*] for information about [*nutrient requiring disclosure*] and other nutrients," (e.g., "See side panel for information about total fat and other nutrients.")
- The following foods trigger the disclosure requirements:
  - Foods with more than 13.0 g of fat, 4.0 g of saturated fat, 60 milligrams (mg) of cholesterol, or 480 mg of sodium per RACC, per labeled serving (except food intended specifically for use by infants and children less than 2 years of age, or foods in the special cases listed below)
  - For foods with RACC of 30 g or less, or 2 tablespoons or less, the nutrient levels are also per 50 grams
  - For dehydrated foods that must be reconstituted before consumption with water or a diluent containing an insignificant amount of all nutrients per RACC, the per 50-g criterion refers to the 'as prepared' form)
  - If a food is a meal product as defined in §101.13(l), the trigger levels are more than 26 g of fat, 8.0 g of saturated fat, 120 mg of cholesterol, or 960 mg of sodium per labeled serving
  - For a food that is a main dish product as defined in §101.13(m), the trigger levels are more than 19.5 g of fat, 6.0 g of saturated fat, 90 mg of cholesterol, or 720 mg of sodium per labeled serving

* This summarizes FDA requirements. While type-size and other requirements do apply, FSIS rules do not require the "See referral" statements identified above.

Source: 21 *CFR* §101.13(f)-(h), 9 *CFR* §§317.313(f)-(g), 381.413(f)-(g)

Specific Claims Allowed

As noted previously, some claims describe the level of a certain nutrient ("low fat"), while others compare the labeled food to a reference product. Whether the first type of claim is allowed depends on if absolute criteria are met for the nutrient in question. Other claims compare nutrient levels in the labeled product to levels in a comparable food (e.g., "less fat")—these are called relative claims, and may be made only if numerous criteria are met pertaining to the content of the labeled food and the selection of the reference food.

Nutrients

As shown in **Table 5.3**, the regulations define a framework for making claims on the following:

- Calorie content,
- Sugar,
- Sodium content,
- Saturated fat,
- Total fat, and
- Cholesterol.

These rules define criteria for using specific terms, including "free," "low," "reduced," and "less." For example, to make a "free" claim, a food must have a nutrient content below specified levels. "Low" claims are also allowed based on absolute criteria for nutrient content, as shown in the table.

Synonyms

Certain variations of the listed terms are also specified in the regulations — in addition to "calorie free", the phrase "free of calories" may be used, for example. However, manufacturers have criticized the regulations for being inflexible with respect to allowed synonyms; FDA has considered revising its rules, as discussed later in this chapter.

TABLE 5.3
Requirements for Claiming Specific Nutrients*

Nutrient being claimed	— Level being claimed —		
	Free	Low	Reduced/Less
Calorie content	< 5 calories per RACC and per labeled serving	≤ 40 calories per RACC (and per 50 g if the RACC is ≤ 30 g or 2 Tbsp)	at least 25% less calories per RACC than reference food
Sugar	< 0.5 g sugars per RACC and per labeled serving	no basis provided	at least 25% less sugar per RACC than reference food
Sodium content [1]	< 5 mg per RACC and per labeled serving, and no ingredient is added that is understood to contain sodium [2]	Very low: ≤ 35 mg per RACC (and per 50 g if the RACC is ≤ 30 g or 2 Tbsp) Low: ≤ 140 mg per RACC (and per 50 g if the RACC is ≤ 30 g or 2 Tbsp)	at least 25% less sodium per RACC than reference food
Saturated fat (fatty acid claims)	< 0.5 g saturated fat and < 0.5 g trans fatty acid per RACC and per labeled serving, and no ingredient is added that is understood to contain saturated fat [2]	≤ 1 g saturated fatty acids per RACC and no more than 15% of calories from saturated fatty acids	at least 25% less saturated fat per RACC than reference food

TABLE 5.3
Requirements for Claiming Specific Nutrients*

Nutrient being claimed	— Level being claimed —		
	Free	Low	Reduced/Less
Total fat	< 0.5 g per RACC and per labeled serving, and no ingredient is added that is understood to contain fat [2, 3]	≤ 3 g per RACC (and per 50 g if the RACC is ≤ 30 g or 2 Tbsp)	at least 25% less fat per RACC than reference food
Cholesterol	< 2 mg cholesterol per RACC and per labeled serving, in foods with 1) ≤ 2 g saturated fatty acids per RACC, and 2) ≤ 13 g total fat per RACC (and per 50 g if the RACC is ≤ 30 g or 2 Tbsp); where no ingredient is added that is understood to contain cholesterol [2, 4]	≤ 20 mg cholesterol per RACC (and per 50 g if the RACC is ≤ 30 g or 2 Tbsp), in foods with 1) ≤ 2 g saturated fatty acids per RACC, and 2) ≤ 13 g total fat per RACC (and per 50 g if the RACC is ≤ 30 g or 2 Tbsp) [4]	at least 25% less fat per RACC than reference food that has a significant (≥ 5%) market share; in foods with 1) ≤ 2 g saturated fatty acids per RACC, and 2) ≤ 13 g total fat per RACC (and per 50 g if the RACC is ≤ 30 g or 2 Tbsp) [4]

*One of the terms defined in the regulations must be used (e.g., "low calorie" or "nonfat" or "sodium free"), and rules for making relative claims apply, including restrictions on "reduced" claims when the reference food also qualifies for a "low" claim. Special restrictions apply to meal and main-dish products. Also, all products making these claims must satisfy general rules in 21 *CFR* §101.13, 9 *CFR* §§317.313, or 381.413.

[1] See Table 5.4 entry for "light" claims, which sometimes allow claims of "light in sodium"
[2] Except if the ingredient is noted in the ingredient statement with an asterisk and a footnote stating " * adds a trivial amount of _____ *[sodium, fat, etc.]*" or other footnote specified in the regulations
[3] A 100% "fat free" claim may be made only on foods that meet the criteria for fat free, contain less than 0.5 g of fat per 100 g, and contain no added fat
[4] If the 13-g total fat limit is exceeded, special disclosure requirements apply; disclosures must be placed in immediate proximity to the claim

---

Q: When is a formulated food considered to be specially processed, and thus able to make a "low" or "free" claim for a particular nutrient?

A: If a similar food would normally be expected to contain the nutrient, and the labeled food is made in such a manner that it has little or none of the nutrient, then the food is considered specially processed. For example, sodium in canned pears would normally be expected, so a product without sodium could claim "sodium-free" or "low-sodium".

Table 5.4 lists criteria for additional claims. Rather than identifying specific nutrients, these involve statements about nutrient content in relation to other foods (e.g., "more fiber"). Additionally, a claim that a product is "lean," based on fat content, is also allowed, as shown below.

TABLE 5.4
Requirements for Claiming "Good Source", "High", "More", "Light", and "Lean" [1]

Type of claim	Requirements for use of claim
More (*or* fortified, enriched, added, extra, plus)	• Allowed to describe protein, vitamins, minerals, dietary fiber, or potassium (except as limited by §101.13) if the product contains **10% more** of the RDI or DRV per RACC than an appropriate reference food [2] • For meal products or main dishes, the food must contain at least 10% more of the RDI or DRV per 100 g of food when compared to the reference (fortification and relative-claim requirements apply also) [21 *CFR* §101.54(e), 9 *CFR* §§317.354(e), 381.454(e)]
Good source (*or* contains, provides)	• May be used if the food contains **10-19% or more** of the RDI or DRV per RACC than an appropriate reference food for the claimed nutrient [2] • For meal products or main dishes, the food that is the subject of the claim must be identified and the item must contain a nutrient that meets the definition of "good source" [21 *CFR* §101.54(c), 9 *CFR* §§317.354(c), 381.454(c)]
Light (*or* lite)	• If the food derives 50% or more of its calories from fat, its fat content must be reduced by 50% or more per RACC compared to an appropriate reference [2] • If the food derives less than 50% of its calories from fat: 1) calories are reduced by at least 33 1/3% per RACC compared to a reference food, or 2) fat content is reduced by 50% or more per RACC compared to the reference food that it resembles • Products for which the reference food contains ≤ 40 calories and ≤ 3 g of fat per RACC may use the term "light" or "light in sodium" if the sodium content is reduced when compared to the reference food [see §101.56(c), 317.365(c), 381.456(c)] • Meal and main-dish products have special requirements; relative-claim rules apply also
High (*or* rich in, excellent source of)	• May be used if the food contains **20% or more** of the RDI or DRV per RACC than an appropriate reference food for the claimed nutrient [2] • For meal products, main dishes, and dietary supplements, the product must contain a nutrient that meets the definition of "high" [21 *CFR* §101.54(b), 9 *CFR* §§317.354(b), 381.454(b)]
Lean, extra lean	• FDA allows these claims on seafood, game meat, main-dish, and meal products • FSIS allows these claims also, although the FSIS rules specify separate criteria for meal-type, but not main-dish, products *Lean:* – No more than 10 g total fat, 4.5 g saturated fat, and 95 mg cholesterol per RACC and per 100 g (per 100 g and per labeled serving for meals and main dishes) *Extra lean:* – No more than 5 g total fat, 2 g saturated fat, and 95 mg cholesterol per RACC and per 100 g (per 100 g and per labeled serving for meal and main-dish products)

[1] If one of these claims is made with respect to the level of dietary fiber, but the product is not low in total fat as defined in the regulations, the level of total fat must be disclosed.
[2] Fortification and relative-claim requirements may apply also. See Figure 5.2 concerning relative claims.
Note: All products making these claims must also satisfy general rules in 21 *CFR* §101.13, 9 *CFR* §317.313, or 9 *CFR* §381.413. Claims for "healthy" are discussed in Chapter 7; Chapter 8 reviews "high potency" claims and antioxidants.

Regulatory Conflict: Nutrient Content Claims and the "General Definition"

In some cases, other federal regulations governing food products can conflict with these relatively new nutrition labeling requirements. For instance, many FDA and USDA regulations specify so-called standards of identity for various products. These require conformity with specified standards in order for product labels to bear a "standardized" name, such as peanut butter or sausage. Much of the motivation behind these standards if identity has been a desire to ensure that products bearing commonly used names meet consumer expectations regarding composition and other characteristics. However, some standards of identity allow use of names or terms that overlap the NLEA-based rules, leading to inconsistency in nomenclature or labeling.

As discussed earlier, in §§101.54–101.67, FDA has established definitions for specific nutrient content claims together with principles for their use — these rules specify when product labels may claim when a product is "low" in fat, is a "good source" of calcium, etc. The agency has also published a rule in 21 CFR §130.10 concerning "Food standards: Requirements for foods named by use of a nutrient content claim and a standardized term."

The §130.10 "general definition" regulation is intended to allow the manufacturer of a product conforming to a standard of identity to use the standardized product name in conjunction with one of the terms authorized in the nutrient content claims rules. For example, a manufacturer might want to offer "reduced fat peanut butter"—in this case, the product would be subject to FDA's standards of identity, which govern the naming of this standardized food. However, under §130.10, the manufacturer would be allowed to use the "reduced fat" nutrient content claim in conjunction with the "peanut butter" standardized term as long as all regulatory requirements were satisfied.

Milk Standards Revoked

With respect to dairy products, when FDA established the "general definition" rules in §130.10, it allowed manufacturers to use terms such as "nonfat," "reduced fat," and "light" in conjunction with *standardized* terms such as "sour cream." This is where the conflict arose — certain standards of identity for dairy products incorporate terms such as "nonfat," "light," and "lowfat" directly into the standardized names of the foods. Consequently, use of terms such as "nonfat," "light," and "lowfat" in keeping with the standardized dairy product names was inconsistent with the definitions for these same terms under the nutrient content claims rules. For example, where "lowfat" foods complying with §101.62(b) generally must contain less than 3 grams of fat per RACC, standards for "lowfat" milk products in 21 CFR §131.135(a) allowed these foods to contain as much as 2% milkfat, or 5 grams of fat per RACC. Thus, the term "lowfat" had multiple meanings under the regulations.

In light of these observed difficulties, several petitions were submitted to FDA requesting that the agency correct regulatory inconsistencies to promote more consistent dairy labeling and nomenclature and remove product specifications that conflict with authorized nutrient content claims. On November 9, 1995 (60 FR 56541), FDA agreed with the requests and proposed to remove several dairy-product standards, noting that they are 1) inconsistent with food labeling regulations established under NLEA, and 2) unnecessary in light of the general standard in §130.10.

One key topic was the widespread use of percent fat declarations in conjunction with product names (e.g., "2% milk"); consumers rely heavily on these numbers to differentiate among milk products. Thus, some expressed concern about how these declarations would be made once the standards of identity for lowfat milk and other products had been revoked. FDA responded by noting that the nutrient content claim requirements in §101.13(i) allow declarations of percent fat content as long as they are not misleading.

On November 20, 1996, FDA revoked 12 milk and dairy standards of identity and made certain other regulatory changes (61 FR 58991). The following table summarizes the agency's actions. Note that FDA acknowledged consumer reliance on the term "skim," and thus is allowing its use as a synonym for "nonfat" on milk product labels.

TABLE 5.5
FDA Actions on Dairy Product Standards

Product standard or requirement	Citation (21 CFR)	FDA action
sweetened condensed skimmed milk	§131.122	
lowfat dry milk	§131.123	
evaporated skimmed milk	§131.132	
lowfat milk	§131.135	
acidified lowfat milk	§131.136	
cultured lowfat milk	§131.138	standards of identity revoked
skim milk	§131.143	
acidified skim milk	§131.144	
cultured skim milk	§131.146	
sour half-and-half	§131.185	
acidified sour half-and-half	§131.187	
lowfat cottage cheese	§133.131	
dry cream	§131.149	removed reference to the lowfat milk standard
rules for nutrient content claims on fat, fatty acid, and cholesterol content	§101.62(b)(1)	provided for use of "skim" as synonym for "nonfat" on labels of milk products
lowfat yogurt	§131.203	deferred decision on revoking standards
nonfat yogurt	§131.206	

Labeling Applications

FSIS policies differ from those of FDA in that FSIS requires prior approval of labels placed on meat and poultry products. However, to streamline the preapproval process, generic labeling procedures have been established—in many cases, labels do not require FSIS approval, such as when an approved label undergoes minor modification. Additionally, in late 1995 the agency expanded the scope of this policy to provide more flexibility for the regulated community. More types of labels now qualify under the generic approval process.

With respect to nutrient content claims, FSIS has separate rules for labeling applications; these appear in §§317.369 and 381.469 for meat and poultry processors, respectively. Basically, the application rules pertain to:

- New (heretofore unauthorized) claims,
- Synonyms, and
- Implied claims in brand names.

These labeling procedures specify documentation requirements, including data from clinical investigations, and spell out how FSIS will respond after receipt of an application.

Possible Changes Coming

As detailed below, industry has petitioned FDA to fine-tune some of its rules governing nutrient content claims. The manufacturers argue that, if more claims are made, consumers will benefit from the additional labeling information.

In some cases, manufacturers request formal revision of the nutrient content claim regulations to accommodate a given term or phrase. For instance, one such request was granted on June 9, 1997: FDA approved the use of "plus" as a synonym for "added" (62 *FR* 31338). In other cases, petitioners want leeway to make ad hoc labeling decisions.

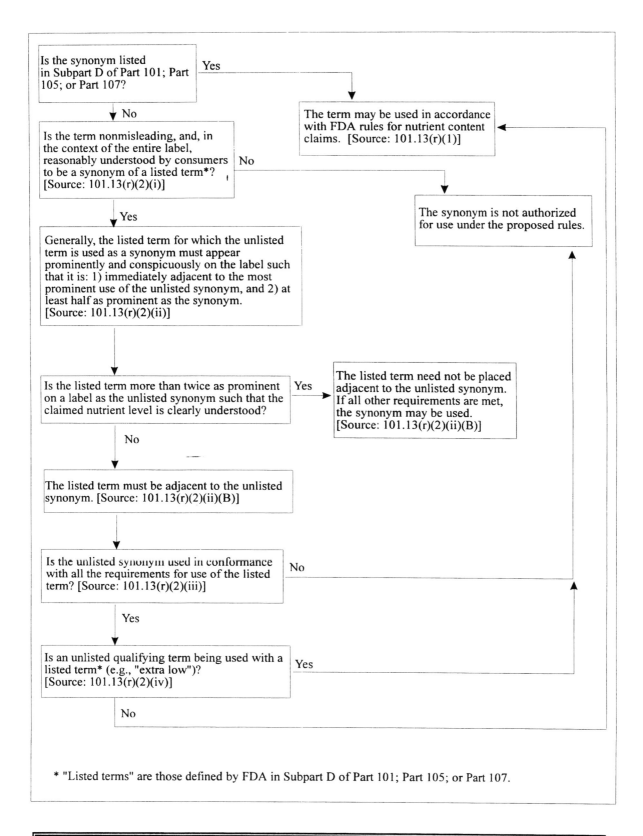

FIGURE 5.3
Using Synonyms Under Proposed Revisions to 21 *CFR* §101.13

Use of Synonyms Is Stringently Limited

Currently, manufacturers must obtain premarket clearance to use terminology not defined in the regulations. Industry petitioners have requested that the agency make the rules more flexible to allow the use of synonyms and implied nutrient content claims on product labels that are understood by consumers to have the same meaning as a defined term — where the defined term would also appear on the labels. This would allow manufacturers more freedom to write creative messages.

In its 1993 nutrient content claims final rule, FDA included a limited number of specific synonyms — but did not promulgate long lists of synonyms or conditions for use of unevaluated terms, because FDCA does not provide for either. Additionally, the agency considered, but rejected, the suggestion that implied nutrition claims defined on the label be permitted. However, it did provide for implied claims on products satisfying definitions for certain expressed claims (e.g., foods that are a good source of fiber could bear "high in oat bran" on the label).

On December 21, 1995, FDA responded to petitioners by proposing to 1) add circumstances under which synonyms describing nutrient content may be used on food labels, and 2) authorize the use of unlisted synonyms as long as they are properly "anchored" to a listed term (60 *FR* 66206). The agency proposed to amend the labeling regulations in §101.13; FDA agreed that the public can benefit if manufacturers have more flexibility in making claims, which after all are intended to assist consumers in maintaining healthy diets.

FDA agrees that more synonyms for defined terms would provide manufacturers with a larger array of nutrient information terms to catch the attention of consumers. However, it is also concerned that an abundance of uncontrolled terms could confuse consumers. The agency believes that an "anchoring" concept (i.e., printing the synonym immediately adjacent to a listed term) may prevent such confusion. Furthermore, FDA proposes that an unlisted term must be unambiguous and clear (e.g., "without any *[nutrient]*" is clearly synonymous with "free of *[nutrient]*").

The decision diagram in **Figure 5.3** (page 5.15) depicts the process for determining if an unlisted synonym may be used on a food product label under the 1995 proposal. The figure also shows how synonyms would need to be formatted.

FDA's anchoring concept would require that an unlisted synonym appear adjacent to a defined term; the defined term would have to be at least half as prominent as the unlisted synonym. This provision would prevent an unlisted term from appearing prominently on a product label while its associated, defined phrase was part of fine print elsewhere on the label.

Fat Substitutes

Another type of rule change being considered by FDA involves nutrient content claims based on reduced availability of fat to the body because of the use of a fat substitute ingredient in a food product. A December 1996 FDA proposal responded to a citizen petition concerning the use of so-called digestibility coefficients in determining the quantity of fat declared on a food label (61 *FR* 67243). The agency has said that it "is undertaking this action to encourage innovation on the part of food manufacturers and to foster a situation that will provide increased product choices for consumers in achieving dietary goals."

This action would add a new 21 *CFR* §101.63 to the FDA regulations. According to agency staff, FDA does plan to complete this rulemaking, although the project is currently low priority.

Counting Calories

Among other things, the NLEA requires that nutrition labels include information on calories from fat and on the quantitative amounts of specified nutrients (e.g., total fat, saturated fat, total carbohydrate) per serving. FDA applies default values of 4, 4, and 9 calories per gram for the purpose of declaring protein, carbohydrate, and fat content, respectively. However, the agency recognizes that many ingredients have caloric values substantially different from these general factors. FDA has provided a number of options for calculating the energy value of foods; for example, calories may be calculated:

- Under 21 *CFR* §101.9(c)(1)(i)(A), by using specific Atwater factors given in Table 13 "Energy Value of Foods-Basis and Derivation," U.S. Department of Agriculture (USDA) Handbook No. 74;

- By multiplying the general factor of 4 calories per g by the amount of total carbohydrate less the amount of insoluble dietary fiber under §101.9(c)(1)(i)(C);

- Under §101.9(c)(1)(i)(D), by using data for specific energy factors for particular foods or ingredients approved by FDA through the food additive or GRAS petition processes in Parts 170 and 171 and provided in Parts 172 or 184; or

- By using bomb calorimetry data under §101.9(c)(1)(i)(E).

FDA rules also define the basic nutrients that are to be declared on labels. These include "total fat" as total lipid fatty acids expressed as triglycerides and "saturated fat" as the sum of all fatty acids containing no double bonds [§101.9(c)(2)].

In late 1994, Nabisco petitioned FDA to amend its food labeling regulations to permit 1) the use of a "digestibility coefficient" or "food factor" in determining the quantity of fat to be declared on the nutrition label, and 2) nutrient content claims to be based on the quantity of fat declared. This action would permit claims on a class of products that contain significantly less available fat compared to an appropriate reference food, but that may not qualify to bear a calorie or fat claim based on the total analytically determined amount of fat in the food. The petition asserted that the nutritional benefit of foods with reduced *available* fat is similar to that of foods with reduced *total* fat, and that these claims would further FDA's goal of promoting healthier diets by encouraging product innovation.

Specifically, the petition requested that FDA amend §101.9(c)(2) by inserting the following language at the end of the first paragraph in that section:

> "Fat content may be calculated by applying a food factor to the actual amount of fat present per serving, using specific food factors for particular foods or ingredients approved by FDA and provided in Parts 172 or 184 of this chapter, or by other means as appropriate."

The change would allow the amount of total fat present per serving to be multiplied by a specific factor approved by FDA, to yield the quantity of fat that is to be declared in nutrition labeling, even though the declared value may be less than the actual amount of fat in the food. FDA notes that the suggested approach (i.e., that the factor used to calculate available fat content be approved by FDA), is similar to the approach taken in §101.9(c)(1)(i)(D), which provides that specific food factors may be used to calculate total caloric content if they have been approved by FDA and provided for in Part 172, Part 184 (or by other means as appropriate). The petitioner also suggested that the agency could permit "self-determination" of a food factor for calculating nutrient availability by a manufacturer, pending agency review of a GRAS petition for the ingredient to which the factor applies.

In essence, the petition suggested 1) that the amount of available (i.e., absorbed/ digestible) fat in an ingredient should be reflected in its "food factor" or "digestibility coefficient," and 2) that manufacturers be permitted to make fat-reduction claims based on the available fat as opposed to the chemically analyzed quantity of fat in the food. Additionally, the petitioner requested that FDA amend §101.9(c)(2) to provide that a food factor be used to calculate the quantity of *all* fatty acids (i.e., saturated fat, polyunsaturated fat, and monounsaturated fat) declared.

Apparently, the agency has anticipated including specific digestibility coefficients that could be used in determining the quantitative declaration of, and the caloric contribution from, these "substitute" fats as a component of their statement of identity in Part 172 or in the GRAS regulations in Part 184. As another mechanism for bringing these issues before the agency, FDA suggested using §101.9(g)(9) as a possible means of requesting the use of specific digestibility coefficients.

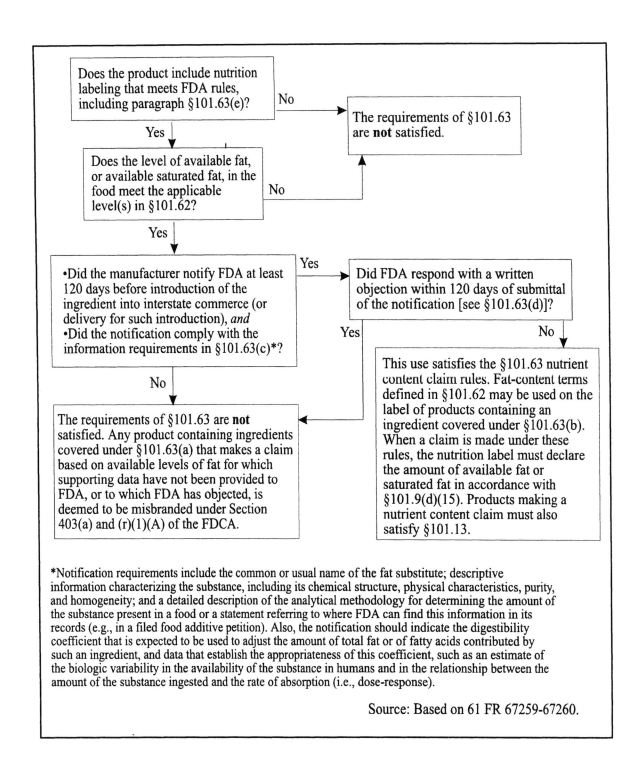

FIGURE 5.4
Proposed Criteria for Making Claims Reflecting "Available" Fat

In the original nutrition labeling rules, FDA was not persuaded to exclude any compounds from the definition of saturated fat on the basis of their physiologic effects. The agency defined saturated fat as "the sum of all fatty acids containing no double bonds" (58 FR 2089). FDA did not address the issue of digestibility or availability of individual fatty acids in its discussion, but did note that an inclusive chemical definition avoids controversy about which saturated fatty acids are associated with increases in blood.

Factors & Coefficients

Dietary fats consist of one, two, or three fatty acid molecules attached to a glycerol "backbone" (i.e., mono-, di-, or triglycerides). Manufactured fat substitutes are designed to lower caloric content through mechanisms such as controlling certain absorption processes or lowering the energy value of certain fatty acids.

This reduction in caloric value could then be incorporated into calculations of fat content for nutrition labeling purposes. For example, if a given substitute reduced calorie levels to 55% of the energy value of conventional fats, it would have a "food factor" of 55%, or 5/9. The digestibility coefficient, on the other hand, would address the *availability* of fat (e.g., it might consider the incomplete absorption of stearic acid from the substitute ingredient).

FDA finds merit in providing a *generic* means of allowing for the digestibility of fat substitutes, rather than in addressing this issue on a case-by-case basis (as it was in the Olestra final rule). The agency acknowledges that technology pertaining to fat-based substitutes is advancing, and that these products appear to offer significant advantages to consumers. **Figure 5.4** illustrates the criteria proposed by FDA.

The agency's 1996 proposal would authorize claims describing the level of *available* fat in a food product. The rules would define nutrient content claims for fat and fatty acids to allow claims for foods containing fat substitutes that have been formulated to limit the amount of fat and fatty acids that can be absorbed and digested by the body, thereby reducing the availability of the fat. Among other things, the proposal defines the circumstances in which the new claim could be used. These requirements are summarized in the diagram. FDA is proposing that claims may be made in food labeling if:

- Appropriate notification procedures are followed, and the agency has not objected to the digestibility coefficient suggested by the manufacturer;

- The food meets the criteria for a fat content claim as specified in §101.62; and

- The food bears appropriate nutrition labeling under §101.63(e).

Notification Required

In a sense, the proposal represents a compromise between maintaining a case-by-case policy and establishing generic procedures: Manufacturers may petition for use of alternative energy factors in nutrition labeling through established procedures for food additive or GRAS petitions. Ideally, FDA believes that it would be useful if factors such as food factors and digestibility coefficients were always listed in the food additive or GRAS regulations — this would allow all information about a compound to be located in one place.

However, not all ingredients used in food are listed there (FDCA does not preclude the use of an ingredient based on a self-determination that a use is GRAS). As a solution, FDA proposes a notification scheme, intended to improve communication, understanding of manufacturer's data, and evaluation of analytical methodology.

Footnotes and Other Changes

FDA has considered how one of these new claims might appear on a package. The agency proposed that manufacturers mark the fat declaration with an asterisk, referring to a footnote informing the consumer that the nutrition information reflects the adjustment of fat content for reduced availability of fat from a substitute ingredient:

> "A number of different possible footnote statements could be used to signal the fact that 'total fat' and any fatty acid content declarations have

been adjusted to reflect reduced availability. For example, direct reference to the adjustment could be made with statements such as 'Fat content adjusted for reduced availability of fat from [*name of ingredient*],' 'Adjusted for reduced absorption of [name of ingredient],' or 'Represents an amount adjusted for absorption of [*name of ingredient*].' (61 *FR* 67252)

To accommodate the new rules, some conforming changes were also proposed. For example, the definition of "total fat" would require that nutrition labels declare total lipid fatty acids, expressed as triglycerides, except that foods making claims under §101.63 may instead state the grams of available fat.

Should This Apply to *All* Claims?

In explaining the proposal, FDA noted that providing for claims based on fat availability raises the question of whether claims for *all* fats should be based on this criterion. The agency "is aware that certain conventional food fats are less available than others (e.g., fats rich in stearic acid, e.g., cocoa butter, are not well-absorbed relative to some other fats) (61 *FR* 67247)." However, FDA is reluctant to include conventional fats under proposed §101.63(a) because few, if any, such fats have undergone testing to determine a digestibility coefficient, i.e., availability.

Moreover, including such fats in the coverage of the proposed regulation would create inconsistencies among nutrition label values, standard food composition tables, and data bases used by consumers and health professionals. If some products continued to declare total analytically determined fat levels, while similar ones chose to declare only available fat, additional inconsistencies would become apparent.

Modernizing Claim Approval

During November 1997, Congress adopted the FDA Modernization Act of 1997. While many components of this bill amended provisions pertaining to medical devices or pharmaceuticals, several sections of the act amend the FDCA food labeling and additive provisions.

Congress' intent was to streamline the claim-approval process. The full text of Sections 301–371, which pertain to nutrient content claims, health claims, and disclosure of food irradiation, are reproduced on the following pages.

Premarket Notification, Scientific Basis

Sections 303 and 304 of the Modernization Act revised FDCA Sections 403(r)(3) and (r)(2) with respect to health claims and nutrient content claims. Both provisions allow a manufacturer to submit premarket notification to FDA within 120 days of introducing a product bearing a new labeling claim.

This permits manufacturers to make statements not previously authorized in FDA's regulations. However, such a claim must be based on an authoritative statement that is currently in effect and is published by a scientific body within the federal government having official responsibility for public health protection or research (e.g., the National Institutes of Health or the Centers for Disease Control and Prevention).

- For **nutrient content claims**, that authoritative statement must identify the nutrient level to which the claim refers.

- **Health claims** made under this statutory provision must rely on an authoritative statement about the relationship between a nutrient and the disease or health-related condition to which the claim refers.

Effect of Claim

If the claim satisfies all applicable requirements (e.g., is stated clearly and accurately represents the authoritative statement on which it relies), it may be made on product labels until 1) FDA prohibits or modifies it through regulation, 2) the agency determines that the manufacturer failed to meet the statutory requirements, or 3) during an enforcement proceeding, a federal court finds that the applicable legal requirements were not met.

FDA Modernization Act of 1997 (Public Law 105-115)

SEC. 301. FLEXIBILITY FOR REGULATIONS REGARDING CLAIMS.
Section 403(r) (21 U.S.C. 343(r)) is amended by adding at the end the following: "(7) The Secretary may make proposed regulations issued under this paragraph effective upon publication pending consideration of public comment and publication of a final regulation if the Secretary determines that such action is necessary—
"(A) to enable the Secretary to review and act promptly on petitions the Secretary determines provide for information necessary to- "(i) enable consumers to develop and maintain healthy dietary practices; "(ii) enable consumers to be informed promptly and effectively of important new knowledge regarding nutritional and health benefits of food; or "(iii) ensure that scientifically sound nutritional and health information is provided to consumers as soon as possible; or
"(B) to enable the Secretary to act promptly to ban or modify a claim under this paragraph. Such proposed regulations shall be deemed final agency action for purposes of judicial review.".

SEC. 302. PETITIONS FOR CLAIMS.
Section 403(r)(4)(A)(i) (21 U.S.C. 343(r)(4)(A)(i)) is amended—
(1) by adding after the second sentence the following: "If the Secretary does not act within such 100 days, the petition shall be deemed to be denied unless an extension is mutually agreed upon by the Secretary and the petitioner.";
(2) in the fourth sentence (as amended by paragraph (1)) by inserting immediately before the comma the following: "or the petition is deemed to be denied"; and
(3) by adding at the end the following: "If the Secretary does not act within such 90 days, the petition shall be deemed to be denied unless an extension is mutually agreed upon by the Secretary and the petitioner. If the Secretary issues a proposed regulation, the rulemaking shall be completed within 540 days of the date the petition is received by the Secretary. If the Secretary does not issue a regulation within such 540 days, the Secretary shall provide the Committee on Commerce of the House of Representatives and the Committee on Labor and Human Resources of the Senate the reasons action on the regulation did not occur within such 540 days.".

SEC. 303. HEALTH CLAIMS FOR FOOD PRODUCTS.
Section 403(r)(3) (21 U.S.C. 343(r)(3)) is amended by adding at the end thereof the following:
"(C) Notwithstanding the provisions of clauses (A)(i) and (B), a claim of the type described in subparagraph (1)(B) which is not authorized by the Secretary in a regulation promulgated in accordance with clause (B) shall be authorized and may be made with respect to a food if-
"(i) a scientific body of the United States Government with official responsibility for public health protection or research directly relating to human nutrition (such as the National Institutes of Health or the Centers for Disease Control and Prevention) or the National Academy of Sciences or any of its subdivisions has published an authoritative statement, which is currently in effect, about the relationship between a nutrient and a disease or health-related condition to which the claim refers;
"(ii) a person has submitted to the Secretary, at least 120 days (during which the Secretary may notify any person who is making a claim as authorized by clause (C) that such person has not submitted all the information required by such clause) before the first introduction into interstate commerce of the food with a label containing the claim, (I) a notice of the claim, which shall include the exact words used in the claim and shall include a concise description of the basis upon which such person relied for determining that the requirements of subclause (i) have been satisfied, (II) a copy of the statement referred to in subclause (i) upon which such person relied in making the claim, and (III) a balanced representation of the scientific literature relating to the relationship between a nutrient and a disease or health-related condition to which the claim refers;
"(iii) the claim and the food for which the claim is made are in compliance with clause (A)(ii) and are otherwise in compliance with paragraph (a) and section 201(n); and
"(iv) the claim is stated in a manner so that the claim is an accurate representation of the authoritative statement referred to in subclause (i) and so that the claim enables the public to comprehend the information provided in the claim and to understand the relative significance of such information in the context of a total daily diet. For purposes of this clause, a statement shall be regarded as an authoritative statement of a scientific body described in sub-clause (i) only if the statement is published by the scientific body and shall not include a statement of an employee of the scientific body made in the individual capacity of the employee.
"(D) A claim submitted under the requirements of clause (C) may be made until- "(i) such time as the Secretary issues a regulation under the standard in clause (B)(i): "(I) prohibiting or modifying the claim and the regulation has become effective, or "(II) finding that the requirements of clause (C) have not been met, including finding that the petitioner has

not submitted all the information required by such clause; or "(ii) a district court of the United States in an enforcement proceeding under chapter III has determined that the requirements of clause (C) have not been met.".

SEC. 304. NUTRIENT CONTENT CLAIMS.

Section 403(r)(2) (21 U.S.C. 343(r)(2)) is amended by adding at the end the following:

"(G) A claim of the type described in subparagraph (1)(A) for a nutrient, for which the Secretary has not promulgated a regulation under clause (A)(i), shall be authorized and may be made with respect to a food if- "(i) a scientific body of the United States Government with official responsibility for public health protection or research directly relating to human nutrition (such as the National Institutes of Health or the Centers for Disease Control and Prevention) or the National Academy of Sciences or any of its subdivisions has published an authoritative statement, which is currently in effect, which identifies the nutrient level to which the claim refers;

"(ii) a person has submitted to the Secretary, at least 120 days (during which the Secretary may notify any person who is making a claim as authorized by clause (C) that such person has not submitted all the information required by such clause) before the first introduction into interstate commerce of the food with a label containing the claim, (I) a notice of the claim, which shall include the exact words used in the claim and shall include a concise description of the basis upon which such person relied for determining that the requirements of subclause (i) have been satisfied, (II) a copy of the statement referred to in subclause (i) upon which such person relied in making the claim, and (III) a balanced representation of the scientific literature relating to the nutrient level to which the claim refers;

"(iii) the claim and the food for which the claim is made are in compliance with clauses (A) and (B), and are otherwise in compliance with paragraph (a) and section 201(n); and

"(iv) the claim is stated in a manner so that the claim is an accurate representation of the authoritative statement referred to in subclause (i) and so that the claim enables the public to comprehend the information provided in the claim and to understand the relative significance of such information in the context of a total daily diet. For purposes of this clause, a statement shall be regarded as an authoritative statement of a scientific body described in sub-clause (i) only if the statement is published by the scientific body and shall not include a statement of an employee of the scientific body made in the individual capacity of the employee.

"(H) A claim submitted under the requirements of clause (G) may be made until- "(i) such time as the Secretary issues a regulation- "(I) prohibiting or modifying the claim and the regulation has become effective, or "(II) finding that the requirements of clause (G) have not been met, including finding that the petitioner had not submitted all the information required by such clause; or "(ii) a district court of the United States in an enforcement proceeding under chapter III has determined that the requirements of clause (G) have not been met.".

SEC. 305. REFERRAL STATEMENTS.

Section 403(r)(2)(B) (21 U.S.C. 343(r)(2)(B)) is amended to read as follows:

"(B) If a claim described in subparagraph (1)(A) is made with respect to a nutrient in a food and the Secretary makes a determination that the food contains a nutrient at a level that increases to persons in the general population the risk of a disease or health-related condition that is diet related, the label or labeling of such food shall contain, prominently and in immediate proximity to such claim, the following statement: 'See nutrition information for ___ content.' The blank shall identify the nutrient associated with the increased disease or health-related condition risk. In making the determination described in this clause, the Secretary shall take into account the significance of the food in the total daily diet.".

SEC. 306. DISCLOSURE OF IRRADIATION.

Chapter IV (21 U.S.C. 341 et seq.) is amended by inserting after section 403B the following:"DISCLOSURE "SEC. 403C. (a) No provision of section 201(n), 403(a), or 409 shall be construed to require on the label or labeling of a food a separate radiation disclosure statement that is more prominent than the declaration of ingredients required by section 403(i)(2). "(b) In this section, the term 'radiation disclosure statement' means a written statement that discloses that a food has been intentionally subject to radiation.".

SEC. 307. IRRADIATION PETITION.

Not later than 60 days following the date of the enactment of this Act, the Secretary of Health and Human Services shall make a final determination on any petition pending with the Food and Drug Administration that would permit the irradiation of red meat under section 409(b)(1) of the Federal Food, Drug, and Cosmetic Act. If the Secretary does not make such determination, the Secretary shall, not later than 60 days following the date of the enactment of this Act, provide the Committee on Commerce of the House of Representatives and the Committee on Labor and Human Resources of the Senate an explanation of the process followed by the Food and Drug Administration in reviewing the petition referred to in paragraph (1) and the reasons action on the petition was delayed.

Chapter 6—HEALTH CLAIMS

> ## In This Chapter...
>
> - What is a **health claim**? How do "medical foods" differ from foods requiring health claims?
> - Which claims are **allowed** by FDA? Does FSIS plan to allow health claims?
> - What is **required** when making these claims?
> - Is it possible to **petition** for approval of a new claim? What **changes** are on the horizon?

One of the more controversial components of the NLEA (and of FDA's implementing regulations) concerns the ability of food manufacturers to make statements regarding relationships between their products and various diseases and other health problems. Only when FDA has specifically authorized a particular claim may a food label include such statements. The current regulations, and some efforts to revise them, are discussed in this chapter.

Regulating Health Claims

Under the FDA rules, a "health claim" is a statement that explicitly or implicitly characterizes the relationship between any substance and a disease or health-related condition—thus, these are sometimes referred to as disease-prevention claims. Dietary supplements are also subject to these rules; special nutrition labeling requirements for those products are discussed in Chapter 8.

Regulatory History

For many years, disease-prevention claims were prohibited from food labeling by FDA. After all, those claims were more traditionally made on drugs, not foods—possible links between diet and health were not always well-understood. Eventually, FDA relaxed its policy in 1985, and in 1987 formally proposed rules that would allow "health messages" concerning the value of the product in reducing disease risk. These messages were required to be

- Consistent with generally recognized medical and nutritional principles,
- Truthful and not misleading, and
- Accompanied by nutritional labeling.

Although the health-message regulations were never actually finalized, FDA essentially assured manufacturers they would be considered to be in compliance if their label statements complied with the proposal.

The 1987 proposal was withdrawn in 1990, and new rules were proposed. This proposal focused on a handful of specific health messages (claims). FDA promised to evaluate scientific findings and develop "model label statements" for manufacturers to use. However, because the NLEA was passed in November 1990, the agency reproposed its 1990 regulations to conform to specific NLEA provisions. Among other requirements, under the revised FDCA, food products are considered misbranded if they bear label claims that characterize the relationship of a nutrient to a disease or health-related condition unless the claims are made in accordance with agency regulations.

When it issued the bulk of its nutrition labeling rules in 1993, FDA also finalized a set of health-claim rules. Using a process similar to the one governing nutrient content claims, the agency sets out general requirements applying to all products making claims (see 21 *CFR* §101.14); then, FDA authorizes individual claims and establishes specific requirements for each.

FDA has considered allowing health claims for several different disease-nutrient relationships. They are aimed at helping consumers purchase products that are part of a healthy diet or that help ward off a diet- related health condition. Since the original rules were published, the agency has authorized a handful of additional claims. **Table 6.1** lists the ten claims currently authorized — FDA also specifically prohibits five claims.

"Modernization"

Congress recently passed the FDA Modernization Act, allowing a manufacturer to make additional health and nutrient content claims under certain conditions — these claims must be based on an authoritative statement from an official government scientific body. Turn to **page 5.20** for a discussion of these new provisions and the full text of the act.

Where is FSIS?

Although FSIS proposed its own set of health-claim regulations on May 25, 1994, the agency never finalized those rules (59 *FR* 27144). In the mean time, that proposal became obsolete as FDA pursued subsequent rulemaking projects.

On April 22, 1998, FSIS withdrew its 1994 health claim proposal (63 *FR* 19852). Now, it intends to propose "a more comprehensive document on health claims regulations for meat and poultry products that will parallel FDA's rules for other foods." The action is scheduled for October 1998.

Important Definitions

Many of FDA's requirements are based on the following terms. Note, however, that "healthy" nutrient content claims, as described in Chapter 7, are not health claims because they do not include statements about disease prevention.

- *Health claim* means any claim made on the label of a food, including a dietary supplement, that expressly or by implication characterizes the relationship of any substance to a disease or health-related condition. These include "third party" references, written statements (e.g., a brand name including a term such as "heart"), symbols (e.g., a heart symbol), and vignettes.

- *Implied health claims* include those statements, symbols, vignettes, or other forms of communication that suggest, within the context in which they are presented, that a relationship exists between the presence or level of a substance in the food and a disease or health-related condition.

- A *substance* is a specific food or component of food, regardless of whether the food is in conventional food form or a dietary supplement that includes vitamins, minerals, herbs, or other similar nutritional substances.

- *Nutritive value* means a value in sustaining human existence by such processes as promoting growth, replacing loss of essential nutrients, or providing energy.

- A *dietary supplement* is a food, not in conventional food form, that supplies a component to supplement the diet by increasing the total dietary intake of that component.

- *Disqualifying nutrient levels* means the levels of total fat, saturated fat, cholesterol, or sodium in a food above which the food will be disqualified from making a health claim. (These are listed later in this chapter.)

- *Disease or health-related condition* means damage to an organ, part, structure, or system of the body such that it does not function properly (e.g., cardiovascular disease), or a state of health leading to such dysfunction (e.g., hypertension). However, diseases resulting from essential nutrient deficiencies (e.g., scurvy, pellagra) are not included in this definition (claims pertaining to such diseases are thereby not subject to §101.14 or §101.70).

TABLE 6.1
Health Claims Either Allowed or Specifically Prohibited by FDA*

Nutrient/disease relationship	Citation in 21 CFR
Claims specifically allowed—	
Calcium and osteoporosis	§101.72
Dietary lipids and cancer	§101.73
Sodium and hypertension	§101.74
Dietary saturated fat and cholesterol and risk of coronary heart disease	§101.75
Fiber-containing grain products, fruits, and vegetables and cancer.	§101.76
Fruits, vegetables, and grain products that contain fiber, particularly soluble fiber, and risk of coronary heart disease	§101.77
Fruits and vegetables and cancer	§101.78
Folate and neural tube defects	§101.79
Dietary sugar alcohols and dental caries	§101.80
Soluble fiber from certain foods and risk of coronary heart disease	§101.81
Specifically prohibited—	
Dietary fiber and cancer	§101.71(a)
Dietary fiber and cardiovascular disease	§101.71(b)
Antioxidant vitamins and cancer	§101.71(c)
Zinc and immune function in the elderly	§101.71(d)
Omega-3 fatty acids and coronary heart disease	§101.71(e)

* FSIS does not currently authorize health claims. The agency proposed in 1994 to allow certain claims, and intended to publish a final rule in 1996. However, in the mean time the proposal became obsolete, and was withdrawn. FSIS now plans to propose a more comprehensive document.

Note: As FDA approves new claims, they are added to Part 101, Subpart E. Products bearing health claims must have "Nutrition Facts" labeling.

Source: 21 CFR §§101.14, 101.70-101.80

General Concepts

As mentioned previously, FDA has published a set of general principles for health claims in §101.14, which sets forth:

- Circumstances in which a substance in a food product is eligible for a health claim,

- Standard the agency applies in deciding whether to allow a claim about a substance-disease relationship,

- General rules on how authorized claims are to be made in food labeling, and

- Limitations on the circumstances in which health claims can be made.

Disqualification

FDA's general rules require that any product making a health claim also comply with the "Nutrition Facts" labeling requirements. Also, only claims specifically defined by FDA may be made — the agency outlines detailed requirements for each allowed claim, such as requiring the use of phrases like "may reduce" instead of "will reduce". For example, a health claim on the label of a fiber-containing grain product can state, "Low fat diets rich in fiber-containing grain products, fruits, and vegetables may reduce the risk of some types of cancer, a disease associated with many factors" [21 CFR §101.76(e)(1)]. (Individual claims are addressed in more detail later in this chapter.)

Furthermore, the labeled food product may not exceed the disqualifying nutrient levels specified in §101.14(a); these are summarized below in **Table 6.2**. Foods exceeding one or more of these levels may not make a health claim unless the FDA rules make a specific exception.

Making the Claim

Key components of FDA's general provisions, which apply to all food products making health claims, are illustrated in **Figure 6.1** (pages 6.5-6.6) and discussed further below.

TABLE 6.2
Disqualifying Nutrient Levels Under the Health Claim Rules

Product	Allowed nutrient level (per RACC and labeled serving size)			
	Total fat	*Saturated fat*	*Cholesterol*	*Sodium*
Foods with RACC > 30 g or 2 Tbsp	13.0 g	4.0 g	60 mg	480 mg
Foods with RACC ≤ 30 g or 2 Tbsp [1]	13.0 g	4.0 g	60 mg	480 mg
Meal product	26.0 g	8.0 g	120 mg	960 mg
Main-dish product	19.5 g	6.0 g	90 mg	720 mg

[1] Levels must be met on a per-50 gram basis also. For dehydrated foods, the per-50 gram criterion refers to the as-prepared form.

Note: Any one of these levels will disqualify a food product from making a health claim unless an exception is provided in the regulations.

Source: 21 CFR §101.14(a)(5)

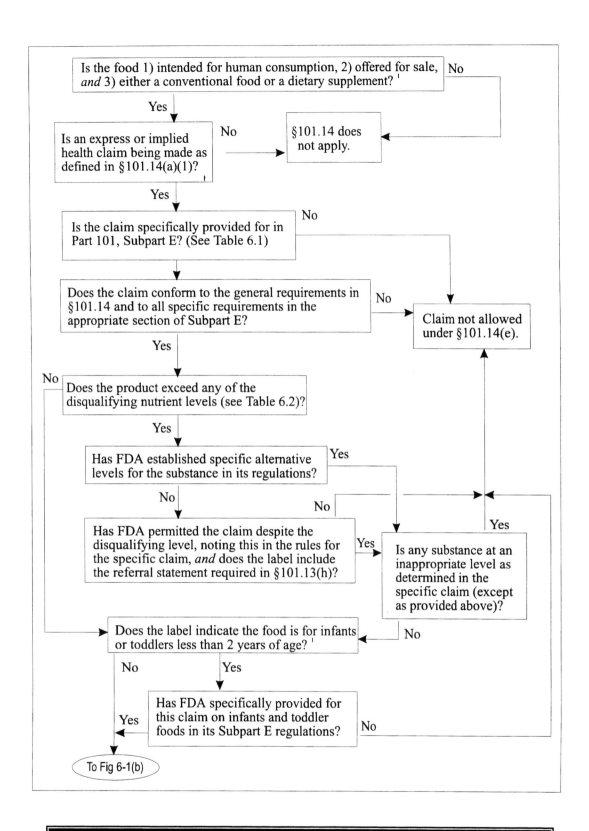

FIGURE 6.1(a)
General Health Claim Requirements

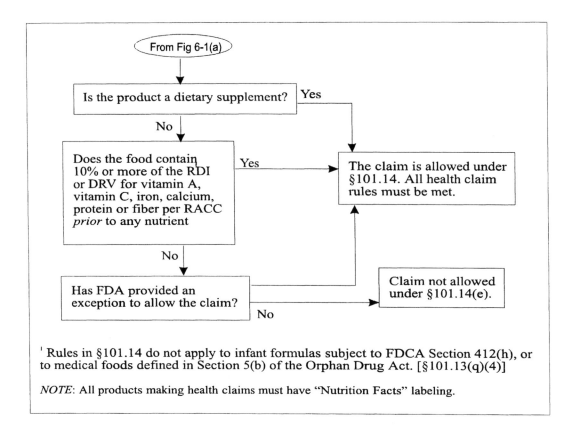

FIGURE 6.1(b)
General Health Claim Requirements

The "Jelly Bean" Rule

An important aspect of these general provisions is the 10% nutrient contribution requirement [see Figure 6.1(b)]. The rules for prohibited claims in §101.14(e)(6) require that, except for dietary supplements, a product making a health claim must contain 10% of more of the RDI or DRV for vitamins A or C, iron, calcium, protein, or fiber per RACC *prior* to any nutrient addition. (If the FDA rules specifically make an exception, the claim will be allowed.) The provision has been nick-named the "jelly bean" rule because it is intended to prevent health claims on products that typically do not have significant nutritional value (at least in the absence of fortification).

This has been a controversial component of the health claim regulations; FDA has in some cases added relief from this requirement, as discussed later in this chapter.

Additional Requirements

Other general health claim labeling requirements are specified in §101.14, as outlined below:

- *Value of product*— A claim must be limited to describing the value that ingestion (or reduced ingestion) of the substance, as part of a total dietary pattern, may have on a particular disease or health- related condition.

- *Truthfulness*— The claim is complete, truthful, and not misleading. (Where factors other than dietary intake affect the relationship between the substance and the disease or health- related condition, FDA often requires that these factors be addressed in the claim.)

- *Completeness*— Information required to be included in the claim must appear in one place, without other intervening material. However, the principal display panel may bear the reference statement, "See *[location of the labeling containing the health claim]* for information about the relationship between *[name of substance]* and *[the disease or health-related condition]* (e.g., "See attached pamphlet for information about calcium and osteoporosis'); the entire claim would then appear elsewhere on the labeling. Note that, where any graphic material (such as a heart symbol) constituting an explicit or implied health claim is used on the label, the reference statement or the complete claim must appear in immediate proximity to that graphic material.

- *Clarity*— Claims must enable the public to comprehend the information provided and understand the relative significance of the information in the context of a total daily diet.

- *Decreased dietary levels*— If the claim concerns the effects of consuming a substance at decreased levels, the level of the substance in the food must be sufficiently low to justify the claim. (If a definition for use of the term "low" has been established for the substance, it must be present at a level that meets the requirements for use of that term, unless a specific alternative level has been established. If no definition for "low" has been established, the substance must meet the level established in the regulation authorizing the claim.)

- *Sufficient levels*— If the claim is about the effects of consuming a substance at other than decreased dietary levels, the level of the substance must be sufficiently high, and in an appropriate form, to justify the claim. (If a definition for use of the term "high" for the substance has been established, it must be present at a level that meets the requirements for use of that term, unless a specific alternative level has been established. If no definition has been established (e.g., where the claim pertains to a food either as a whole food or as an ingredient in another food), the claim must specify the daily dietary intake necessary to achieve the claimed effect, as established in the regulation authorizing the claim.

- *RACC*— If the product meets the above requirements based on its RACC, but the labeled serving size differs from that amount, the claim must be followed by a statement explaining that the claim is based on the reference amount, rather than the labeled serving size (e.g., "Diets low in sodium may reduce the risk of high blood pressure, a disease associated with many factors. A serving of ___ ounces of this product conforms to such a diet.").

- *Restaurant-type foods*— If the food is sold in a restaurant or in other establishments in which food that is ready for immediate human consumption is sold, it can meet the above requirements if the firm selling the food has a reasonable basis on which to believe that the product meets these requirements, rather than by performing analytical testing. The firm must provide that basis upon request.

- *"Nutrition Facts"*— Nutrition labels must be provided on any food for which a health claim is made in accordance with §101.9 (for restaurant foods, §101.10; for dietary supplements of vitamins or minerals, §101.36).

Which Substances Might Qualify for a New Claim?

In §101.70, FDA outlines procedures for petitioning the agency to allow a specific claim. Requirements for scientific data and other documentation are given. The agency has also summarized its policy on evaluating a given substance for a possible health claim.

Eligibility

FDA lists in §101.14 the criteria used to determine whether a substance qualifies for a new health claim. For a substance to be eligible for a health claim:

1. It must be associated with a disease or health-related condition for which the general U.S. population, or an identified subgroup, is at risk (or the proponent of the claim must otherwise explain the prevalence of the disease or health-related condition and the relevance of the claim).

2. If the substance is to be consumed as a component of a conventional food at *decreased* dietary levels, the substance must be a nutrient

listed in the FDCA, or one that FDA has required to be included in the label.

3. If the substance is to be consumed at *other than decreased* dietary levels, it must 1) regardless of whether the product is a conventional food or dietary supplement, contribute taste, aroma, or nutritive value, or any other technical effect listed in §170.3(o), and retain that attribute when consumed at levels necessary to justify a claim; and 2) be a food, ingredient, or component of a food whose use, at the levels necessary to justify a claim, has been demonstrated to FDA's satisfaction to be safe and lawful under FDCA.

Validity

After considering a petition, FDA issues regulations authorizing a health claim only when it determines, based on the "totality of publicly available scientific evidence," that the claim is valid. This evidence includes data from well-designed studies conducted in a manner consistent with generally recognized scientific procedures and principles. Additionally, the agency must find that there is "significant scientific agreement" among experts, qualified by scientific training and experience to evaluate potential claims, that a new claim is supported by the available evidence.

Once FDA determines that a health claim meets its validity requirements, the agency will propose to add a new regulation to Subpart E authorizing the use of that claim. If the claim pertains to a substance not already provided for in the labeling rules in §101.9 or §101.36, FDA will propose to allow declaration of that substance. After a final rule has been adopted, firms may make health claims based on the new regulations.

FDA Faces New Deadlines

Responding to a federal court decision, FDA recently amended §101.70(j) to specify a schedule for publishing final determinations on potential new health claims (62 FR 28232; May 22, 1997).

In explaining its action, FDA noted that FDCA Section 403(r), which was added by the NLEA, authorizes the agency to provide for claims on food labeling that characterize the relationship of one or more nutrients to a disease or other health-related condition. In providing for these health claims, the law treats conventional foods differently than dietary supplements. For conventional foods, the FDCA sets out the standards that FDA must use when deciding whether to authorize a particular claim — for dietary supplements, however, it states that health claims are subject to procedures and standards established by the agency.

Anyone may petition FDA to authorize a claim about a particular nutrient-disease relationship (§101.70). The agency's stated procedure is intended to parallel requirements in FDCA Section 403(r)(4)(A)(i):

- Within 100 days of receiving a petition, FDA will notify the petitioner by letter that the petition has either been filed for comprehensive review or denied.

- If the agency files the petition, it will, within 90 days of filing, either issue a denial or advise the petitioner that a proposal to authorize the health claim will be published in the *Federal Register*.

However, in publishing those rules, FDA made no mention of when final regulations for the health claim will be issued. The agency was sued on several occasions by dietary supplement trade associations, manufacturers, and retailers on the grounds that the agency regulations violate the First Amendment to the U.S. Constitution. One of these cases, *Nutritional Health Alliance v. Shalala* was decided on January 31, 1997 by the district court for the Southern District of New York.

The court found that — once FDA has proposed to allow a particular health claim — the absence of a time frame for issuing a final ruling fails to satisfy certain legal tests and therefore violates the First Amendment to the U.S. Constitution. Accordingly, FDA was ordered to establish a reasonable time limit for issuing final rules for a new health claim.

After evaluating its "typical" rulemaking process, and considering public comments, FDA determined that a 9-month time frame is appropriate. Under 21 CFR §101.70(j)(4)(i), within 270 days of publication of a health claim proposal, the agency will publish a final rule either authorizing its use or explaining why FDA has decided against it. (The agency notes that this time period is approximately 90 days shorter than what was required in the oat-product and dental-caries rulemakings.)

The court recognized that FDA may receive information during a comment period that could require the agency to rethink whether to authorize a health claim. Such circumstances could be handled by an extension, based on a showing of cause. Consequently, FDA is allowing extensions beyond the 270 days if circumstances justify such a delay.

Specific Claims

As noted earlier, ten specific health claims are currently authorized by FDA. Recent actions affecting the requirements for specific claims are discussed beginning on page 6.12; full details for each of the ten claims are summarized in **Table 6.3** (page 6.18).

As shown in Table 6.3, for each allowed claim, FDA describes the nature of the claim being authorized. The agency also specifies certain restrictions, such as what may or may not be said about the ability of the substance to reduce risk of disease. Frequently, the rules also identify additional information that may be included on an optional basis. Also, the agency provides "model" terminology for each authorized health claim.

What Is a "Medical Food"?

One possible source of confusion is the regulatory distinction between so-called medical foods, foods formulated for special dietary use, and foods qualifying for health claims. The term "medical food" means a food which is intended to be consumed enterally under a physician's supervision, and which is formulated for the specific dietary management of a disease or condition with distinctive nutritional requirements.

FDA now believes its policy toward regulation of these foods should be reevaluated due to a number of developments, including: 1) enactment of a statutory definition of "medical food," 2) the rapid increase in the variety of products marketed as medical foods, 3) possible safety problems associated with the manufacture and quality control of these products, and 4) the potential for fraudulent claims. The agency issued an advance notice on November 29, 1996 requesting public comment (61 FR 60661).

The agency is therefore reevaluating its approach to the regulation of the broad group of heterogeneous products marketed as medical foods. FDA's stated goal is "to arrive at a regulatory regime that will ensure that: these products are safe for their intended uses, especially because they are likely to be the sole or a major source of nutrients for sick and otherwise vulnerable people; claims for these products are truthful, not misleading, and supported by sound science; and the labeling of these products is adequate to inform consumers about how to use them in a safe and appropriate manner."

Foods for Special Dietary Use, Labeling Exemptions Develop

Infant formula was one of the first types of medical foods marketed in the United States; some of the early products were developed for use in the management of phenylketonuria (PKU). Before 1972, because of their role in mitigating serious adverse effects of diseases, FDA regulated such products as drugs under Section 201(g)(1)(B) of the FDCA. At that time, however, the nutritional formulation requirements for this type of product were believed to be well-established by the medical community; medical foods were manufactured by a limited number of firms.

After reassessing its regulatory approach, FDA in 1972 stated that PKU formulas and similar products would now be regulated as "foods for special dietary use" rather than as drugs. In doing so, the agency recognized that use of these products by healthy individuals could be hazardous (e.g., a product formulated for infants with PKU would be nutritionally inadequate for a normal infant). Thus, the agency saw that it was important to differentiate these products from foods for general use.

When FDA made nutrition labeling mandatory for many products in 1973, the agency exempted certain types of foods for special dietary use from this requirement. In the preamble to the January 19, 1973 final rule (38 FR at 2126), FDA noted that nutrition labeling developed for foods intended for nutrition labeling developed for foods intended for consumption by the general population was not well-suited for some products, including two types of foods for special dietary use: 1) any food represented for use as the sole item of the diet, and 2) foods represented for use solely under medical supervision in the dietary management of specific diseases and disorders. Therefore, the regulation exempted these two types of products from the general requirements for nutrition labeling; instead, they were to be labeled in compliance with regulations that the agency intended to include in 21 *CFR* Part 125 (later redesignated as Part 105).

Generally, medical foods are consumed enterally under the guidance of a physician. [According to FDA's terminology, enteral nutrition is nutrition provided through the gastrointestinal tract, taken by mouth or provided through a tube or catheter that delivers nutrients beyond the oral cavity (i.e., directly to the stomach or small intestine).] Such products have proliferated; more than 200 products were sold as "medical foods" in the U.S. by 1990. Examples include protein products for kidney and liver diseases and high-fat, low-carbohydrate products for hospitalized patients with certain lung diseases.

Statutory Definitions, More Exemptions

Medical foods are now common in the United States. When the Orphan Drug Amendments of 1988 were enacted, a statutory definition of "medical food" was provided for the first time:

"The term 'medical food' means a food which is formulated to be consumed or administered enterally under the supervision of a physician and which is intended for the specific dietary management of a disease or condition for which distinctive nutritional requirements, based on recognized scientific principles, are established by medical evaluation."

Although Congress provided a statutory definition, the legislative history of the Orphan Drug Amendments does not provide any further information regarding the types of products that the medical food definition was intended to cover. In the 1990 NLEA, Congress incorporated this definition into FDCA Section 403(q)(5)(A)(iv). The NLEA exempted medical foods from the nutrition labeling, health claim, and nutrient content claim requirements applicable to most other foods.

FDA's 1993 rules implementing the NLEA provisions, specifically the requirements for mandatory nutrition labeling (58 *FR* 2151), exempted medical foods and incorporated the new statutory definition into 21 *CFR* §101.9(j)(8). The agency enumerated criteria intended to clarify the characteristics of medical foods. FDA's regulation provides that a food may claim the exemption from nutrition labeling requirements only if it is:

- A specially formulated and processed product (as opposed to a naturally occurring foodstuff used in its natural state) for the partial or exclusive feeding of a patient by means of oral intake or enteral feeding by tube;

- Intended for the dietary management of a patient who, because of therapeutic or chronic medical needs, has limited or impaired capacity to ingest, digest, absorb, or metabolize ordinary foodstuffs or certain nutrients, or who has other special medically determined nutrient requirements, the management of which cannot be achieved by the modification of the normal diet alone;

- Providing nutritional support specifically modified for the management of the unique nutrient needs that result from the specific disease or condition, as determined by medical evaluation (*Note*: In a future action, FDA intends to replace "unique" with "distinctive" since that word should have been used here);

- Intended to be used under medical supervision; and

- Intended only for a patient receiving ongoing supervision wherein the patient requires medical care on a recurring basis for, among other things, instructions on use of the medical food.

Medical Foods vs. Special Dietary Use

In 1990, the Life Sciences Research Office of the Federation of American Societies for Experimental Biology (LSRO/FASEB) published "Guidelines for the Scientific Review of Enteral Food Products for Special Medical Purposes." This report defined medical foods as products distinct from foods for special dietary use in that they "demonstrate greater suitability for nutritional management of a specific disease than standard enteral formulas" and are intended for patients with "special medically determined nutrient requirements, the dietary management of whom cannot be achieved by the modification of the normal diet alone, by other foods for special dietary uses, or by a combination thereof." The report proposed criteria that would establish a strict standard for a product to be considered a medical food. The proposed definition of medical food did *not* include all foods that might be useful for persons with a disease or medical condition.

In its 1996 advance notice, FDA noted the paradox currently found in federal policy toward these products:

"The statutory definitions of 'medical food' . . . and food for special dietary use (see Section 411(c)(3) of the act), and the differing treatment of these two categories of products under the 1990 amendments to the act [i.e., medical foods are exempted under section 403(q)(5)(A)(iv) and (r)(5)(A) of the act, while there is no special treatment of foods for special dietary use], establish that *Congress intended that medical foods and foods for special dietary use be viewed and regulated as separate and distinct categories* of products.

"Foods for special dietary use are subject to the same nutrition labeling requirements and requirements for health claims and nutrient content claims established for most other foods by the 1990 amendments. Thus, foods for special dietary use, like ordinary foods, must be labeled with certain nutrition information in a prescribed format to ensure that such information is presented in an informative and understandable fashion. Moreover, any nutrient content claims or health claims on the label or in the labeling of a food for special dietary use must have been authorized by FDA to ensure that the claim is scientifically valid and is presented in such a way that it is truthful and not misleading.

"In contrast, under the 1990 amendments, medical foods are specifically exempted from the requirements for nutrition labeling, nutrient content claims, and health claims. Thus, a *medical food* that is intended for the specific dietary management of a disease or condition for which distinctive nutritional requirements have been established *may be sold without any nutrition information on its label* or labeling, and it may *bear claims that have not been evaluated* under the 1990 amendments to ensure that they are scientifically valid. Moreover, there is no assurance that the formulation of a medical food has been evaluated prior to sale to ensure that it is suitable for the intended patient population.

"The exemption from the requirements of the 1990 amendments, therefore, creates a *troubling paradox*: Medical *foods intended for use by sick people are subject to much less scrutiny* than virtually all other foods, which are intended for the healthy general population. This lack of scrutiny creates a situation that could have adverse public health consequences if these products bear claims that are not scientifically valid, or if their labeling does not disclose nutrition or other information that is necessary for the safe and effective use of the food." (61 *FR* 60663-60664; emphasis added)

Safety Problems

FDA has determined that, as the number of medical food manufacturers has grown, the level of industry experience in the current good manufacturing practices (CGMPs) and quality control procedures necessary to produce products that contain nutrients within a narrow range of declared label values has become more variable. Medical foods are complex formulated products, generally requiring sophisticated and exacting technology comparable to that used in the manufacture of infant formulas and drugs. Moreover, the populations that consume these products, often as the sole or a major source of nutrition, are extremely vulnerable (e.g., pediatric patients in periods of rapid growth, the elderly, etc.).

Specific Claims Allowed

As detailed in **Table 6.3** (beginning on page 6.18), ten specific health claims are currently authorized. Agency actions pertaining to the most recently approved claims are reviewed below.

Folate and Neural Tube Defects

Manufacturers may link folate (i.e., the entire group of folate vitamin forms, including folic acid) with reduced risk of neural tube defects (NTDs). FDA's requirements for making such a product claim are illustrated in the decision diagram in **Figure 6.2** (pages 6.13–6.14).

Preventing Tooth Decay

Another authorized claim allows manufacturers to point out the benefits of the sugar alcohol in their products for reducing the likelihood of tooth decay (as compared to other carbohydrates). These rules, published at 21 *CFR* §101.80, were originally issued on August 23, 1996 (61 *FR* 43433). Since then, FDA has also allowed products containing the sugar alcohol erythritol (December 2, 1997; 62 *FR* 63653).

Oats, Psyllium, and CHD

A 1997 (62 *FR* 3584) action allowed products such as oat bran to use claims that alert consumers about the relationship between oat products and the reduction in risk of developing coronary heart disease (CHD). According to 1985 data reported by FDA, total direct costs related to CHD are estimated at $13 billion, with indirect costs totaling another $36 billion. Note that the original rules were clarified on March 31, 1997 (62 *FR* 15343).

In addition, on February 18, 1998, FDA revised its requirements to allow similar claims on products containing soluble fiber derived from psyllium seed husk (63 *FR* 8103). Current criteria are summarized beginning on page 6.18 and are reviewed below. Note that the rules for this claim are no longer specific to one food type — §101.81 now refers to "certain foods," rather than oats, psyllium, etc., putting into place terminology that could accommodate still other soluble fiber sources.

The original soluble fiber/CHD rules marked the first time the agency authorized a health claim for a specific food type; the specific substance that was the subject of the original claim is "β-glucan soluble fiber from whole oats" (FDA had originally proposed to authorize claims on oat bran and rolled oats, but later clarified the precise nature of the beneficial food constituent).

While the agency has previously denied the use of claims relating dietary fiber to reduced risk of cardiovascular disease (CVD), it has authorized claims relating diets low in saturated fat and cholesterol and high in fruits, vegetables, and grain products that contain fiber (particularly soluble fiber) and risk of CHD. (CHD is the most common and most serious form of CVD.) In denying the dietary fiber/CVD health claim, FDA concluded that dietary fiber consists of too diverse a group of chemical substances, each with a different physiological function.

On the other hand, FDA did recognize scientific evidence documenting the effectiveness of naturally occurring dietary fibers in specific food products in reducing the risk of CVD, and in particular, CHD. Therefore, manufacturers are encouraged to petition for a health claim for a particular product if they can 1) document the effects of dietary consumption of soluble fiber in their food to lowering low density lipoprotein-cholesterol [(LDL)-cholesterol] in the blood (since lower blood cholesterol is known to decrease risk of CHD), and 2) show that the product has no adverse effects on other heart disease risk factors, such as high density lipoprotein-cholesterol [(HDL)-cholesterol].

The following issues were addressed by FDA when it first published the "oats" rule:

- *Subject of claim:* The agency found significant scientific agreement that β-glucan soluble fiber from whole oats, as part of a diet low in saturated fat and cholesterol, may reduce the risk of CHD. FDA reached this conclusion based on evidence that there is a dose response between the level of β-glucan soluble fiber from whole oats and the level of reduction in blood total-and LDL-cholesterol, and that intake at or above 3 g per day was more effective in lowering serum lipids than were lower intakes. Therefore, the subject of this claim was "soluble fiber from whole oats." Any product meeting the eligibility requirements qualified for the claim. Because

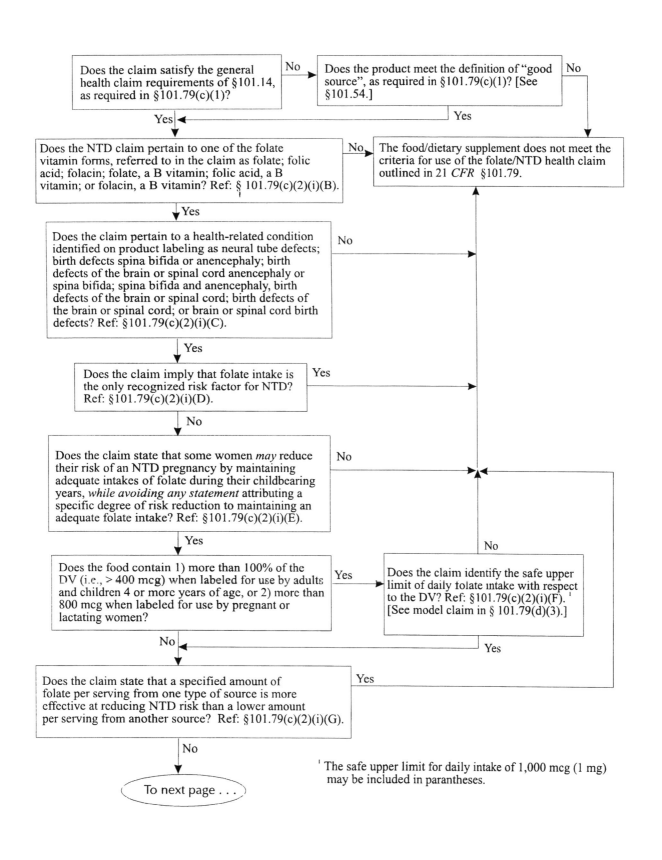

FIGURE 6.2(a)
Criteria for Making a Claim Relating Folate to NTD Risk

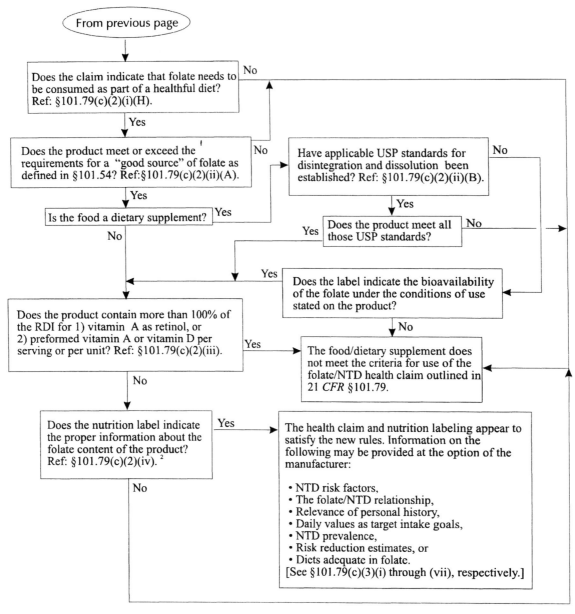

[2] Nutritional information must be declared after the declaration for iron if only the levels of vitamin A, vitamin C, calcium, and iron are provided. Otherwise, if optional vitamins or minerals are declared, the information must be provided in accordance with §101.9(c)(8) and (c)(9).

Abbreviations:
DV = daily value; mcg = micrograms; NTD = neural tube defect;
RDI = reference daily intake; Ref = Regulatory reference in Part 21 of the
Code of Federal Regulations; USP = United States Pharmacopeia.

Source: Based on §101.79 published March 5, 1996 (61 *FR* 8779).

FIGURE 6.2(b)
Criteria for Making a Claim Relating Folate to NTD Risk

"rolled oats" is more commonly used to describe the dry form of the food, FDA uses this phrase rather than "oatmeal."

- *Qualifying criteria:* In developing these rules, FDA assumed a desired consumption of 3 g β-glucan soluble fiber/day. Because this intake of 3 grams is expected to be distributed over 4 eating occasions, the qualifying level is 0.75 g/RACC.

- *Mixtures:* FDA has noted that it should not matter if fiber is derived from a mixture, as long as the requisite amount of β-glucan soluble fiber is present. The regulations state that the product must provide the required level of soluble fiber per RACC from the *eligible sources* of whole oat soluble fiber listed in §101.81(c)(2)(ii). Therefore, a mixture of oat bran, rolled oats, and whole oat flour may be used.

- *Fiber declarations:* FDA believes that using "β-glucan" as a subcategory of soluble fiber would likely be confusing since it is a technical term unfamiliar to consumers. Therefore, the agency does not require declaration of β-glucan on the nutrition label for oats products.

Psyllium & Certain Other Foods

As noted earlier, FDA has expanded the "oats" claim to include soluble fiber derived from psyllium seed husk.. To satisfy the claim criteria, the psyllium must comply with §101.81(c)(2)(ii)(B) and other applicable requirements. The food must contain at least 1.7 g of soluble fiber per RACC.

Currently, the minimum daily intake levels for fiber sources authorized under this CHD claim are as follows: 3 g/day of β-glucan soluble fiber from whole oats, and 7 g/day of soluble fiber from psyllium seed husk. FDA uses these intake levels to determine the required fiber content per RACC. Finally, the agency has added a warning/notice requirement for psyllium husk in §101.17.

Note that the statement "diets low in saturated fat and cholesterol and *high* in soluble fiber from ***" cannot be used at this time because the term "high" and its synonyms have been defined under §101.54(b) as meaning that the food contains 20% or more of the DRV per RACC for a particular substance; there is now no DRV for soluble fiber. FDA has expressed an interest in proposing a DRV value for soluble fiber (although this might depend on the fiber source).

What's Next?

It would be an understatement to say that industry has had concerns about some of the NLEA health claim provisions. Since 1990, FDA has received numerous requests to revise certain regulatory requirements. Among other things, FDA has been asked to grant manufacturers permission to:

- Use abbreviated or implied health claims with a referral statement directing consumers to a complete claim elsewhere on the label.

- Make a health claim in some cases even when a nutrient is present at a disqualifying level (under one such request, food containing a nutrient at a disqualifying level would be barred from health claims only if that nutrient were directly and adversely associated with the disease to which the claim referred).

- Place claims on products that contain a nutrient above certain threshold levels if disclosure of that fact is printed on the labels.

Catch-22

Additionally, some have requested exemptions from the 10% nutrient contribution requirement (i.e., the "jelly bean" rule). For example, one petitioner asked that enriched cereal-grain products (conforming to standards in 21 *CFR* Parts 136, 137, or 139) and bread (conforming to the standard in §136.115, except containing whole wheat or other grains not permitted under that standard) be permitted to bear health claims even if they do not satisfy the 10% requirement. Ironically, most enriched grain products cannot meet the 10% rule for any of the six listed nutrients because their standards of identity preclude them from doing so. Thus, enriched grain products were currently excluded from health claims, even though they are help reduce the risk of certain diet-related diseases, and current guidelines recommend increasing their consumption.

As an alternative, one petitioner has suggested that FDA 1) expand the list of qualifying nutrients to include complex carbohydrates, niacin, or thiamin; or 2) allow the 10% contribution requirement to apply based on the summation of applicable nutrients.

The 10% Requirement

At least one petitioner has argued that the 10% requirement precludes some truthful, non-misleading health claims by setting an arbitrary nutritional contribution in order for a food to qualify for any claim. Additionally, this approach has prohibited some common fruits, vegetables, and other wholesome foods from making health claims. Finally, some argue that fortified nutrients are just as healthful as indigenous ones.

FDA Response

Since 1993, FDA has proposed several revisions in an effort to encourage manufacturers to use health claims. These include changes to: 1) the 10% nutrient contribution requirement for health claims, 2) specific requirements for individual claims, and 3) disqualifying levels for health claims.

FDA has acknowledged that the 10% nutrient contribution requirement may have unintentionally excluded certain beneficial foods from bearing health claims. However, it is concerned that eliminating it will permit misleading claims on foods with little nutritional value or encourage over-fortification. While the agency has considered prohibiting health claims on specific foods such as confections or snack foods, FDA believes this would not satisfy the intent of the FDCA.

The agency believes the 10% nutrient contribution requirement is a necessary component of the health claims provisions and should not be revoked. It has also concluded that expanding the list of nutrients required to be present at the 10% level would not sufficiently address the concern that the current regulation precludes certain healthful foods from making claims. Similarly, permitting the 10% rule to be based on the daily consumption of an entire food group would not enhance the likelihood of consumers achieving dietary goals. As for the fortification requirement, FDA believes that fortification solely to bear a health claim could result in deceptive or misleading labeling.

Proposed Fruit, Vegetable, and Grain Product Exemption

On December 21, 1995, FDA proposed to exempt certain products from the "jelly bean" rule (60 *FR* 66206). Section 101.14(e)(6) would be amended to exempt the following products from the 10% rule (as long as all other requirements were met):

1. Fruit and vegetable products comprised solely of fruits and vegetables,

2. Enriched-grain products conforming to a standard of identity, and

3. Bread conforming to the standard of identity for enriched bread (except that it contains whole wheat or other grain products not permitted under that standard)

Abbreviations

Under the current rules, abbreviated health claims are not permitted. However, a referral statement is permitted that states,"See *[location on the label containing the health claim]* for information about the relationship between *[name of the substance in the food]* and *[disease or health-related condition]*"—the complete health claim is then included elsewhere.

FDA has proposed increased latitude in an effort to encourage industry to use applicable health claims more extensively — it recognizes that an abbreviated claim on the principal display panel designed to capture the consumer's attention will likely facilitate more extensive use. Therefore, the agency also proposed to permit a shorter health claim on a PDP as long as it was accompanied by a referral to the complete claim.

Abbreviated health claims could become a moot point, since the proposal also included provisions for shortening health claims by making optional some of the elements currently required. Under the current scheme, health claims tend to be long statements. Industry representatives have requested that these be modified to permit simplified, nondeceptive claims that are likely to be more easily understood by consumers. The agency's 1995 responses are summarized below.

Calcium/Osteoporosis

Regarding the calcium/osteoporosis health claim, several revisions have been proposed, including plans to require these claims to:

- Make it clear that adequate calcium intake as part of a healthful diet throughout life is essential to reduce the risk of osteoporosis;

- Identify the population at particular risk for development of osteoporosis as women in their bone-forming years from approximately ages 11 to 35;

- Not attribute any degree to which maintaining adequate calcium intake throughout life may reduce the risk of osteoporosis; and

- State that total dietary intake of calcium greater than 2,000 mg per day provides no further benefit to bone health in reducing osteoporosis (2,000 mg per day represents 200% of the calcium daily value for adults and children 4 years of age or older or 154% of the daily value for pregnant or lactating women). (Under the proposal, the limited benefit statement would be required only for foods that provide more than 1,500 mg calcium per day.)

FDA also proposed several optional statements that would not be required. A manufacturer making a calcium/osteoporosis claim could:

- List specific risk factors for osteoporosis, identifying them among the multifactorial risks for the disease (e.g., a person's sex, age, and race, and adequate exercise);

- Identify the population particularly at risk, including Caucasian and Asian women in their bone-forming years, menopausal women, persons with a family history of osteoporosis, and elderly men and women; and

- State that adequate calcium intake throughout life is linked to reduced risk of osteoporosis through the mechanism of optimizing peak bone mass during adolescence and early adulthood (the phrases "build and maintain good bone health" or "slowing the rate of bone loss" may be used to convey this concept.

Other Claims

A common requirement for the other authorized health claims is a statement that development of a particular disease depends on many factors. However, upon review, FDA found that consumers have become aware that major chronic diseases, such as cancer and coronary heart disease, are caused by a number of factors. For this reason, the agency proposed to make the multi-factor statement optional. Instead, FDA proposed to substitute a requirement that the claim not imply that the noted substance is the only recognized risk factor for the corresponding disease/health-related condition.

Disclosure Versus Disqualifying Nutrient Levels for Health Claims

As discussed earlier, FDA has established disqualifying levels for different types of foods, depending on the role they play in the daily diet. Foods exceeding the values listed in Table 6.2 may not make health claims.

One petitioner requested that the disqualifying levels be converted to "disclosure" levels under certain circumstances. FDA rejected this suggestion, because it considers the current rules to be consistent with congressional intent. However, the agency agrees that, in certain situations, disclosure rather than disqualification may be appropriate. This determination would be made on a case-by-case basis, with petitioners requesting a specific exemption. Factors FDA may consider in evaluating such petitions might include whether:

1. The risk of the condition is of such significance that disqualification is not appropriate, but disclosure would be appropriate,

2. Availability of foods qualifying for a claim is adequate to address health concern, and

3. There is evidence that the target population is not at risk for the health-related condition associated with the disqualifying nutrient.

TABLE 6.3
Criteria for Making Specific Health Claims Under FDA Rules*

— Overview of FDA requirements —

Claim being made	Nature of food	Nature of claim [1]	Model claim(s) [2]
Calcium and osteoporosis	• "high" level of calcium as defined in §101.54(c) • calcium must be assimilable • dietary supplements must meet standards for disintegration and dissolution under the conditions of use stated on the product label • a food or total daily recommended supplement intake may not contain more phosphorus than calcium on a weight per weight basis	A claim associating calcium with a reduced risk of osteoporosis may be made if it: • makes clear that adequate calcium intake throughout life is not the only recognized risk factor in this multifactorial bone disease by listing specific factors, that place persons at risk and stating that an adequate level of exercise and a healthful diet are also needed • does not state or imply that the risk of osteoporosis is equally applicable to the general United States population; the claim must identify the populations at particular risk • states that adequate calcium intake throughout life is linked to reduced risk of osteoporosis through the mechanism of optimizing peak bone mass during adolescence and early adulthood • does not attribute any degree of risk reduction to maintaining an adequate calcium intake throughout life • states that a total dietary intake greater than 200% of the RDI (2,000 mg) has no further known benefit to bone health (this requirement does not apply to foods that contain less than 40% of the RDI of 1,000 mg per RACC or per total daily recommended supplement intake	*Appropriate for most conventional foods:* "Regular exercise and a healthy diet with enough calcium helps teen and young adult white and Asian women maintain good bone health and may reduce their high risk of osteoporosis later in life." *Appropriate for foods exceptionally high in calcium and most calcium supplements:* "Regular exercise and a healthy diet with enough calcium helps teen and young adult white and Asian women maintain good bone health and may reduce their high risk of osteoporosis later in life. Adequate calcium intake is important, but daily intakes above about 2,000 mg are not likely to provide any additional benefit."

(more claims on following pages . . .)

* Products making these claims must satisfy general rules in 21 *CFR* §101.14.

[1] This information is required. The rules for some claims also list information that may be included at the option of the manufacturer.

[2] These examples are provided in the regulations.

TABLE 6.3
Criteria for Making Specific Health Claims Under FDA Rules*

— Overview of FDA requirements —

Claim being made	Nature of food	Nature of claim [1]	Model claim(s) [2]
Dietary lipids and cancer	• meet all of the nutrient content requirements of §101.62 for a "low fat" food • fish and game meats (i.e., deer, bison, rabbit, quail, wild turkey, geese, ostrich) may meet the requirements for "extra lean" in §101.62	A claim associating diets low in fat with reduced risk of cancer may be made if it: • states that diets low in fat "may" or "might" reduce the risk of some cancers • uses "some types of cancer" or "some cancers" • uses the term "total fat" or "fat" • does not specify types of fat or fatty acid that may be related to the risk of cancer • does not attribute any degree of cancer risk reduction to diets low in fat • indicates that the development of cancer depends on many factors	"Development of cancer depends on many factors. A diet low in total fat may reduce the risk of some cancers." "Eating a healthful diet low in fat may help reduce the risk of some types of cancers. Development of cancer is associated with many factors, including a family history of the disease, cigarette smoking, and what you eat."
Sodium and hypertension	• meet all of the nutrient content requirements of §101.61 for a "low sodium" food	A claim associating diets low in sodium with reduced risk of high blood pressure may be made if it: • states that diets low in sodium "may" or "might" reduce the risk of high blood pressure • uses the term "high blood pressure" • In specifying the nutrient, uses the term "sodium" • does not attribute any degree of reduction in risk of high blood pressure to diets low in sodium • indicates that development of high blood pressure depends on many factors	"Diets low in sodium may reduce the risk of high blood pressure, a disease associated with many factors." "Development of hypertension or high blood pressure depends on many factors. [This product] can be part of a low sodium, low salt diet that might reduce the risk of hypertension or high blood pressure."

TABLE 6.3
Criteria for Making Specific Health Claims Under FDA Rules*

— Overview of FDA requirements —

Claim being made	Nature of food	Nature of claim [1]	Model claim(s) [2]
Dietary saturated fat and cholesterol and risk of coronary heart disease	• meet all of the nutrient content requirements of §101.62 for a "low saturated fat," "low cholesterol," and "low fat" food; • fish and game meats (i.e., deer, bison, rabbit, quail, wild turkey, geese, ostrich) may meet the requirements for "extra lean" in §101.62.	A health claim associating diets low in saturated fat and cholesterol with reduced risk of coronary heart disease may be made if it: • states that diets low in saturated fat and cholesterol "may" or "might" reduce the risk of heart disease • uses the terms "heart disease" or "coronary heart disease" • In specifying the nutrient, uses "saturated fat" and "cholesterol" and lists both • does not attribute any degree of risk reduction for coronary heart disease to diets low in dietary saturated fat and cholesterol • states that coronary heart disease risk depends on many factors	"While many factors affect heart disease, diets low in saturated fat and cholesterol may reduce the risk of this disease." "Development of heart disease depends upon many factors, but its risk may be reduced by diets low in saturated fat and cholesterol and healthy lifestyles." "Development of heart disease depends upon many factors, including a family history of the disease, high blood LDL-cholesterol, diabetes, high blood pressure, being overweight, cigarette smoking, lack of exercise, and the type of dietary pattern. A healthful diet low in saturated fat, total fat, and cholesterol, as part of a healthy lifestyle, may lower blood cholesterol levels and may reduce the risk of heart disease." "Many factors, such as a family history of the disease, increased blood- and LDL-cholesterol levels, high blood pressure, cigarette smoking, diabetes, and being overweight, contribute to developing heart disease. A diet low in saturated fat, cholesterol, and total fat may help reduce the risk of heart disease." "Diets low in saturated fat, cholesterol, and total fat may reduce the risk of heart disease. Heart disease is dependent upon many factors, including diet, a family history of the disease, elevated blood LDL-cholesterol levels, and physical inactivity."

TABLE 6.3
Criteria for Making Specific Health Claims Under FDA Rules*

— Overview of FDA requirements —

Claim being made	Nature of food	Nature of claim [1]	Model claim(s) [2]
Fruits and vegetables and cancer	• the food must be, or contain, a fruit or vegetable • meet the nutrient requirements of §101.62 for a "low fat" food • meet, without fortification, the nutrient content requirements of §101.54 for a "good source" of at least one of the following: vitamin A, vitamin C, or dietary fiber.	A claim associating substances in diets low in fat and high in fruits and vegetables with reduced risk of cancer may be made if it: • states that diets low in fat and high in fruits and vegetables "may" or "might" reduce the risk of some cancers • uses the following terms: "some types of cancer," or "some cancers" • characterizes fruits and vegetables as foods that are low in fat and may contain vitamin A, vitamin C, and dietary fiber • characterizes the food bearing the claim as containing one or more of the following, for which the food is a "good source" under §101.54: dietary fiber, vitamin A, or vitamin C • does not attribute any degree of cancer risk reduction to diets low in fat and high in fruits and vegetables • in specifying the fat component of the labeled food, uses "total fat" or "fat" • does not specify types of fats or fatty acids that may be related to risk of cancer • uses "fiber," "dietary fiber," or "total dietary fiber" • does not specify types of dietary fiber that may be related to risk of cancer • indicates that development of cancer depends on many factors	"Low fat diets rich in fruits and vegetables (foods that are low in fat and may contain dietary fiber, vitamin A, and vitamin C) may reduce the risk of some types of cancer, a disease associated with many factors. Broccoli is high in vitamins A and C, and it is a good source of dietary fiber." "Development of cancer depends on many factors. Eating a diet low in fat and high in fruits and vegetables, foods that are low in fat and may contain vitamin A, vitamin C, and dietary fiber, may reduce your risk of some cancers. Oranges, a food low in fat, are a good source of fiber and vitamin C."

HEALTH CLAIMS   6.21

TABLE 6.3
Criteria for Making Specific Health Claims Under FDA Rules*

Claim being made	— Overview of FDA requirements —		
	Nature of food	Nature of claim [1]	Model claim(s) [2]
Folate and neural tube defects	• meet or exceed the requirements for a "good source" of folate as defined in §101.54 • dietary supplements must meet standards for disintegration and dissolution (if there are no applicable USP standards, show bioavailability under conditions of use) • may *not* contain more than 100% of the RDI for vitamin A (as retinol or preformed vitamin A or vitamin D) per serving or per unit • the nutrition label must include folate content; this must be declared after the declaration for iron, if only the levels of vitamin A, vitamin C, calcium, and iron are provided, or in accordance with §101.9(c)(8)–(9), if other optional vitamins or minerals are declared	A claim that women who are capable of becoming pregnant and who consume adequate amounts of folate daily during their childbearing years may reduce their risk of having a pregnancy affected by spina bifida or other neural tube defects may be made if it: • uses the terms "folate," "folic acid," "folacin," "folate, a B vitamin," "folic acid, a B vitamin," or "folacin, a B vitamin" • identifies the birth defects as "neural tube defects," "birth defects spina bifida or anencephaly," "birth defects of the brain or spinal cord anencephaly or spina bifida," "spina bifida and anencephaly, birth defects of the brain or spinal cord," "birth defects of the brain or spinal cord," or "brain or spinal cord birth defects", *and* does *not* imply that folate intake is the only recognized risk factor for neural tube defects • does not attribute any specific degree of reduction in risk from maintaining an adequate folate intake throughout the childbearing years (the claim must state that some women may reduce their risk by maintaining adequate intakes of folate during their childbearing years) • for foods that contain more than 100% of the Daily Value (DV) [400 micrograms (mcg) when labeled for use by adults and children 4 or more years of age, or 800 mcg when labeled for use by pregnant or lactating women] identifies the safe upper limit of daily intake with respect to the DV [the upper limit of 1,000 mcg (1 mg) may be included in parentheses] • does not state that a specified amount of folate from one source more effectively reduces the risk of neural tube defects than a lower amount from another source • states that folate needs to be consumed as part of a healthful diet	*Examples 1 and 2: Appropriate for foods containing 100% or less of the DV (general population) for folate per serving or per unit* (these examples contain only the required elements): 1) "Healthful diets with adequate folate may reduce a woman's risk of having a child with a brain or spinal cord birth defect." 2) "Adequate folate in healthful diets may reduce a woman's risk of having a child with a brain or spinal cord birth defect." *Example 3. Appropriate for foods containing 100% or less of the DV for folate per serving or per unit* (contains all required elements, plus optional information): "Women who consume healthful diets with adequate folate throughout their childbearing years may reduce their risk of having a child with a birth defect of the brain or spinal cord. Sources of folate include fruits, vegetables, whole grain products, fortified cereals, and dietary supplements." *Example 4. Appropriate for foods intended for use by the general population and containing more than 100% of the DV of folate per serving or per unit:* "Women who consume healthful diets with adequate folate may reduce their risk of having a child with birth defects of the brain or spinal cord. Folate intake should not exceed 250% of the DV (1,000 mcg)."

TABLE 6.3
Criteria for Making Specific Health Claims Under FDA Rules*

— Overview of FDA requirements —

Claim being made	Nature of food	Nature of claim [1]	Model claim(s) [2]
Fruits, vegetables, and grain products that contain fiber, particularly soluble fiber, and risk of coronary heart disease	<ul><li>the food must be, or must contain, a fruit, vegetable, or grain product</li><li>meet the nutrient requirements of §101.62 for a "low saturated fat," "low cholesterol," and "low fat" food</li><li>must contain, without fortification, at least 0.6 g of soluble fiber per RACC</li><li>soluble fiber content must be declared in the nutrition information panel, consistent with §101.9(c)(6)(i)(A)</li></ul>	A claim associating diets low in saturated fat and cholesterol and high in fruits, vegetables, and grain products that contain fiber, particularly soluble fiber, with reduced risk of heart disease may be made if it: <ul><li>states that diets low in saturated fat and cholesterol and high in fruits, vegetables, and grain products that contain fiber "may" or "might" reduce the risk of heart disease</li><li>uses the following terms: "heart disease" or "coronary heart disease"</li><li>is limited to those fruits, vegetables, and grains that contain fiber</li><li>uses the term "fiber," "dietary fiber," "some types of dietary fiber," "some dietary fibers," or "some fibers;" the term "soluble fiber" may be used in addition</li><li>in specifying the fat component, uses the terms "saturated fat" and "cholesterol"</li><li>indicates that development of heart disease depends on many factors</li><li>does not attribute any degree of risk reduction for coronary heart disease to diets low in saturated fat and cholesterol and high in fruits, vegetables, and grain products that contain fiber</li></ul>	"Diets low in saturated fat and cholesterol and rich in fruits, vegetables, and grain products that contain some types of dietary fiber, particularly soluble fiber, may reduce the risk of heart disease, a disease associated with many factors." "Development of heart disease depends on many factors. Eating a diet low in saturated fat and cholesterol and high in fruits, vegetables, and grain products that contain fiber may lower blood cholesterol levels and reduce your risk of heart disease."

TABLE 6.3
Criteria for Making Specific Health Claims Under FDA Rules*

— Overview of FDA requirements —

Claim being made	Nature of food	Nature of claim [1]	Model claim(s) [2]
Dietary sugar alcohols and dental caries (tooth decay)	• the food must meet requirements in §101.60(c)(1)(i) with respect to sugars content • the sugar alcohol in the food shall be xylitol, sorbitol, mannitol, maltitol, isomalt, lactitol, hydrogenated starch hydrolysates, hydrogenated glucose syrups, or a combination of these • when fermentable carbohydrates are present, the food shall not lower plaque pH below 5.7 by fermentation either during consumption or up to 30 minutes thereafter, as measured by the indwelling plaque pH test specified in §101.80(c)(2)(ii)(C)	A claim relating sugar alcohols, compared to other carbohydrates, and the nonpromotion of dental caries may be made if it: • states that frequent between-meal consumption of foods high in sugars and starches can promote tooth decay (see note below) • states that the sugar alcohol present in the food "does not promote," "may reduce the risk of," "useful [or is useful] in not promoting," or "expressly [or is expressly] for not promoting" dental caries • in specifying the nutrient, uses "sugar alcohol," "sugar alcohols," or the name or names of the sugar alcohols (e.g., "sorbitol") (see note below) • uses the terms "dental caries" or "tooth decay" • does not attribute any degree of reduction in the risk of dental caries to the use of the sugar alcohol-containing food • does not imply that consuming sugar alcohol-containing foods is the only recognized means of achieving a reduced risk of dental caries *Note:* Under §101.80(c)(2)(i)(G), packages with less than 15 square inches of surface area available for labeling are exempt from §101.80(c)(2)(i)(A) and (C). Also, the general health claim requirements in §101.14 must be met, except sugar alcohol-containing foods are exempt from §101.14(e)(6).	*Example 1 — the full claim:* 1) "Frequent eating of foods high in sugars and starches as between-meal snacks can promote tooth decay. The sugar alcohol *[name, optional]* used to sweeten this food may reduce the risk of dental caries." 2) "Frequent between-meal consumption of foods high in sugars and starches promotes tooth decay. The sugar alcohols in *[name of food]* do not promote tooth decay." *Example 2 — the shortened claim for small packages:* 1) "Does not promote tooth decay." 2) "May reduce the risk of tooth decay."

TABLE 6.3
Criteria for Making Specific Health Claims Under FDA Rules*

Claim being made	— Overview of FDA requirements —		
	Nature of food	Nature of claim [1]	Model claim(s) [2]
Soluble fiber from certain foods and risk of coronary heart disease (CHD)	• the product must include fiber sources (whole oats or psyllium seed husk) identified in §101.81(c)(2)(ii) • whole oat foods must contain at least 0.75 g per RACC, while psyllium foods must contain at least 1.7 g soluble fiber per RACC of the food product • soluble fiber must be declared in the nutrition label, consistent with §101.9(c)(6)(i)(A) • meet the nutrient requirements of §101.62 for a "low saturated fat," "low cholesterol," and "low fat" food	A claim associating diets low in saturated fat and cholesterol that include soluble fiber from certain foods with reduced risk of heart disease may be made if it: • states that diets low in saturated fat and cholesterol that include soluble fiber from certain foods "may" or "might" reduce the risk of heart disease • uses the following terms: "heart disease" or "coronary heart disease" • uses the term "soluble fiber" qualified by the name of the eligible source of soluble fiber from §101.81(c)(2)(ii) (the name of the food product containing the eligible source of soluble fiber may also be used) • in specifying the fat component, uses the terms "saturated fat" and "cholesterol" • does not attribute any degree of risk reduction for CHD to diets that are low in saturated fat and cholesterol and that include soluble fiber from whole oats • does not imply that consumption of diets that are low in saturated fat and cholesterol and that include soluble fiber from eligible sources is the only recognized means of achieving a reduced risk of CHD • specifies the daily dietary intake of the soluble fiber source that is necessary to reduce CHD risk and the contribution one product serving makes to that intake level (daily levels for eligible sources are currently as follows: 3 g or more/day of β-glucan soluble fiber from whole oats, or 7 g or more/day of soluble fiber from psyllium seed husk)	"Soluble fiber from foods such as *[name of soluble fiber source from paragraph (c)(2)(ii) of this section and, if desired, the name of food product]*, as part of a diet low in saturated fat and cholesterol, may reduce the risk of heart disease. A serving of *[name of food]* supplies ___ grams of the *[grams of soluble fiber specified in paragraph (c)(2)(i)(G) of this section]* soluble fiber from *[name of the soluble fiber source from paragraph (c)(2)(ii) of this section]* necessary per day to have this effect." Diets low in saturated fat and cholesterol that include *[___ grams of soluble fiber specified in paragraph (c)(2)(i)(G) of this section]* of soluble fiber per day from *[name of soluble fiber source from paragraph (c)(2)(ii) of this section and, if desired, the name of food product]* may reduce the risk of heart disease. One serving of *[name of food]* provides ___ grams of this soluble fiber."

TABLE 6.3
Criteria for Making Specific Health Claims Under FDA Rules*

Claim being made	— Overview of FDA requirements —		
	Nature of food	Nature of claim [1]	Model claim(s) [2]
Fiber-containing grain products, fruits, and vegetables and cancer.	• the food must be, or must contain, a grain product, fruit, or vegetable • meet the nutrient requirements of §101.62 for a "low fat" food • meet, without fortification, the nutrient content requirements of §101.54 for a "good source" of dietary fiber	A claim associating diets low in fat and high in fiber-containing grain products, fruits, and vegetables with reduced risk of cancer may be made if it: • states that diets low in fat and high in fiber-containing grain products, fruits, and vegetables "may" or "might" reduce the risk of some cancers • uses the following terms: "some types of cancer," or "some cancers" • is limited to grain products, fruits, and vegetables that contain dietary fiber • indicates that development of cancer depends on many factors • does not attribute any degree of cancer risk reduction to diets low in fat and high in fiber-containing grain products, fruits, and vegetables • in specifying the dietary fiber component, uses "fiber", "dietary fiber", or "total dietary fiber" • does not specify types of dietary fiber that may be related to risk of cancer	"Low fat diets rich in fiber-containing grain products, fruits, and vegetables may reduce the risk of some types of cancer, a disease associated with many factors." "Development of cancer depends on many factors. Eating a diet low in fat and high in grain products, fruits, and vegetables that contain dietary fiber may reduce your risk of some cancers."

Source: 21 *CFR* §§101.72-101.81

* Products making these claims must satisfy general rules in 21 *CFR* §101.14.

[1] This information is required. The rules for some claims also list information that may be included at the option of the manufacturer.

[2] These examples are provided in the regulations.

Chapter 7—IMPLIED & DESCRIPTIVE NUTRIENT CONTENT CLAIMS

In This Chapter...

- What is an "implied" **nutrient content claim**?
- Which claims are **allowed** by FDA and FSIS, and what are the criteria?

As explained in Chapter 6, statements such as "low in cholesterol and saturated fat" or "sodium free" may be made only in accordance with established criteria—these are known as *expressed* nutrient content claims. Under the NLEA, FDA has been required by law to develop regulations governing this type of statement; FSIS has adopted rules that parallel the FDA regulations.

What about *implied* claims, such as "high in oat bran" or "healthy"? These statements are also regulated because they suggest that a particular nutrient is absent or present at a certain level. The rules that have been adopted by FDA and/or FSIS governing implied claims are outlined in this chapter.

By way of review, most nutrient content claims are one of two distinct types. An *expressed* claim is any direct statement about the level or range of a nutrient in a food, such as "low sodium" or "contains 100 calories." A less specific type of statement is an *implied* nutrient content claim. For instance, a product label may indicate that the contents are "healthy;" while this statement does not provide information on a specific ingredient, it nevertheless implies that the product somehow contributes to a diet that satisfies available guidelines.

Again, keep in mind that these claims include statements only about nutrient levels. Claims concerning how a particular food or nutrient relates to a disease or health-related condition are known as "health claims", and are discussed in Chapter 6.

Current Regulations

With respect to these claims, the FDA and FSIS rules are structured differently. Listed in **Table 7.1** are the regulatory provisions currently in place for implied nutrient content claims. Specific agency definitions and criteria are reviewed below.

FSIS Requirements

In its general principles for nutrient content claims, FSIS defines "implied claim" as a statement that 1) describes the product or an ingredient in a manner that suggests that a nutrient is absent or present in a certain amount (e.g., "high in oat bran"); or 2) suggests that the product, because of its nutrient content, may be useful in maintaining healthy dietary practices and is made in conjunction with an explicit claim about a nutrient (e.g., "healthy, contains 3 grams of fat") [§§317.313(b)(2), 381.413(b)(2)].

As shown in the following table, the FSIS regulations do not include separate provisions for implied claims per se. Rather, the rules simply specify criteria for making a nutrient content claim that a product is "healthy".

TABLE 7.1
Overview of Rules for Implied Nutrient Content Claims

Regulatory requirement	Citation
FDA provisions—	
definition of "implied nutrient content claim"	21 *CFR* §101.13(b)(2)
implied nutrient content claims and related label statements	21 *CFR* §101.65
statements that are *not* implied claims	21 *CFR* §101.65(b)
specific implied nutrient content claims	21 *CFR* §101.65(c)
claims for "healthy"	21 *CFR* §101.65(d)
FSIS provisions—	
definition of "implied nutrient content claim"	9 *CFR* §§317.313(b)(2), 381.413(b)(2)
claims for "healthy"	9 *CFR* §§317.363, 381.463

"Healthy" Meat and Poultry Products

Table 7.2 shows the criteria specified by FSIS for use of the term "healthy" (any product making this claim must bear "Nutrition Facts" labeling). These are similar to the criteria for other claims, in that determining whether a product satisfies required limits depends on the size of the RACC.

FDA Rules

FDA's nutrient content claim regulations define implied claims according to the same definition given above for FSIS products [see 21 *CFR* §101.13(b)(2)]. Under §101.65(b), the following types of statements about the nature of a product are generally *not* considered implied claims, and thus are typically not subject to the regulatory requirements for nutrient content claims:

- Claims that a specific ingredient or food component is absent from a product, when the purpose is to facilitate avoidance of the substance because of allergies, intolerance, religious beliefs, or dietary practices such as vegetarianism (e.g., "100% milk-free").

- Statements about a substance that does not have a nutritive function, (e.g., "contains no preservatives" or "no artificial colors").

- Claims about the presence of an ingredient that is perceived to add value to the product (e.g., "made with real butter").

- Statements of identity in which an ingredient constitutes essentially 100% of the food, (e.g, "corn oil").

- A statement of identity that names, as a characterizing ingredient, a substance associated with a nutrient benefit (e.g., "corn oil margarine") unless the claim is made in a context in which symbols or other forms of communication suggest that a nutrient is absent or present in a certain amount.

- A label statement on a food for special dietary use, made in compliance with 21 *CFR* Part 105, where the claim identifies the special diet of which the food is intended to be a part.

TABLE 7.2
Criteria for Making a "Healthy" Claim*

Regulatory requirement	nutrient per RACC and per labeled serving size			
	fat	cholesterol	sodium	other nutrients
FSIS provisions (9 *CFR* §§317.363 and 381.463)—				
foods with RACC > 30 g or > 2 Tbsp [1,2]	meet rules for "low fat" and "low saturated fat"	60 mg	360 mg [3]	10% of RDI or DRV for 1 nutrient [4]
meal-type products, including those weighing > 10 oz/serving (container)	meet rules for "low fat" and "low saturated fat"	90 mg	480 [5]	10% of RDI or DRV for 3 nutrients [4]
meal-type products, including those weighing > 6 but ≤ 10 oz/serving	meet rules for "low fat" and "low saturated fat"	90 mg	480 [5]	10% of RDI or DRV for 2 nutrients [4]
FDA provisions [21 *CFR* §101.65(d)]—				
foods with RACC > 30 g or 2 Tbsp [2]	meet rules for "low fat" and "low saturated fat"	disclosure levels in §101.13(h)	480 mg [6]	10% of RDI or DRV for 1 nutrient [4,7]
raw, single-ingredient seafood or game meat [2]	5 g total, 2 g saturated per RACC and per 100 g	95 mg per RACC and per 100 g	480 mg [6]	10% of RDI or DRV for 1 nutrient [4]
main-dish and meal products	meet rules for "low fat" and "low saturated fat"	90 mg per labeled serving	600 mg per labeled serving [8]	10% of RDI or DRV for 3 nutrients [4,9]

[1] Single-ingredient, raw products may meet the total fat, saturated fat, or cholesterol criteria for "extra lean." Also, sodium requirements do not apply to these products.
[2] If the RACC is ≤ 30 g or 2 Tbsp, some of these criteria also must be met on a per-50 gram basis.
[3] Sodium levels may be 480 mg during the first 24 months of this regulatory program (FSIS extended this to January 1, 2000 on February 13, 1998).
[4] The nutrients are: vitamins A and C, iron, calcium, protein, and/or fiber.
[5] Sodium levels may be 600 mg during the first 24 months of this regulatory program (FSIS extended this to January 1, 2000 on February 13, 1998).
[6] After January 1, 1998, this level is 360 mg (FDA stayed this requirement until January 1, 2000 on April 1, 1997).
[7] Other than raw fruits and vegetables; frozen or canned single-ingredient fruits and vegetables and mixtures thereof; and enriched cereal-grain products that conform to a standard of identity in Part 136, 137, or 139.
[8] After January 1, 1998, this level is 480 mg (FDA stayed this requirement until January 1, 2000 on April 1, 1997).
[9] Two nutrients for main-dish products.

*FSIS defines meal-type products in 9 *CFR* §§317.313(l) and 381.413(l). FDA defines meal-type and main-dish products in 21 *CFR* §101.13(l) and (m), respectively. All products must have nutrition labeling and also must comply with applicable definitions and declaration requirements in the rules governing specific nutrient content claims. For FDA-regulated products, fortification requirements may also apply.

Specific Requirements

Under FDA's rules, statements suggesting that a nutrient is absent or present in a certain amount (e.g., "high in oat bran") must 1) make proper use of one of the terms described in §101.65, 2) comply with general nutrient content claim rules, and 3) provide nutrition labeling.

The agency also allows the phrases "contains the same amount of *[nutrient]* as *[food]*" and "as much *[nutrient]* as *[food]*" as long as the nutrient in the reference food qualifies as a "good source" and the labeled food is an equivalent "good source" on a per-serving basis. For example, a label could include the implied claim "as much fiber as an apple".

Finally, claims may indicate that a food contains, or is made with, an ingredient known to contain a particular nutrient, or is prepared in a way that affects the content of a particular nutrient. These statements are allowed if the finished product is either "low" in or a "good source" of the nutrient that is associated with the ingredient or type of preparation. If a more specific level is claimed, such as "high", that level also must be present. For example, a claim that a food is made only with vegetable oil is considered to be a claim that the product is low in saturated fat.

The Trouble With "Healthy"

FDA also provides criteria for general nutritional claims, including "healthy" (and synonyms such as "healthful"). The current regulations for using a statement suggesting that a food is useful in maintaining a healthful diet are summarized in **Table 7.2** (products making this claim must be labeled per §§101.9, 101.10, or 101.36). As discussed below, FDA recently relaxed some of its requirements for these implied claims.

FDA Adds Flexibility

In a March 25, 1998 final rule (63 *FR* 14349), the agency revised §101.65(d)(2)(iv) to allow use of "healthy" on additional products. The term was defined to permit certain processed fruits and vegetables and enriched cereal-grain products to bear a "healthy" claim even if they do not satisfy the agency requirement that all "healthy" products contain 10% of the RDI or DRV for vitamins A and C, iron, calcium, protein, or fiber.

Beforehand, with the exception of raw fruits and vegetables, foods were required to qualify as a "good source" of one of the six listed nutrients. (The foods also must meet other requirements, such as restrictions on fat and sodium content, in order to meet the "healthy" definition.)

FDA had earlier allowed the "healthy" claim on raw fruits and vegetables because these products can contribute significantly to a healthy diet — even if particular items, such as cucumbers, do not always contain 10% of the daily value of one of the six identified nutrients. However, the agency was not prepared to extend the exemption to all fruit and vegetable products because it did not have adequate information on the effects of processing (e.g., exposure to liquid packing media).

Industry Petitions

After publication of the nutrient content claims rule in 1994, industry petitioned FDA to reconsider its decision, contending that 1) there was an adequate basis for evaluating the effects of freezing on the nutritional profile of fruits and vegetables, and 2) data submitted to FDA showed that, in some cases, the nutrient levels in frozen products sometimes exceed corresponding levels in fresh products.

Another petitioner requested that FDA amend its definition to permit certain enriched cereal-grain products to use "healthy" in their labeling. Although these products typically are believed to play a key role in a balanced diet, and thus are considered "healthy," manufacturers of products conforming to FDA's standards of identity were essentially in a regulatory Catch-22. While some breads might meet the 10% nutrient contribution for fiber, most enriched grain products cannot meet the requirement for any of the six listed nutrients because they are specifically precluded from doing so by FDA's standards of identity. (Similar requests have been made to relax certain health claim requirements, based on the same reasoning.)

FDA relented in 1996, announcing its intent to revise the definition of "healthy" (61 *FR* 5349).

More than two years later, the agency completed that rulemaking. As shown in footnote 7 to Table 7.2, the following items are now excused from the 10% nutrient requirement:

- Raw fruits and vegetables;

- Frozen or canned single-ingredient fruits and vegetables, and mixtures thereof — except that ingredients whose addition does not change the nutrient profile may be added; and

- Enriched cereal-grain products that conform to a standard of identity in Parts 136, 137, or 139.

Sodium Requirement Also Questioned

The FDA and FSIS "healthy" provisions impose a two-tiered requirement for sodium content. The agencies intended that, in order for a label to continue bearing this term, compliance with the second, lower tier would be required after a specified period of time had elapsed. However, a citizen petition has raised concerns and questions about how these sodium levels were established.

- As shown in Table 7.2, footnotes 3 and 5, FSIS recently issued an interim rule delaying the effective date of the lower sodium requirements until January 1, 2000 (63 *FR* 7279).

- Previously, FDA had stayed its parallel provisions in §§101.65(d)(2)(ii)(C) and 101.65(d)(4)(ii)(B) until January 1, 2000 (April 1, 1997; 62 FR 15391) — see footnotes 6 and 8.

Descriptive Claims

The FDA regulations in §101.95 also address the use of terms such as "fresh" and "frozen fresh". These are not actually part of the agency's nutrition labeling program, but are included in FDA's overall labeling requirements.

Among other things, these rules specify that a "fresh" product is generally expected to be raw and unprocessed.

Blurring of Regulatory "Lines"

The line between so-called implied claims and other types of nutrition claims can often blur. In some cases, long-standing federal regulations governing food products conflict with the relatively new nutrition labeling requirements. As explained in Chapter 5 and in the discussion of "healthy" above, many FDA and USDA regulations specify so-called standards of identity for various products.

These require conformity with specified standards in order for product labels to bear a "standardized" name, such as peanut butter or sausage. Much of the motivation behind these standards of identity has been a desire to ensure that products bearing commonly used names meet consumer expectations regarding composition and other characteristics. However, some standards require use of names or terms that overlap the NLEA-based rules, leading to inconsistency in nomenclature or labeling.

- FDA has addressed the use of specific nutrient content claim terms with specified standard product names in 21 *CFR* §130.10: "Food standards: Requirements for foods named by use of a nutrient content claim and a standardized term." That "general definition" is intended to allow manufacturers of products conforming to a standard of identity to use the standardized name in conjunction with one of the authorized nutrient content claims.

- FSIS proposed to establish a similar "expressed nutrient content claim" rule in 1995 (60 *FR* 67474). It has announced plans to complete that action during June 1998.

For example, a manufacturer might want to offer "reduced fat peanut butter"— in this case, the product would be subject to a standard of identity, which governs the product name, and does not provide for use of the phrase "reduced fat." However, under §130.10, the "reduced fat" nutrient content claim could be used in conjunction with the "peanut butter" standardized name as long as all regulatory requirements were satisfied.

Part D
NUTRITION LABELING UNDER DSHEA

- *Chapter 8 —* DIETARY SUPPLEMENTS

Chapter 8—DIETARY SUPPLEMENTS

> ### In This Chapter...
>
> - What is a **dietary supplement**?
>
> - What **statutes** and regulations govern these products?
>
> - How do the FDA nutrition labeling rules **differ** for these products? What is the current status of this regulatory program?

Generally speaking, the nutrition labeling concepts discussed in Chapters 1–7 apply to all food products, whether they are in conventional food form or dietary supplement form. FDA has noted in several instances, however, that the regulations apply differently to dietary supplements — some key differences were noted in the discussions of specific rules, such as the health-claim requirement for a 10% nutrient contribution for "conventional" foods only. Clearly, some special requirements are based on the need for practical considerations and on the nature of the various products being regulated. FDA has developed special regulations for dietary supplements in 21 *CFR* §101.36.

As FDA begin implementing NLEA requirements, a number of issues arose with respect to nutrition labeling for supplements of vitamins and minerals. Since then, Congress has passed two bills — in 1992 and again in 1994 — concerning how FDA regulates supplement products. These actions, and the current status of the agency's program, are explained in this chapter.

A Fuzzy Line

Approximately 50% of Americans regularly consume dietary supplements including vitamins, minerals, or herbs to improve nutrition. Perhaps nowhere else is the difficulty of regulating nutrition information more evident than in the handling of dietary supplements. These products raise questions about the separation between "drugs" and "foods" as well as the appropriateness of disease-prevention claims on food products. As discussed in Chapter 6, federal policy toward health claims has evolved significantly over the past 10 years, and more changes may be on the horizon.

Foods for Special Dietary Use

Many different categories of products are intended to address particular dietary requirements. These include infant formulas and so-called medical foods; by necessity, the FDA rules make special provisions for these products, as noted on numerous occasions in the preceding chapters. Furthermore, the agency has issued special regulations in Parts 105 and 107. However, those types of foods are not the focus of this discussion. This chapter reviews FDA's labeling regulations for products intended to supplement the diets of the general population.

Statutory and Regulatory History

Since 1990, FDA has had to change course as it mapped out a plan to regulate nutrition labeling on all food products, including dietary supplements. For example, two 1991 proposals tentatively defined nutrient content claims for food labeling. FDA intended that these proposals would apply both to dietary supplements *and* to conventional foods. However, as noted earlier, a series of actions has delayed completion of rulemaking for dietary supplements:

- Congress passed the Dietary Supplement Act (DSA) on October 6, 1992, placing a moratorium until December 15, 1993 on NLEA implementation with respect to dietary supplements not in the form of conventional food. DSA Section 202(a)(2)(A) directed FDA to issue new proposed regulations applicable to dietary supplements of vitamins, minerals, herbs, and other similar nutritional substances.

- Meanwhile, food labeling and nutrient content claims rules for conventional foods were finalized on January 6, 1993.

- On June 18, 1993, FDA proposed a rule governing nutrition labeling of dietary supplements, which was subsequently finalized on January 4, 1994. That rule established separate labeling provisions for dietary supplements of vitamins or minerals, codified in 21 *CFR* §101.36; the agency also indicated that supplements of herbs or other nutritional substances were subject to regulations in §101.9.

- Still more changes came about on October 25, 1994, when Congress enacted the Dietary Supplement Health and Education Act of 1994 (DSHEA). The DSHEA amended the 1992 DSA, changing labeling requirements for dietary supplements, including provisions for the following:

    - The order in which dietary ingredients are to be listed,

    - The listing of the quantity of each dietary ingredient,

    - The optional listing of the source of a dietary ingredient, and

    - The listing of other ingredients in the dietary supplement.

DSHEA Requirements

Among other things, the DSHEA: 1) provided a statutory definition for "dietary supplements," 2) allowed some flexibility in ingredient and nutrition labeling of these products, and 3) provided for statements characterizing the level of ingredients for which no RDI or DRV has been established. With respect to the statutory definition, FDA later noted that:

> "In the 1994 nutrient content claims final rule, FDA used the terms 'dietary supplements of vitamins, minerals, herbs, and other similar nutritional substances' and 'food in conventional food form.' With the passage of the DSHEA, however, Congress has defined the term 'dietary supplement' and has modified the Act . . . to make clear that the form of the food is not necessarily determinative of whether it is a dietary supplement or not. Therefore, . . . FDA will use the more simple terms 'dietary supplement' and 'conventional food.' " (60 *FR* 67184-67185)

Structure/Function Claims

Moreover, the DSHEA provided that dietary supplements may bear statements claiming benefits related to nutrient-deficient diseases, describing and characterizing the role of the nutrients in human structure or function, and/or describing general well-being from consumption of such nutrients. However, to bear these statements, dietary supplement manufacturers must have evidence to substantiate the claim and must place a disclaimer on the label. (A recent FDA proposal concerning these claims is discussed at the end of this chapter.)

§101.36

FDA has issued regulations specifically for dietary supplements in 21 *CFR* §101.36. Other special requirements or exceptions for dietary supplements are interwoven with the rest of FDA's food labeling rules — such as §101.54, which governs nutrient content claims for "good source," "high," "more," and "high potency" on both conventional foods and dietary supplement products.

Table 8.1 lists the key provisions provided in §101.36 as revised in 1997. Details of the 1997 actions, which revamped the FDA program, are provided below.

TABLE 8.1
Primary Provisions of FDA's Dietary Supplement Rules

Regulatory provision—	21 CFR §101.36
Nutrition labeling required	(a)
Label formats	(b)-(e)
Nutrient listings, amounts by weight for dietary ingredients that have an established RDI or DRV	(b)(2)
Nutrient listings, amounts by weight for dietary ingredients for which no RDI or DRV has been established	(b)(3)
Labeling a proprietary blend of dietary ingredients	(b)(4)
Type-size requirements, sample labels	(e)
Applying for alternative means of compliance	(f)
Small business exemption	(h)(1)
Low-volume products	(h)(2)
Special labeling provisions and exceptions	(h)

A State of Flux

To complete its implementation of NLEA and DSHEA, FDA published several significant proposals on December 28, 1995 — these were finalized on September 23, 1997.

With the five recent actions, FDA created a full set of requirements for dietary supplements — and met its statutory obligations. Four of the 1997 rules revised 21 *CFR* Part 101. The fifth — which concerns premarket submittal of safety information — added a new Part 190 to the *CFR*. The subject of each of these actions was as follows:

1. Statement of identity, nutrition labeling and ingredient labeling of dietary supplements; and compliance policy guide revocation (62 *FR* 49826).

2. Food labeling; requirements for nutrient content claims, health claims, and statements of nutritional support for dietary supplements (62 *FR* 49859).

3. Nutrient content claims: definition for "high potency" and "antioxidant" for use in nutrient content claims for dietary supplements and conventional foods (62 *FR* 49868).

4. Notification procedures for statements on dietary supplements (62 *FR* 49883).

5. Premarket notification for a new dietary ingredient (62 *FR* 49886).

Among other things, FDA outlined requirements for how dietary supplements are to be identified on labels — these products will be distinguished from conventional foods, and will clearly use the term "supplement." The agency also set standards for how nutrition information is displayed. In finalizing this action, FDA modified its initial proposal to provide more flexibility for manufacturers, while ensuring that nutrition labels conform to up-to-date scientific views.

Three other rules cover requirements for nutrient content claims, health claims, and statements of nutritional support for dietary supplements, while a fifth requires submittal of safety information to FDA. Specifically, the agency amended the terminology used to describe dietary supplements; provided for statements to characterize percentage levels of dietary ingredients that do not have established reference daily intakes (RDIs) or daily reference values (DRVs); and withdrew the provision that dietary supplements may not give prominence on a label to any ingredient that is not a vitamin or mineral. The agency has also changed how industry must display disclaimers on labels.

Furthermore, FDA included definitions and standards for using the terms "high potency" and "antioxidant." Finally, the agency has established notification procedures for manufacturers, packers, and distributors of dietary supplements who make nutrition or health statements in marketing about their products, or market new dietary ingredients.

Although the new notification procedures were effective on October 23, 1997, the remainder of the new requirements take effect March 23, 1999.

Statutory and Regulatory Background

Among other things, DSHEA: 1) provided statutory definition for "dietary supplements," 2) allowed some flexibility in ingredient and nutrition labeling of these products, and 3) provided for statements characterizing the level of ingredients for which no RDIs or DRVs have been established.

On December 28, 1995, FDA published three proposals to finish implementing NLEA and DSHEA. The first two rules dealt directly with new dietary supplement provisions, while the third proposed nutrient content claim requirements for "antioxidant" and "high potency." These were finalized in September 1997, along with the two actions establishing notification procedures.

Statement of Identity and Labeling Requirements

The first in the series of new final rules established regulations for: 1) identification of dietary supplements, and 2) nutrition labeling and ingredient requirements for dietary supplements so that they more clearly convey nutrition information. Since DSHEA mandates that dietary supplements clearly bear those words on the label, FDA has added a new paragraph [21 *CFR* §101.3(g)]. To allow for increased flexibility, the agency is allowing the name of dietary ingredients to replace the word "dietary" on product labels. The new regulation reads:

> "Dietary supplements shall be identified by the term 'dietary supplement' as part of the statement of identity, except that the word 'dietary' may be deleted and replaced by the name of the dietary ingredients in the product (e.g., calcium supplement) or an appropriately descriptive term indicating the type of dietary ingredients that are in the product (e.g., herbal supplement with vitamins)."

In addition, the agency revised §101.36. This change affects the listings on supplement labels for:

- Serving size;

- Dietary ingredients both *with* and *without* established RDIs or DRVs;

- The quantity and percent daily value (%DV) of certain nutrients or their subcomponents (e.g., total calories, cholesterol, etc.);

- Proprietary blends; and

- The source of dietary ingredients (e.g., oyster shell powder).

The final rule also addressed label format and location requirements, as well as how ingredients are designated (§101.4).

"Supplement Facts"

As of the effective date, the agency is requiring that dietary supplement labels have a "Supplement Facts" heading to distinguish these products from conventional "Nutrition Facts" labels. The term "Serving Size" must be placed as a subheading under "Supplement Facts" and aligned on the left side of the nutrition label [(§101(b)(1)].

Serving size must be expressed using a term appropriate for the particular supplement (e.g., tablet, capsule, packet, or teaspoonful). Another subheading, "Servings Per Container," must appear directly below the "Serving Size" heading.

Figures 8.1–8.4 (pages 8.6–8.7) depict sample

labels for multiple vitamins, supplements containing ingredients both with and without established RDIs or DRVs, products declaring an herb ingredient, and products using proprietary blends, respectively. The text of §101.36 provides other example labels, such as alternative formats.

Ingredients With RDIs or DRVs

RDIs for vitamins and minerals are listed in §101.9(c), including vitamin A, vitamin C, calcium, and iron. DRVs have been established for several other nutrients, including sodium and potassium. When added to dietary supplements, these nutrients and their subcomponents are called dietary ingredients. Because FDA lists these 14 components in 21 *CFR* §101.36(b)(2), they are referred to as "(b)(2)-dietary ingredients." The agency is requiring their declaration as follows:

TABLE 8.2
(b)(2)-Dietary Ingredients Required on Dietary Supplement Labels

Total calories *(listed first in the column of names under a bar separating "Amount Per Serving" [or serving unit, such as Tablet] from the list of names)*
Calories from fat *(indented under "Calories")*
Total fat
Saturated fat
Cholesterol
Sodium
Total carbohydrate
Dietary fiber
Sugars
Protein
Vitamin A
Vitamin C
Calcium
Iron

These ingredients do not need to be declared on dietary supplement labels if present in amounts that can be declared as zero under §101.9(c) (e.g., for vitamins and minerals, in amounts of less than 2% of the RDI).

Other (b)(2)-dietary ingredients include calories from saturated, polyunsaturated, and monounsaturated fats; soluble and insoluble fiber; sugar alcohol; and other carbohydrates. These are required only when a claim is made about them, but they may be declared in the absence of such a claim. Any other vitamins or minerals listed in §§101.9(c)(8)(iv) or 101.9(c)(9) may be declared — but only when they are added for purposes of supplementation or when a claim is made.

When ingredients with established daily values are present, FDA is requiring that they be identified on the left side of the nutrition label with quantitative amounts by weight in the following order under the heading "Amount Per Serving:"

- Vitamin A,
- Vitamin C,
- Vitamin D,
- Vitamin E,
- Vitamin K,
- Thiamin,
- Riboflavin,
- Niacin,
- Vitamin B_6,
- Folate,
- Vitamin B_{12},
- Biotin,
- Pantothenic acid,
- Calcium,
- Iron,
- Phosphorus,
- Iodine,
- Magnesium,
- Zinc,
- Selenium,
- Copper,
- Manganese,
- Chromium,
- Molybdenum,
- Chloride,
- Sodium, and
- Potassium.

Supplement Facts

Serving Size 1 Tablet

Amount Per Serving		% Daily Values for Children Under 4 Years of Age	% Daily Value for Adults and Children 4 or more Years of Age
Calories	5		
Total Carbohydrate	1 g	†	< 1%*
Sugars	1 g	†	†
Vitamin A (50% as beta-carotene)	2500 IU	100%	50%
Vitamin C	40 mg	100%	67%
Vitamin D	400 IU	100%	100%
Vitamin E	15 IU	150%	50%
Thiamin	1.1 mg	157%	73%
Riboflavin	1.2 mg	150%	71%
Niacin	14 mg	156%	70%
Vitamin B_6	1.1 mg	157%	55%
Folate	300 mcg	150%	75%
Vitamin B_{12}	5 mcg	167%	83%

* Percent Daily Values are based on a 2,000 calorie diet.
† Daily Value not established.

Other ingredients: Sucrose, sodium ascorbate, stearic acid, gelatin, maltodextrine, artificial flavors, d-alpha tocopheryl acetate, niacinamide, magnesium stearate, Yellow 6, artificial colors, stearic acid, palmitic acid, pyridoxine hydrochloride, thiamin mononitrate, vitamin A acetate, beta-carotene, folic acid, cholecalciferol, and cyanocobalamin.

FIGURE 8.1
Vitamins for Children & Adults

Supplement Facts

Serving Size 1 Capsule

Amount Per Capsule		% Daily Value
Calories 20		
Calories from Fat 20		
Total Fat 2 g		3%*
Saturated Fat 0.5 g		3%*
Polyunsaturated Fat 1 g		†
Monounsaturated Fat 0.5 g		†
Vitamin A 4250 IU		85%
Vitamin D 425 IU		106%
Omega-3 fatty acids 0.5 g		†

* Percent Daily Values are based on a 2,000 calorie diet.
† Daily Value not established.

Ingredients: Cod liver oil, gelatin, water, and glycerin.

FIGURE 8.2
Supplement Containing Ingredients Both With and Without RDIs and DRVs

Supplement Facts

Serving Size 1 Capsule

Amount Per Capsule

Oriental Ginseng, powdered (root) — 250 mcg*

* Daily Value not established.

Other ingredients: Gelatin, water, and glycerin.

FIGURE 8.3
Supplement Declaring an Herb Ingredient

Supplement Facts

Serving Size 1 tsp (3 g) (makes 8 fl oz prepared)
Servings Per Container 24

	Amount Per Teaspoon	% Daily Value
Calories	10	
Total Carbohydrate	2 g	< 1%*
Sugars	2 g	†
Proprietary blend	0.7 g	
German Chamomile (flower)		†
Hyssop (leaves)		†

* Percent Daily Values are based on a 2,000 calorie diet.
† Daily Value not established.

Other ingredients: Fructose, lactose, starch, and stearic acid.

FIGURE 8.4
Declaring a Proprietary Blend
of Dietary Ingredients

FDA is allowing manufacturers to add synonyms in parentheses immediately following the name of some of these "(b)(2)-dietary ingredients." The following synonyms are approved by the agency: Vitamin C (ascorbic acid), thiamin (vitamin B_1), riboflavin (vitamin B_2), folate (folacin or folic acid), and calories (energy). Also, the term "folic acid" or "folacin" may be listed without parentheses in place of "folate." Additionally, beta-carotene may be declared as the percent of vitamin A that is present as beta-carotene — the declaration is *required* only when a claim is made about beta-carotene.

Quantitative amounts by weight of each ingredient may be presented in one of two ways: either in a separate column aligned to the right of the column of "(b)(2)-dietary ingredients," or immediately following the names within the same column. In either case, quantitative amounts must be expressed in increments specified in §101.9(c)(1)-(c)(7). The agency points out that the separate column method of expressing quantitative amounts allows space between the dietary ingredient names and the quantitative amounts so that dietary supplement manufacturers may provide information on the source of the ingredients.

Percent Daily Value

Another column to the right of "Amount Per Serving" must bear the heading "% Daily Value." When known for any of the (b)(2)-dietary ingredients, the percent daily value must be declared (based on RDI and DRV values for children 4 or more years of age). FDA is requiring that this value be calculated by dividing the quantitative amount of each (b)(2)-dietary ingredient by the RDI and multiplying by 100. The percentages must be expressed in the column to the nearest whole percent. When the quantitative amount of the (b)(2)-dietary ingredients is significant enough to be listed on the label, but when rounded to the nearest whole percent still equals zero, then the term "Less than 1%" or "<1%" must be used. The percent daily value for protein must be calculated as specified in §101.9(c)(7)(ii).

If the percent daily value is declared for total fat, saturated fat, total carbohydrate, dietary fiber, or protein, a symbol (such as an asterisk) must follow the value listed for those nutrients. At the bottom of the label, the same symbol must appear with the words, "Percent Daily Values are based on a 2,000 calorie diet."

For some products on which the unit amount may be of interest to consumers, FDA is allowing dietary supplement labels to bear quantitative information on a "per unit" basis. However, "per unit" information must be presented *in addition* to the required "per serving" information and must be presented in additional columns that have clearly identified, appropriately named headings.

Dietary Ingredients Without Established RDIs or DRVs

FDA is requiring that a dietary ingredient for which there is no established RDI or DRV also be listed on the nutrition label of a dietary supplement. These substances must be identified by their common or usual names, and must be listed either in a column under the (b)(2)-dietary ingredients or in a linear display under a heavy bar that separates them from the (b)(2)-dietary ingredients. However, the agency is not proposing any particular order for their listing. If the column format is used, then a symbol (e.g., an asterisk) must appear under the "% Daily Value" column, referring to a statement in the bottom portion of the label indicating "Daily Value not established." When the linear display format is used, the quantitative amounts must follow the name of each dietary ingredient with the symbol following the weight.

For any dietary ingredient that is a liquid extract from which the solvent has not been removed, the quantity listed must be the weight of the total extract. Information on the concentration of the dietary ingredient, solvent used, and the condition of the starting material (i.e., fresh or dried) must also be listed. If the solvent has been removed, the weight of the dried extract must be listed. But the dried extract must be described by an appropriate term that identifies the solvent used (e.g., dried hexane extract of X).

Proprietary Blends

Dietary supplements that contain proprietary blends of ingredients are also regulated under the new rule. The term "Proprietary Blend" or other appropriately descriptive term or fanciful name must be listed under dietary ingredients; the term may be highlighted in bold type. Then, ingredients contained in the blend must be declared in descending order of predominance by weight in a column or in linear fashion under the term, "Proprietary Blend."

The quantitative weight of the blend must appear in the "% Daily Value" column to refer to a statement at the bottom of the nutrition label—"Daily Value not established." If the blend contributes any significant amount of the "(b)(2)-dietary ingredients," then those must be listed along with a quantitative amount and percent daily value.

Sources of Ingredients

FDA has added a new section to §101.36 to allow room on the supplement nutrition labels for information on the source of dietary ingredients. Under new section (d), source ingredients may be identified within parentheses immediately following or indented beneath the name of a dietary ingredient and preceded by the words "as" or "from" [e.g., Calcium (as calcium carbonate)]. When the source of an ingredient is identified in this manner, then it is not required in the list of ingredients that appears outside the nutrition label.

In accordance with §101.4, source ingredients must be listed by common or usual name as found in the reference, *Herbs of Commerce*. If the common or usual name is not listed in this reference, then the complete Latin binomials are required. In addition, the listing of botanicals must specify the part of the plant from which the ingredient is derived.

Small Packages

To accommodate the nutrition label requirements on small packages, FDA is providing for some flexibility. The agency is allowing 4.5-type size on packages with: 1) less than 12 square inches of label space, 2) less than 20 sq in of label space and more than 8 dietary ingredients, and 3) 20-40 sq in of label space if more than 16 dietary ingredients are listed. Additionally, the agency is allowing the nutrient names to be connected to the quantitative amounts with dots rather than hairlines between the label rows on small and intermediate-sized supplement packages. However, FDA is allowing dots only when it is not possible to meet the minimum type size requirement of 4.5 if hairlines are used.

Compliance Requirements

Dietary ingredients, when added to supplements, must be present in amounts at least equal to the values declared on the nutrition label. When ingredients occur naturally, they must be present in amounts at least equaling 80% of the value declared. Reasonable excesses are permitted within current good manufacturing practices.

In addition, if the amount of calories, sugars, total fat, cholesterol, and/or sodium is found to be more than 20% in excess of the amounts declared on the label, a dietary supplement is deemed misbranded. These requirements parallel those already in place for nutrition labels.

To provide greater flexibility for industry, FDA has eliminated the requirement that packages for sampling analysis come from 12 different shipping cases. The agency is now requiring only that packages come from the same inspection lot. The sample for analysis must consist of a composite of 12 consumer packages or 10% of the packages in the inspection lot, whichever is smaller. This approach allows FDA to ensure compliance with the regulations without being impeded by the low availability of certain dietary supplement products.

Exemptions

The agency is exempting small businesses from this rule if total gross annual sales are not more than $500,000 or annual food product sales are not more than $50,000. However, these small business may not provide nutrition information or make health or nutrient content claims on their dietary supplement labels or in advertising. Additionally, foods shipped in bulk form that are not intended for distribution to consumers would be exempted from these labeling requirements.

Nutrient Content Claims: Conforming to DSHEA

In an effort to develop regulations conforming to the NLEA and the DSHEA, FDA has also finalized two other amendments to the dietary supplement regulations. Specifically, the agency is:

- Changing some of the regulatory terminology that describes dietary supplements,

- Allowing dietary supplement labels to bear statements characterizing the percentage level of dietary ingredients for which no RDIs or DRVs are established, and

- Requiring a disclaimer on labels of dietary supplements that bear nutritional support statements.

FDA also withdrew the provision prohibiting dietary supplement manufacturers from giving prominence to ingredients that are not vitamins or minerals.

Terminology

The dietary supplement regulations govern nutrient content claims for supplements of vitamins, minerals, herbs, and other similar nutritional substances. As noted by FDA, the phrase "other similar nutritional substances" suggests that the regulations are broad in scope. The DSHEA provided an amended definition for "dietary supplement" that essentially explains "other similar nutritional substances:"

> "[A] product, other than tobacco, intended to supplement the diet that bears or contains one or more of the following dietary ingredients: a vitamin; a mineral; an herb or other botanical; an amino acid; a dietary substance for use by man to supplement the diet by increasing the total dietary intake; or a concentrate, metabolite, constituent, extract, or combination of any of the aforementioned dietary ingredients." [60 FR 67177; proposed rule published December 28, 1995]

To ensure that the regulatory terminology reflects this statutory language, FDA amended the nutrient content claim regulations in §101.13 by 1) referring to the dietary supplement nutrition labeling regulations (§101.36), and 2) dropping reference to "other similar nutritional substances" in the introductory paragraph. In other words, the agency is stating that nutrient content claims in §101.13 apply to "conventional foods and dietary supplements." Similar changes were made to the health claim regulations found at §101.14(b)(3)(i) and (d)(3).

Substances Without an RDI or DRV

Under NLEA, food labels may include nutrient content claims only if the nutrient level characterizations use terms that are defined in the regulations. However, the DSHEA amended this provision for supplements. The amended language permits the use of statements in the labeling of a dietary supplement characterizing the percentage level of dietary ingredients for which FDA has *not* established RDIs or DRVs.

To conform with this DSHEA provision, FDA is authorizing claims on supplement labels that disclose the percentage level of dietary ingredients for which no RDI or DRV has been established.

- Whenever a simple percentage claim is made, the actual amount of the dietary ingredient in a serving of the product must also be declared (e.g., "40% omega-3 fatty acids, 10 mg per capsule").

- Percentage statements also may compare the amount of a dietary ingredient to the amount found in another product *if* the actual ingredient amounts in the product and in the comparative food are declared [e.g., "twice the omega-3 fatty acids per capsule (80 mg) as in 100 mg of menhaden oil (40 mg)"].

These changes are included in §101.13(q)(3) and apply only to dietary supplements. The decision diagram in **Figure 8.5** outlines how nutrient content claims may be included on dietary supplements.

Prominence of Non-Vitamin and Non-Mineral Ingredients

The NLEA stated that the labeling and advertising of dietary supplements of vitamins and minerals could not emphasize ingredients that are not represented as a source of vitamins or minerals. This restricted the use of nutrient content claims such as "more fiber" or "high protein" from use on dietary supplement labels. However, Section 7(d) of the

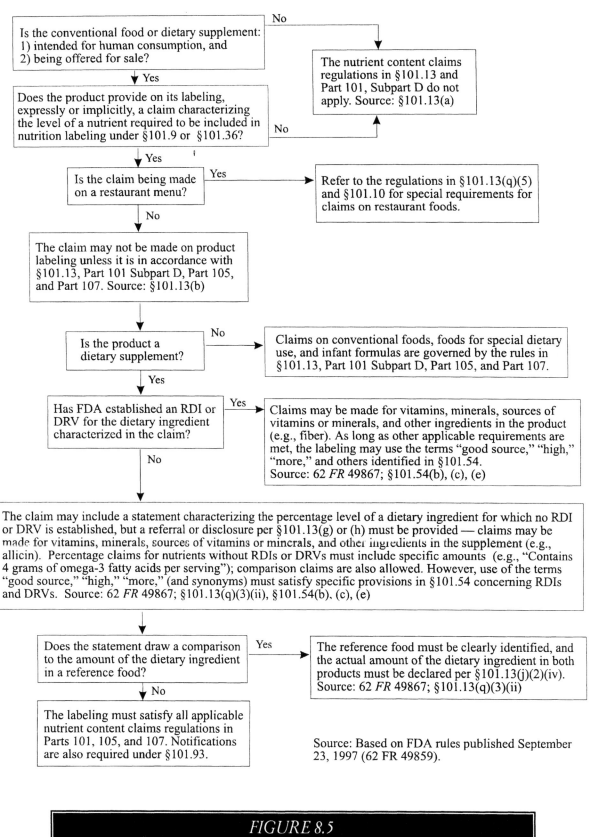

FIGURE 8.5
MAKING A NUTRIENT CONTENT CLAIM ON A DIETARY SUPPLEMENT

DSHEA removed this restriction.

In response to the amended law, FDA removed certain prohibitions from §101.54 (the regulations for "high," "good source," and "more") to accommodate dietary supplements. Under the revised rules:

- To use the term "high" on a label, the food must contain 20% or more of the RDI or the DRV per serving.

- "Good source," "contains," or "provides" may be used on dietary supplements provided that the food contains 10-19% of the RDI or DRV per serving.

- Claims for "more" (e.g., "more," "fortified," "enriched," and "added") may be used to describe levels of protein, vitamins, minerals, dietary fiber, or potassium except as limited by §101.54(e) and §101.13(j)(1)(i).

Nutritional Support Statement Disclaimer

The DSHEA provides for use of statements of nutritional support on dietary supplement labels. Such statements alert the consumer to the role of various nutrients or dietary ingredients in supporting human structure or function. However, the law mandates that manufacturers have evidence substantiating their nutritional support statements in order to demonstrate that the claim is truthful and not misleading.

To implement this requirement, FDA is requiring that manufacturers include a disclaimer to accompany the nutritional support statement on supplement labels. The text of the disclaimer must read as follows:

This statement has not been evaluated by the Food and Drug Administration. This product is not intended to diagnose, treat, cure, or prevent any disease.

The disclaimer must be placed immediately adjacent to the statement of nutritional support with no intervening material. If more than one statement is included on a label, a symbol (such as an asterisk) must follow each statement, and the disclaimer, preceded by the symbol, must be placed elsewhere on the same label.(The disclaimer must be in plural form, i.e., "These statements . . . ."). The agency is also specifying type size for the disclaimer. These provisions are included in §101.93(b)-(e).

Nutrient Content Claims Finalized

Nutrient claims apply to dietary supplements as well as to other products. In another action, FDA defined two new nutrient content claims — "high potency" and "antioxidant." Although these rules are targeted toward dietary supplements, they also apply to conventional foods. (Note that this rulemaking does not respond to any of the DSHEA requirements, but instead is a part of ongoing NLEA implementation.) FDA also corrected an omission pertaining to the use of "sugar free" claims on dietary supplements.

"High Potency"

FDA has acknowledged that dietary supplements differ from conventional foods in several respects. In particular, they can contain much higher levels of nutrients than conventional foods; currently defined nutrient content claims such as "good source" and "high" are not always adequate to describe nutrient levels found in dietary supplements. Therefore, the agency has concluded that nutrient content claims specifically applicable to dietary supplements are appropriate. "High potency" is just such a term.

Accordingly, the agency is allowing the term "high potency" on dietary supplement and conventional food labels to describe a nutrient that is present at 100% or more of the RDI for a vitamin or mineral. If the term is used, it must clearly identify which vitamin or mineral it describes (e.g., "Botanical X with high potency vitamin E"). The term "high potency" may also be used on the label of a multi-ingredient food product to describe the product as a whole if the product contains 100% or more of the RDI for at least two-thirds of the vitamins and minerals listed.

Of note, the term "high potency" may be used only on vitamins and minerals with established RDIs, not DRVs. For this reason, "high potency" claims may not be made for protein and fiber — the claim is limited to use with vitamins and minerals only.

FDA has concluded that 100% of the RDI per serving is a reasonable definition of "high potency" because this level is high enough to meet the needs of most healthy people. Furthermore, the agency believes that this level is consistent with current industry practice and is a matter of common sense; providing 100% of the RDI for a vitamin or mineral is an amount that can be considered highly potent.

Note that the "high potency" claim may be used on a conventional food product that has been fortified. However, this provision does not extend to dietary supplements.

"Antioxidant" Defined for *All* Products

As part of the final rule, FDA defined the word "antioxidant" so that it can be used in a clear and consistent manner in conjunction with currently defined nutrient content terms such as "good source" or "high." Because claims using this term often refer to a variety of nutrients or dietary ingredients, the agency is allowing these nutrient content claims on all food products, not just dietary supplements. A claim characterizing the level of antioxidant nutrients may be used on product labeling when:

- An RDI is established for each of the nutrients;

- The nutrients have recognized antioxidant activity;

- The level of each nutrient claimed on the label qualifies for a claim of "high," "good source," or "more" as found in §101.54(b), (c), and (e); and

- The names of the nutrients are included as part of the claim (e.g., "high in antioxidant vitamins C and E").

For example, to claim "high in antioxidant vitamin C," the product must contain 20% or more of the RDI for vitamin C. Also, beta-carotene may be claimed if the level of vitamin A in the product present as beta-carotene qualifies for the claim: to claim "good source of antioxidant beta-carotene," 10%–19% of the RDI for vitamin A must be present as beta-carotene per RACC.

When used as part of a nutrient content claim, the term "antioxidant" may be linked by a symbol (such as an asterisk) that refers to the same symbol elsewhere on the same label panel. The symbol must be followed by the names of the nutrients with recognized antioxidant activity.

Antioxidant Nutrients

To define "antioxidant," FDA first tried to determine which nutrients should be included within coverage of the term. Initially, the agency proposed that only vitamins C and E and beta-carotene possessed direct antioxidant activity. However, new literature questions the relevance of direct and indirect antioxidants. Therefore, FDA is not limiting the "antioxidant" definition to vitamin C, vitamin E, and beta-carotene. Instead, it broadened the number of substances that can claim antioxidant levels:

> "[A]ntioxidant may be used for a substance for which there is scientific evidence that, following absorption from the gastrointestinal tract, the substance participates in physiological, biochemical, or cellular processes that inactivate free radicals or that prevent free radical-initiated chemical reactions." [62 *FR* 49874]

In the future, FDA will consider proposing an affirmative list of antioxidant nutrients. Until then, the final rule allows manufacturers to determine what nutrients in a product meet the definition for antioxidants found in §101.54(g).

"Sugar free" Clarified

FDA has also clarified use of the term "sugar free" in nutrient content claims on dietary supplements. In the current regulations, FDA states that "calorie free" and "low calorie" claims may not be made on dietary supplements unless an equivalent amount of a dietary supplement that the labeled food resembles and for which it substitutes normally exceeds the definition for "low calorie."

This had the unintended effect of limiting the use of "sugar free" and "no sugar" claims on dietary supplements that would otherwise meet the requirements for the "low calorie" claim but were prohibited from bearing the claim because they do not substitute for a similar dietary supplement that normally exceeds the definition for "low calorie."

FDA sees no reason why its position toward these claims should be any different for supplements than it is for conventional foods. Thus, the agency revised §101.60(c)(1)(iii)(A) to excuse dietary supplement products that meet the definition of "low calorie," but are otherwise prohibited from bearing that claim, from the limitation.

Effective Date

All three rules discussed above become effective 18 months after publication — March 23, 1999. FDA believes this is sufficient time for manufacturers, packers, and distributors of dietary supplements to deplete existing label stores and obtain new labels that conform to the regulations.

Economic Impact

Although these actions could financially impact much of the dietary supplement industry, FDA believes that providing an 18-month compliance period will alleviate some of the burden. Furthermore, the agency asserts that the industry has been aware of these impending changes since at least January 1994, and that the majority of firms have been taking the necessary steps to reduce their label inventories.

In determining the impact of these actions, FDA had difficulty quantifying all the benefits of the labeling changes. But the agency believes, in general, that the new regulations will benefit consumers by assuring that adequate and complete nutrition information is provided accurately and consistently.

Notification Procedures

Effective October 23, 1997, if a manufacturer, packer, or distributor of dietary supplements makes a health or nutrient content claim, or includes a statement of nutrition support, FDA must be notified of the statement. Notification must occur no later than 30 days after first marketing that product.

The DSHEA added Section 403(r)(6) to the FDCA, allowing for a statement on the label of a dietary supplement that does the following:

- Claims a benefit related to a classical nutrient deficiency disease and discloses the prevalence of such disease in the United States,

- Describes the role of a nutrient or dietary ingredient intended to affect the structure or function in humans,

- Characterizes the documented mechanism by which a nutrient or dietary ingredient acts to maintain such structure or function, or

- Describes general well-being from consumption of a nutrient or dietary ingredient if the statements are made in accordance with certain requirements.

Procedures

As part of notification, the manufacturer, packer, or distributor must include:

- Name and address of the manufacturer, packer, or distributor of the dietary supplement that bears the statement,

- The text of the statement,

- Name of the dietary ingredient or supplement that is the subject of the statement,

- Name of the dietary supplement (including brand name), and

- The signature of the person who can certify the accuracy of the information presented in the statement and notice.

The notification must be sent to:

Office of Special Nutritionals (HFS-450)
Center for Food Safety and Applied Nutrition
Food and Drug Administration
200 C Street, SW
Washington, DC 20204.

In addition, manufacturers/distributors and other covered entities must 1) make the required disclaimer and 2) submit information on product safety (see below).

Requirements for Submitting Supplement Safety Information

Finally, FDA has also developed procedures outlining how dietary supplement manufacturers and/or distributors marketing new products — or products containing new dietary ingredients — must submit to the agency information demonstrating that their products can reasonably be expected to be safe. This action also took effect October 23, 1997.

The rule is somewhat of a companion to the other rule requiring notification when a supplement label bears a statement of nutrition support, a health claim, or a nutrient content claim (reviewed above). FDA issued this regulation to enable industry to comply with underlying DSHEA requirements in FDCA Section 413(a)(2). FDA originally proposed this action on September 27, 1996 (61 *FR* 50774) — the completed regulations added a new Part 190 to Title 21 of the *CFR*.

Requirements

In response to comments on the proposal, FDA has clarified which dietary supplements are covered by this action. The proposed language was revised by incorporating the phrase "that has not been present in the food supply as an article used for food in a form in which the food has not been chemically altered" to define which new dietary ingredients are subject to these notification requirements.

In "New Dietary Ingredient Notification," added as Subpart B of the new Part 190, FDA outlines the following requirements:

- At least 75 days before introducing or delivering into interstate commerce a dietary supplement that contains a new dietary ingredient (i.e., one "that has not been present in the food supply as an article used for food in a form in which the food has not been chemically altered"), the manufacturer or distributor of that supplement or ingredient shall submit to FDA the information that is the basis on which the manufacturer or distributor has concluded that a dietary supplement containing such dietary ingredient will reasonably be expected to be safe. This information may include citation to published articles.

- The notification must include:

    — Name and address of the manufacturer or distributor;

    — The name of the new dietary ingredient that is the subject of the premarket notification, including the Latin binomial name (including the author) of any herb or other botanical; and

    — A description of the dietary supplement(s) that contain the new ingredient, including: 1) the level of the new ingredient in the dietary supplement, and 2) the conditions of use recommended or suggested in the labeling of the supplement product (if no conditions of use are recommended, the ordinary conditions of use should be provided).

- Provide a history of use or other evidence of safety establishing that the dietary ingredient, when used under the conditions recommended or suggested on the label, will reasonably be expected to be safe. Include any citation to published articles or other evidence that is the basis for the distributor's or manufacturer's safety decision. Any reference to published information must be accompanied by reprints or copies of such references (and accompanied by an accurate and complete English translation, if applicable).

- Include the signature of the designated representative of the manufacturer or distributor.

Acknowledgment, Publication by FDA

FDA will acknowledge receipt of these so-called FDCA Section 413 submittals by notifying the submitter of the date of receipt; this becomes the "filing date" for the notification. For 75 days thereafter, the manufacturer or distributor may not introduce, or deliver for introduction, the supplement containing the new dietary ingredient into interstate commerce.

If the submitter later provides additional information in support of the new dietary ingredient notification, the agency will review all pertinent material, including responses to its inquiries, to determine whether they are substantive and whether FDA should reset the 75-day period (i.e., assign a new filing date). FDA will acknowledge receipt of any additional information and, when applicable, notify the manufacturer of a new filing date (this would be the date the agency received the new substantive information).

FDA will not disclose "the existence of, or the information contained in, the new dietary ingredient notification for 90 days after the filing date. [62 *FR* 49892]" After the 90th day, all information will be placed on public display, except for trade secrets or otherwise confidential commercial information. Lastly, "failure of the agency to respond to a notification does not constitute a finding that the new dietary ingredient or the dietary supplement that contains the new dietary ingredient is safe or is not adulterated" under FDCA Section 402. (62 *FR* 49892)

Submittal

An original and two copies should be submitted. Notifications should be sent to:

Office of Special Nutritionals (HFS–450)
Center for Food Safety and Applied Nutrition
Food and Drug Administration
200 C St SW
Washington, DC 20204.

Ephedrine Alkaloid Restrictions

FDA has begun looking at the content of certain dietary supplements where evidence indicates that significant health problems may result from product misuse. On June 4, 1997, the agency proposed restrictions on the use of ephedrine alkaloids in supplement products (62 *FR* 30678).

Among other things, this action would prohibit labeling claims indicating that long-term intake would be necessary to achieve the purported effect. FDA will add a new Subpart D, "Restricted Dietary Ingredients," to the rules in 21 *CFR* Part 111 if this proposal is completed.

More Changes Still to Come

FDA is planning additional regulations concerning dietary supplements. On April 29, 1998, it proposed regulations to clarify the distinction between drugs and dietary supplements (63 *FR* 23624).

The key issue is how to define criteria for acceptable label statements concerning the effect of a dietary supplement product on the structure or function of the human body. The agency plans to define in the regulations the types of statements that can be made; the proposal also would establish criteria for determining when a statement about a dietary supplement is a claim to diagnose, cure, mitigate, treat, or prevent disease. The full text of FDA's regulatory proposal is reproduced on **page 8.21**.

If the proposal becomes final, unacceptable claims will have to be removed from labeling, or the product will have to be approved as a drug under FDCA.

Types of Statements and Claims

FDA wants to make allowed claims more uniform. In taking this action, it intends to provide direction to the dietary supplement industry and to respond to guidance provided by the Commission on Dietary Supplement Labels. It is no secret that the line between function and disease claims is a fuzzy one:

- *Function*— FDA points out that currently, certain types of permitted statements represent claims that, prior to enactment of the DSHEA, might have rendered the product a "drug" under FDCA. Specifically, Section 403(r)(6) allows labeling to bear a statement that "describes the role of a nutrient or dietary ingredient intended to affect the structure or function in humans" or that "characterizes the documented mechanism by which a nutrient or dietary ingredient acts to maintain such structure or function." These types of claims are generally referred to as structure/function claims. However, the agency notes that a supplement for which only structure/function claims are made may nevertheless be subject to regulation as a drug if FDA has other evidence that the intended use of the product is for the diagnosis, cure, mitigation, treatment, or prevention of disease (see 21 *CFR* §201.128).

- *Disease*— Certain other types of statements continue, even under the DSHEA, to cause a product to be regulated as a drug. Statements "may not claim to diagnose, mitigate, treat, cure, or prevent a specific disease or class of diseases" under Section 403(r)(6), except that they may claim a benefit related to a classical nutrient deficiency disease — provided they also disclose the prevalence of the disease in the United States. These are typically known as disease claims.

A manufacturer who wishes to make a structure or function statement must have substantiation that the statement is truthful and not misleading, and must include the following disclaimer:

This statement has not been evaluated by the Food and Drug Administration. This product is not intended to diagnose, treat, cure, or prevent any disease.

Moreover, the DSHEA requires the manufacturer of a product bearing a statement under Section 403(r)(6) to notify FDA no later than 30 days after the first marketing of a product with the statement. New regulations implementing these requirements were published in 1997, and are discussed above.

Health Claims O.K.

FDA also acknowledges that another type of statement pertaining to diseases or health-related conditions is currently permitted on food products, including dietary supplements — as discussed in Chapter 6, these are known as health claims, and are authorized under FDCA Section 403(r)(1)(B). Health claims describe the relationship between a nutrient and a disease or health-related condition. Unlike structure/function claims, however, these must be formally authorized before they may be used on the label of a food or dietary supplement.

Consequently, certain claims about disease may be made for foods and dietary supplements without causing these products to be regulated as drugs, provided the claim has been authorized for use in accordance with the applicable regulations.

FDA Review

The DSHEA does not authorize the use of all claims that describe the effect of a dietary supplement on the structure or function of the body. Instead, Section 403(r)(6) authorizes only those structure/function statements that describe an effect of a product on the body but that are not also disease claims. Because the distinction between the two types of claims is not always obvious, the dietary supplement industry requested clarification from FDA on which structure/function claims may be made under Section 403(r)(6).

Since DSHEA, FDA has received approximately 2,300 notifications about structure/function claims for various products. The agency informed manufacturers that roughly 150 of these were problematic; FDA estimates that another 60 claims would not be permitted under the proposed criteria.

To develop criteria for evaluating claims, FDA reviewed the notification letters submitted previously. It also reviewed the recent final report of the Commission on Dietary Supplement Labels, which was established by the DSHEA to provide guidance and recommendations for the regulation of label claims and statements for these products.

The Commission issued a draft report in June 1997; among other things, it included views on what constitutes an acceptable statement of the "structure/function type." After reviewing public comments, a final report was released November 24, 1997. (Note that FDA issued a formal response to that report at the same time it published this proposal — 63 *FR* 23633; April 29, 1998).

The Commission's report includes findings, guidance, and recommendations.

- *Guidance* represents advice to specific agencies, groups, or individuals that will hopefully be considered by the identified recipients as they take action related to the availability of dietary supplements in the marketplace.

- *Findings* represent conclusions reached by the Commission during its deliberations and are based on the information and data received and reviewed.

- *Recommendations* are indicated as such and identify the intended recipients. These call for consideration of changes in existing regulations, development of new rules or policies, or legislative action as indicated.

Commission Guidance

Among other things, the Commission offered the following guidance on the scope of permissible structure/function claims:

"While the Commission recognizes that the context

of a claim has to be considered on a case-by-case basis, the Commission proposes the following general guidelines:

1. Statements of nutritional support should provide useful information to consumers about the intended use of a product.

2. Statements of nutritional support should be supported by scientifically valid evidence substantiating that the statements are truthful and not misleading.

3. Statements indicating the role of a nutrient or dietary ingredient in affecting the structure or function of humans may be made when the statements do not suggest disease prevention or treatment.

4. Statements that mention a body system, organ, or function affected by the supplement using terms such as 'stimulate,' 'maintain,' 'support,' 'regulate,' or 'promote' can be appropriate when the statements do not suggest disease prevention or treatment or use for a serious health condition that is beyond the ability of the consumer to evaluate.

5. Statements should not be made that products 'restore' normal or 'correct' abnormal function when the abnormality implies the presence of disease. An example might be a claim to 'restore' normal blood pressure when the abnormality implies hypertension.

6. Health claims are specifically defined under NLEA as statements that characterize the relationship between a nutrient or a food component and a specific disease or health-related condition. Statements of nutritional support should be distinct from NLEA health claims in that they do not state or imply a link between a supplement and prevention of a specific disease or health-related condition.

7. Statements of nutritional support are not to be drug claims. They should not refer to specific diseases, disorders, or classes of diseases and should not use drug-related terms such as 'diagnose,' 'treat,' 'prevent,' 'cure,' or 'mitigate.'" (pp. 38-39)

Note that the Commission report sometimes refers to statements made under Section 403(r)(6) as "statements of nutritional support." However, as explained by FDA in its September 23, 1997 final rule regarding labeling claims for dietary supplements, the agency no longer uses that phrasing — "because many of the substances that can be the subject of this type of claim have no nutritional value. (63 FR 23625)" Thus, the term "statement of nutritional support" is not accurate in all instances.

The guidance focuses on the distinction between allowable structure/function statements and claims that a product can diagnose, treat, prevent, cure, or mitigate disease (i.e., disease claims), and makes clear that structure/ function claims on dietary supplements should not imply treatment or prevention of disease. The Commission report also provides examples of types of structure/function claims that do and do not imply disease claims.

FDA also reviewed the Commission Report's guidance on distinguishing structure/function and disease claims. Eventually, the agency developed a definition of "disease," relying on standard medical and legal definitions as a basis for a proposed regulatory definition. FDA then applied the proposed definition of disease to explore potential criteria, testing the preliminary definition on a wide variety of statements currently being made by dietary supplement manufacturers. Based on those reviews, the agency developed its proposed general criteria.

Defining "Disease"

To assist in describing what constitutes a disease claim, the proposal defines "disease" as any deviation from, impairment of, or interruption of the normal structure or function of any part, organ, or system (or combination thereof) of the body that is manifested by a characteristic set of one or more signs or symptoms. For purposes of this definition, "signs or symptoms" include laboratory or clinical measurements that are characteristic of a disease, such as elevated cholesterol fraction, uric acid, blood sugar, and glycosylated hemoglobin, and characteristic signs of disease, such as elevated blood pressure or intraocular pressure.

TABLE 8.3
Distinguishing Between Drugs and Dietary Supplements

Examples of prohibited claim(s)	Reason prohibited	Examples of allowable statement(s)
"Protects against the development of cancer"	indicates that the product has an effect on a specific disease or class of diseases	"Helps promote urinary tract health"
Description of a product as an antibiotic, antiseptic, laxative, antidepressant, vaccine, or diuretic	product would be in a drug class, intended to be used to diagnose, mitigate, treat, prevent, etc.	Product description such as "energizer"
"Lowers cholesterol" or "Reduces joint pain"	product has effect on one or more symptoms recognizable to health professionals or consumers as characteristic of a specific disease	"Reduces stress and frustration" or "Improves absentmindedness" (symptoms do not identify a disease)
Product names such as "Herbal Prozac" or a suggestion that a product should be used "as part of your diet when taking insulin to help maintain a healthy blood sugar level"	the product is a substitute for, or augmentation of, a drug or other medical therapy	
"supports the body's antiviral capabilities" or "supports the body's ability to resist infection"	product has a role in the body's response to a disease or carrier of disease	"supports the immune system" (effect on a normal body function)
"treats toxemia of pregnancy" or treats "decreased sexual function," "hot flashes," or "Alzheimer's disease"	the product has an effect on a consequence of a natural state (e.g., aging or pregnancy) that presents characteristic symptoms recognizable as an abnormality of the body	
"helps avoid diarrhea associated with antibiotic use," "reduces nausea associated with chemotherapy"	this product treats, prevents or mitigates adverse events associated with a medical therapy or procedure	"helps maintain healthy intestinal flora" (no mention of a therapy or procedure)
Product name: -"Raynaudin" -"Hepatacure" -"Carpaltum"	-Raynaud's phenomenon -liver problems -implies effect on carpal tunnel syndrome	Allowable names: "Cardiohealth" or "Heart Tabs"
Statements about product formulation, including a claim that it contains an ingredient that is a well-known drug, such as aspirin, laetrile or digoxin		
Citation of a publication with a title referring to a disease or use — e.g., for vitamin E, an article entitled *Serial Coronary Angiographic Evidence That Antioxidant Vitamin Intake Reduces Progression of Coronary Artery Atherosclerosis*		
Use of "disease" or "diseased"		
Suggestions of an effect on disease through pictures, symbols and other means		
		Source: FDA Fact Sheet dated April 27, 1998

Examples

FDA has provided examples of various types of prohibited disease claims, and when possible or appropriate, contrasted them with examples of statements that are allowable, as shown in **Table 8.3** on the previous page.

Proposed Changes

To eliminate any inconsistency between the definitions of "disease" and "disease or health-related condition" found in the current regulations, FDA will amend §101.14(a)(6). That section defines "disease or health-related condition" as damage to an organ, part, structure, or system of the body such that it does not function properly (e.g., cardiovascular disease), or a state of health leading to such dysfunctioning (e.g. hypertension); except that diseases resulting from essential nutrient deficiencies (e.g., scurvy, pellagra) are not included.

Under the proposal, "disease or health-related condition" would be defined, in relevant part, as any deviation from, impairment of, or interruption of the normal structure or function of any part, organ, or system (or combination thereof) of the body that is manifested by a characteristic set of one or more signs or symptoms (including laboratory or clinical measurements that are characteristic of a disease), or a state of health leading to such deviation, impairment, or interruption; except that diseases resulting from essential nutrient deficiencies (e.g., scurvy, pellagra) are not included.

All of the regulatory changes proposed by FDA are listed on the following page.

Regulatory Changes Proposed by FDA Concerning Structure/Function and Disease-Related Statements on Dietary Supplements:

Amend 21 *CFR* §101.14 by revising paragraph (a)(6) to read as follows:

§101.14 Health claims: general requirements.
(a)(6) Disease or health-related condition means any deviation from, impairment of, or interruption of the normal structure or function of any part, organ, or system (or combination thereof) of the body that is manifested by a characteristic set of one or more signs or symptoms (including laboratory or clinical measurements that are characteristic of a disease), or a state of health leading to such deviation, impairment, or interruption; except that diseases resulting from essential nutrient deficiencies (e.g., scurvy, pellagra) are not included in this definition (claims pertaining to such diseases are thereby not subject to this section or §101.70).

Amend §101.93 by revising the heading and adding paragraphs (f) and (g) to read as follows:

§101.93 Certain types of statements for dietary supplements.
(f) *Permitted structure/function statements.*
    (1) Dietary supplement labels or labeling may, subject to the requirements of this section, bear statements that describe the role of a nutrient or dietary ingredient intended to affect the structure or function in humans or that characterize the documented mechanism by which a nutrient or dietary ingredient acts to maintain such structure or function, but may not bear statements that are disease claims under paragraph (g) of this section.

(g) *Disease claims.*
    (1) *Definition of disease.* For purposes of 21 U.S.C. 343(r)(6), a "disease" is any deviation from, impairment of, or interruption of the normal structure or function of any part, organ, or system (or combination thereof) of the body that is manifested by a characteristic set of one or more signs or symptoms, including laboratory or clinical measurements that are characteristic of a disease.

    (2) *Disease claims.* FDA will find that a statement about a product claims to diagnose, mitigate, treat, cure, or prevent disease (other than a classical nutrient deficiency disease) under section 403(r)(6) of the act if it meets one or more of the criteria listed in this paragraph (g)(2). In determining whether a statement is a disease claim under these criteria, FDA will consider the context in which the claim is presented. A statement claims to diagnose, mitigate, treat, cure, or prevent disease if it claims, explicitly or implicitly, that the product:

      (i) Has an effect on a specific disease or class of diseases;
      (ii) Has an effect, using scientific or lay terminology, on one or more signs or symptoms that are recognizable to health care professionals or consumers as being characteristic of a specific disease or of a number of different specific diseases;
      (iii) Has an effect on a consequence of a natural state that presents a characteristic set of signs or symptoms recognizable to health care professionals or consumers as constituting an abnormality of the body;
      (iv) Has an effect on disease through one or more of the following factors: (A) The name of the product; (B) A statement about the formulation of the product, including a claim that the product contains an ingredient that has been regulated by FDA as a drug and is well known to consumers for its use in preventing or treating a disease; (C) Citation of the title of a publication or reference, if the title refers to a disease use; (D) Use of the term "disease" or "diseased"; or (E) Use of pictures, vignettes, symbols, or other means;
      (v) Belongs to a class of products that is intended to diagnose, mitigate, treat, cure, or prevent a disease;
      (vi) Is a substitute for a product that is a therapy for a disease;
      (vii) Augments a particular therapy or drug action;
      (viii) Has a role in the body's response to a disease or to a vector of disease;
      (ix) Treats, prevents, or mitigates adverse events associated with a therapy for a disease and manifested by a characteristic set of signs or symptoms; or
      (x) Otherwise suggests an effect on a disease or diseases.

Source: 63 *FR* 23631-23632, April 29, 1998

Part E
REGULATORY TEXT

- *Chapter 9* — FDA FOOD LABELING RULES IN 21 *CFR* PART 101

- *Chapter 10* — FSIS LABELING REGULATIONS FOR MEAT PRODUCTS IN 9 *CFR* PART 317, SUBPARTS A AND B

- *Chapter 11* — FSIS POULTRY LABELING RULES IN 9 *CFR* PART 381, SUBPARTS N AND Y

Chapter 9—FDA LABELING RULES

> **In This Chapter...**
>
> **Full text of FDA's nutrition labeling regulations** (as of May 1, 1998):
>
> - General labeling provisions: 21 CFR §§101.1–101.8 — **page 9.2**
>
> - Basic nutrition labeling rules: 21 CFR §101.9 — **page 9.9**
>
> - Claims and other requirements: 21 CFR §§101.10–101.108 — **page 9.33**

PART 101-FOOD LABELING
Authority: 15 U.S.C. 1453, 1454, 1455; 21 U.S.C. 321, 331, 342, 343, 348, 371.

Subpart A-General Provisions
- §101.1 Principal display panel of package form food.
- §101.2 Information panel of package form food.
- §101.3 Identity labeling of food in packaged form.
- §101.4 Food; designation of ingredients.
- §101.5 Food; name and place of business of manufacturer, packer, or distributor.
- §101.8 [Removed]
- §101.9 Nutrition labeling of food.
- §101.10 Nutrition labeling of restaurant foods.
- §101.12 Reference amounts customarily consumed per eating occasion.
- §101.13 Nutrient content claims-general principles.
- §101.14 Health claims: general requirements.
- §101.15 Food; prominence of required statements.
- §101.17 Food labeling warning and notice statements.
- §101.18 Misbranding of food.

Subpart B-Specific Food Labeling Requirements
- §101.22 Foods; labeling of spices, flavorings, colorings and chemical preservatives.
- §101.29 [Removed]
- §101.30 Percentage juice declaration for foods purporting to be beverages that contain fruit or vegetable juice.

Subpart C-Specific Nutrition Labeling Requirements and Guidelines
- §101.36 Nutrition labeling of dietary supplements.
- §101.42 Nutrition labeling of raw fruit, vegetables, and fish.
- §101.43 Substantial compliance of food retailers with the guidelines for the voluntary nutrition labeling of raw fruit, vegetables, and fish.
- §101.44 Identification of the 20 most frequently consumed raw fruit, vegetables, and fish in the U.S.
- §101.45 Guidelines for the voluntary nutrition labeling of raw fruit, vegetables, and fish.

21 CFR §101.1

Subpart D-Specific Requirements for Nutrient Content Claims
- §101.54 Nutrient content claims for "good source," "high," "more," and "high potency."
- §101.56 Nutrient content claims for "light" or "lite."
- §101.60 Nutrient content claims for the calorie content of foods.
- §101.61 Nutrient content claims for the sodium content of foods.
- §101.62 Nutrient content claims for fat, fatty acid, and cholesterol content of foods.
- §101.65 Implied nutrient content claims and related label statements.
- §101.67 Use of nutrient content claims for butter.
- §101.69 Petitions for nutrient content claims.

Subpart E-Specific Requirements for Health Claims
- §101.70 Petitions for health claims.
- §101.71 Health claims: claims not authorized.
- §101.72 Health claims: calcium and osteoporosis.
- §101.73 Health claims: dietary lipids and cancer.
- §101.74 Health claims: sodium and hypertension.
- §101.75 Health claims: dietary saturated fat and cholesterol and risk of coronary heart disease.
- §101.76 Health claims: fiber-containing grain products, fruits, and vegetables and cancer.
- §101.77 Health claims: fruits, vegetables, and grain products that contain fiber, particularly soluble fiber, and risk of coronary heart disease.
- §101.78 Health claims: fruits and vegetables and cancer.
- §101.79 Health claims: folate and neural tube defects.
- §101.80 Health claims: dietary sugar alcohols and dental caries.
- §101.81 Health claims: soluble fiber from certain foods and risk of coronary heart disease (CHD).

Subpart F-Specific Requirements for Descriptive Claims that are Neither Nutrient Contents Claims nor Health Claims
- §101.93 Notification procedures for certain types of statements on dietary supplements.
- §101.95 "Fresh," "freshly frozen," "fresh frozen," "frozen fresh."

Subpart G-Exemptions From Food Labeling Requirements
- §101.100 Food; exemptions from labeling.
- §101.105 Declaration of net quantity of contents when exempt.
- §101.108 Temporary exemptions for purposes of conducting authorized food labeling experiments.

Appendix A to Part 101-Monier-Williams Procedure (With Modifications) for Sulfites in Food, Center for Food Safety and
Applied Nutrition, Food and Drug Administration (November 1985)
Appendix B to Part 101-Graphic Enhancements Used by the FDA
Appendix C to Part 101-Nutrition Facts for Raw Fruits and Vegetables
Appendix D to Part 101-Nutrition Facts for Cooked Fish

Subpart A-General Provisions

§101.1 Principal display panel of package form food.

The term "principal display panel" as it applies to food in package form and as used in this part, means the part of a label that is most likely to be displayed, presented, shown, or examined under customary conditions of display for retail sale. The principal display panel shall be large enough to accommodate all the mandatory label information required to be placed thereon by this part with clarity and conspicuousness and without obscuring design, vignettes,

or crowding. Where packages bear alternate principal display panels, information required to be placed on the principal display panel shall be duplicated on each principal display panel. For the purpose of obtaining uniform type size in declaring the quantity of contents for all packages of substantially the same size, the term "area of the principal display panel" means the area of the side or surface that bears the principal display panel, which area shall be:
(a) In the case of a rectangular package where one entire side properly can be considered to be the principal display panel side, the product of the height times the width of that side;
(b) In the case of a cylindrical or nearly cylindrical container, 40 percent of the product of the height of the container times the circumference;
(c) In the case of any otherwise shaped container, 40 percent of the total surface of the container: *Provided*, however, That where such container presents an obvious "principal display panel" such as the top of a triangular or circular package of cheese, the area shall consist of the entire top surface. In determining the area of the principal display panel, exclude tops, bottoms, flanges at tops and bottoms of cans, and shoulders and necks of bottles or jars. In the case of cylindrical or nearly cylindrical containers, information required by this part to appear on the principal display panel shall appear within that 40 percent of the circumference which is most likely to be displayed, presented, shown, or examined under customary conditions of display for retail sale.

§101.2 Information panel of package form food.

(a) The term "information panel" as it applies to packaged food means that part of the label immediately contiguous and to the right of the principal display panel as observed by an individual facing the principal display panel with the following exceptions:
   (1) If the part of the label immediately contiguous and to the right of the principal display panel is too small to accommodate the necessary information or is otherwise unusable label space, e.g., folded flaps or can ends, the panel immediately contiguous and to the right of this part of the label may be used.
   (2) If the package has one or more alternate principal display panels, the information panel is immediately contiguous and to the right of any principal display panel.
   (3) If the top of the container is the principal display panel and the package has no alternate principal display panel, the information panel is any panel adjacent to the principal display panel.
(b) All information required to appear on the label of any package of food under §§101.4, 101.5, 101.9, 101.13, 101.17, 101.36, subpart D of part 101, and part 105 of this chapter shall appear either on the principal display panel or on the information panel, unless otherwise specified by regulations in this chapter.
(c) All information appearing on the principal display panel or the information panel pursuant to this section shall appear prominently and conspicuously, but in no case may the letters and/or numbers be less than one-sixteenth inch in height unless an exemption pursuant to paragraph (f) of this section is established. The requirements for conspicuousness and legibility shall include the specifications of §§101.105(h) (1) and (2) and 101.15.
   (1)(i) Soft drinks packaged in bottles manufactured before October 31, 1975 shall be exempt from the requirements prescribed by this section to the extent that information which is blown, lithographed, or formed onto the surface of the bottle is exempt from the size and placement requirements of this section.
   (ii) Soft drinks packaged in bottles shall be exempt from the size and placement requirements prescribed by this section if all of the following conditions are met:
   (*a*) If the soft drink is packaged in a bottle bearing a paper, plastic foam jacket, or foil label, or is packaged in a nonreusable bottle bearing a label lithographed onto the surface of the bottle or is packaged in metal cans, the product shall not be exempt from any requirement of this section other than the exemptions created by §1.24(a)(5) (ii) and (v) of this chapter and the label shall bear all required information in the specified minimum type size, except the label will not be required to bear the information required by §101.5 if this information appears on the bottle closure or on the lid of the can in a type size not less than one-sixteenth inch in height, or if embossed on the lid of the can in a type size not less than one-eighth inch in height.
   (*b*) If the soft drink is packaged in a bottle which does not bear a paper, plastic foam jacket or foil label, or is packaged in a reusable bottle bearing a label lithographed onto the surface of the bottle:
   (*1*) Neither the bottle nor the closure is required to bear nutrition labeling in compliance with §101.9, except that any multiunit retail package in which it is contained shall bear nutrition labeling if required by §101.9; and any

21 *CFR* §101.2

vending machine in which it is contained shall bear nutrition labeling if nutrition labeling is not present on the bottle or closure, if required by §101.9.

(2) All other information pursuant to this section shall appear on the top of the bottle closure prominently and conspicuously in letters and/or numbers no less than one thirty-second inch in height, except that if the information required by §101.5 is placed on the side of the closure in accordance with §1.24(a)(5)(ii) of this chapter, such information shall appear in letters and/or numbers no less than one-sixteenth inch in height.

(3) Upon the petition of any interested person demonstrating that the bottle closure is too small to accommodate this information, the Commissioner may by regulation establish an alternative method of disseminating such information. Information appearing on the closure shall appear in the following priority:

(*i*) The statement of ingredients.

(*i*) The name and address of the manufacturer, packer, or distributor.

(*iii*) The statement of identity.

(2) Individual serving-size packages of food served with meals in restaurants, institutions, and on board passenger carriers, and not intended for sale at retail, are exempt from type-size requirements of this paragraph, provided:

(i) The package has a total area of 3 square inches or less available to bear labeling;

(ii) There is insufficient area on the package available to print all required information in a type size of 1/16 inch in height;

(iii) The information required by paragraph (b) of this section appears on the label in accordance with the provisions of this paragraph, except that the type size is not less than 1/32 inch in height.

(d)

(1) Except as provided by §101.9(j)(13) and (j)(17) and 101.36(i)(2) and (i)(5), all information required to appear on the principal display panel or on the information panel under this section shall appear on the same panel unless there is insufficient space. In determining the sufficiency of the available space, except as provided by §101.9(j)(17) and 101.36(i)(5), any vignettes, designs, and other nonmandatory label information shall not be considered. If there is insufficient space for all of this information to appear on a single panel, it may be divided between these two panels, except that the information required under any given section or part shall all appear on the same panel. A food whose label is required to bear the ingredient statement on the principal display panel may bear all other information specified in paragraph (b) of this section on the information panel.

(2) Any food, not otherwise exempted in this section, if packaged in a container consisting of a separate lid and body, and bearing nutrition labeling pursuant to §101.9, and if the lid qualifies for and is designed to serve as a principal display panel, shall be exempt from the placement requirements of this section in the following respects:

(i) The name and place of business information required by §101.5 shall not be required on the body of the container if this information appears on the lid in accordance with this section.

(ii) The nutrition information required by §101.9 shall not be required on the lid if this information appears on the container body in accordance with this section.

(iii) The statement of ingredients required by §101.4 shall not be required on the lid if this information appears on the container body in accordance with this section. Further, the statement of ingredients is not required on the container body if this information appears on the lid in accordance with this section.

(e) All information appearing on the information panel pursuant to this section shall appear in one place without other intervening material.

(f) If the label of any package of food is too small to accommodate all of the information required by §§101.4, 101.5, 101.9, 101.13, 101.17, 101.36, subpart D of part 101, and part 105 of this chapter, the Commissioner may establish by regulation an acceptable alternative method of disseminating such information to the public, e.g., a type size smaller than one-sixteenth inch in height, or labeling attached to or inserted in the package or available at the point of purchase. A petition requesting such a regulation, as an amendment to this paragraph shall be submitted under part 10 of this chapter.

[42 FR 14308, Mar. 15, 1977, as amended at 42 FR 15673, Mar. 22, 1977; 42 FR 45905, Sept. 13, 1977; 42 FR 47191, Sept. 20, 1977; 44 FR 16006, Mar. 16, 1979; 49 FR 13339, Apr. 4, 1984; 53 FR 16068, May 5, 1988; 58 FR 44030, Aug. 18, 1993; 60 FR 17205, Apr. 5, 1995; 62 *FR* 43074, Aug. 12, 1997; 62 *FR* 49847, Sept. 23, 1997; 63 FR 14817, Mar. 27, 1998]

21 CFR §101.3

§101.3 Identity labeling of food in packaged form.

(a) The principal display panel of a food in package form shall bear as one of its principal features a statement of the identity of the commodity.

(b) Such statement of identity shall be in terms of:

(1) The name now or hereafter specified in or required by any applicable Federal law or regulation; or, in the absence thereof,

(2) The common or usual name of the food; or, in the absence thereof,

(3) An appropriately descriptive term, or when the nature of the food is obvious, a fanciful name commonly used by the public for such food.

(c) Where a food is marketed in various optional forms (whole, slices, diced, etc.), the particular form shall be considered to be a necessary part of the statement of identity and shall be declared in letters of a type size bearing a reasonable relation to the size of the letters forming the other components of the statement of identity; except that if the optional form is visible through the container or is depicted by an appropriate vignette, the particular form need not be included in the statement. This specification does not affect the required declarations of identity under definitions and standards for foods promulgated pursuant to section 401 of the act.

(d) This statement of identity shall be presented in bold type on the principal display panel, shall be in a size reasonably related to the most prominent printed matter on such panel, and shall be in lines generally parallel to the base on which the package rests as it is designed to be displayed.

(e) Under the provisions of section 403(c) of the Federal Food, Drug, and Cosmetic Act, a food shall be deemed to be misbranded if it is an imitation of another food unless its label bears, in type of uniform size and prominence, the word "imitation" and, immediately thereafter, the name of the food imitated.

(1) A food shall be deemed to be an imitation and thus subject to the requirements of section 403(c) of the act if it is a substitute for and resembles another food but is nutritionally inferior to that food.

(2) A food that is a substitute for and resembles another food shall not be deemed to be an imitation provided it meets each of the following requirements:

(i) It is not nutritionally inferior to the food for which it substitutes and which it resembles.

(ii) Its label bears a common or usual name that complies with the provisions of §102.5 of this chapter and that is not false or misleading, or in the absence of an existing common or usual name, an appropriately descriptive term that is not false or misleading. The label may, in addition, bear a fanciful name which is not false or misleading.

(3) A food for which a common or usual name is established by regulation (e.g., in a standard of identity pursuant to section 401 of the act, in a common or usual name regulation pursuant to part 102 of this chapter, or in a regulation establishing a nutritional quality guideline pursuant to part 104 of this chapter), and which complies with all of the applicable requirements of such regulation(s), shall not be deemed to be an imitation.

(4) Nutritional inferiority includes:

(i) Any reduction in the content of an essential nutrient that is present in a measurable amount, but does not include a reduction in the caloric or fat content provided the food is labeled pursuant to the provisions of §101.9, and provided the labeling with respect to any reduction in caloric content complies with the provisions applicable to caloric content in part 105 of this chapter.

(ii) For the purpose of this section, a measurable amount of an essential nutrient in a food shall be considered to be 2 percent or more of the Daily Reference Value (DRV) of protein listed under §101.9(c)(7)(iii) and of potassium listed under §101.9(c)(9) per reference amount customarily consumed and 2 percent or more of the Reference Daily Intake (RDI) of any vitamin or mineral listed under §101.9(c)(8)(iv) per reference amount customarily consumed, except that selenium, molybdenum, chromium, and chloride need not be considered.

(iii) If the Commissioner concludes that a food is a substitute for and resembles another food but is inferior to the food imitated for reasons other than those set forth in this paragraph, he may propose appropriate revisions to this regulation or he may propose a separate regulation governing the particular food.

(f) A label may be required to bear the percentage(s) of a characterizing ingredient(s) or information concerning the presence or absence of an ingredient(s) or the need to add an ingredient(s) as part of the common or usual name of the food pursuant to subpart B of part 102 of this chapter.

(g) Dietary supplements shall be identified by the term "dietary supplement" as a part of the statement of identity, except that the word "dietary" may be deleted and replaced by the name of the dietary ingredients in the product

21 CFR §101.4

(e.g., calcium supplement) or an appropriately descriptive term indicating the type of dietary ingredients that are in the product (e.g., herbal supplement with vitamins).
[42 FR 14308, Mar. 15, 1977, as amended at 48 FR 10811, Mar. 15, 1983; 58 FR 2227, Jan. 6, 1993; 60 FR 67164, Dec. 28, 1995; 62 *FR* 49847, Sept. 23, 1997]

§101.4 Food; designation of ingredients.

(a)
   (1) Ingredients required to be declared on the label or labeling of a food, including foods that comply with standards of identity, except those ingredients exempted by §101.100, shall be listed by common or usual name in descending order of predominance by weight on either the principal display panel or the information panel in accordance with the provisions of §101.2, except that ingredients in dietary supplements that are listed in the nutrition label in accordance with §101.36 need not be repeated in the ingredient list. Paragraph (g) of this section describes the ingredient list on dietary supplement products.
   (2) The descending order of predominance requirements of paragraph (a)(1) of this section do not apply to ingredients present in amounts of 2 percent or less by weight when a listing of these ingredients is placed at the end of the ingredient statement following an appropriate quantifying statement, e.g., "Contains ___ percent or less of ___," or "Less than ___ percent of ___." The blank percentage within the quantifying statement shall be filled in with a threshold level of 2 percent, or, if desired, 1.5 percent, 1.0 percent, or 0.5 percent, as appropriate. No ingredient to which the quantifying phrase applies may be present in an amount greater than the stated threshold.

(b) The name of an ingredient shall be a specific name and not a collective (generic) name, except that:
   (1) Spices, flavorings, colorings and chemical preservatives shall be declared according to the provisions of §101.22.
   (2) An ingredient which itself contains two or more ingredients and which has an established common or usual name, conforms to a standard established pursuant to the Meat Inspection or Poultry Products Inspection Acts by the U.S. Department of Agriculture, or conforms to a definition and standard of identity established pursuant to section 401 of the Federal Food, Drug, and Cosmetic Act, shall be designated in the statement of ingredients on the label of such food by either of the following alternatives:
   (i) By declaring the established common or usual name of the ingredient followed by a parenthetical listing of all ingredients contained therein in descending order of predominance except that, if the ingredient is a food subject to a definition and standard of identity established in subchapter B of this chapter that has specific labeling provisions for optional ingredients, optional ingredients may be declared within the parenthetical listing in accordance with those provisions.
   (ii) By incorporating into the statement of ingredients in descending order of predominance in the finished food, the common or usual name of every component of the ingredient without listing the ingredient itself.
   (3) Skim milk, concentrated skim milk, reconstituted skim milk, and nonfat dry milk may be declared as "skim milk" or "nonfat milk".
   (4) Milk, concentrated milk, reconstituted milk, and dry whole milk may be declared as "milk".
   (5) Bacterial cultures may be declared by the word "cultured" followed by the name of the substrate, e.g., "made from cultured skim milk or cultured buttermilk".
   (6) Sweetcream buttermilk, concentrated sweetcream buttermilk, reconstituted sweetcream buttermilk, and dried sweetcream buttermilk may be declared as "buttermilk".
   (7) Whey, concentrated whey, reconstituted whey, and dried whey may be declared as "whey".
   (8) Cream, reconstituted cream, dried cream, and plastic cream (sometimes known as concentrated milk fat) may be declared as "cream".
   (9) Butteroil and anhydrous butterfat may be declared as "butterfat".
   (10) Dried whole eggs, frozen whole eggs, and liquid whole eggs may be declared as "eggs".
   (11) Dried egg whites, frozen egg whites, and liquid egg whites may be declared as "egg whites".
   (12) Dried egg yolks, frozen egg yolks, and liquid egg yolks may be declared as "egg yolks".
   (13) [Reserved]

21 CFR §101.4

(14) Each individual fat and/or oil ingredient of a food intended for human consumption shall be declared by its specific common or usual name (e.g., "beef fat", "cottonseed oil") in its order of predominance in the food except that blends of fats and/or oils may be designated in their order of predominance in the foods as "-- shortening" or "blend of -- oils", the blank to be filled in with the word "vegetable", "animal", "marine", with or without the terms "fat" or "oils", or combination of these, whichever is applicable if, immediately following the term, the common or usual name of each individual vegetable, animal, or marine fat or oil is given in parentheses, e.g., "vegetable oil shortening (soybean and cottonseed oil)". For products that are blends of fats and/or oils and for foods in which fats and/or oils constitute the predominant ingredient, i.e., in which the combined weight of all fat and/or oil ingredients equals or exceeds the weight of the most predominant ingredient that is not a fat or oil, the listing of the common or usual names of such fats and/or oils in parentheses shall be in descending order of predominance. In all other foods in which a blend of fats and/or oils is used as an ingredient, the listing of the common or usual names in parentheses need not be in descending order of predominance if the manufacturer, because of the use of varying mixtures, is unable to adhere to a constant pattern of fats and/or oils in the product. If the fat or oil is completely hydrogenated, the name shall include the term "hydrogenated", or if partially hydrogenated, the name shall include the term "partially hydrogenated". If each fat and/or oil in a blend or the blend is completely hydrogenated, the term "hydrogenated" may precede the term(s) describing the blend, e.g., "hydrogenated vegetable oil (soybean, cottonseed, and palm oils)", rather than preceding the name of each individual fat and/or oil; if the blend of fats and/or oils is partially hydrogenated, the term "partially hydrogenated" may be used in the same manner. Fat and/or oil ingredients not present in the product may be listed if they may sometimes be used in the product. Such ingredients shall be identified by words indicating that they may not be present, such as "or", "and/or", "contains one or more of the following:", e.g., "vegetable oil shortening (contains one or more of the following: cottonseed oil, palm oil, soybean oil)". No fat or oil ingredient shall be listed unless actually present if the fats and/or oils constitute the predominant ingredient of the product, as defined in this paragraph (b)(14).

(15) When all the ingredients of a wheat flour are declared in an ingredient statement, the principal ingredient of the flour shall be declared by the name(s) specified in §§137.105, 137.200, 137.220 and 137.225 of this chapter, i.e., the first ingredient designated in the ingredient list of flour, or bromated flour, or enriched flour, or self-rising flour is "flour", "white flour", "wheat flour", or "plain flour"; the first ingredient designated in the ingredient list of durum flour is "durum flour"; the first ingredient designated in the ingredient list of whole wheat flour, or bromated whole wheat flour is "whole wheat flour", "graham flour", or "entire wheat flour"; and the first ingredient designated in the ingredient list of whole durum wheat flour is "whole durum wheat flour".

(16) Ingredients that act as leavening agents in food may be declared in the ingredient statement by stating the specific common or usual name of each individual leavening agent in parentheses following the collective name "leavening", e.g., "leavening (baking soda, monocalcium phosphate, and calcium carbonate)". The listing of the common or usual name of each individual leavening agent in parentheses shall be in descending order of predominance: Except, That if the manufacturer is unable to adhere to a constant pattern of leavening agents in the product, the listing of individual leavening agents need not be in descending order of predominance. Leavening agents not present in the product may be listed if they are sometimes used in the product. Such ingredients shall be identified by words indicating that they may not be present, such as "or", "and/or", "contains one or more of the following:".

(17) Ingredients that act as yeast nutrients in foods may be declared in the ingredient statement by stating the specific common or usual name of each individual yeast nutrient in parentheses following the collective name "yeast nutrients", e.g., "yeast nutrients (calcium sulfate and ammonium phosphate)". The listing of the common or usual name of each individual yeast nutrient in parentheses shall be in descending order of predominance: Except, That if the manufacturer is unable to adhere to a constant pattern of yeast nutrients in the product, the listing of the common or usual names of individual yeast nutrients need not be in descending order of predominance. Yeast nutrients not present in the product may be listed if they are sometimes used in the product. Such ingredients shall be identified by words indicating that they may not be present, such as "or", "and/or", or "contains one or more of the following:".

(18) Ingredients that act as dough conditioners may be declared in the ingredient statement by stating the specific common or usual name of each individual dough conditioner in parentheses following the collective name "dough conditioner", e.g., "dough conditioners (L-cysteine, ammonium sulfate)". The listing of the common or usual name of each dough conditioner in parentheses shall be in descending order of predominance:

21 CFR §101.4

Except, That if the manufacturer is unable to adhere to a constant pattern of dough conditioners in the product, the listing of the common or usual names of individual dough conditioners need not be in descending order of predominance. Dough conditioners not present in the product may be listed if they are sometimes used in the product. Such ingredients shall be identified by words indicating that they may not be present, such as "or", "and/or", or "contains one or more of the following:".

(19) Ingredients that act as firming agents in food (e.g., salts of calcium and other safe and suitable salts in canned vegetables) may be declared in the ingredient statement, in order of predominance appropriate for the total of all firming agents in the food, by stating the specific common or usual name of each individual firming agent in descending order of predominance in parentheses following the collective name "firming agents". If the manufacturer is unable to adhere to a constant pattern of firming agents in the food, the listing of the individual firming agents need not be in descending order of predominance. Firming agents not present in the product may be listed if they are sometimes used in the product. Such ingredients shall be identified by words indicating that they may not be present, such as "or", "and/or", "contains one or more of the following:".

(20) For purposes of ingredient labeling, the term "sugar" shall refer to sucrose, which is obtained from sugar cane or sugar beets in accordance with the provisions of §184.1854 of this chapter.

(21) [Reserved]

(22) Wax and resin ingredients on fresh produce when such produce is held for retail sale, or when held for other than retail sale by packers or repackers shall be declared collectively by the phrase "coated with food-grade animal-based wax, to maintain freshness" or the phrase "coated with food-grade vegetable-, petroleum-, beeswax-, and/or shellac-based wax or resin, to maintain freshness" as appropriate. The terms "food- grade" and "to maintain freshness" are optional. The term "lac-resin" may be substituted for the term "shellac."

(c) When water is added to reconstitute, completely or partially, an ingredient permitted by paragraph (b) of this section to be declared by a class name, the position of the ingredient class name in the ingredient statement shall be determined by the weight of the unreconstituted ingredient plus the weight of the quantity of water added to reconstitute that ingredient, up to the amount of water needed to reconstitute the ingredient to single strength. Any water added in excess of the amount of water needed to reconstitute the ingredient to single strength shall be declared as "water" in the ingredient statement.

(d) When foods characterized on the label as "nondairy" contain a caseinate ingredient, the caseinate ingredient shall be followed by a parenthetical statement identifying its source. For example, if the manufacturer uses the term "nondairy" on a creamer that contains sodium caseinate, it shall include a parenthetical term such as "a milk derivative" after the listing of sodium caseinate in the ingredient list.

(e) If the percentage of an ingredient is included in the statement of ingredients, it shall be shown in parentheses following the name of the ingredient and expressed in terms of percent by weight. Percentage declarations shall be expressed to the nearest 1 percent, except that where ingredients are present at levels of 2 percent or less, they may be grouped together and expressed in accordance with the quantifying guidance set forth in paragraph (a)(2) of this section.

(f) Except as provided in §101.100, ingredients that must be declared on labeling because there is no label for the food, including foods that comply with standards of identity, shall be listed prominently and conspicuously by common or usual name in the manner prescribed by paragraph (b) of this section.

(g) When present, the ingredient list on dietary supplement products shall be located immediately below the nutrition label, or, if there is insufficient space below the nutrition label, immediately contiguous and to the right of the nutrition label and shall be preceded by the word "Ingredients," unless some ingredients (i.e., sources) are identified within the nutrition label in accordance with §101.36(d), in which case the ingredients listed outside the nutrition label shall be in a list preceded by the words "Other ingredients." Ingredients in dietary supplements that are not dietary ingredients or that do not contain dietary ingredients, such as excipients, fillers, artificial colors, artificial sweeteners, flavors, or binders, shall be included in the ingredient list.

(h) The common or usual name of ingredients of dietary supplements that are botanicals (including fungi and algae) shall be consistent with the names standardized in *Herbs of Commerce*, 1992 edition, which is incorporated by reference in accordance with 5 U.S.C. 552(a) and 1 CFR part 51. Copies may be obtained from the American Herbal Products Association, 4733 Bethesda Ave., suite 345, Bethesda, MD 20814, or may be examined at the Center for Food Safety and Applied Nutrition's Library, 200 C St. SW., rm. 3321, Washington, DC, or at the Office of the Federal Register, 800 Capital St. NW., suite 700, Washington, DC. The listing of these names on the label shall be followed by statements of:

(1) The part of the plant (e.g., root, leaves) from which the dietary ingredient is derived (e.g., "Garlic bulb" or "Garlic (bulb)"), except that this designation is not required for algae. The name of the part of the plant shall be expressed in English (e.g., "flower" rather than "flos");

(2) The Latin binomial name of the plant, in parentheses, except that this name is not required when it is available in the reference entitled: *Herbs of Commerce* for the common or usual name listed on the label, and, when required, the Latin binomial name may be listed before the part of the plant. Any name in Latin form shall be in accordance with internationally accepted rules on nomenclature, such as those found in the *International Code of Botanical Nomenclature* and shall include the designation of the author or authors who published the Latin name, when a positive identification cannot be made in its absence. The *International Code of Botanical Nomenclature* (Tokyo Code), 1994 edition, a publication of the International Association for Plant Taxonomy, is incorporated by reference in accordance with 5 U.S.C. 552(a) and 1 CFR part 51. Copies of the *International Code of Botanical Nomenclature* may be obtained from Koeltz Scientific Books, D–61453 Konigstein, Germany, and University Bookstore, Southern Illinois University, Carbondale, IL 62901–4422, 618–536–3321, FAX 618–453–5207, or may be examined at the Center for Food Safety and Applied Nutrition's Library, 200 C St. SW, Rm 3321, Washington DC, or at the Office of the Federal Register, 800 North Capitol St. NW, Suite 700, Washington DC.

(3) On labels of single-ingredient dietary supplements that do not include an ingredient list, the identification of the Latin binomial name, when needed, and the part of the plant may be prominently placed on the principal display panel or information panel, or included in the nutrition label.

[42 FR 14308, Mar. 15, 1977, as amended at 43 FR 12858, Mar. 28, 1978; 43 FR 24519, June 6, 1978; 48 FR 8054, Feb. 25, 1983; 55 FR 17433, Apr. 25, 1990; 58 FR 2875, Jan. 6, 1993; 62 *FR* 49847, Sept. 23, 1997]

§101.5 Food; name and place of business of manufacturer, packer, or distributor.

(a) The label of a food in packaged form shall specify conspicuously the name and place of business of the manufacturer, packer, or distributor.

(b) The requirement for declaration of the name of the manufacturer, packer, or distributor shall be deemed to be satisfied, in the case of a corporation, only by the actual corporate name, which may be preceded or followed by the name of the particular division of the corporation. In the case of an individual, partnership, or association, the name under which the business is conducted shall be used.

(c) Where the food is not manufactured by the person whose name appears on the label, the name shall be qualified by a phrase that reveals the connection such person has with such food; such as "Manufactured for ———", "Distributed by ———", or any other wording that expresses the facts.

(d) The statement of the place of business shall include the street address, city, State, and ZIP code; however, the street address may be omitted if it is shown in a current city directory or telephone directory. The requirement for inclusion of the ZIP code shall apply only to consumer commodity labels developed or revised after the effective date of this section. In the case of nonconsumer packages, the ZIP code shall appear either on the label or the labeling (including invoice).

(e) If a person manufactures, packs, or distributes a food at a place other than his principal place of business, the label may state the principal place of business in lieu of the actual place where such food was manufactured or packed or is to be distributed, unless such statement would be misleading.

§101.8 Labeling of food with number of servings.

[Removed August 12, 1997; 62 *FR* 43074]

§101.9 Nutrition labeling of food.

(a) Nutrition information relating to food shall be provided for all products intended for human consumption and

21 CFR §101.9

offered for sale unless an exemption is provided for the product in paragraph (j) of this section.

(1) When food is in package form, the required nutrition labeling information shall appear on the label in the format specified in this section.

(2) When food is not in package form, the required nutrition labeling information shall be displayed clearly at the point of purchase (e.g., on a counter card, sign, tag affixed to the product, or some other appropriate device). Alternatively, the required information may be placed in a booklet, looseleaf binder, or other appropriate format that is available at the point of purchase.

(3) Solicitation of requests for nutrition information by a statement "For nutrition information write to _____" on the label or in the labeling or advertising for a food, or providing such information in a direct written reply to a solicited or unsolicited request, does not subject the label or the labeling of a food exempted under paragraph (j) of this section to the requirements of this section if the reply to the request conforms to the requirements of this section.

(4) If any vitamin or mineral is added to a food so that a single serving provides 50 percent or more of the Reference Daily Intake (RDI) for the age group for which the product is intended, as specified in paragraph (c)(8)(iv) of this section, of any one of the added vitamins or minerals, unless such addition is permitted or required in other regulations, e.g., a standard of identity or nutritional quality guideline, or is otherwise exempted by the Commissioner, the food shall be considered a food for special dietary use within the meaning of §105.3(a)(1)(iii) of this chapter.

(b) Except as provided in §101.9(h)(3), all nutrient and food component quantities shall be declared in relation to a serving as defined in this section.

(1) The term "serving" or "serving size" means an amount of food customarily consumed per eating occasion by persons 4 years of age or older which is expressed in a common household measure that is appropriate to the food. When the food is specially formulated or processed for use by infants or by toddlers, a serving or serving size means an amount of food customarily consumed per eating occasion by infants up to 12 months of age or by children 1 through 3 years of age, respectively.

(2) Except as provided in paragraphs (b)(3), (b)(4), and (b)(6) of this section and for products that are intended for weight control and are available only through a weight-control or weight-maintenance program, serving size declared on a product label shall be determined from the "Reference Amounts Customarily Consumed Per Eating Occasion * * *" (reference amounts) that appear in §101.12(b) using the procedures described below. For products that are both intended for weight control and available only through a weight-control program, a manufacturer may determine the serving size that is consistent with the meal plan of the program. Such products must bear a statement, "for sale only through the ---- program" (fill in the blank with the name of the appropriate weight-control program, e.g., Smith's Weight Control), on the principal display panel. However, the reference amounts in §101.12(b) shall be used for purposes of evaluating whether weight-control products that are available only through a weight-control program qualify for nutrient content claims or health claims.

(i) For products in discrete units (e.g., muffins, sliced products, such as sliced bread, or individually packaged products within a multiserving package) and for products which consist of two or more foods packaged and presented to be consumed together where the ingredient represented as the main ingredient is in discrete units (e.g., pancakes and syrup), the serving size shall be declared as follows:

(A) If a unit weighs 50 percent or less of the reference amount, the serving size shall be the number of whole units that most closely approximates the reference amount for the product category;

(B) If a unit weighs more than 50 percent, but less than 67 percent of the reference amount, the manufacturer may declare one unit or two units as the serving size;

(C) If a unit weighs 67 percent or more, but less than 200 percent of the reference amount, the serving size shall be one unit;

(D) If a unit weighs 200 percent or more of the reference amount, the manufacturer may declare one unit as the serving size if the whole unit can reasonably be consumed at a single- eating occasion.

(E) For products that have reference amounts of 100 grams (g) (or milliliter (mL)) or larger and are individual units within a multiserving package, if a unit contains more than 150 percent but less than 200 percent of the reference amount, the manufacturer may decide whether to declare the individual unit as 1 or 2 servings.

(F) The serving size for maraschino cherries shall be expressed as 1 cherry with the parenthetical metric measure equal to the average weight of a medium size cherry.

(G) The serving size for products that naturally vary in size (e.g., pickles, shellfish, whole fish, and fillet of fish)

21 *CFR* §101.9

may be the amount in ounces that most closely approximates the reference amount for the product category. Manufacturers shall adhere to the requirements in paragraph (b)(5)(vi) of this section for expressing the serving size in ounces.

(H) For products which consist of two or more foods packaged and presented to be consumed together where the ingredient represented as the main ingredient is in discrete units (e.g., pancakes and syrup), the serving size may be the number of discrete units represented as the main ingredient plus proportioned minor ingredients used to make the reference amount for the combined product determined in §101.12(f).

(I) For packages containing several individual single-serving containers, each of which is labeled with all required information including nutrition labeling as specified in §101.9 (that is, are labeled appropriately for individual sale as single-serving containers), the serving size shall be 1 unit.

(ii) For products in large discrete units that are usually divided for consumption (e.g., cake, pie, pizza, melon, cabbage), for unprepared products where the entire contents of the package is used to prepare large discrete units that are usually divided for consumption (e.g., cake mix, pizza kit), and for products which consist of two or more foods packaged and presented to be consumed together where the ingredient represented as the main ingredient is a large discrete unit usually divided for consumption (e.g., prepared cake packaged with a can of frosting), the serving size shall be the fractional slice of the ready-to-eat product (e.g., 1/12 cake, 1/8 pie, 1/4 pizza, 1/4 melon, 1/6 cabbage) that most closely approximates the reference amount for the product category, and may be the fraction of the package used to make the reference amount for the unprepared product determined in §101.12(c) or the fraction of the large discrete unit represented as the main ingredient plus proportioned minor ingredients used to make the reference amount for the combined product determined in §101.12(f). In expressing the fractional slice, manufacturers shall use 1/2, 1/3, 1/4, 1/5, 1/6, or smaller fractions that can be generated by further division by 2 or 3.

(iii) For nondiscrete bulk products (e.g., breakfast cereal, flour, sugar, dry mixes, concentrates, pancake mixes, macaroni and cheese kits), and for products which consist of two or more foods packaged and presented to be consumed together where the ingredient represented as the main ingredient is a bulk product (e.g., peanut butter and jelly), the serving size shall be the amount in household measure that most closely approximates the reference amount for the product category and may be the amount of the bulk product represented as the main ingredient plus proportioned minor ingredients used to make the reference amount for the combined product determined in §101.12(f).

(3) The serving size for meal products and main dish products as defined in §101.13(l) and (m) that comes in single-serving containers as defined in paragraph (b)(6) of this section shall be the entire content (edible portion only) of the package. Serving size for meal products and main dish products in multiserving containers shall be based on the reference amount applicable to the product in §101.12(b) if the product is listed in §101.12(b). Serving size for meal products and main dish products in multiserving containers that are not listed in §101.12(b) shall be based on the reference amount according to §101.12(f).

(4) A variety pack, such as a package containing several varieties of single-serving units as defined in paragraph (b)(2)(i) of this section, and a product having two or more compartments with each compartment containing a different food, shall provide nutrition information for each variety or food per serving size that is derived from the reference amount in §101.12(b) applicable for each variety or food and the procedures to convert the reference amount to serving size in paragraph (b)(2) of this section.

(5) For labeling purposes, the term "common household measure" or "common household unit" means cup, tablespoon, teaspoon, piece, slice, fraction (e.g., 1/4 pizza), ounce (oz), fluid ounce (fl oz), or other common household equipment used to package food products (e.g., jar, tray). In expressing serving size in household measures, except as specified in paragraphs (b)(5)(iv), (b)(5)(v), (b)(5)(vi), and (b)(5)(vii) of this section, the following rules shall be used:

(i) Cups, tablespoons, or teaspoons shall be used wherever possible and appropriate except for beverages. For beverages, a manufacturer may use fluid ounces. Cups shall be expressed in 1/4- or 1/3-cup increments, tablespoons in whole number of tablespoons for quantities less than 1/4 cup but greater than or equal to 2 tablespoons (tbsp), 1, 1 1/3, 1 1/2, or 1 2/3 tbsp for quantities less than 2 tbsp but greater than or equal to 1 tbsp, and teaspoons in whole number of teaspoons for quantities less than 1 tbsp but greater than or equal to 1 teaspoon (tsp), and in 1/4-tsp increments for quantities less than 1 tsp.

(ii) If cups, tablespoons or teaspoons are not applicable, units such as piece, slice, tray, jar, and fraction shall be used.

21 CFR §101.9

(iii) If paragraphs (b)(5)(i) and (b)(5)(ii) of this section are not applicable, ounces may be used with an appropriate visual unit of measure such as a dimension of a piece, e.g., 1 oz (28 g/about 1/2 pickle). Ounce measurements shall be expressed in 0.5 oz increments most closely approximating the reference amount.

(iv) A description of the individual container or package shall be used for single serving containers and for individually packaged products within multiserving containers (e.g., can, box, package). A description of the individual unit shall be used for other products in discrete units (e.g., piece, slice, cracker, bar).

(v) For unprepared products where the entire contents of the package is used to prepare large discrete units that are usually divided for consumption (e.g., cake mix, pizza kit), the fraction or portion of the package may be used.

(vi) Ounces with an appropriate visual unit of measure, as described in paragraph (b)(5)(iii) of this section, may be used for products that naturally vary in size as provided for in paragraph (b)(2)(i)(G) of this section.

(vii) As provided for in §101.9(h)(1), for products that consist of two or more distinct ingredients or components packaged and presented to be consumed together (e.g. dry macaroni and cheese mix, cake and muffin mixes with separate ingredient packages, pancakes and syrup), nutrition information may be declared for each component or as a composite. The serving size may be provided in accordance with the provisions of paragraphs (b)(2)(i), (b)(2)(ii), and (b)(2)(iii) of this section, or alternatively in ounces with an appropriate visual unit of measure, as described in paragraph (b)(5)(iii) of this section (e.g., declared as separate components: "3 oz dry macaroni (84 g/about 2/3 cup)" and "1 oz dry cheese mix (28 g/about 2 tbsp);" declared as a composite value: "4 oz (112 g/about 2/3 cup macaroni and 2 tbsp dry cheese mix)").

(viii) For nutrition labeling purposes, a teaspoon means 5 milliliters (mL), a tablespoon means 15 mL, a cup means 240 mL, 1 fl oz means 30 mL, and 1 oz in weight means 28 g.

(ix) When a serving size, determined from the reference amount in §101.12(b) and the procedures described in this section, falls exactly half way between two serving sizes, e.g., 2.5 tbsp, manufacturers shall round the serving size up to the next incremental size.

(6) A product that is packaged and sold individually and that contains less than 200 percent of the applicable reference amount shall be considered to be a single-serving container, and the entire content of the product shall be labeled as one serving except for products that have reference amounts of 100 g (or mL) or larger, manufacturers may decide whether a package that contains more than 150 percent but less than 200 percent of the reference amount is 1 or 2 servings. Packages sold individually that contain 200 percent or more of the applicable reference amount may be labeled as a single-serving if the entire content of the package can reasonably be consumed at a single-eating occasion.

(7) A label statement regarding a serving shall be the serving size expressed in common household measures as set forth in paragraphs (b)(2) through (b)(6) of this section and shall be followed by the equivalent metric quantity in parenthesis (fluids in milliliters and all other foods in grams) except for single- serving containers.

(i) For a single-serving container, the parenthetical metric quantity, which will be presented as part of the net weight statement on the principal display panel, is not required except where nutrition information is required on a drained weight basis according to §101.9(b)(9). However, if a manufacturer voluntarily provides the metric quantity on products that can be sold as single servings, then the numerical value provided as part of the serving size declaration must be identical to the metric quantity declaration provided as part of the net quantity of contents statement.

(ii) The gram or milliliter quantity equivalent to the household measure should be rounded to the nearest whole number except for quantities that are less than 5 g (mL). The gram (mL) quantity between 2 and 5 g (mL) should be rounded to the nearest 0.5 g (mL) and the g (mL) quantity less than 2 g (mL) should be expressed in 0.1-g (mL) increments.

(iii) In addition, serving size may be declared in ounce and fluid ounce, in parenthesis, following the metric measure separated by a slash where other common household measures are used as the primary unit for serving size, e.g., 1 slice (28 g/1 oz) for sliced bread. The ounce quantity equivalent to the metric quantity should be expressed in 0.1 oz increments.

(iv) If a manufacturer elects to use abbreviations for units, the following abbreviations shall be used: tbsp for tablespoon, tsp for teaspoon, g for gram, mL for milliliter, oz for ounce, and fl oz for fluid ounce.

(v) For products that only require the addition of water or another ingredient that contains insignificant amounts of nutrients in the amount added and that are prepared in such a way that there are no significant changes to the nutrient profile, the amount of the finished product may be declared in parentheses at the end of the serving size declaration (e.g., 1/2 cup (120 mL) concentrated soup (makes 1 cup prepared)).

(vi) To promote uniformity in label serving sizes in household measures declared by different manufacturers, FDA has provided a guideline entitled, "Guidelines for Determining the Gram Weight of the Household Measure." The guideline can be obtained from the Office of Food Labeling (HFS-150), Center for Food Safety and Applied Nutrition, Food and Drug Administration, 200 C St. SW., Washington, DC 20204.

(8) Determination of the number of servings per container shall be based on the serving size of the product determined by following the procedures described in this section.

(i) The number of servings shall be rounded to the nearest whole number except for the number of servings between 2 and 5 servings and random weight products. The number of servings between 2 and 5 servings shall be rounded to the nearest 0.5 serving. Rounding should be indicated by the use of the term "about" (e.g., about 2 servings, about 3.5 servings).

(ii) When the serving size is required to be expressed on a drained solids basis and the number of servings varies because of a natural variation in unit size (e.g., maraschino cherries, pickles), the manufacturer may state the typical number of servings per container (e.g., usually 5 servings).

(iii) For random weight products, a manufacturer may declare "varied" for the number of servings per container provided the nutrition information is based on the reference amount expressed in ounces. The manufacturer may provide the typical number of servings in parenthesis following the "varied" statement.

(iv) For packages containing several individual single-serving containers, each of which is labeled with all required information including nutrition labeling as specified in §101.9 (that is, are labeled appropriately for individual sale as single-serving containers), the number of servings shall be the number of individual packages within the total package.

(v) For packages containing several individually packaged multiserving units, the number of servings shall be determined by multiplying the number of individual multiserving units in the total package by the number of servings in each individual unit.

(9) The declaration of nutrient and food component content shall be on the basis of food as packaged or purchased with the exception of raw fish covered under §101.42 (see 101.44), packaged single-ingredient products that consist of fish or game meat as provided for in paragraph (j)(11) of this section, and of foods that are packed or canned in water, brine, or oil but whose liquid packing medium is not customarily consumed (e.g., canned fish, maraschino cherries, pickled fruits, and pickled vegetables). Declaration of nutrient and food component content of raw fish shall follow the provisions in §101.45. Declaration of the nutrient and food component content of foods that are packed in liquid which is not customarily consumed shall be based on the drained solids.

(10) Another column of figures may be used to declare the nutrient and food component information:

(i) Per 100 g or 100 mL, or per 1 oz or 1 fl oz of the food as packaged or purchased;

(ii) Per one unit if the serving size of a product in discrete units in a multiserving container is more than 1 unit;

(iii) Per cup popped for popcorn in a multiserving container.

(11) If a product is promoted on the label, labeling, or advertising for a use that differs in quantity by twofold or greater from the use upon which the reference amount in §101.12(b) was based (e.g., liquid cream substitutes promoted for use with breakfast cereals), the manufacturer shall provide a second column of nutrition information based on the amount customarily consumed in the promoted use, in addition to the nutrition information per serving derived from the reference amount in §101.12(b), except that nondiscrete bulk products that are used primarily as ingredients (e.g., flour, sweeteners, shortenings, oils), or traditionally used for multipurposes (e.g., eggs, butter, margarine), and multipurpose baking mixes are exempt from this requirement.

(c) The declaration of nutrition information on the label and in labeling of a food shall contain information about the level of the following nutrients, except for those nutrients whose inclusion, and the declaration of amounts, is voluntary as set forth in this paragraph. No nutrients or food components other than those listed in this paragraph as either mandatory or voluntary may be included within the nutrition label. Except as provided for in paragraphs (f) or (j) of this section, nutrient information shall be presented using the nutrient names specified and in the following order in the formats specified in paragraphs (d) or (e) of this section.

(1) "Calories, total," "Total calories," or "Calories": A statement of the caloric content per serving, expressed to the nearest 5-calorie increment up to and including 50 calories, and 10-calorie increment above 50 calories, except that amounts less than 5 calories may be expressed as zero. Energy content per serving may also be expressed in kilojoule units, added in parentheses immediately following the statement of the caloric content.

21 CFR §101.9

(i) Caloric content may be calculated by the following methods. Where either specific or general food factors are used, the factors shall be applied to the actual amount (i.e., before rounding) of food components (e.g., fat, carbohydrate, protein, or ingredients with specific food factors) present per serving.

(A) Using specific Atwater factors (i.e., the Atwater method) given in Table 13, "Energy Value of Foods-Basis and Derivation," by A. L. Merrill and B. K. Watt, United States Department of Agriculture (USDA) Handbook No. 74 (slightly revised, 1973), which is incorporated by reference in accordance with 5 U.S.C. 552(a) and 1 CFR part 51 and is available from the Office of Food Labeling (HFS-150), Center for Food Safety and Applied Nutrition, Food and Drug Administration, 200 C St. SW., Washington, DC 20204, or may be inspected at the Office of the Federal Register, 800 North Capitol St. NW., suite 700, Washington, DC.;

(B) Using the general factors of 4, 4, and 9 calories per gram for protein, total carbohydrate, and total fat, respectively, as described in USDA Handbook No. 74 (slightly revised 1973) pp. 9-11, which is incorporated by reference in accordance with 5 U.S.C. 552(a) and 1 CFR part 51 (the availability of this incorporation by reference is given in paragraph (c)(1)(i)(A) of this section);

(C) Using the general factors of 4, 4, and 9 calories per gram for protein, total carbohydrate less the amount of insoluble dietary fiber, and total fat, respectively, as described in USDA Handbook No. 74 (slightly revised 1973) pp. 9-11, which is incorporated by reference in accordance with 5 U.S.C. 552(a) and 1 CFR part 51 (the availability of this incorporation by reference is given in paragraph (c)(1)(i)(A) of this section);

(D) Using data for specific food factors for particular foods or ingredients approved by the Food and Drug Administration (FDA) and provided in parts 172 or 184 of this chapter, or by other means, as appropriate; or

(E) Using bomb calorimetry data and subtracting 1.25 calories per gram protein to correct for incomplete digestibility, as described in USDA Handbook No. 74 (slightly revised 1973) p. 10, which is incorporated by reference in accordance with 5 U.S.C. 552(a) and 1 CFR part 51 (the availability of this incorporation by reference is given in paragraph (c)(1)(i)(A) of this section).

(ii) "Calories from fat": A statement of the caloric content derived from total fat as defined in paragraph (c)(2) of this section in a serving, expressed to the nearest 5-calorie increment, up to and including 50 calories, and the nearest 10-calorie increment above 50 calories, except that label declaration of "calories from fat" is not required on products that contain less than 0.5 gram of fat in a serving and amounts less than 5 calories may be expressed as zero. This statement shall be declared as provided in paragraph (d)(5) of this secton. Except as provided for in paragraph (f) of this section, if "Calories from fat" is not required and, as a result, not declared, the statement "Not a significant source of calories from fat" shall be placed at the bottom of the table of nutrient values in the same type size.

(iii) "Calories from saturated fat" or "Calories from saturated" (VOLUNTARY): A statement of the caloric content derived from saturated fat as defined in paragraph (c)(2)(i) of this section in a serving may be declared voluntarily, expressed to the nearest 5-calorie increment, up to and including 50 calories, and the nearest 10-calorie increment above 50 calories, except that amounts less than 5 calories may be expressed as zero. This statement shall be indented under the statement of calories from fat as provided in paragraph (d)(5) of this section.

(2) "Fat, total" or "Total fat": A statement of the number of grams of total fat in a serving defined as total lipid fatty acids and expressed as triglycerides. Amounts shall be expressed to the nearest 0.5 (1/2) gram increment below 5 grams and to the nearest gram increment above 5 grams. If the serving contains less than 0.5 gram, the content shall be expressed as zero.

(i) "Saturated fat," or "Saturated": A statement of the number of grams of saturated fat in a serving defined as the sum of all fatty acids containing no double bonds, except that label declaration of saturated fat content information is not required for products that contain less than 0.5 gram of total fat in a serving if no claims are made about fat or cholesterol content, and if "calories from saturated fat" is not declared. Except as provided for in paragraph (f) of this section, if a statement of the saturated fat content is not required and, as a result, not declared, the statement "Not a significant source of saturated fat" shall be placed at the bottom of the table of nutrient values in the same type size. Saturated fat content shall be indented and expressed as grams per serving to the nearest 0.5 (1/2) gram increment below 5 grams and to the nearest gram increment above 5 grams. If the serving contains less than 0.5 gram, the content shall be expressed as zero.

(ii) "Polyunsaturated fat" or "Polyunsaturated" (VOLUNTARY): A statement of the number of grams of polyunsaturated fat in a serving defined as cis,cis-methylene-interrupted polyunsaturated fatty acids may be declared voluntarily, except that when monounsaturated fat is declared, or when a claim about fatty acids or cholesterol is made on the label or in labeling of a food other than one that meets the criteria in §101.62(b)(1) for a claim for "fat free," label declaration of polyunsaturated fat is required. Polyunsaturated fat content shall be indented and expressed as grams per serving to the nearest 0.5 (1/2) gram increment below 5 grams and to the nearest gram increment above 5 grams. If the serving contains less than 0.5 gram, the content shall be expressed as zero.

(iii) "Monounsaturated fat" or "Monounsaturated" (VOLUNTARY): A statement of the number of grams of monounsaturated fat in a serving defined as cis-monounsaturated fatty acids may be declared voluntarily except that when polyunsaturated fat is declared, or when a claim about fatty acids or cholesterol is made on the label or in labeling of a food other than one that meets the criteria in §101.62(b)(1) for a claim for "fat free," label declaration of monounsaturated fat is required. Monounsaturated fat content shall be indented and expressed as grams per serving to the nearest 0.5 (1/2) gram increment below 5 grams and to the nearest gram increment above 5 grams. If the serving contains less than 0.5 gram, the content shall be expressed as zero.

(3) "Cholesterol": A statement of the cholesterol content in a serving expressed in milligrams to the nearest 5-milligram increment, except that label declaration of cholesterol information is not required for products that contain less than 2 milligrams cholesterol in a serving and make no claim about fat, fatty acids, or cholesterol content, or such products may state the cholesterol content as zero. Except as provided for in paragraph (f) of this section, if cholesterol content is not required and, as a result, not declared, the statement "Not a significant source of cholesterol" shall be placed at the bottom of the table of nutrient values in the same type size. If the food contains 2 to 5 milligrams of cholesterol per serving, the content may be stated as "less than 5 milligrams."

(4) "Sodium": A statement of the number of milligrams of sodium in a specified serving of food expressed as zero when the serving contains less than 5 milligrams of sodium, to the nearest 5- milligram increment when the serving contains 5 to 140 milligrams of sodium, and to the nearest 10-milligram increment when the serving contains greater than 140 milligrams.

(5) "Potassium" (VOLUNTARY): A statement of the number of milligrams of potassium in a specified serving of food may be declared voluntarily, except that when a claim is made about potassium content, label declaration shall be required. Potassium content shall be expressed as zero when the serving contains less than 5 milligrams of potassium, to the nearest 5-milligram increment when the serving contains less than or equal to 140 milligrams of potassium, and to the nearest 10-milligram increment when the serving contains more than 140 milligrams.

(6) "Carbohydrate, total" or "Total carbohydrate": A statement of the number of grams of total carbohydrate in a serving expressed to the nearest gram, except that if a serving contains less than 1 gram, the statement "Contains less than 1 gram" or "less than 1 gram" may be used as an alternative, or if the serving contains less than 0.5 gram, the content may be expressed as zero. Total carbohydrate content shall be calculated by subtraction of the sum of the crude protein, total fat, moisture, and ash from the total weight of the food. This calculation method is described in A. L. Merrill and B. K. Watt, "Energy Value of Foods-Basis and Derivation," USDA Handbook 74 (slightly revised 1973) pp. 2 and 3, which is incorporated by reference in accordance with 5 U.S.C. 552(a) and 1 CFR part 51 (the availability of this incorporation by reference is given in paragraph (c)(1)(i)(A) of this section).

(i) "Dietary fiber": A statement of the number of grams of total dietary fiber in a serving, indented and expressed to the nearest gram, except that if a serving contains less than 1 gram, declaration of dietary fiber is not required or, alternatively, the statement "Contains less than 1 gram" or "less than 1 gram" may be used, and if the serving contains less than 0.5 gram, the content may be expressed as zero. Except as provided for in paragraph (f) of this section, if dietary fiber content is not required and as a result, not declared, the statement "Not a significant source of dietary fiber" shall be placed at the bottom of the table of nutrient values in the same type size.

(A) "Soluble fiber" (VOLUNTARY): A statement of the number of grams of soluble dietary fiber in a serving may be declared voluntarily except when a claim is made on the label or in labeling about soluble fiber, label declaration shall be required. Soluble fiber content shall be indented under dietary fiber and expressed to the nearest gram, except that if a serving contains less than 1 gram, the statement "Contains less than 1 gram" or "less than 1 gram" may be used as an alternative, and if the serving contains less than 0.5 gram, the content may be expressed as zero.

(B) "Insoluble fiber" (VOLUNTARY): A statement of the number of grams of insoluble dietary fiber in a serving may be declared voluntarily except that when a claim is made on the label or in labeling about insoluble fiber, label declaration shall be required. Insoluble fiber content shall be indented under dietary fiber and expressed to the nearest gram except that if a serving contains less than 1 gram, the statement "Contains less than 1 gram" or "less than 1 gram" may be used as an alternative, and if the serving contains less than 0.5 gram, the content may be expressed as zero.

(ii) "Sugars": A statement of the number of grams of sugars in a serving, except that label declaration of sugars content is not required for products that contain less than 1 gram of sugars in a serving if no claims are made about sweeteners, sugars, or sugar alcohol content. Except as provided for in paragraph (f) of this section, if a statement of the sugars content is not required and, as a result, not declared, the statement "Not a significant source of sugars" shall be placed at the bottom of the table of nutrient values in the same type size. Sugars shall be defined as the sum of all free mono- and disaccharides (such as glucose, fructose, lactose, and sucrose). Sugars content shall be indented and expressed to the nearest gram, except that if a serving contains less than 1 gram, the statement "Contains less then 1 gram" or "less than 1 gram" may be used as an alternative, and if the serving contains less than 0.5 gram, the content may be expressed as zero.

(iii) "Sugar alcohol" (VOLUNTARY): A statement of the number of grams of sugar alcohols in a serving may be declared voluntarily on the label, except that when a claim is made on the label or in labeling about sugar alcohol or sugars when sugar alcohols are present in the food, sugar alcohol content shall be declared. For nutrition labeling purposes, sugar alcohols are defined as the sum of saccharide derivatives in which a hydroxyl group replaces a ketone or aldehyde group and whose use in the food is listed by FDA (e.g., mannitol or xylitol) or is generally recognized as safe (e.g., sorbitol). In lieu of the term "sugar alcohol," the name of the specific sugar alcohol (e.g., "xylitol") present in the food may be used in the nutrition label provided that only one sugar alcohol is present in the food. Sugar alcohol content shall be indented and expressed to the nearest gram, except that if a serving contains less than 1 gram, the statement "Contains less then 1 gram" or "less than 1 gram" may be used as an alternative, and if the serving contains less than 0.5 gram, the content may be expressed as zero.

(iv) "Other carbohydrate" (VOLUNTARY): A statement of the number of grams of other carbohydrates may be declared voluntarily. Other carbohydrates shall be defined as the difference between total carbohydrate and the sum of dietary fiber, sugars, and sugar alcohol, except that if sugar alcohol is not declared (even if present), it shall be defined as the difference between total carbohydrate and the sum of dietary fiber and sugars. Other carbohydrate content shall be indented and expressed to the nearest gram, except that if a serving contains less than 1 gram, the statement "Contains less than 1 gram" or "less than 1 gram" may be used as an alternative, and if the serving contains less than 0.5 gram, the content may be expressed as zero.

(7) "Protein": A statement of the number of grams of protein in a serving, expressed to the nearest gram, except that if a serving contains less than 1 gram, the statement "Contains less than 1 gram" or "less than 1 gram" may be used as an alternative, and if the serving contains less than 0.5 gram, the content may be expressed as zero. When the protein in foods represented or purported to be for adults and children 4 or more years of age has a protein quality value that is a protein digestibility-corrected amino acid score of less than 20 expressed as a percent, or when the protein in a food represented or purported to be for children greater than 1 but less than 4 years of age has a protein quality value that is a protein digestibility- corrected amino acid score of less than 40 expressed as a percent, either of the following shall be placed adjacent to the declaration of protein content by weight: The statement "not a significant source of protein," or a listing aligned under the column headed "Percent Daily Value" of the corrected amount of protein per serving, as determined in paragraph (c)(7)(ii) of this section, calculated as a percentage of the Daily Reference Value (DRV) or Reference Daily Intake (RDI), as appropriate, for protein and expressed as Percent of Daily Value. When the protein quality in a food as measured by the Protein Efficiency Ratio (PER) is less than 40 percent of the reference standard (casein) for a food represented or purported to be for infants, the statement "not a significant source of protein" shall be placed adjacent to the declaration of protein content. Protein content may be calculated on the basis of the factor of 6.25 times the nitrogen content of the food as determined by the appropriate method of analysis as given in the "Official Methods of Analysis of the AOAC International" (formerly the Association of Official Analytical Chemists), 15th Ed. (1990), which is incorporated by reference in accordance with 5 U.S.C. 552(a) and 1 CFR part 51, except when the official procedure for a specific food requires another factor. Copies may be obtained from Association of Official Analytical Chemists International, 481 North Frederick Ave., suite 500, Gaithersburg, MD 20877-2504, or may be inspected at the Office of the Federal Register, 800 North Capitol St. NW., suite 700, Washington, DC.

21 CFR §101.9

(i) A statement of the corrected amount of protein per serving, as determined in paragraph (c)(7)(ii) of this section, calculated as a percentage of the RDI or DRV for protein, as appropriate, and expressed as Percent of Daily Value, may be placed on the label, except that such a statement shall be given if a protein claim is made for the product, or if the product is represented or purported to be for use by infants or children under 4 years of age. When such a declaration is provided, it shall be placed on the label adjacent to the statement of grams of protein and aligned under the column headed "Percent Daily Value," and expressed to the nearest whole percent. However, the percentage of the RDI for protein shall not be declared if the food is represented or purported to be for use by infants and the protein quality value is less than 40 percent of the reference standard.

(ii) The "corrected amount of protein (gram) per serving" for foods represented or purported for adults and children 1 or more years of age is equal to the actual amount of protein (gram) per serving multiplied by the amino acid score corrected for protein digestibility. If the corrected score is above 1.00, then it shall be set at 1.00. The protein digestibility-corrected amino acid score shall be determined by methods given in sections 5.4.1, 7.2.1, and 8.00 in "Protein Quality Evaluation, Report of the Joint FAO/WHO Expert Consultation on Protein Quality Evaluation," Rome, 1990, except that when official AOAC procedures described in section (c)(7) of this paragraph require a specific food factor other than 6.25, that specific factor shall be used. The "Report of the Joint FAO/WHO Expert Consultation on Protein Quality Evaluation" as published by the Food and Agriculture Organization of the United Nations/World Health Organization is incorporated by reference in accordance with 5 U.S.C. 552(a) and 1 CFR part 51. Copies are available from the Center for Food Safety and Applied Nutrition (HFS-150), Food and Drug Administration, 200 C St. SW., Washington, DC 20204, or may be inspected at the Office of the Federal Register, 800 North Capitol St. NW., suite 700, Washington, DC. For foods represented or purported for infants, the corrected amount of protein (grams) per serving is equal to the actual amount of protein (grams) per serving multiplied by the relative protein quality value. The relative protein quality value shall be determined by dividing the subject food protein PER value by the PER value for casein. If the relative protein value is above 1.00, it shall be set at 1.00.

(iii) For the purpose of labeling with a percent of the Daily Reference Value (DRV) or RDI, a value of 50 grams of protein shall be the DRV for adults and children 4 or more years of age, and the RDI for protein for children less than 4 years of age, infants, pregnant women, and lactating women shall be 16 grams, 14 grams, 60 grams, and 65 grams, respectively.

(8) Vitamins and minerals: A statement of the amount per serving of the vitamins and minerals as described in this paragraph, calculated as a percent of the RDI and expressed as percent of Daily Value.

(i) For purposes of declaration of percent of Daily Value as provided for in paragraphs (d), (e), and (f) of this section, foods represented or purported to be for use by infants, children less than 4 years of age, pregnant women, or lactating women shall use the RDI's that are specified for the intended group. For foods represented or purported to be for use by both infants and children under 4 years of age, the percent of Daily Value shall be presented by separate declarations according to paragraph (e) of this section based on the RDI values for infants from birth to 12 months of age and for children under 4 years of age. Similarly, the percent of Daily Value based on both the RDI values for pregnant women and for lactating women shall be declared separately on foods represented or purported to be for use by both pregnant and lactating women. When such dual declaration is used on any label, it shall be included in all labeling, and equal prominence shall be given to both values in all such labeling. All other foods shall use the RDI for adults and children 4 or more years of age.

(ii) The declaration of vitamins and minerals as a percent of the RDI shall include vitamin A, vitamin C, calcium, and iron, in that order, and shall include any of the other vitamins and minerals listed in paragraph (c)(8)(iv) of this section when they are added as a nutrient supplement, or when a claim is made about them. Other vitamins and minerals need not be declared if neither the nutrient nor the component is otherwise referred to on the label or in labeling or advertising and the vitamins and minerals are:

(A) Required or permitted in a standardized food (e.g., thiamin, riboflavin, and niacin in enriched flour) and that standardized food is included as an ingredient (i.e., component) in another food; or

(B) Included in a food solely for technological purposes and declared only in the ingredient statement. The declaration may also include any of the other vitamins and minerals listed in paragraph (c)(8)(iv) of this section when they are naturally occurring in the food. The additional vitamins and minerals shall be listed in the order established in paragraph (c)(8)(iv) of this section.

(iii) The percentages for vitamins and minerals shall be expressed to the nearest 2-percent increment up to and including the 10-percent level, the nearest 5-percent increment above 10 percent and up to and including the

50-percent level, and the nearest 10-percent increment above the 50-percent level. Amounts of vitamins and minerals present at less than 2 percent of the RDI are not required to be declared in nutrition labeling but may be declared by a zero or by the use of an asterisk (or other symbol) that refers to another asterisk (or symbol) that is placed at the bottom of the table and that is followed by the statement "Contains less than 2 percent of the Daily Value of this (these) nutrient (nutrients)" or "Contains < 2 percent of the Daily Value of this (these) nutrient (nutrients)." Alternatively, except as provided for in paragraph (f) of this section, if vitamin A, vitamin C, calcium, or iron is present in amounts less than 2 percent of the RDI, label declaration of the nutrient(s) is not required if the statement "Not a significant source of _____ (listing the vitamins or minerals omitted)" is placed at the bottom of the table of nutrient values. Either statement shall be in the same type size as nutrients that are indented.

(iv) The following RDI's and nomenclature are established for the following vitamins and minerals which are essential in human nutrition:

Vitamin A, 5,000 International Units
Vitamin C, 60 milligrams
Calcium, 1,000 milligrams
Iron, 18 milligrams
Vitamin D, 400 International Units
Vitamin E, 30 International Units
Vitamin K, 80 micrograms
Thiamin, 1.5 milligrams
Riboflavin, 1.7 milligrams
Niacin, 20 milligrams
Vitamin B_6, 2.0 milligrams
Folate, 400 micrograms
Vitamin B_{12}, 6 micrograms
Biotin, 300 micrograms
Pantothenic acid, 10 milligrams
Phosphorus, 1,000 milligrams
Iodine, 150 micrograms
Magnesium, 400 milligrams
Zinc, 15 milligrams
Selenium, 70 micrograms
Copper, 2.0 milligrams
Manganese, 2.0 milligrams
Chromium, 120 micrograms
Molybdenum, 75 micrograms
Chloride, 3,400 milligrams.

(v) The following synonyms may be added in parentheses immediately following the name of the nutrient or dietary component:

Calories—Energy,
Vitamin C—Ascorbic acid,
Thiamin—Vitamin B_1,
Riboflavin—Vitamin B_2,
Folate—Folic acid or Folacin.
Alternatively, folic acid or folacin may be listed without parentheses in place of folate.

(vi) A statement of the percent of vitamin A that is present as beta-carotene may be declared voluntarily. When the vitamins and minerals are listed in a single column, the statement shall be indented under the information on vitamin A. When vitamins and minerals are arrayed horizontally, the statement of percent shall be presented in parenthesis following the declaration of vitamin A and the percent DV of vitamin A in the food (e.g., "Percent

21 *CFR* §101.9

Daily Value: Vitamin A 50 (90 percent as beta-carotene)"). When declared, the percentages shall be expressed in the same increments as are provided for vitamins and minerals in paragraph (c)(8)(iii) of this section.

(9) For the purpose of labeling with a percent of the DRV, the following DRV's are established for the following food components based on the reference caloric intake of 2,000 calories:

Food component	Unit of measurement	DRV
Fat.................	gram (g)	65
Saturated fatty acids......	do	20
Cholesterol...................	milligrams (mg)	300
Total carbohydrate........	grams (g)	300
Fiber............................	do	25
Sodium........................	milligrams (mg)	2,400
Potassium.....................	do	3,500
Protein.........................	grams (g)	50

(d)

(1) Nutrient information specified in paragraph (c) of this section shall be presented on foods in the following format, as shown in paragraph (d)(12) of this section, except on foods on which dual columns of nutrition information are declared as provided for in paragraph (e) of this section, on those food products on which the simplified format is required to be used as provided for in paragraph (f) of this section, on foods for infants and children less than 4 years of age as provided for in paragraph (j)(5) of this section, and on foods in small or intermediate-sized packages as provided for in paragraph (j)(13) of this section. In the interest of uniformity of presentation, FDA urges that the nutrition information be presented using the graphic specifications set forth in appendix B to part 101.

(i) The nutrition information shall be set off in a box by use of hairlines and shall be all black or one color type, printed on a white or other neutral contrasting background whenever practical.

(ii) All information within the nutrition label shall utilize:

(A) A single easy-to-read type style,

(B) Upper and lower case letters,

(C) At least one point leading (i.e., space between two lines of text) except that at least four points leading shall be utilized for the information required by paragraphs (d)(7) and (d)(8) of this section as shown in paragraph (d)(12), and

(D) Letters should never touch.

(iii) Information required in paragraphs (d)(3), (d)(5), (d)(7), and (d)(8) of this section shall be in type size no smaller than 8 point. Except for the heading "Nutrition Facts," the information required in paragraphs (d)(4), (d)(6), and (d)(9) of this section and all other information contained within the nutrition label shall be in type size no smaller than 6 point. When provided, the information described in paragraph (d)(10) of this section shall also be in type no smaller than 6 point.

(iv) The headings required by paragraphs (d)(2), (d)(4), and (d)(6) of this section (i.e., "Nutrition Facts," "Amount per Serving," and "% Daily Value*"), the names of all nutrients that are not indented according to requirements of paragraph (c) of this section (i.e., "Calories," "Total Fat," "Cholesterol," "Sodium," "Total Carbohydrate," and "Protein"), and the percentage amounts required by paragraph (d)(7)(ii) of this section shall be highlighted by bold or extra bold type or other highlighting (reverse printing is not permitted as a form of highlighting) that prominently distinguishes it from other information. No other information shall be highlighted.

(v) A hairline rule that is centered between the lines of text shall separate "Amount Per Serving" from the calorie statements required in paragraph (d)(5) of this section and shall separate each nutrient and its corresponding percent Daily Value required in paragraphs (d)(7)(i) and (d)(7)(ii) of this section from the nutrient and percent Daily Value above and below it, as shown in paragraph (d)(12) of this section.

(2) The information shall be presented under the identifying heading of "Nutrition Facts" which shall be set in a type size larger than all other print size in the nutrition label and, except for labels presented according to the format provided for in paragraph (d)(11) of this section, unless impractical, shall be set the full width of the information provided under paragraph (d)(7) of this section, as shown in paragraph (d)(12) of this section.

(3) Information on serving size shall immediately follow the heading as shown in paragraph (d)(12) of this section. Such information shall include:

(i) "Serving Size": A statement of the serving size as specified in paragraph (b)(7) of this section.

(ii) "Servings Per Container": The number of servings per container, except that this statement is not required on single serving containers as defined in paragraph (b)(6) of this section.

(4) A subheading "Amount Per Serving" shall be separated from serving size information by a bar as shown in paragraph (d)(12) of this section.

(5) Information on calories shall immediately follow the heading "Amount Per Serving" and shall be declared in one line, leaving sufficient space between the declaration of "Calories" and "Calories from fat" to allow clear differentiation, or, if "Calories from saturated fat" is declared, in a column with total "Calories" at the top, followed by "Calories from fat" (indented), and "Calories from saturated fat" (indented).

(6) The column heading "% Daily Value," followed by an asterisk (e.g., "% Daily Value*"), shall be separated from information on calories by a bar as shown in paragraph (d)(12) of this section. The position of this column heading shall allow for a list of nutrient names and amounts as described in paragraph (d)(7) of this section to be to the left of, and below, this column heading. The column headings "Percent Daily Value," "Percent DV," or "% DV" may be substituted for "% Daily Value."

(7) Except as provided for in paragraph (j)(13) of this section, nutrient information for both mandatory and any voluntary nutrients listed in paragraph (c) of this section that are to be declared in the nutrition label, except vitamins and minerals, shall be declared as follows:

(i) The name of each nutrient, as specified in paragraph (c) of this section, shall be given in a column and followed immediately by the quantitative amount by weight for that nutrient appended with a "g" for grams or "mg" for milligrams as shown in paragraph (d)(12) of this section. The symbol "<" may be used in place of "less than."

(ii) A listing of the percent of the DRV as established in paragraphs (c)(7)(iii) and (c)(9) of this section shall be given in a column aligned under the heading "% Daily Value" established in paragraph (d)(6) of this section with the percent expressed to the nearest whole percent for each nutrient declared in the column described in paragraph (d)(7)(i) of this section for which a DRV has been established, except that the percent for protein may be omitted as provided in paragraph (c)(7) of this section. The percent shall be calculated by dividing either the amount declared on the label for each nutrient or the actual amount of each nutrient (i.e., before rounding) by the DRV for the nutrient, except that the percent for protein shall be calculated as specified in paragraph (c)(7)(ii) of this section. The numerical value shall be followed by the symbol for percent (i.e., %).

(8) Nutrient information for vitamins and minerals shall be separated from information on other nutrients by a bar and shall be arrayed horizontally (e.g., Vitamin A 4%, Vitamin C 2%, Calcium 15%, Iron 4%) or may be listed in two columns as shown in paragraph (d)(12) of this section, except that when more than four vitamins and minerals are declared, they may be declared vertically with percentages listed under the column headed "% Daily Value."

(9) A footnote, preceded by an asterisk, shall be placed beneath the list of vitamins and minerals and shall be separated from that list by a hairline.

(i) The footnote shall state:

Percent Daily Values are based on a 2,000 calorie diet. Your daily values may be higher or lower depending on your calorie needs.

21 CFR §101.9

	Calories:	2,000	2,500
Total fat	Less than	65 g	80 g
Saturated fat	Less than	20 g	25 g
Cholesterol	Less than	300 mg	300 mg
Sodium	Less than	2,400 mg	2,400 mg
Total carbohydrate		300 g	375 g
Dietary fiber		25 g	30 g

(ii) If the percent of Daily Value is given for protein in the Percent of Daily Value column as provided in paragraph (d)(7)(ii) of this section, protein shall be listed under dietary fiber, and a value of 50 g shall be inserted on the same line in the column headed "2,000" and a value of 65 g in the column headed "2,500".

(iii) If potassium is declared in the column described in paragraph (d)(7)(i) of this section, potassium shall be listed under sodium and the DRV established in paragraph (c)(9) of this section shall be inserted on the same line in the numeric columns.

(iv) The abbreviations established in paragraph (j)(13)(ii)(B) of this section may be used within the footnote.

(10) Caloric conversion information on a per gram basis for fat, carbohydrate, and protein shall be presented beneath the information required in paragraph (d)(9) and shall be separated from that information by a hairline. This information may be presented horizontally as shown in paragraph (d)(12) of this section (i.e., "Calories per gram: fat 9, carbohydrate 4, protein 4") or vertically in columns.

(11)(i) If the space beneath the information on vitamins and minerals is not adequate to accommodate the information required in paragraph (d)(9) of this section, the information required in paragraph (d)(9) may be moved to the right of the column required in paragraph (d)(7)(ii) of this section and set off by a line that distinguishes it and sets it apart from the percent Daily Value information. The caloric conversion information provided for in paragraph (d)(10) of this section may be presented beneath either side or along the full length of the nutrition label.

(ii) If the space beneath the mandatory declaration of iron is not adequate to accommodate any remaining vitamins and minerals to be declared or the information required in paragraph (d)(9) of this section, the remaining information may be moved to the right and set off by a line that distinguishes it and sets it apart from the nutrients and the percent DV information given to the left. The caloric conversion information provided for in paragraph (d)(10) of this section may be presented beneath either side or along the full length of the nutrition label.

(iii) If there is not sufficient continuous vertical space (i.e., approximately 3 in) to accommodate the required components of the nutrition label up to and including the mandatory declaration of iron, the nutrition label may be presented in a tabular display as shown below.

21 CFR §101.9

(12) The following sample label illustrates the provisions of paragraph (d) of this section.

```
Nutrition Facts
Serving Size 1 cup (228g)
Servings Per Container 2

Amount Per Serving
Calories 260    Calories from Fat 120

                              % Daily Value*
Total Fat 13g                      20%
  Saturated Fat 5g                 25%
Cholesterol 30mg                   10%
Sodium 660mg                       28%
Total Carbohydrate 31g             10%
  Dietary Fiber 0g                  0%
  Sugars 5g
Protein 5g

Vitamin A 4%        •     Vitamin C 2%
Calcium 15%         •     Iron 4%

* Percent Daily Values are based on a 2,000
  calorie diet. Your daily values may be higher
  or lower depending on your calorie needs:
                    Calories:  2,000     2,500
  Total Fat      Less than     65g       80g
  Sat Fat        Less than     20g       25g
  Cholesterol    Less than     300mg     300mg
  Sodium         Less than     2,400mg   2,400mg
  Total Carbohydrate           300g      375g
    Dietary Fiber              25g       30g

Calories per gram:
Fat 9  •  Carbohydrate 4  •  Protein 4
```

(13)(i) Nutrition labels on the outer label of packages of products that contain two or more separately packaged foods that are intended to be eaten individually (e.g., variety packs of cereals or snack foods) or of packages that are used interchangeably for the same type of food (e.g., round ice cream containers) may use an aggregate display.

(ii) Aggregate displays shall comply with the format requirements of paragraph (d) of this section to the maximum extent possible, except that the identity of each food shall be specified immediately under the "Nutrition Facts" title, and both the quantitative amount by weight (i.e., g/mg amounts) and the percent Daily Value for each nutrient shall be listed in separate columns under the name of each food. The following sample label illustrates an aggregate display.

Nutrition Facts	Wheat Squares Sweetened		Corn Flakes Not Sweetened		Mixed Grain Flakes Sweetened	
Serving Size 1 Box	(35g)		(19g)		(27g)	
Servings Per Container	1		1		1	
Amount Per Serving						
Calories	120		70		100	
Calories from Fat	0		0		0	
	% Daily Value*		% Daily Value*		% Daily Value*	
Total Fat	0g	0%	0g	0%	0g	0%
Saturated Fat	0g	0%	0g	0%	0g	0%
Cholesterol	0mg	0%	0mg	0%	0mg	0%
Sodium	0mg	0%	200mg	8%	120mg	5%
Potassium	125mg	4%	25mg	1%	30mg	1%
Total Carbohydrate	29g	10%	17g	6%	24g	8%
Dietary Fiber	3g	12%	1g	4%	1g	4%
Sugars	8g		6g		13g	
Protein	4g		1g		1g	
Vitamin A		0%		10%		10%
Vitamin C		0%		15%		90%
Calcium		0%		0%		0%
Iron		10%		6%		20%
Thiamin		30%		15%		20%
Riboflavin		30%		15%		20%
Niacin		30%		15%		20%
Vitamin B₆		30%		15%		20%

*Percent Daily Values are based on a 2,000 calorie diet. Your daily values may be higher or lower depending on your calorie needs:

	Calories:	2,000	2,500
Total Fat	Less than	65g	80g
Sat Fat	Less than	20g	25g
Cholesterol	Less than	300mg	300mg
Sodium	Less than	2,400mg	2,400mg
Total Carbohydrate		300g	375g
Dietary Fiber		25g	30g

Calories per gram:
Fat 9 • Carbohydrate 4 • Protein 4

(14) In accordance with §101.15(c)(2), when nutrition labeling must appear in a second language, the nutrition information may be presented in a separate nutrition label for each language or in one nutrition label with the information in the second language following that in English. Numeric characters that are identical in both languages need not be repeated (e.g., "Protein/Proteinas 2 g"). All required information must be included in both languages.

(e) Nutrition information may be presented for two or more forms of the same food (e.g., both "as purchased" and "as prepared") or for common combinations of food as provided for in paragraph (h)(4) of this section, for different units (e.g., slices of bread or per 100 grams) as provided for in paragraph (b) of this section, or for two or more groups for which RDI's are established (e.g., both infants and children less than 4 years of age) as shown in paragraph (e)(5) of this section. When such dual labeling is provided, equal prominence shall be given to both sets of values. Information shall be presented in a format consistent with paragraph (d) of this section, except that:

(1) Following the subheading of "Amount Per Serving," there shall be two or more column headings accurately describing the forms of the same food (e.g., "Mix" and "Baked"), the combinations of food, the units, or the RDI groups that are being declared. The column representing the product as packaged and according to the label serving size based on the reference amount in §101.12(b) shall be to the left of the numeric columns.

(2) When the dual labeling is presented for two or more forms of the same food, for combinations of food, or for different units, total calories and calories from fat (and calories from saturated fat, when declared) shall be listed in a column and indented as specified in paragraph (d)(5) of this section with quantitative amounts declared in columns aligned under the column headings set forth in paragraph (e)(1) of this section.

(3) Quantitative information by weight required in paragraph (d)(7)(i) of this section shall be specified for the form of the product as packaged and according to the label serving size based on the reference amount in

§101.12(b).

(i) Quantitative information by weight may be included for other forms of the product represented by the additional column(s) either immediately adjacent to the required quantitative information by weight for the product as packaged and according to the label serving size based on the reference amount in §101.12(b) or as a footnote.

(A) If such additional quantitative information is given immediately adjacent to the required quantitative information, it shall be declared for all nutrients listed and placed immediately following and differentiated from the required quantitative information (e.g., separated by a comma). Such information shall not be put in a separate column.

(B) If such additional quantitative information is given in a footnote, it shall be declared in the same order as the nutrients are listed in the nutrition label. The additional quantitative information may state the total nutrient content of the product identified in the second column or the nutrient amounts added to the product as packaged for only those nutrients that are present in different amounts than the amounts declared in the required quantitative information. The footnote shall clearly identify which amounts are declared. Any subcomponents declared shall be listed parenthetically after principal components (e.g., 1/2 cup skim milk contributes an additional 40 calories, 65 mg sodium, 6 g total carbohydrate (6 g sugars), and 4 g protein).

(ii) Total fat and its quantitative amount by weight shall be followed by an asterisk (or other symbol) (e.g., "Total fat (2 g)*") referring to another asterisk (or symbol) at the bottom of the nutrition label identifying the form(s) of the product for which quantitative information is presented.

(4) Information required in paragraphs (d)(7)(ii) and (d)(8) of this section shall be presented under the subheading "% DAILY VALUE" and in columns directly under the column headings set forth in paragraph (e)(1) of this section.

(5) The following sample label illustrates the provisions of paragraph (e) of this section:

Nutrition Facts

Serving Size 1/12 package (44g, about 1/4 cup dry mix)
Servings Per Container 12

Amount Per Serving	Mix	Baked
Calories	190	280
Calories from Fat	45	140
	% Daily Value**	
Total Fat 5g*	8%	24%
Saturated Fat 2g	10%	13%
Cholesterol 0mg	0%	23%
Sodium 300mg	13%	13%
Total Carbohydrate 34g	11%	11%
Dietary Fiber 0g	0%	0%
Sugars 18g		
Protein 2g		
Vitamin A	0%	0%
Vitamin C	0%	0%
Calcium	6%	8%
Iron	2%	4%

* Amount in Mix
** Percent Daily Values are based on a 2,000 calorie diet. Your daily values may be higher or lower depending on your calorie needs:

	Calories:	2,000	2,500
Total Fat	Less than	65g	80g
Sat Fat	Less than	20g	25g
Cholesterol	Less than	300mg	300mg
Sodium	Less than	2,400mg	2,400mg
Total Carbohydrate		300g	375g
Dietary Fiber		25g	30g

Calories per gram:
Fat 9 • Carbohydrate 4 • Protein 4

21 CFR §101.9

(f) The declaration of nutrition information may be presented in the simplified format set forth herein when a food product contains insignificant amounts of seven or more of the following: Calories, total fat, saturated fat, cholesterol, sodium, total carbohydrate, dietary fiber, sugars, protein, vitamin A, vitamin C, calcium, and iron; except that for foods intended for children less than 2 years of age to which §101.9(j)(5)(i) applies, nutrition information may be presented in the simplified format when a food product contains insignificant amounts of six or more of the following: Calories, total fat, sodium, total carbohydrate, dietary fiber, sugars, protein, vitamin A, vitamin C, calcium, and iron.

(1) An "insignificant amount" shall be defined as that amount that allows a declaration of zero in nutrition labeling, except that for total carbohydrate, dietary fiber, and protein, it shall be an amount that allows a declaration of "less than 1 gram."

(2) The simplified format shall include information on the following nutrients:

(i) Total calories, total fat, total carbohydrate, protein, and sodium;

(ii) Calories from fat and any other nutrients identified in paragraph (f) of this section that are present in the food in more than insignificant amounts; and

(iii) Any vitamins and minerals listed in paragraph (c)(8)(iv) of this section when they are required to be added as a nutrient supplement to foods for which a standard of identity exists.

(iv) Any vitamins or minerals listed in paragraph (c)(8)(iv) of this section voluntarily added to the food as nutrient supplements.

(3) Other nutrients that are naturally present in the food in more than insignificant amounts may be voluntarily declared as part of the simplified format.

(4) If any nutrients are declared as provided in paragraphs (f)(2)(iii), (f)(2)(iv), or (f)(3) of this section as part of the simplified format or if any nutrition claims are made on the label or in labeling, the statement "Not a significant source of _____ " (with the blank filled in with the name(s) of any nutrient(s) identified in §101.9(f) and calories from fat that are present in insignificant amounts) shall be included at the bottom of the nutrition label.

(5) Except as provided for in paragraphs (j)(5) and (j)(13) of this section, nutrient information declared in the simplified format shall be presented in the same manner as specified in paragraphs (d) or (e) of this section, except that the footnote required in paragraph (d)(9) of this section is not required. When the footnote is omitted, an asterisk shall be placed at the bottom of the label followed by the statement "Percent Daily Values are based on a 2,000 calorie diet" and, if the term "Daily Value" is not spelled out in the heading, a statement that "DV" represents "Daily Value."

(g) Compliance with this section shall be determined as follows:

(1) A collection of primary containers or units of the same size, type, and style produced under conditions as nearly uniform as possible, designated by a common container code or marking, or in the absence of any common container code or marking, a day's production, constitutes a "lot."

(2) The sample for nutrient analysis shall consist of a composite of 12 subsamples (consumer units), taken 1 from each of 12 different randomly chosen shipping cases, to be representative of a lot. Unless a particular method of analysis is specified in paragraph (c) of this section, composites shall be analyzed by appropriate methods as given in the "Official Methods of Analysis of the AOAC International," 15th Ed. (1990), which is incorporated by reference in accordance with 5 U.S.C. 552(a) or 1 CFR part 51 or, if no AOAC method is available or appropriate, by other reliable and appropriate analytical procedures. The availability of this incorporation by reference is given in paragraph (c)(7) of this section.

(3) Two classes of nutrients are defined for purposes of compliance:

(i) *Class I.* Added nutrients in fortified or fabricated foods; and

(ii) *Class II.* Naturally occurring (indigenous) nutrients. If any ingredient which contains a naturally occurring (indigenous) nutrient is added to a food, the total amount of such nutrient in the final food product is subject to class II requirements unless the same nutrient is also added.

(4) A food with a label declaration of a vitamin, mineral, protein, total carbohydrate, dietary fiber, other carbohydrate, polyunsaturated or monounsaturated fat, or potassium shall be deemed to be misbranded under section 403(a) of the Federal Food, Drug, and Cosmetic Act (the act) unless it meets the following requirements:

(i) *Class I vitamin, mineral, protein, dietary fiber, or potassium.* The nutrient content of the composite is at least equal to the value for that nutrient declared on the label.

(ii) *Class II vitamin, mineral, protein, total carbohydrate, dietary fiber, other carbohydrate, polyunsaturated or*

21 *CFR* §101.9

monounsaturated fat, or potassium. The nutrient content of the composite is at least equal to 80 percent of the value for that nutrient declared on the label.

Provided, That no regulatory action will be based on a determination of a nutrient value that falls below this level by a factor less than the variability generally recognized for the analytical method used in that food at the level involved.

(5) A food with a label declaration of calories, sugars, total fat, saturated fat, cholesterol, or sodium shall be deemed to be misbranded under section 403(a) of the act if the nutrient content of the composite is greater than 20 percent in excess of the value for that nutrient declared on the label. *Provided*, That no regulatory action will be based on a determination of a nutrient value that falls above this level by a factor less than the variability generally recognized for the analytical method used in that food at the level involved.

(6) Reasonable excesses of a vitamin, mineral, protein, total carbohydrate, dietary fiber, other carbohydrate, polyunsaturated or monounsaturated fat, or potassium over labeled amounts are acceptable within current good manufacturing practice. Reasonable deficiencies of calories, sugars, total fat, saturated fat, cholesterol, or sodium under labeled amounts are acceptable within current good manufacturing practice.

(7) Compliance will be based on the metric measure specified in the label statement of serving size.

(8) Compliance with the provisions set forth in paragraphs (g)(1) through (g)(6) of this section may be provided by use of an FDA approved data base that has been computed following FDA guideline procedures and where food samples have been handled in accordance with current good manufacturing practice to prevent nutrition loss. FDA approval of a data base shall not be considered granted until the Center for Food Safety and Applied Nutrition has agreed to all aspects of the data base in writing. The approval will be granted where a clear need is presented (e.g., raw produce and seafood). Approvals will be in effect for a limited time, e.g., 10 years, and will be eligible for renewal in the absence of significant changes in agricultural or industry practices. Approval requests shall be submitted in accordance with the provisions of §10.30 of this chapter. Guidance in the use of data bases may be found in the "FDA Nutrition Labeling Manual- A Guide for Developing and Using Data Bases," available from the Office of Food Labeling (HFS-150), Center for Food Safety and Applied Nutrition, Food and Drug Administration, 200 C St. SW., Washington, DC 20204.

(9) When it is not technologically feasible, or some other circumstance makes it impracticable, for firms to comply with the requirements of this section (e.g., to develop adequate nutrient profiles to comply with the requirements of paragraph (c) of this section), FDA may permit alternative means of compliance or additional exemptions to deal with the situation. Firms in need of such special allowances shall make their request in writing to the Center for Food Safety and Applied Nutrition (HFS-150), Food and Drug Administration, 200 C St. SW., Washington, DC 20204.

(h) Products with separately packaged ingredients or foods, with assortments of food, or to which other ingredients are added by the user may be labeled as follows:

(1) If a product consists of two or more separately packaged ingredients enclosed in an outer container or of assortments of the same type of food (e.g., assorted nuts or candy mixtures) in the same retail package, nutrition labeling shall be located on the outer container or retail package (as the case may be) to provide information for the consumer at the point of purchase. However, when two or more food products are simply combined together in such a manner that no outer container is used, or no outer label is available, each product shall have its own nutrition information, e.g., two boxes taped together or two cans combined in a clear plastic overwrap. When separately packaged ingredients or assortments of the same type of food are intended to be eaten at the same time, the nutrition information may be specified per serving for each component or as a composite value.

(2) If a product consists of two or more separately packaged foods that are intended to be eaten individually and that are enclosed in an outer container (e.g., variety packs of cereals or snack foods), the nutrition information shall:

(i) Be specified per serving for each food in a location that is clearly visible to the consumer at the point of purchase; and

(ii) Be presented in separate nutrition labels or in one aggregate nutrition label with separate columns for the quantitative amount by weight and the percent Daily Value for each food.

(3) If a package contains a variety of foods, or an assortment of foods, and is in a form intended to be used as a gift, the nutrition labeling shall be in the form required by paragraphs (a) through (f) of this section, but it may be modified as follows:

(i) Nutrition information may be presented on the label of the outer package or in labeling within or attached to the outer package.

(ii) In the absence of a reference amount customarily consumed in §101.12(b) that is appropriate for the variety or assortment of foods in a gift package, 1 ounce for solid foods, 2 fluid ounces for nonbeverage liquids (e.g., syrups), and 8 fluid ounces for beverages may be used as the standard serving size for purposes of nutrition labeling of foods subject to this paragraph. However, the reference amounts customarily consumed in §101.12(b) shall be used for purposes of evaluating whether individual foods in a gift package qualify for nutrient content claims or health claims.

(iii) The number of servings per container may be stated as "varied."

(iv) Nutrition information may be provided per serving for individual foods in the package, or, alternatively, as a composite per serving for reasonable categories of foods in the package having similar dietary uses and similar significant nutritional characteristics. Reasonable categories of foods may be used only if accepted by FDA. In determining whether a proposed category is reasonable, FDA will consider whether the values of the characterizing nutrients in the foods proposed to be in the category meet the compliance criteria set forth in paragraphs (g)(3) through (g)(6) of this section. Proposals for such categories may be submitted in writing to the Office of Food Labeling (HFS-150), Center for Food Safety and Applied Nutrition, Food and Drug Administration, 200 C St. SW., Washington, DC 20204.

(v) If a food subject to paragraph (j)(13) of this section because of its small size is contained in a gift package, the food need not be included in the determination of nutrition information under paragraph (h) of this section if it is not specifically listed in a promotional catalogue as being present in the gift package, and:

(A) It is used in small quantities primarily to enhance the appearance of the gift package; or

(B) It is included in the gift package as a free gift or promotional item.

(4) If a food is commonly combined with other ingredients or is cooked or otherwise prepared before eating, and directions for such combination or preparations are provided, another column of figures may be used to declare nutrition information on the basis of the food as consumed in the format required in paragraph (e) of this section (e.g., a dry ready-to-eat cereal may be described with one set of Percent Daily Values for the cereal as sold (e.g., per ounce), and another set for the cereal and milk as suggested in the label (e.g., per ounce of cereal and 1/2 cup of vitamin D fortified skim milk); and a cake mix may be labeled with one set of Percent Daily Values for the dry mix (per serving) and another set for the serving of the final cake when prepared): *Provided*, That, the type and quantity of the other ingredients to be added to the product by the user and the specific method of cooking and other preparation shall be specified prominently on the label.

(i) Except as provided in paragraphs (j)(13) and (j)(17) of this section, the location of nutrition information on a label shall be in compliance with §101.2.

(j) The following foods are exempt from this section or are subject to special labeling requirements:

(1) (i) Food offered for sale by a person who makes direct sales to consumers (e.g., a retailer) who has annual gross sales made or business done in sales to consumers that is not more than $500,000 or has annual gross sales made or business done in sales of food to consumers of not more than $50,000, *Provided*, That the food bears no nutrition claims or other nutrition information in any context on the label or in labeling or advertising. Claims or other nutrition information subject the food to the provisions of this section.

(ii) For purposes of this paragraph, calculation of the amount of sales shall be based on the most recent 2-year average of business activity. Where firms have been in business less than 2 years, reasonable estimates must indicate that annual sales will not exceed the amounts specified. For foreign firms that ship foods into the United States, the business activities to be included shall be the total amount of food sales, as well as other sales to consumers, by the firm in the United States.

(2) Food products which are:

(i) Served in restaurants, *Provided*, That the food bears no nutrition claims or other nutrition information in any context on the label or in labeling or advertising. Claims or other nutrition information subject the food to the provisions of this section;

(ii) Served in other establishments in which food is served for immediate human consumption (e.g., institutional food service establishments, such as schools, hospitals, and cafeterias; transportation carriers, such as trains and airplanes; bakeries, delicatessens, and retail confectionery stores where there are facilities for immediate consumption on the premises; food service vendors, such as lunch wagons, ice cream shops, mall cookie counters, vending machines, and sidewalk carts where foods are generally consumed immediately where purchased or while the consumer is walking away, including similar foods sold from convenience stores; and food delivery systems or establishments where ready-to-eat foods are delivered to homes or offices), *Provided*, That the food bears no nutrition claims or other nutrition information in any context on the label or in labeling or advertising. Claims or other nutrition information subject the food to the provisions of this section;

(iii) Sold only in such facilities, *Provided*, That the food bears no nutrition claims or other nutrition information in any context on the label or in labeling or advertising. Claims or other nutrition information subject the food to the provisions of this section;

(iv) Used only in such facilities and not served to the consumer in the package in which they are received (e.g., foods that are not packaged in individual serving containers); or

(v) Sold by a distributor who principally sells food to such facilities; *Provided*, That:

(A) This exemption shall not be available for those foods that are manufactured, processed, or repackaged by that distributor for sale to any persons other than restaurants or other establishments that serve food for immediate human consumption, and

(B) The manufacturer of such products is responsible for providing the nutrition information on the products if there is a reasonable possibility that the product will be purchased directly by consumers.

(3) Food products that are:

(i) Of the type of food described in paragraphs (j)(2)(i) and (j)(2)(ii) of this section,

(ii) Ready for human consumption,

(iii) Offered for sale to consumers but not for immediate human consumption,

(iv) Processed and prepared primarily in a retail establishment, and

(v) Not offered for sale outside of that establishment (e.g., ready-to-eat foods that are processed and prepared on-site and sold by independent delicatessens, bakeries, or retail confectionery stores where there are no facilities for immediate human consumption; by in-store delicatessen, bakery, or candy departments; or at self-service food bars such as salad bars), Provided, That the food bears no nutrition claims or other nutrition information in any context on the label or in labeling or advertising. Claims or other nutrition information subject the food to the provisions of this section.

(4) Foods that contain insignificant amounts of all of the nutrients and food components required to be included in the declaration of nutrition information under paragraph (c) of this section, *Provided*, That the food bears no nutrition claims or other nutrition information in any context on the label or in labeling or advertising. Claims or other nutrition information subject the food to the provisions of this section. An insignificant amount of a nutrient or food component shall be that amount that allows a declaration of zero in nutrition labeling, except that for total carbohydrate, dietary fiber, and protein, it shall be an amount that allows a declaration of "less than 1 gram." Examples of foods that are exempt under this paragraph include coffee beans (whole or ground), tea leaves, plain unsweetened instant coffee and tea, condiment-type dehydrated vegetables, flavor extracts, and food colors.

(5)(i) Foods, other than infant formula, represented or purported to be specifically for infants and children less than 2 years of age shall bear nutrition labeling, except as provided in paragraph (j)(5)(ii) and except that such labeling shall not include calories from fat (paragraph (c)(1)(ii) of this section), calories from saturated fat ((c)(1)(iii)), saturated fat ((c)(2)(i)), polyunsaturated fat ((c)(2)(ii)), monounsaturated fat ((c)(2)(iii)), and cholesterol ((c)(3)).

(ii) Foods, other than infant formula, represented or purported to be specifically for infants and children less than 4 years of age shall bear nutrition labeling, except that:

(A) Such labeling shall not include declarations of percent of Daily Value for total fat, saturated fat, cholesterol, sodium, potassium, total carbohydrate, and dietary fiber;

(B) Nutrient names and quantitative amounts by weight shall be presented in two separate columns.

(C) The heading "Percent Daily Value" required in paragraph (d)(6) of this section shall be placed immediately below the quantitative information by weight for protein;

(D) Percent of Daily Value for protein, vitamins, and minerals shall be listed immediately below the heading "Percent Daily Value"; and

(E) Such labeling shall not include the footnote specified in paragraph (d)(9) of this section.

(6) Dietary supplements, except that such foods shall be labeled in compliance with §101.36.

(7) Infant formula subject to section 412 of the act, as amended, except that such foods shall be labeled in compliance with part 107 of this chapter.

21 CFR §101.9

(8) Medical foods as defined in section 5(b) of the Orphan Drug Act (21 U.S.C. 360ee(b)(3)). A medical food is a food which is formulated to be consumed or administered enterally under the supervision of a physician and which is intended for the specific dietary management of a disease or condition for which distinctive nutritional requirements, based on recognized scientific principles, are established by medical evaluation. A food is subject to this exemption only if:

(i) It is a specially formulated and processed product (as opposed to a naturally occurring foodstuff used in its natural state) for the partial or exclusive feeding of a patient by means of oral intake or enteral feeding by tube;

(ii) It is intended for the dietary management of a patient who, because of therapeutic or chronic medical needs, has limited or impaired capacity to ingest, digest, absorb, or metabolize ordinary foodstuffs or certain nutrients, or who has other special medically determined nutrient requirements, the dietary management of which cannot be achieved by the modification of the normal diet alone;

(iii) It provides nutritional support specifically modified for the management of the unique nutrient needs that result from the specific disease or condition, as determined by medical evaluation;

(iv) It is intended to be used under medical supervision; and

(v) It is intended only for a patient receiving active and ongoing medical supervision wherein the patient requires medical care on a recurring basis for, among other things, instructions on the use of the medical food.

(9) Food products shipped in bulk form that are not for distribution to consumers in such form and that are for use solely in the manufacture of other foods or that are to be processed, labeled, or repacked at a site other than where originally processed or packed.

(10) Raw fruits, vegetables, and fish subject to section 403(q)(4) of the act, except that the labeling of such foods should adhere to guidelines in §101.45. This exemption is contingent on the food bearing no nutrition claims or other nutrition information in any context on the label or in labeling or advertising. Claims or other nutrition information subject the food to nutrition labeling in accordance with §101.45. The term "fish" includes freshwater or marine fin fish, crustaceans, and mollusks, including shellfish, amphibians, and other forms of aquatic animal life.

(11) Packaged single-ingredient products that consist of fish or game meat (i.e., animal products not covered under the Federal Meat Inspection Act or the Poultry Products Inspection Act, such as flesh products from deer, bison, rabbit, quail, wild turkey, or ostrich) subject to this section may provide required nutrition information for a 3-ounce cooked edible portion (i.e., on an "as prepared" basis), except that:

(i) Such products that make claims that are based on values as packaged must provide nutrition information on an as packaged basis, and

(ii) Nutrition information is not required for custom processed fish or game meats.

(12) Game meats (i.e., animal products not covered under the Federal Meat Inspection Act or the Poultry Products Inspection Act, such as flesh products from deer, bison, rabbit, quail, wild turkey, or ostrich) may provide required nutrition information on labeling in accordance with the provisions of paragraph (a)(2) of this section.

(13)(i) Foods in small packages that have a total surface area available to bear labeling of less than 12 square inches, *Provided*, That the labels for these foods bear no nutrition claims or other nutrition information in any context on the label or in labeling or advertising. Claims or other nutrition information subject the food to the provisions of this section.

(A) The manufacturer, packer, or distributor shall provide on the label of packages that qualify for and use this exemption an address or telephone number that a consumer can use to obtain the required nutrition information (e.g., "For nutrition information, call 1-800-123-4567").

(B) When such products bear nutrition labeling, either voluntarily or because nutrition claims or other nutrition information is provided, all required information shall be in type size no smaller than 6 point or all uppercase type of 1/16 inches minimum height, except that individual serving-size packages of food served with meals in restaurants, institutions, and on board passenger carriers, and not intended for sale at retail, may comply with §101.2(c)(5).

(ii) Foods in packages that have a total surface area available to bear labeling of 40 or less square inches may modify the requirements of paragraphs (c) through (f) and (i) of this section by one or more of the following means:

(A) Presenting the required nutrition information in a tabular or, as provided below, linear (i.e., string) fashion rather than in vertical columns if the product has a total surface area available to bear labeling of less than 12 square inches, or if the product has a total surface area available to bear labeling of 40 or less square inches and the package shape or size cannot accommodate a standard vertical column or tabular display on any label panel. Nutrition information may be given in a linear fashion only if the label will not accommodate a tabular display.

(*1*) The following sample label illustrates the tabular display.

```
Nutrition Facts
Serv. Size ⅓ cup (56g)
Servings about 3
Calories 80
Fat Cal. 10
*Percent Daily Values (DV) are based on a 2,000 calorie diet.

              Amount/serving  %DV*        Amount/serving  %DV*
              Total Fat 1g    2%          Total Carb. 0g  0%
              Sat.Fat 0g      0%          Fiber 0g        0%
              Cholest. 10mg   3%          Sugars 0g
              Sodium 200mg    8%          Protein 17g
              Vitamin A 0% • Vitamin C 0% • Calcium 0% • Iron 6%
```

(*2*) The following sample label illustrates the linear display. When nutrition information is given in a linear fashion, bolding is required only on the title "Nutrition Facts" and is allowed voluntarily for the nutrient names for "Calories," "Total fat," "Cholesterol," "Sodium," "Total carbohydrate," and "Protein."

```
Nutrition Facts Serv size: 1 package, Amount Per Serving: Calories 45, Fat Cal. 10, Total Fat 1g (2% DV), Sat. Fat 1g (5% DV), Cholest. 0mg (0% DV), Sodium 50mg (2% DV), Total carb. 8g (3% DV), Fiber 1g (4% DV), Sugars 4g, Protein 1g, Vitamin A (8% DV), Vitamin C (8% DV), Calcium (0% DV), Iron (2 % DV). Percent Daily Values (DV) are based on a 2,000 calorie diet.
```

(B) Using any of the following abbreviations:

Serving size—Serv size
Servings per container—Servings
Calories from fat—Fat cal
Calories from saturated fat—Sat fat cal
Saturated fat—Sat fat
Monounsaturated fat—Monounsat fat
Polyunsaturated fat—polyunsat fat
Cholesterol—Cholest
Total carbohydrate—Total carb
Dietary fiber—Fiber
Soluble fiber—Sol fiber
Insoluble fiber—Insol fiber
Sugar alcohol—Sugar alc
Other carbohydrate—Other carb

(C) Omitting the footnote required in paragraph (d)(9) of this section and placing another asterisk at the bottom of the label followed by the statement "Percent Daily Values are based on a 2,000 calorie diet" and, if the term "Daily Value" is not spelled out in the heading, a statement that "DV" represents "Daily Value."
(D) Presenting the required nutrition information on any label panel.
(14) Shell eggs packaged in a carton that has a top lid designed to conform to the shape of the eggs are exempt from outer carton label requirements where the required nutrition information is clearly presented immediately beneath the carton lid or in an insert that can be clearly seen when the carton is opened.
(15) The unit containers in a multiunit retail food package where:
(i) The multiunit retail food package labeling contains all nutrition information in accordance with the requirements of this section;
(ii) The unit containers are securely enclosed within and not intended to be separated from the retail package under conditions of retail sale; and
(iii) Each unit container is labeled with the statement "This Unit Not Labeled For Retail Sale" in type size not less than 1/16-inch in height, except that this statement shall not be required when the inner unit containers bear no labeling at all. The word "individual" may be used in lieu of or immediately preceding the word "Retail" in the statement.
(16) Food products sold from bulk containers: *Provided*, That nutrition information required by this section be displayed to consumers either on the labeling of the bulk container plainly in view or in accordance with the provisions of paragraph (a)(2) of this section.
(17) Foods in packages that have a total surface area available to bear labeling greater than 40 square inches but whose principal display panel and information panel do not provide sufficient space to accommodate all required information may use any alternate panel that can be readily seen by consumers for the nutrition label. The space needed for vignettes, designs, and other nonmandatory label information on the principal display panel may be considered in determining the sufficiency of available space on the principal display panel for the placement of the nutrition label. Nonmandatory label information on the information panel shall not be considered in determining the sufficiency of available space for the placement of the nutrition label.
(18) Food products that are low-volume (that is, they meet the requirements for units sold in paragraphs (j)(18)(i) or (j)(18)(ii) of this section); that, except as provided in paragraph (j)(18)(iv) of this section, are the subject of a claim for an exemption that provides the information required under paragraph (j)(18)(iv) of this section, that is filed before the beginning of the time period for which the exemption is claimed, and that is filed by a person, whether it is the manufacturer, packer, or distributor, that qualifies to claim the exemption under the requirements for average full-time equivalent employees in paragraphs (j)(18)(i) or (j)(18)(ii) of this section; and whose labels, labeling, and advertising do not provide nutrition information or make a nutrient content or health claim.
(i) For food products first introduced into interstate commerce before May 8, 1994, the product shall be exempt for the period:
(A) Between May 8, 1995, and May 7, 1996, if, for the period between May 8, 1994, and May 7, 1995, the person claiming the exemption employed fewer than an average of 300 full-time equivalent employees and fewer than 400,000 units of that product were sold in the United States; and
(B) Between May 8, 1996, and May 7, 1997, if for the period between May 8, 1995, and May 7, 1996, the person claiming the exemption employed fewer than an average of 200 full-time equivalent employees and fewer than 200,000 units of that product were sold in the United States.
(ii) For all other food products, the product shall be eligible for an exemption for any 12-month period if, for the preceding 12 months, the person claiming the exemption employed fewer than an average of 100 full-time equivalent employees and fewer than 100,000 units of that product were sold in the United States, or in the case of a food product that was not sold in the 12-month period preceding the period for which exemption is claimed, fewer than 100,000 units of such product are reasonably anticipated to be sold in the United States during the period for which exemption is claimed.
(iii) If a person claims an exemption under paragraphs (j)(18)(i) or (j)(18)(ii) of this section for a food product and then, during the period of such exemption, the number of full-time equivalent employees of such person exceeds the appropriate number, or the number of food products sold in the United States exceeds the appropriate number, or, if at the end of the period of such exemption, the food product no longer qualifies for an exemption under the provisions of paragraphs (j)(18)(i) or (j)(18)(ii) of this section, such person shall have 18

months from the date that the product was no longer qualified as a low-volume product of a small business to comply with this section.

(iv) A notice shall be filed with the Office of Food Labeling (HFS-150), Center for Food Safety and Applied Nutrition, Food and Drug Administration, 200 C St. SW., Washington, DC 20204 and contain the following information, except that if the person is not an importer and has fewer than 10 full-time equivalent employees, that person does not have to file a notice for any food product with annual sales of fewer than 10,000 total units:

(A) Name and address of person requesting exemption. This should include a telephone number or FAX number that can be used to contact the person along with the name of a specific contact;

(B) Names of the food products (including the various brand names) for which exemption is claimed;

(C) Name and address of the manufacturer, distributor, or importer of the food product for which an exemption is claimed, if different than the person that is claiming the exemption;

(D) The number of full-time equivalent employees. Provide the average number of full-time equivalent individuals employed by the person and its affiliates for the 12 months preceding the period for which a small business exemption is claimed for a product. The average number of full-time equivalent employees is to be determined by dividing the total number of hours of salary or wages paid to employees of the person and its affiliates by the number of hours of work in a year, 2,080 hours (i.e., 40 hours x 52 weeks);

(E) Approximate total number of units of the food product sold by the person in the United States in the 12-month period preceding that for which a small business exemption is claimed. Provide the approximate total number of units sold, or expected to be sold, in a 12- month period for each product for which an exemption is claimed. For products that have been in production for 1 year or more prior to the period for which exemption is claimed, the 12- month period is the period immediately preceding the period for which an exemption is claimed. For other products, the 12-month period is the period for which an exemption is claimed; and

(F) The notice shall be signed by a responsible individual for the person who can certify the accuracy of the information presented in the notice. The individual shall certify that the information contained in the notice is a complete and accurate statement of the average number of full-time equivalent employees of this person and its affiliates and of the number of units of the product for which an exemption is claimed sold by the person. The individual shall also state that should the average number of full-time equivalent employees or the number of units of food products sold in the United States by the person exceed the applicable numbers for the time period for which exemption is claimed, the person will notify FDA of that fact and the date on which the number of employees or the number of products sold exceeded the standard.

(v) FDA may by regulation lower the employee or units of food products requirements of paragraph (j)(18)(ii) of this section for any food product first introduced into interstate commerce after May 8, 2002, if the agency determines that the cost of compliance with such lower requirement will not place an undue burden on persons subject to it.

(vi) For the purposes of this paragraph, the following definitions apply:

(A) *Unit* means the packaging or, if there is no packaging, the form in which a food product is offered for sale to consumers.

(B) *Food product* means food in any sized package which is manufactured by a single manufacturer or which bears the same brand name, which bears the same statement of identity, and which has similar preparation methods.

(C) *Person* means all domestic and foreign affiliates, as defined in 13 CFR 121.401, of the corporation, in the case of a corporation, and all affiliates, as defined in 13 CFR 121.401, of a firm or other entity, when referring to a firm or other entity that is not a corporation.

(D) *Full-time equivalent employee* means all individuals employed by the person claiming the exemption. This number shall be determined by dividing the total number of hours of salary or wages paid directly to employees of the person and of all of its affiliates by the number of hours of work in a year, 2,080 hours (i.e., 40 hours x 52 weeks).

(k) A food labeled under the provisions of this section shall be deemed to be misbranded under sections 201(n) and 403(a) of the act if its label or labeling represents, suggests, or implies:

(1) That the food, because of the presence or absence of certain dietary properties, is adequate or effective in the prevention, cure, mitigation, or treatment of any disease or symptom. Information about the relationship of a dietary property to a disease or health-related condition may only be provided in conformance with the requirements of §101.14 and part 101, subpart E.

(2) That the lack of optimum nutritive quality of a food, by reason of the soil on which that food was grown, is or may be responsible for an inadequacy or deficiency in the quality of the daily diet.

(3) That the storage, transportation, processing, or cooking of a food is or may be responsible for an inadequacy or deficiency in the quality of the daily diet.

(4) That a natural vitamin in a food is superior to an added or synthetic vitamin.

[58 FR 2175, Jan. 6, 1993, as amended at 58 FR 2227, 2291, 2533, Jan. 6, 1993; 58 FR 17086, 17104, Apr. 1, 1993; 58 FR 17328- 17331, Apr. 2, 1993; 58 FR 44048, 44076, Aug. 18, 1993; 58 FR 59363, Nov. 9, 1993; 58 FR 60109, Nov. 15, 1993; 59 FR 371, Jan. 4, 1994; 59 FR 62317, Dec. 5, 1994; 60 FR 17205, Apr. 5, 1995; 60 FR 67164, Dec. 28, 1995; 61 FR 8779, Mar. 5, 1996; 61 FR 14479, Apr. 2, 1996; 61 FR 40978-40979, Aug. 7, 1996; 62 FR 15342, Mar. 31, 1997; 62 *FR* 49848, Sept. 23, 1997]

§101.10 Nutrition labeling of restaurant foods.

Nutrition labeling in accordance with §101.9 shall be provided upon request for any restaurant food or meal for which a nutrient content claim (as defined in §101.13 or in subpart D of this part) or a health claim (as defined in §101.14 and permitted by a regulation in subpart E of this part) is made, except that information on the nutrient amounts that are the basis for the claim (e.g., "low fat, this meal provides less than 10 grams of fat") may serve as the functional equivalent of complete nutrition information as described in §101.9. Nutrient levels may be determined by nutrient data bases, cookbooks, or analyses or by other reasonable bases that provide assurance that the food or meal meets the nutrient requirements for the claim. Presentation of nutrition labeling may be in various forms, including those provided in §101.45 and other reasonable means.
[58 FR 2410, Jan. 6, 1993; 58 FR 17341, Apr. 2, 1993; 61 FR 40332, Aug. 2, 1996]

§101.12 Reference amounts customarily consumed per eating occasion.

(a) The general principles and factors that the Food and Drug Administration (FDA) considered in arriving at the reference amounts customarily consumed per eating occasion (reference amounts) which are set forth in paragraph (b) of this section, are that:

(1) FDA calculated the reference amounts for persons 4 years of age or older to reflect the amount of food customarily consumed per eating occasion by persons in this population group. These reference amounts are based on data set forth in appropriate national food consumption surveys.

(2) FDA calculated the reference amounts for an infant or child under 4 years of age to reflect the amount of food customarily consumed per eating occasion by infants up to 12 months of age or by children 1 through 3 years of age, respectively. These reference amounts are based on data set forth in appropriate national food consumption surveys. Such reference amounts are to be used only when the food is specially formulated or processed for use by an infant or by a child under 4 years of age.

(3) An appropriate national food consumption survey includes a large sample size representative of the demographic and socioeconomic characteristics of the relevant population group and must be based on consumption data under actual conditions of use.

(4) To determine the amount of food customarily consumed per eating occasion, FDA considered the mean, median, and mode of the consumed amount per eating occasion.

(5) When survey data were insufficient, FDA took various other sources of information on serving sizes of food into consideration. These other sources of information included:

(i) Serving sizes used in dietary guidance recommendations or recommended by other authoritative systems or organizations;

(ii) Serving sizes recommended in comments;

(iii) Serving sizes used by manufacturers and grocers; and

(iv) Serving sizes used by other countries.

(6) Because they reflect the amount customarily consumed, the reference amount and, in turn, the serving size declared on the product label are based on only the edible portion of food, and not bone, seed, shell, or other

21 CFR §101.12

inedible components.

(7) The reference amount is based on the major intended use of the food (e.g., milk as a beverage and not as an addition to cereal).

(8) The reference amounts for products that are consumed as an ingredient of other foods, but that may also be consumed in the form in which they are purchased (e.g., butter), are based on use in the form purchased.

(9) FDA sought to ensure that foods that have similar dietary usage, product characteristics, and customarily consumed amounts have a uniform reference amount.

(b) The following reference amounts shall be used as the basis for determining serving sizes for specific products:

Table 1: Reference Amounts Customarily Consumed Per Eating Occasion: Infant and Toddler Foods [1,2,3,4]

Product category	Reference amount	Label statement[5]
Cereals, dry instant................................	15 g..	--- cup (--- g)
Cereals, prepared, ready-to-serve...........	110 g..	--- cup(s) (--- g)
Other cereal and grain products, dry ready-to-eat, e.g., ready-to-eat cereals, cookies, teething biscuits, and toasts.	7 g for infants and 20 g for toddlers for ready-to-eat cereals; 7 g for all others.	--- cup(s) (--- g) for ready-to-eat cereals; eat cereals; --- piece(s) (--- g) for others.
Dinners, desserts, fruits, vegetables or soups, dry mix.	15 g..	--- tbsp(s) (--- g); --- cup(s) (--- g)
Dinners, desserts, fruits, vegetables or soups, ready-to-serve, junior type.	110 g..	--- cup(s) (--- g); --- cup(s) (--- mL)
Dinners, desserts, fruits, vegetables or soups, ready-to-serve, strained type.	60 g..	--- cup(s) (--- g); --- cup(s) (--- mL)
Dinners, stews or soups for toddlers, ready-to-serve.	170 g..	--- cup(s) (--- g); --- cup(s) (--- mL)
Fruits for toddlers, ready-to-serve.	125 g..	--- cup(s) (--- g)
Vegetables for toddlers, ready-to-serve.	70 g..	--- cup(s) (--- g)
Eggs/egg yolks, ready-to-serve................	55 g..	--- cup(s) (--- g)
Juices, all varieties................................	120 mL..	4 fl oz (120 mL)

{1} These values represent the amount of food customarily consumed per eating occasion and were primarily derived from the 1977-1978 and the 1987-1988 Nationwide Food Consumption Surveys conducted by the U.S. Department of Agriculture.

{2} Unless otherwise noted in the Reference amount column, the reference amounts are for the ready-to-serve or almost ready-to-serve form of the product (i.e., heat and serve, brown and serve). If not listed separately, the reference amount for the unprepared form (e.g., dry cereal) is the amount required to make the reference amount of the prepared form. Prepared means prepared for consumption (e.g., cooked).

{3} Manufacturers are required to convert the reference amount to the label serving size in a household measure most appropriate to their specific product using the procedures in 21 CFR 101.9(b).

{4} Copies of the list of products for each product category are available from the Office of Food Labeling (HFS-150), Center for Food Safety and Applied Nutrition, Food and Drug Administration, 200 C St. SW., Washington, DC 20204.

{5} The label statements are meant to provide guidance to manufacturers on the presentation of serving size information on the label, but they are not required. The term "piece" is used as a generic description of a discrete unit. Manufacturers should use the description of a unit that is most appropriate for the specific product (e.g., sandwich for sandwiches, cookie for cookies, and bar for frozen novelties).

Table 2: Reference Amounts Customarily Consumed per Eating Occasion: General Food Supply[1,2,3,4]

Product category	Reference amount	Label statement[5]
BAKERY PRODUCTS:		
Biscuits, croissants, bagels, tortillas, soft bread sticks, soft pretzels, corn bread, hush puppies.	55 g	--- piece(s) (--- g)
Breads (excluding sweet quick type), rolls.	50 g	--- piece(s) (--- g) for sliced bread and distinct pieces (e.g., rolls); 2 oz (56 g/--- inch slice) for unsliced bread
Bread sticks-see crackers		
Toaster pastries-see coffee cakes		
Brownies	40 g	--- piece(s) (--- g) for distinct pieces; fractional slice (--- g) for bulk
Cakes, heavy weight (cheese cake; pineapple upside-down cake; fruit, nut, and vegetable cakes with more than or equal to 35 percent of the finished weight as fruit, nuts, or vegetables or any of these combined)[6].	125 g	--- piece(s) (--- g) for distinct pieces (e.g., sliced or individually packaged products); --- fractional slice (--- g) for large discrete units
Cakes, medium weight (chemically leavened cake with or without icing or filling except those classified as light weight cake; fruit, nut, and vegetable cake with less than 35 percent of the finished weight as fruit, nuts, or vegetables or any of these combined; light weight cake with icing; Boston cream pie; cupcake; eclair; cream puff)[7].	80 g	--- piece(s) (--- g) for distinct pieces (e.g., cupcake); ---fractional slice (--- g) for large discrete units
Cakes, light weight (angel food, chiffon, or sponge cake without icing or filling)[8].	55 g	--- piece(s) (--- g) for distinct pieces (e.g., sliced or individually packaged products); --- fractional slice (--- g) for large discrete units
Coffee cakes, crumb cakes, doughnuts, Danish, sweet rolls, sweet quick type breads, muffins, toaster pastries.	55 g	--- piece(s) (--- g) for sliced bread and distinct pieces (e.g., doughnut); 2 oz (56 g/visual unit of measure) for bulk products (e.g., unsliced bread)
Cookies	30 g	--- piece(s) (--- g)
Crackers that are usually not used as snack, melba toast, hard bread sticks, ice cream cones[9].	15 g	--- piece(s) (--- g)
Crackers that are usually used as snacks.	30 g	--- piece(s) (--- g)
Croutons	7 g	--- tbsp(s) (--- g); --- cup(s) (--- g); --- piece(s) (--- g) for large pieces

French toast, pancakes, variety mixes.	110 g prepared for french toast and pancakes; 40 g dry mix for variety mixes	--- piece(s) (--- g); --- cup(s) (--- g) for dry mix
Grain-based bars with or without filling or coating, e.g., breakfast bars, granola bars, rice cereal bars.	40 g...............................	--- piece(s) (--- g)
Ice cream cones-see crackers		
Pies, cobblers, fruit crisps, turnovers, other pastries.	125 g............................	--- piece(s) (---g) for distinct pieces; --- fractional slice (---g) for large discrete units
Pie crust.......................................	1/6 of 8 inch crust; 1/8 of 9 inch crust	1/6 of 8 inch crust (--- g); 1/8 of 9 inch crust (--- g)
Pizza crust.....................	55 g............................	--- fractional slice (--- g)
Taco shells, hard...............................	30 g............................	--- shell(s) (--- g)
Waffles...	85 g............................	--- piece(s) (--- g)
BEVERAGES:		
Carbonated and noncarbonated beverages, wine coolers, water.	240 mL	8 fl oz (240 mL)
Coffee or tea, flavored and sweetened.	240 mL prepared..........................	8 fl oz (240 mL)
CEREALS AND OTHER GRAIN PRODUCTS:		
Breakfast cereals (hot cereal type), hominy grits.	1 cup prepared; 40 g plain dry cereal; 55 g flavored, sweetened dry cereal	--- cup(s) (--- g)
Breakfast cereals, ready-to-eat, weighing less than 20 g per cup, e.g., plain puffed cereal grains.	15 g.............................	--- cup(s) (--- g)
Breakfast cereals, ready-to-eat weighing 20 g or more but less than 43 g per cup; high fiber cereals containing 28 g or more of fiber per 100 g.	30 g.............................	--- cup(s) (--- g)
Breakfast cereals, ready-to-eat, weighing 43 g or more per cup; biscuit types.	55 g.............................	--- piece(s) (--- g) for large distinct pieces (e.g., biscuit type);--- cup(s) (---g) for all others
Bran or wheat germ.............................	15 g.............................	--- tbsp(s) (--- g) or --- cup(s) (---g)
Flours or cornmeal.............................	30 g.............................	--- tbsp(s) (--- g) or --- cup(s) (---g)
Grains, e.g., rice, barley, plain.	140 g prepared; 45 g dry.............	--- cup(s) (--- g)

21 CFR §101.12

Product	Reference Amount	Label Statement
Pastas, plain	140 g prepared; 55 g dry	--- cup(s) (--- g); --- piece(s) (--- g) for large pieces (e.g., large shells or lasagna noodles) or 2 oz (56 g/visual unit of measure) for dry bulk products (e.g., spaghetti)
Pastas, dry, ready-to-eat, e.g., fried canned chow mein noodles.	25 g	--- cup(s) (--- g)
Starches, e.g., cornstarch, potato starch, tapioca, etc.	10 g	--- tbsp (--- g)
Stuffing	100 g	--- cup(s) (--- g)
DAIRY PRODUCTS AND SUBSTITUTES:		
Cheese, cottage	110 g	--- cup (--- g)
Cheese used primarily as ingredients, e.g., dry cottage cheese, ricotta cheese.	55 g	--- cup (--- g)
Cheese, grated hard, e.g., Parmesan, Romano.	5 g	--- tbsp (--- g)
Cheese, all others except those listed as separate categories-includes cream cheese and cheese spread.	30 g	--- piece(s) (--- g) for distinct pieces;--- tbsp(s) (--- g) for cream cheese and cheese spread; 1oz (28 g/visual unit of measure) for bulk
Cheese sauce-see sauce category		
Cream or cream substitutes, fluid.	15 mL	1 tbsp (15 mL)
Cream or cream substitutes, powder	2 g	--- tbsp (--- g)
Cream, half & half	30 mL	2 tbsp (30 mL)
Eggnog	120 mL	1/2 cup (120 mL); 4 fl oz (120 mL)
Milk, condensed, undiluted	30 mL	2 tbsp (30 mL)
Milk, evaporated, undiluted	30 mL	2 tbsp (30 mL)
Milk, milk-based drinks, e.g., instant breakfast, meal replacement, cocoa.	240 mL	1 cup (240 mL); 8 fl oz (240 mL)
Shakes or shake substitutes, e.g., dairy shake mixes, fruit frost mixes.	240 mL	1 cup (240 mL); 8 fl oz (240 mL)
Sour cream	30 g	--- tbsp (--- g)
Yogurt	225 g	--- cup (--- g)
DESSERTS:		
Ice cream, ice milk, frozen yogurt, sherbet: all types, bulk and novelties (e.g., bars, sandwiches, cones).	1/2 cup-includes the volume for coatings and wafers for the novelty type varieties	--- piece(s) (--- g) for individually wrapped or packaged products; 1/2 cup (--- g) for others

21 CFR §101.12

Frozen flavored and sweetened ice and pops, frozen fruit juices: all types, bulk and novelties (e.g., bars, cups).	85 g	--- piece(s) (--- g) for individually wrapped or packaged products; --- cup(s) (--- g) for others
Sundae	1 cup	1 cup (--- g)
Custards, gelatin or pudding	1/2 cup	--- piece(s) (--- g) for distinct unit (e.g., individually packaged products); 1/2 cup (--- g) for bulk
DESSERT TOPPINGS AND FILLINGS:		
Cake frostings or icings	35 g	--- tbsp(s) (--- g)
Other dessert toppings, e.g., fruits, syrups, spreads, marshmallow cream, nuts, dairy and nondairy whipped toppings.	2 tbsp	2 tbsp (--- g); 2 tbsp (30 mL)
Pie fillings	85 g	--- cup(s) (--- g)
EGG AND EGG SUBSTITUTES:		
Egg mixtures, e.g., egg foo young, scrambled eggs, omelets.	110 g	--- piece(s) (--- g) for discrete pieces; --- cup(s) (--- g)
Eggs (all sizes)[9]	50 g	1 large, medium, etc. (--- g)
Egg substitutes	An amount to make 1 large (50 g) egg	--- cup(s) (--- g); --- cup(s) (--- mL)
FATS AND OILS:		
Butter, margarine, oil, shortening	1 tbsp	1 tbsp (--- g); 1 tbsp (15 mL)
Butter replacement, powder	2 g	--- tsp(s) (--- g)
Dressings for salads	30 g	--- tbsp(s) (--- g); --- tbsp (--- mL)
Mayonnaise, sandwich spreads, mayonnaise-type dressings.	15 g	--- tbsp (--- g)
Spray types	0.25 g	About --- seconds spray (--- g)
FISH, SHELLFISH, GAME MEATS[10], AND MEAT OR POULTRY SUBSTITUTES:		
Bacon substitutes, canned anchovies,[11] anchovy pastes, caviar.	15 g	--- piece(s) (--- g) for discrete pieces; --- tbsp(s) (--- g) for others
Dried, e.g., jerky	30 g	--- piece(s) (--- g)
Entrees with sauce, e.g., fish with cream sauce, shrimp with lobster sauce.	140 g cooked	--- cup(s) (--- g); 5 oz (140 g/visual unit of measure) if not measurable by cup
Entrees without sauce, e.g., plain or fried fish and shellfish, fish and shellfish cake.	85 g cooked; 110 g uncooked[12]	--- piece(s) (--- g) for discrete pieces; --- cup(s) (--- g); --- oz (--- g/visual unit of measure) if not measurable by cup[13]

21 CFR §101.12

Fish, shellfish or game meat[10], canned[11].	55 g.	--- piece(s) (--- g) for discrete pieces; --- cup(s) (--- g); 2 oz (56 g/--- cup) for products that are difficult to measure the g weight of cup measure (e.g., tuna); 2 oz (56 g/--- pieces) for products that naturally vary in size (e.g., sardines)
Substitute for luncheon meat, meat spreads, Canadian bacon, sausages and frankfurters.	55 g.	--- piece(s) (--- g) for distinct pieces (e.g., slices, links); --- cup(s) (--- g); 2 oz (56 g/visual unit of measure) for nondiscrete bulk product
Smoked or pickled[11] fish, shellfish, or game meat[10]; fish or shellfish spread.	55 g.	--- piece(s) (--- g) for distinct pieces (e.g., slices, links) or --- cup(s) (--- g); 2 oz (56 g/ visual unit of measure) for nondiscrete bulk product
Substitutes for bacon bits-see miscellaneous category.		
FRUITS AND FRUIT JUICES:		
Candied or pickled[11].	30 g.	--- piece(s) (--- g)
Dehydrated fruits-see snacks category.		
Dried.	40 g.	--- piece(s) (--- g) for large pieces (e.g., dates, figs, prunes); --- cup(s) (--- g) for small pieces (e.g., raisins)
Fruits for garnish or flavor, e.g., maraschino cherries[11].	4 g.	1 cherry (--- g)
Fruit relishes, e.g., cranberry sauce, cranberry relish.	70 g.	--- cup(s) (--- g)
Fruits used primarily as ingredients, avocado.	30 g.	See footnote 13
Fruits used primarily as ingredients, others (cranberries, lemon, lime).	55 g.	--- piece(s) (--- g) for large fruits; --- cup(s) (--- g) for small fruits measurable by cup[13]
Watermelon.	280 g.	See footnote 13
All other fruits (except those listed as separate categories), fresh, canned, or frozen.	140 g.	--- piece(s) (--- g) for large pieces (e.g., strawberries, prunes, apricots, etc.); --- cup(s) (--- g) for small pieces (e.g., blueberries, raspberries, etc.)[13]
Juices, nectars, fruit drinks.	240 mL.	8 fl oz (240 mL)
Juices used as ingredients, e.g., lemon juice, lime juice.	5 mL.	1 tsp (5 mL)

21 *CFR* §101.12

LEGUMES:		
Bean cake (tofu)[11], tempeh..................	85 g...................................	--- piece(s) (--- g) for discrete pieces; 3 oz (84 g/visual unit of measure) for bulk products
Beans, plain or in sauce...........................	130 g for beans in sauce or canned in liquid and refried beans prepared; 35 g dry	--- cup (--- g)
MISCELLANEOUS CATEGORY:		
Baking powder, baking soda, pectin	1 g...................................	--- tsp (--- g)
Baking decorations, e.g., colored sugars and sprinkles for cookies, cake decorations.	1/4 tsp or 4 g if not measurable by teaspoon	--- piece(s) (--- g) for discrete pieces; 1 tsp (--- g)
Batter mixes, bread crumbs.....................	30 g...................................	--- tbsp(s) (--- g) or--- cup(s) (--- g)
Cooking wine...	30 mL...................................	2 tbsp (30 mL)
******* Dietary supplements	The maximum amount recommended, as appropriate, on the label for consumption per eating occasion, or, in the absence of recommendations, 1 unit, e.g., tablet, capsule, packet, teaspoonful, etc.	--- tablet(s), --- capsule(s), ---packet(s), ---tsp(s), (--- g), etc.
Drink mixers (without alcohol)................	Amount to make 240 mL drink (without ice)	--- fl oz (--- mL)
Chewing gum[9]	3 g...................................	--- piece(s) (--- g)
Meat, poultry and fish coating mixes, dry; seasoning mixes, dry, e.g., chili seasoning mixes, pasta salad seasoning mixes.	Amount to make one reference amount of final dish	--- tsp(s) (--- g); --- tbsp(s) (--- g)
Salad and potato toppers, e.g., salad crunchies, salad crispins, substitutes for bacon bits.	7 g...................................	--- tbsp(s) (--- g)
Salt, salt substitutes, seasoning salts (e.g., garlic salt).	1/4 tsp.............................	1/4 tsp (-- g); --- piece(s) (-- g) for discrete pieces (e.g., individually packaged products)
Spices, herbs (other than dietary supplements).	1/4 tsp or 0.5 g if not measurable by teaspoon	1/4 tsp (--- g); --- piece(s) (---g) if not measurable by teaspoons (e.g., bay leaf)
MIXED DISHES:		
Measurable with cup, e.g., casseroles, hash, macaroni and cheese, pot pies, spaghetti with sauce, stews, etc.	1 cup	1 cup (--- g)

21 CFR §101.12

Not measurable with cup, e.g., burritos, egg rolls, enchiladas, pizza, pizza rolls, quiche, all types of sandwiches.	140 g, add 55 g for products with gravy or sauce topping, e.g., enchilada with cheese sauce, crepe with white sauce[14]	--- piece(s) (--- g) for discrete pieces; --- fractional slice (---g) for large discrete units
NUTS AND SEEDS:		
Nuts, seeds, and mixtures, all types: sliced, chopped, slivered, and whole.	30 g..............................	--- piece(s) (--- g) for large pieces (e.g., unshelled nuts); --- tbsp(s) (--- g) or--- cup(s) (---g) for small pieces (e.g., peanuts, sunflower seeds)
Nut and seed butters, pastes, or creams.	2 tbsp.........................	2 tbsp (--- g)
Coconut, nut and seed flours....................	15 g..............................	--- tbsp(s) (--- g); --- cup (---g)
POTATOES AND SWEET POTATOES/YAMS:		
French fries, hash browns, skins, or pancakes.	70 g prepared; 85 g for frozen unprepared french fries.	--- piece(s) (--- g) for large distinct pieces (e.g., patties, skins); 2.5 oz (70 g/--- pieces) for prepared fries; 3 oz (84 g/---pieces) for unprepared fries
Mashed, candied, stuffed, or with sauce.	140 g............................	--- piece(s) (--- g) for discrete pieces (e.g., stuffed potato); --- cup(s) (--- g)
Plain, fresh, canned, or frozen..................	110 g for fresh or frozen; 125 g for vacuum packed; 160 g for canned in liquid	--- piece(s) (--- g) for discrete pieces;--- cup(s) (--- g) for sliced or chopped products
SALADS:		
Gelatin salad............................	120 g............................	--- cup (--- g)
Pasta or potato salad.................	140 g............................	--- cup(s) (--- g)
All other salads, e.g., egg, fish, shellfish, bean, fruit, or vegetable salads.	100 g............................	--- cup(s) (--- g)
SAUCES, DIPS, GRAVIES AND CONDIMENTS:		
Barbecue sauce, hollandaise sauce, tartar sauce, other sauces for dipping (e.g., mustard sauce, sweet and sour sauce), all dips (e.g., bean dips, dairy-based dips, salsa).	2 tbsp...........................	2 tbsp (--- g); 2 tbsp (30 mL)
Major main entree sauces, e.g., spaghetti sauce.	125 g............................	--- cup (--- g); --- cup (--- mL)
Minor main entree sauces (e.g., pizza sauce, pesto sauce), other sauces used as toppings (e.g., gravy, white sauce, cheese sauce), cocktail sauce.	1/4 cup.........................	1/4 cup (--- g); 1/4 cup (60 mL)

21 CFR §101.12

Major condiments, e.g., catsup, steak sauce, soy sauce, vinegar, teriyaki sauce, marinades.	1 tbsp	1 tbsp (--- g); 1 tbsp (15 mL)
Minor condiments, e.g., horseradish, hot sauces, mustards, worcestershire sauce.	1 tsp	1 tsp (--- g); 1 tsp (5 mL)
SNACKS:		
All varieties, chips, pretzels, popcorns, extruded snacks, fruit-based snacks (e.g., fruit chips,) grain-based snack mixes.	30 g	--- cup(s) (--- g) for small pieces (e.g., popcorn) --- piece(s) (--- g) for large pieces (e.g., large pretzels; pressed dried fruit sheet); 1 oz (28 g/visual unit of measure) for bulk products (e.g., potato chips)
SOUPS:		
All varieties	245 g	--- cup (--- g); --- cup (--- mL)
SUGARS AND SWEETS:		
Baking candies (e.g., chips)	15 g	--- piece(s) (--- g) for large pieces; --- tbsp(s) (--- g) for small pieces; 1/2 oz (14 g/visual unit of measure) for bulk products
Hard candies, breath mints	2 g	--- piece(s) (--- g)
Hard candies, roll-type, mini-size in dispenser packages.	5 g	--- piece(s) (--- g)
Hard candies, others	15 g	--- piece(s) (--- g) for large pieces;--- tbsp(s) (--- g) for "mini-size" candies measurable by tablespoon; 1/2 oz (14 g/visual unit of measure) for bulk products
All other candies	40 g	--- piece(s) (--- g); 1 1/2 oz (42 g/visual unit of measure) for bulk products
Confectioner's sugar	30 g	--- cup (--- g)
Honey, jams, jellies, fruit butter, molasses.	1 tbsp	1 tbsp (--- g); 1 tbsp (15 mL)
Marshmallows	30 g	--- cup(s) (--- g) for small pieces or --- piece(s) (--- g) for large pieces
Sugar	4 g	--- tsp (--- g); --- piece(s) (--- g) for discrete pieces (e.g., sugar cubes, individually packaged products)

21 CFR §101.12

Sugar substitutes..........................	An amount equivalent to one reference amount for sugar in sweetness	--- tsp(s) (--- g) for solids; ---drop(s) (--- g) for liquid; ---piece(s) (--- g) (e.g., individually packaged products)
Syrups..	30 mL for syrups used primarily as an ingredient (e.g., light or dark corn syrup); 60 mL for all others	2 tbsp (30 mL) for syrups used primarily as an ingredient; 1/4 cup (60 mL) for all others
VEGETABLES:		
Vegetables primarily used for garnish or flavor, e.g., pimento, parsley.	4 g...	--- piece(s) (--- g); --- tbsp(s) (--- g) for chopped products
Chili pepper, green onion	30 g...	--- piece(s) (--- g)[13]; --- tbsp(s) (--- g); --- cup(s) (--- g) for sliced or chopped products
All other vegetables without sauce: fresh, canned, or frozen.	85 g for fresh or frozen; 95 g for vacuum packed; 130 g for canned in liquid, cream-style corn, canned or stewed tomatoes, pumpkin, or winter squash	--- piece(s) (--- g) for large pieces (e.g., brussel sprouts); --- cup(s) (--- g) for small pieces (e.g., cut corn, green peas); 3 oz (84 g/visual unit of measure) if not measurable by cup[13]
All other vegetables with sauce: fresh, canned, or frozen.	110 g...	--- piece(s) (--- g) for large pieces (e.g., brussel sprouts); --- cup(s) (--- g) for small pieces (e.g., cut corn, green peas); 4 oz (112 g/visual unit of measure) if not measurable by cup
Vegetable juice.........................	240 mL......................................	8 fl oz (240 mL)
Olives[11]....................................	15 g...	--- piece(s) (--- g); --- tbsp(s) (--- g) for sliced products
Pickles, all types[11]....................	30 g...	1 oz (28 g/visual unit of measure)
Pickle relishes.........................	15 g...	--- tbsp (--- g)
Vegetable pastes, e.g., tomato paste.	30 g...	--- tbsp (--- g)
Vegetable sauces or purees, e.g, tomato sauce, tomato puree.	60 g...	--- cup (---g); --- cup (--- mL)

{1} These values represent the amount (edible portion) of food customarily consumed per eating occasion and were primarily derived from the 1977-1978 and the 1987-1988 Nationwide Food Consumption Surveys conducted by the United States Department of Agriculture.

{2} Unless otherwise noted in the Reference Amount column, the reference amounts are for the ready-to-serve or almost ready-to-serve form of the product (i.e., heat and serve, brown and serve). If not listed separately, the reference amount for the unprepared form (e.g., dry mixes; concentrates; dough; batter; fresh and frozen pasta) is the amount required to make the reference amount of the prepared form. Prepared means prepared for consumption (e.g., cooked).

{3} Manufacturers are required to convert the reference amount to the label serving size in a household measure most appropriate to their specific product using the procedures in 21 CFR 101.9(b).

{4} Copies of the list of products for each product category are available from the Office of Food Labeling (HFS-150), Center for Food Safety and Applied Nutrition, Food and Drug Administration, 200 C St. SW., Washington, DC 20204.

{5} The label statements are meant to provide guidance to manufacturers on the presentation of serving size information on the label, but they are not required. The term "piece" is used as a generic description of a discrete unit. Manufacturers should use the description of a unit that is most appropriate for the specific product (e.g., sandwich for sandwiches, cookie for cookies, and bar for ice cream bars). The guidance provided is for the label statement of products in ready-to-serve or almost ready-to-serve form. The guidance does not apply to the products which require further preparation for consumption (e.g., dry mixes, concentrates) unless specifically stated in the product category, reference amount, or label statement column that it is for these forms of the product. For products that require further preparation, manufacturers must determine the label statement following the rules in §101.9(b) using the reference amount determined according to §101.12(c).

{6} Includes cakes that weigh 10 g or more per cubic inch.

{7} Includes cakes that weigh 4 g or more per cubic inch but less than 10 g per cubic inch.

{8} Includes cakes that weigh less than 4 g per cubic inch.

{9} Label serving size for ice cream cones and eggs of all sizes will be one unit. Label serving size of all chewing gums that weigh more than the reference amount that can reasonably be consumed at a single-eating occasion will be one unit.

{10} Animal products not covered under the Federal Meat Inspection Act or the Poultry Products Inspection Act, such as flesh products from deer, bison, rabbit, quail, wild turkey, geese, ostrich, etc.

{11} If packed or canned in liquid, the reference amount is for the drained solids, except for products in which both the solids and liquids are customarily consumed (e.g., canned chopped clam in juice).

{12} The reference amount for the uncooked form does not apply to raw fish in §101.45 or to single- ingredient products that consist of fish or game meat as provided for in §101.9(b)(j)(11).

{13} For raw fruit, vegetables, and fish, manufacturers should follow the label statement for the serving size specified in Appendices A and B to the regulation entitled "Food Labeling; Guidelines for Voluntary Nutrition Labeling; and Identification of the 20 Most Frequently Consumed Raw Fruits, Vegetables, and Fish; Definition of Substantial Compliance; Correction" (56 FR 60880 as amended 57 FR 8174, March 6, 1992).

{14} Pizza sauce is part of the pizza and is not considered to be sauce topping.

(c) If a product requires further preparation, e.g., cooking or the addition of water or other ingredients, and if paragraph (b) of this section provides a reference amount for the product in the prepared but not the unprepared form, then the reference amount for the unprepared product shall be determined using the following rules:

(1) Except as provided for in paragraph (c)(2) of this section, the reference amount for the unprepared product shall be the amount of the unprepared product required to make the reference amount for the prepared product as established in paragraph (b) of this section.

(2) For products where the entire contents of the package is used to prepare one large discrete unit usually divided for consumption, the reference amount for the unprepared product shall be the amount of the unprepared product required to make the fraction of the large discrete unit closest to the reference amount for the prepared product as established in paragraph (b) of this section.

(d) The reference amount for an imitation or substitute food or altered food, such as a "low calorie" version, shall be the same as for the food for which it is offered as a substitute.

(e) If a food is modified by incorporating air (aerated), and thereby the density of the food is lowered by 25 percent or more in weight than that of an appropriate reference regular food as described in §101.13(j)(1)(ii)(A), and the reference amount of the regular food is in grams, the manufacturer may determine the reference amount of the aerated food by adjusting for the difference in density of the aerated food relative to the density of the appropriate reference food provided that the manufacturer will show FDA detailed protocol and records of all data that were used to determine the density-adjusted reference amount for the aerated food. The reference amount for the aerated food shall be rounded to the nearest 5-g increment. Such products shall bear a descriptive term indicating that extra air has been incorporated (e.g., whipped, aerated). The density-adjusted reference amounts described in paragraph (b) of this section may not be used for cakes except for cheese cake. The differences in the densities of different types of cakes having different degrees of air incorporation have already been taken into consideration in determining the reference amounts for cakes in §101.12(b). In determining the difference in density of the aerated and the regular food, the manufacturer shall adhere to the following:

(1) The regular and the aerated product must be the same in size, shape, and volume. To compare the densities of products having nonsmooth surfaces (e.g., waffles), manufacturers shall use a device or method that ensures that the volumes of the regular and the aerated products are the same.

(2) Sample selections for the density measurements shall be done in accordance with the provisions in §101.9(g).

(3) Density measurements of the regular and the aerated products shall be conducted by the same trained operator using the same methodology (e.g., the same equipment, procedures, and techniques) under the same conditions.

21 CFR §101.12

(4) Density measurements shall be replicated a sufficient number of times to ensure that the average of the measurements is representative of the true differences in the densities of the regular and the "aerated" products.

(f) For products that have no reference amount listed in paragraph (b) of this section for the unprepared or the prepared form of the product and that consist of two or more foods packaged and presented to be consumed together (e.g., peanut putter and jelly, cracker and cheese pack, pancakes and syrup, cake and frosting), the reference amount for the combined product shall be determined using the following rules:

(1) For bulk products (e.g., peanut butter and jelly), the reference amount for the combined product shall be the reference amount, as established in paragraph (b) of this section, for the ingredient that is represented as the main ingredient plus proportioned amounts of all minor ingredients.

(2) For products where the ingredient represented as the main ingredient is one or more discrete units (e.g., cracker and cheese pack, pancakes and syrup, cake and frosting), the reference amount for the combined product shall be either the number of small discrete units or the fraction of the large discrete unit that is represented as the main ingredient that is closest to the reference amount for that ingredient as established in paragraph (b) of this section plus proportioned amounts of all minor ingredients.

(3) If the reference amounts are in compatible units, they shall be summed (e.g., the reference amount for equal volumes of peanut butter and jelly for which peanut butter is represented as the main ingredient would be 4 tablespoons (tbsp) (2 tbsp peanut butter plus 2 tbsp jelly). If the reference amounts are in incompatible units, the weights of the appropriate volumes should be used (e.g., 110 grams (g) pancakes plus the gram weight of the proportioned amount of syrup).

(g) The reference amounts set forth in paragraphs (b) through (f) of this section shall be used in determining whether a product meets the criteria for nutrient content claims, such as "low calorie," and for health claims. If the serving size declared on the product label differs from the reference amount, and the product meets the criteria for the claim only on the basis of the reference amount, the claim shall be followed by a statement that sets forth the basis on which the claim is made. That statement shall include the reference amount as it appears in paragraph (b) of this section followed, in parenthesis, by the amount in common household measure if the reference amount is expressed in measures other than common household measures (e.g., for a beverage, "Very low sodium, 35 mg or less per 240 mL (8 fl oz)").

(h) The Commissioner of Food and Drugs, either on his or her own initiative or in response to a petition submitted pursuant to part 10 of this chapter, may issue a proposal to establish or amend a reference amount in paragraph (b) of this section. A petition to establish or amend a reference amount shall include:

(1) Objective of the petition;

(2) A description of the product;

(3) A complete sample product label including nutrition label, using the format established by regulation;

(4) A description of the form (e.g., dry mix, frozen dough) in which the product will be marketed;

(5) The intended dietary uses of the product with the major use identified (e.g., milk as a beverage and chips as a snack);

(6) If the intended use is primarily as an ingredient in other foods, list of foods or food categories in which the product will be used as an ingredient with information on the prioritization of the use;

(7) The population group for which the product will be offered for use (e.g., infants, children under 4 years of age);

(8) The names of the most closely related products (or in the case of foods for special dietary use and imitation or substitute foods, the names of the products for which they are offered as substitutes);

(9) The suggested reference amount (the amount of edible portion of food as consumed, excluding bone, seed, shell, or other inedible components) for the population group for which the product is intended with full description of the methodology and procedures that were used to determine the suggested reference amount. In determining the reference amount, general principles and factors in paragraph (a) of this section should be followed.

(10) The suggested reference amount shall be expressed in metric units. Reference amounts for fluids shall be expressed in milliliters. Reference amounts for other foods shall be expressed in grams except when common household units such as cups, tablespoons, and teaspoons, are more appropriate or are more likely to promote uniformity in serving sizes declared on product labels. For example, common household measures would be more appropriate if products within the same category differ substantially in density, such as frozen desserts.

(i) In expressing the reference amounts in milliliters, the following rules shall be followed:

(A) For volumes greater than 30 milliliters (mL), the volume shall be expressed in multiples of 30 mL.
(B) For volumes less than 30 mL, the volume shall be expressed in milliliters equivalent to a whole number of teaspoons or 1 tbsp, i.e., 5, 10, or 15 mL.
(ii) In expressing the reference amounts in grams, the following general rules shall be followed:
(A) For quantities greater than 10 g, the quantity shall be expressed in the nearest 5-g increment.
(B) For quantities less than 10 g, exact gram weights shall be used.
(11) A petition to create a new subcategory of food with its own reference amount shall include the following additional information:
(i) Data that demonstrate that the new subcategory of food will be consumed in amounts that differ enough from the reference amount for the parent category to warrant a separate reference amount. Data must include sample size; and the mean, standard deviation, median, and modal consumed amount per eating occasion for the petitioned product and for other products in the category, excluding the petitioned product. All data must be derived from the same survey data.
(ii) Documentation supporting the difference in dietary usage and product characteristics that affect the consumption size that distinguishes the petitioned product from the rest of the products in the category.
(12) A claim for categorical exclusion under §25.30 or 25.32 of this chapter or an environmental assessment under §25.40 of this chapter; and
(13) In conducting research to collect or process food consumption data in support of the petition, the following general guidelines should be followed.
(i) Sampled population selected should be representative of the demographic and socioeconomic characteristics of the target population group for which the food is intended.
(ii) Sample size (i.e., number of eaters) should be large enough to give reliable estimates for customarily consumed amounts.
(iii) The study protocol should identify potential biases and describe how potential biases are controlled for or, if not possible to control, how they affect interpretation of results. (iv) The methodology used to collect or process data should be fully documented and should include: study design, sampling procedures, materials used (e.g., questionnaire, and interviewer's manual), procedures used to collect or process data, methods or procedures used to control for unbiased estimates, and procedures used to correct for nonresponse.
(14) A statement concerning the feasibility of convening associations, corporations, consumers, and other interested parties to engage in negotiated rulemaking to develop a proposed rule consistent with the Negotiated Rulemaking Act (5 U.S.C. 561).
[58 FR 44051, Aug. 18, 1993; 58 FR 60109, Nov. 15, 1993; as amended at 59 FR 371, Jan. 4, 1994 and 59 FR 24039, May 10, 1994; 62 FR 40598, July 29, 1997; 62 FR 49848, Sept. 23, 1997; 62 FR 63653, Dec. 2, 1997; 63 FR 14818, Mar. 27, 1998]

§101.13 Nutrient content claims—general principles.

(a) This section and the regulations in subpart D of this part apply to foods that are intended for human consumption and that are offered for sale, including conventional foods and dietary supplements.
(b) A claim that expressly or implicitly characterizes the level of a nutrient of the type required to be in nutrition labeling under §101.9 or under §101.36 (that is, a nutrient content claim) may not be made on the label or in labeling of foods unless the claim is made in accordance with this regulation and with the applicable regulations in subpart D of this part or in part 105 or part 107 of this chapter.
(1) An expressed nutrient content claim is any direct statement about the level (or range) of a nutrient in the food, e.g., "low sodium" or "contains 100 calories."
(2) An implied nutrient content claim is any claim that:
(i) Describes the food or an ingredient therein in a manner that suggests that a nutrient is absent or present in a certain amount (e.g., "high in oat bran"); or
(ii) Suggests that the food, because of its nutrient content, may be useful in maintaining healthy dietary practices and is made in association with an explicit claim or statement about a nutrient (e.g., "healthy, contains 3 grams (g) of fat").
(3) Except for claims regarding vitamins and minerals described in paragraph (q)(3) of this section, no nutrient

content claims may be made on food intended specifically for use by infants and children less than 2 years of age unless the claim is specifically provided for in parts 101, 105, or 107 of this chapter.

(4) Reasonable variations in the spelling of the terms defined in part 101 and their synonyms are permitted provided these variations are not misleading (e.g., "hi" or "lo").

(5) For dietary supplements, claims for calories, fat, saturated fat, and cholesterol may not be made on products that meet the criteria in §101.60(b)(1) or (b)(2) for "calorie free" or "low calorie" claims, except, in the case of calorie claims, when an equivalent amount of similar dietary supplement (e.g., another protein supplement) that the labeled food resembles and for which it substitutes, normally exceeds the definition for "low calorie" in §101.60(b)(2).

(c) Information that is required or permitted by §101.9 or §101.36, as applicable, to be declared in nutrition labeling, and that appears as part of the nutrition label, is not a nutrient content claim and is not subject to the requirements of this section. If such information is declared elsewhere on the label or in labeling, it is a nutrient content claim and is subject to the requirements for nutrient content claims.

(d) A "substitute" food is one that may be used interchangeably with another food that it resembles, i.e., that it is organoleptically, physically, and functionally (including shelf life) similar to, and that it is not nutritionally inferior to unless it is labeled as an "imitation."

(1) If there is a difference in performance characteristics that materially limits the use of the food, the food may still be considered a substitute if the label includes a disclaimer adjacent to the most prominent claim as defined in paragraph (j)(2)(iii) of this section, informing the consumer of such difference (e.g., "not recommended for frying").

(2) This disclaimer shall be in easily legible print or type and in a size no less than that required by §101.105(i) for the net quantity of contents statement, except where the size of the claim is less than two times the required size of the net quantity of contents statement, in which case the disclaimer shall be no less than one-half the size of the claim but no smaller than one-sixteenth of an inch, unless the package complies with §101.2(c)(5), in which case the disclaimer may be in type of not less than one thirty-second of an inch.

(e)

(1) Because the use of a "free" or "low" claim before the name of a food implies that the food differs from other foods of the same type by virtue of its having a lower amount of the nutrient, only foods that have been specially processed, altered, formulated, or reformulated so as to lower the amount of the nutrient in the food, remove the nutrient from the food, or not include the nutrient in the food, may bear such a claim (e.g., "low sodium potato chips").

(2) Any claim for the absence of a nutrient in a food, or that a food is low in a nutrient when the food has not been specially processed, altered, formulated, or reformulated to qualify for that claim shall indicate that the food inherently meets the criteria and shall clearly refer to all foods of that type and not merely to the particular brand to which the labeling attaches (e.g., "corn oil, a sodium-free food").

(f) A nutrient content claim shall be in type size no larger than two times the statement of identity and shall not be unduly prominent in type style compared to the statement of identity.

(g) The label or labeling of a food for which a nutrient content claim is made shall contain prominently and in immediate proximity to such claim, the following referral statement: "See _____ for nutrition information" with the blank filled in with the identity of the panel on which nutrition labeling is located, except that when such a claim appears on the panel that bears nutrition information the referral statement may be omitted.

(1) The referral statement "See [*appropriate panel*] for nutrition information" shall be in easily legible boldface print or type, in distinct contrast to other printed or graphic matter, and in a size no less than that required by §101.105(i) for the net quantity of contents statement, except where the size of the claim is less than two times the required size of the net quantity of contents statement, in which case the referral statement shall be no less than one-half the size of the claim but no smaller than one-sixteenth of an inch, unless the package complies with §101.2(c)(5), in which case the referral statement may be in type of not less than one thirty-second of an inch.

(2) The referral statement shall be immediately adjacent to the nutrient content claim and may have no intervening material other than, if applicable, other information in the statement of identity or any other information that is required to be presented with the claim under this section (e.g., see paragraph (j)(2) of this section) or under a regulation in subpart D of this part (e.g., see §§101.54 and 101.62). If the nutrient content claim appears on more than one panel of the label, the referral statement shall be adjacent to the claim on each

panel except for the panel that bears the nutrition information where it may be omitted.

(3) If a single panel of a food label or labeling contains multiple nutrient content claims or a single claim repeated several times, a single referral statement may be made. The statement shall be adjacent to the claim that is printed in the largest type on that panel.

(h) In place of the referral statement described in paragraph (g) of this section,

(1) If a food, except a meal product as defined in §101.13(l), a main dish product as defined in §101.13(m), or food intended specifically for use by infants and children less than 2 years of age, contains more than 13.0 g of fat, 4.0 g of saturated fat, 60 milligrams (mg) of cholesterol, or 480 mg of sodium per reference amount customarily consumed, per labeled serving, or, for a food with a reference amount customarily consumed of 30 g or less or 2 tablespoons or less, per 50 g (for dehydrated foods that must be reconstituted before typical consumption with water or a diluent containing an insignificant amount, as defined in §101.9(f)(1), of all nutrients per reference amount customarily consumed, the per 50-g criterion refers to the 'as prepared' form) then that food must disclose, as part of the referral statement, that the nutrient exceeding the specified level is present in the food as follows: "See [*appropriate panel*] for information about [*nutrient requiring disclosure*] and other nutrients," e.g., "See side panel for information about total fat and other nutrients."

(2) If a food is a meal product as defined in §101.13(l), and contains more than 26 g of fat, 8.0 g of saturated fat, 120 mg of cholesterol, or 960 mg of sodium per labeled serving, then that food must disclose, in accordance with the requirements as provided in paragraph (h)(1) of this section, that the nutrient exceeding the specified level is present in the food.

(3) If a food is a main dish product as defined in §101.13(m), and contains more than 19.5 g of fat, 6.0 g of saturated fat, 90 mg of cholesterol, or 720 mg of sodium per labeled serving, then that food must disclose, in accordance with the requirements as provided in paragraph (h)(1) of this section, that the nutrient exceeding the specified level is present in the food.

(i) Except as provided in §101.9 or §101.36, as applicable, or in paragraph (q)(3) of this section, the label or labeling of a product may contain a statement about the amount or percentage of a nutrient if:

(1) The use of the statement on the food implicitly characterizes the level of the nutrient in the food and is consistent with a definition for a claim, as provided in subpart D of this part, for the nutrient that the label addresses. Such a claim might be, "less than 3 g of fat per serving;"

(2) The use of the statement on the food implicitly characterizes the level of the nutrient in the food and is not consistent with such a definition, but the label carries a disclaimer adjacent to the statement that the food is not "low" in or a "good source" of the nutrient, such as "only 200 mg sodium per serving, not a low sodium food." The disclaimer must be in easily legible print or type and in a size no less than that required by § 101.105(i) for the net quantity of contents statement except where the size of the claim is less than two times the required size of the net quantity of contents statement, in which case the disclaimer shall be no less than one-half the size of the claim but no smaller than one-sixteenth of an inch unless the package complies with §101.2(c)(5), in which case the disclaimer may be in type of not less less than one thirty-second of an inch, or

(3) The statement does not in any way implicitly characterize the level of the nutrient in the food and it is not false or misleading in any respect (e.g., "100 calories" or "5 grams of fat"), in which case no disclaimer is required.

(4) "Percent fat free" claims are not authorized by this paragraph. Such claims shall comply with §101.62(b)(6).

(j) A food may bear a statement that compares the level of a nutrient in the food with the level of a nutrient in a reference food. These statements shall be known as "relative claims" and include "light," "reduced," "less" (or "fewer"), and "more" claims.

(1) To bear a relative claim about the level of a nutrient, the amount of that nutrient in the food must be compared to an amount of nutrient in an appropriate reference food as specified below.

(i) (A) For "less" (or "fewer") and "more" claims, the reference food may be a dissimilar food within a product category that can generally be substituted for one another in the diet (e.g., potato chips as reference for pretzels, orange juice as a reference for vitamin C tablets) or a similar food (e.g., potato chips as reference for potato chips, one brand of multivitamin as reference for another brand of multivitamin).

(B) For "light," "reduced," "added," "extra," "plus," "fortified," and "enriched" claims, the reference food shall be a similar food (potato chips as a reference for potato chips, one brand of multivitamin for another brand of multivitamin), and

(ii) (A) For "light" claims, the reference food shall be representative of the type of food that includes the product

21 *CFR* §101.13

that bears the claim. The nutrient value for the reference food shall be representative of a broad base of foods of that type; e.g., a value in a representative, valid data base; an average value determined from the top three national (or regional) brands, a market basket norm; or, where its nutrient value is representative of the food type, a market leader. Firms using such a reference nutrient value as a basis for a claim, are required to provide specific information upon which the nutrient value was derived, on request, to consumers and appropriate regulatory officials.

(B) For relative claims other than "light," including "less" and "more" claims, the reference food may be the same as that provided for "light" in paragraph (j)(1)(ii)(A) of this section, or it may be the manufacturer's regular product, or that of another manufacturer, that has been offered for sale to the public on a regular basis for a substantial period of time in the same geographic area by the same business entity or by one entitled to use its trade name. The nutrient values used to determine the claim when comparing a single manufacturer's product to the labeled product shall be either the values declared in nutrition labeling or the actual nutrient values, provided that the resulting label is internally consistent to (i.e., that the values stated in the nutrition information, the nutrient values in the accompanying information and the declaration of the percentage of nutrient by which the food has been modified are consistent and will not cause consumer confusion when compared), and that the actual modification is at least equal to the percentage specified in the definition of the claim.

(2) For foods bearing relative claims:

(i) The label or labeling must state the identity of the reference food and the percentage (or fraction) of the amount of the nutrient in the reference food by which the nutrient in the labeled food differs (e.g., "50 percent less fat than (reference food)" or "1/3 fewer calories than (reference food)"),

(ii) This information shall be immediately adjacent to the most prominent claim. The type size shall be in accordance with paragraph (g)(1) of this section.

(iii) The determination of which use of the claim is in the most prominent location on the label or labeling will be made based on the following factors, considered in order:

(A) A claim on the principal display panel adjacent to the statement of identity;

(B) A claim elsewhere on the principal display panel;

(C) A claim on the information panel; or

(D) A claim elsewhere on the label or labeling.

(iv) The label or labeling must also bear:

(A) Clear and concise quantitative information comparing the amount of the subject nutrient in the product per labeled serving with that in the reference food; and

(B) This statement shall appear adjacent to the most prominent claim or to the nutrition label, except that if the nutrition label is on the information panel, the quantitative information may be located elsewhere on the information panel in accordance with §101.2.

(3) A relative claim for decreased levels of a nutrient may not be made on the label or in labeling of a food if the nutrient content of the reference food meets the requirement for a "low" claim for that nutrient (e.g., 3 g fat or less).

(k) The term "modified" may be used in the statement of identity of a food that bears a relative claim that complies with the requirements of this part, followed immediately by the name of the nutrient whose content has been altered (e.g., "Modified fat cheesecake"). This statement of identity must be immediately followed by the comparative statement such as "Contains 35 percent less fat than ____." The label or labeling must also bear the information required by paragraph (j)(2) of this section in the manner prescribed.

(l) For purposes of making a claim, a "meal product" shall be defined as a food that:

(1) Makes a major contribution to the total diet by:

(i) Weighing at least 10 ounces (oz) per labeled serving; and

(ii) Containing not less than three 40-g portions of food, or combinations of foods, from two or more of the following four food groups, except as noted in paragraph (l)(1)(ii)(E) of this section.

(A) Bread, cereal, rice, and pasta group;

(B) Fruits and vegetables group;

(C) Milk, yogurt, and cheese group;

(D) Meat, poultry, fish, dry beans, eggs, and nuts group; except that;

(E) These foods shall not be sauces (except for foods in the above four food groups that are in the sauces), gravies, condiments, relishes, pickles, olives, jams, jellies, syrups, breadings or garnishes; and

21 CFR §101.13

(2) Is represented as, or is in a form commonly understood to be, a breakfast, lunch, dinner, or meal. Such representations may be made either by statements, photographs, or vignettes.

(m) For purposes of making a claim, a "main dish product" shall be defined as a food that:

(1) Makes a major contribution to a meal by

(i) Weighing at least 6 oz per labeled serving; and

(ii) Containing not less than 40 g of food, or combinations of foods, from each of at least two of the following four food groups, except as noted in paragraph (m)(1)(ii)(E) of this section.

(A) Bread, cereal, rice, and pasta group;

(B) Fruits and vegetables group;

(C) Milk, yogurt, and cheese group;

(D) Meat, poultry, fish, dry beans, eggs, and nuts groups; except that:

(E) These foods shall not be sauces (except for foods in the above four food groups that are in the sauces) gravies, condiments, relishes, pickles, olives, jams, jellies, syrups, breadings, or garnishes; and

(2) Is represented as, or is in a form commonly understood to be, a main dish (e.g, not a beverage or a dessert). Such representations may be made either by statements, photographs, or vignettes.

(n) Nutrition labeling in accordance with §101.9, §101.10, or §101.36, as applicable, shall be provided for any food for which a nutrient content claim is made.

(o) Except as provided in §101.10, compliance with requirements for nutrient content claims in this section and in the regulations in subpart D of this part, will be determined using the analytical methodology prescribed for determining compliance with nutrition labeling in §101.9.

(p)

(1) Unless otherwise specified, the reference amount customarily consumed set forth in §101.12(b) through (f) shall be used in determining whether a product meets the criteria for a nutrient content claim. If the serving size declared on the product label differs from the reference amount customarily consumed, and the amount of the nutrient contained in the labeled serving does not meet the maximum or minimum amount criterion in the definition for the descriptor for that nutrient, the claim shall be followed by the criteria for the claim as required by §101.12(g) (e.g., "very low sodium, 35 mg or less per 240 milliliters (8 fl oz.)").

(2) The criteria for the claim shall be immediately adjacent to the most prominent claim in easily legible print or type and in a size in accordance with paragraph (g)(1) of this section.

(q) The following exemptions apply:

(1) Nutrient content claims that have not been defined by regulation and that are contained in the brand name of a specific food product that was the brand name in use on such food before October 25, 1989, may continue to be used as part of that brand name for such product, provided that they are not false or misleading under section 403(a) of the Federal Food, Drug, and Cosmetic Act (the act). However, foods bearing such claims must comply with section 403(f), (g), and (h) of the act;

(2) A soft drink that used the term "diet" as part of its brand name before October 25, 1989, and whose use of that term was in compliance with §105.66 of this chapter as that regulation appeared in the Code of Federal Regulations on that date, may continue to use that term as part of its brand name, provided that its use of the term is not false or misleading under section 403(a) of the act. Such claims are exempt from the requirements of section 403(r)(2) of the act (e.g., the referral statement also required by §101.13(g) and the disclosure statement also required by §101.13(h)). Soft drinks marketed after October 25, 1989, may use the term "diet" provided they are in compliance with the current §105.66 of this chapter and the requirements of §101.13.

(3)(i) A statement that describes the percentage of a vitamin or mineral in the food, including foods intended specifically for use by infants and children less than 2 years of age, in relation to a Reference Daily Intake (RDI) as defined in §101.9 may be made on the label or in labeling of a food without a regulation authorizing such a claim for a specific vitamin or mineral unless such claim is expressly prohibited by regulation under section 403(r)(2)(A)(vi) of the act.

(ii) Percentage claims for dietary supplements. Under section 403(r)(2)(F) of the act, a statement that characterizes the percentage level of a dietary ingredient for which a reference daily intake (RDI) or daily reference value (DRV) has not been established may be made on the label or in labeling of dietary supplements without a regulation that specifically defines such a statement. All such claims shall be accompanied by a referral or disclosure statement in accordance with paragraphs (g) or (h) of this section.

(A) *Simple percentage claims.* Whenever a statement is made that characterizes the percentage level of a dietary ingredient for which there is no RDI or DRV, the statement of the actual amount of the dietary ingredient per serving shall be declared next to the percentage statement (e.g., "40 percent omega-3 fatty acids, 10 mg per capsule").

(B) *Comparative percentage claims.* Whenever a statement is made that characterizes the percentage level of a dietary ingredient for which there is no RDI or DRV and the statement draws a comparison to the amount of the dietary ingredient in a reference food, the reference food shall be clearly identified, the amount of that food shall be identified, and the information on the actual amount of the dietary ingredient in both foods shall be declared in accordance with paragraph (j)(2)(iv) of this section (e.g., "twice the omega- fatty acids per capsule (80 mg) as in 100 mg of menhaden oil (40 mg)").

(4) The requirements of this section do not apply to:

(i) Infant formulas subject to section 412(h) of the act; and

(ii) Medical foods defined by section 5(b) of the Orphan Drug Act.

(5) A nutrient content claim used on food that is served in restaurants or other establishments in which food is served for immediate human consumption or which is sold for sale or use in such establishments shall comply with the requirements of this section and the appropriate definition in subpart D of this part, except that:

(i) Such claim is exempt from the requirements for disclosure statements in paragraphs (g) and (h) of this section and §§101.54(d), 101.62(c), (d)(1)(ii)(D), (d)(2)(iii)(C), (d)(3), (d)(4)(ii)(C), and (d)(5)(ii)(C); and

(ii) In lieu of analytical testing, compliance may be determined using a reasonable basis for concluding that the food that bears the claim meets the definition for the claim. This reasonable basis may derive from recognized data bases for raw and processed foods, recipes, and other means to compute nutrient levels in the foods or meals and may be used provided reasonable steps are taken to ensure that the method of preparation adheres to the factors on which the reasonable basis was determined (e.g., types and amounts of ingredients, cooking temperatures, etc.). Firms making claims on foods based on this reasonable basis criterion are required to provide to appropriate regulatory officials on request the specific information on which their determination is based and reasonable assurance of operational adherence to the preparation methods or other basis for the claim; and

(iii) A term or symbol that may in some contexts constitute a claim under this section may be used, provided that the use of the term or symbol does not characterize the level of a nutrient, and a statement that clearly explains the basis for the use of the term or symbol is prominently displayed and does not characterize the level of a nutrient. For example, a term such as "lite fare" followed by an asterisk referring to a note that makes clear that in this restaurant "lite fare" means smaller portion sizes than normal; or an item bearing a symbol referring to a note that makes clear that this item meets the criteria for the dietary guidance established by a recognized dietary authority would not be considered a nutrient content claim under §101.13.

(6) Nutrient content claims that were part of the common or usual names of foods that were subject to a standard of identity on November 8, 1990, are not subject to the requirements of paragraphs (b), (g), and (h) of this section or to definitions in subpart D of this part.

(7) Implied nutrient content claims may be used as part of a brand name, provided that the use of the claim has been authorized by the Food and Drug Administration. Petitions requesting approval of such a claim may be submitted under §101.69(o).

(8) The term "fluoridated, fluoride added" or "with added fluoride" may be used on the label or in labeling of bottled water that contains added fluoride.

[58 FR 2410, Jan. 6, 1993; 58 FR 17341, 17342, Apr. 2, 1993, as amended at 58 FR 44030, Aug. 18, 1993; 59 FR 393, Jan. 4, 1994; 59 FR 15051, Mar. 31, 1994; 60 FR 17205, Apr. 5, 1995; 61 FR 11731, Mar. 22, 1996; 61 FR 40332, Aug. 2, 1996; 61 FR 67452, Dec. 23, 1996; 62 *FR* 31339; June 9, 1997; 62 *FR* 49867, Sept. 23, 1997; 63 FR 14818, Mar. 27, 1998]

§101.14 Health claims: general requirements.

(a) *Definitions.* For purposes of this section, the following definitions apply:

(1) *Health claim* means any claim made on the label or in labeling of a food, including a dietary supplement,

21 CFR §101.14

that expressly or by implication, including "third party" references, written statements (e.g., a brand name including a term such as "heart"), symbols (e.g., a heart symbol), or vignettes, characterizes the relationship of any substance to a disease or health-related condition. Implied health claims include those statements, symbols, vignettes, or other forms of communication that suggest, within the context in which they are presented, that a relationship exists between the presence or level of a substance in the food and a disease or health-related condition.

(2) *Substance* means a specific food or component of food, regardless of whether the food is in conventional food form or a dietary supplement that includes vitamins, minerals, herbs, or other similar nutritional substances.

(3) *Nutritive value* means a value in sustaining human existence by such processes as promoting growth, replacing loss of essential nutrients, or providing energy.

(4) *Disqualifying nutrient levels* means the levels of total fat, saturated fat, cholesterol, or sodium in a food above which the food will be disqualified from making a health claim. These levels are 13.0 grams (g) of fat, 4.0 g of saturated fat, 60 milligrams (mg) of cholesterol, or 480 mg of sodium, per reference amount customarily consumed, per label serving size, and, only for foods with reference amounts customarily consumed of 30 g or less or 2 tablespoons or less, per 50 g. For dehydrated foods that must have water added to them prior to typical consumption, the per 50-g criterion refers to the as prepared form. Any one of the levels, on a per reference amount customarily consumed, a per label serving size or, when applicable, a per 50 g basis, will disqualify a food from making a health claim unless an exception is provided in subpart E of this part, except that:

(i) The levels for a meal product as defined in §101.13(l) are 26.0 g of fat, 8.0 g of saturated fat, 120 mg of cholesterol, or 960 mg of sodium per label serving size, and

(ii) The levels for a main dish product as defined in §101.13(m) are 19.5 g of fat, 6.0 g of saturated fat, 90 mg of cholesterol, or 720 mg of sodium per label serving size.

(5) *Disease or health-related condition* means damage to an organ, part, structure, or system of the body such that it does not function properly (e.g., cardiovascular disease), or a state of health leading to such dysfunctioning (e.g., hypertension); except that diseases resulting from essential nutrient deficiencies (e.g., scurvy, pellagra) are not included in this definition (claims pertaining to such diseases are thereby not subject to §101.14 or §101.70).

(b) *Eligibility.* For a substance to be eligible for a health claim:

(1) The substance must be associated with a disease or health- related condition for which the general U.S. population, or an identified U.S. population subgroup (e.g., the elderly) is at risk, or, alternatively, the petition submitted by the proponent of the claim otherwise explains the prevalence of the disease or health-related condition in the U.S. population and the relevance of the claim in the context of the total daily diet and satisfies the other requirements of this section.

(2) If the substance is to be consumed as a component of a conventional food at decreased dietary levels, the substance must be a nutrient listed in 21 U.S.C. 343(q)(1)(C) or (q)(1)(D), or one that the Food and Drug Administration (FDA) has required to be included in the label or labeling under 21 U.S.C. 343(q)(2)(A); or

(3) If the substance is to be consumed at other than decreased dietary levels:

(i) The substance must, regardless of whether the food is a conventional food or a dietary supplement, contribute taste, aroma, or nutritive value, or any other technical effect listed in §170.3(o) of this chapter, to the food and must retain that attribute when consumed at levels that are necessary to justify a claim; and

(ii) The substance must be a food or a food ingredient or a component of a food ingredient whose use at the levels necessary to justify a claim has been demonstrated by the proponent of the claim, to FDA's satisfaction, to be safe and lawful under the applicable food safety provisions of the Federal Food, Drug, and Cosmetic Act.

(c) *Validity requirement.* FDA will promulgate regulations authorizing a health claim only when it determines, based on the totality of publicly available scientific evidence (including evidence from well-designed studies conducted in a manner which is consistent with generally recognized scientific procedures and principles), that there is significant scientific agreement, among experts qualified by scientific training and experience to evaluate such claims, that the claim is supported by such evidence.

(d) *General health claim labeling requirements.*

(1) When FDA determines that a health claim meets the validity requirements of paragraph (c) of this section, FDA will propose a regulation in subpart E of this part to authorize the use of that claim. If the claim pertains to a substance not provided for in §101.9 or §101.36, FDA will propose amending that regulation to include declaration of the substance.

(2) When FDA has adopted a regulation in subpart E of this part providing for a health claim, firms may make claims based on the regulation in subpart E of this part, provided that:

(i) All label or labeling statements about the substance-disease relationship that is the subject of the claim are based on, and consistent with, the conclusions set forth in the regulations in subpart E of this part;

(ii) The claim is limited to describing the value that ingestion (or reduced ingestion) of the substance, as part of a total dietary pattern, may have on a particular disease or health- related condition;

(iii) The claim is complete, truthful, and not misleading. Where factors other than dietary intake of the substance affect the relationship between the substance and the disease or health- related condition, such factors may be required to be addressed in the claim by a specific regulation in subpart E of this part;

(iv) All information required to be included in the claim appears in one place without other intervening material, except that the principal display panel of the label or labeling may bear the reference statement, "See _____ for information about the relationship between _____ and_____," with the blanks filled in with the location of the labeling containing the health claim, the name of the substance, and the disease or health-related condition (e.g., "See attached pamphlet for information about calcium and osteoporosis"), with the entire claim appearing elsewhere on the other labeling, Provided that, where any graphic material (e.g., a heart symbol) constituting an explicit or implied health claim appears on the label or labeling, the reference statement or the complete claim shall appear in immediate proximity to such graphic material;

(v) The claim enables the public to comprehend the information provided and to understand the relative significance of such information in the context of a total daily diet; and

(vi) If the claim is about the effects of consuming the substance at decreased dietary levels, the level of the substance in the food is sufficiently low to justify the claim. To meet this requirement, if a definition for use of the term "low" has been established for that substance under this part, the substance must be present at a level that meets the requirements for use of that term, unless a specific alternative level has been established for the substance in subpart E of this part. If no definition for "low" has been established, the level of the substance must meet the level established in the regulation authorizing the claim; or

(vii) If the claim is about the effects of consuming the substance at other than decreased dietary levels, the level of the substance is sufficiently high and in an appropriate form to justify the claim. To meet this requirement, if a definition for use of the term "high" for that substance has been established under this part, the substance must be present at a level that meets the requirements for use of that term, unless a specific alternative level has been established for the substance in subpart E of this part. If no definition for "high" has been established (e.g., where the claim pertains to a food either as a whole food or as an ingredient in another food), the claim must specify the daily dietary intake necessary to achieve the claimed effect, as established in the regulation authorizing the claim; *Provided* That:

(A) Where the food that bears the claim meets the requirements of paragraphs (d)(2)(vi) or (d)(2)(vii) of this section based on its reference amount customarily consumed, and the labeled serving size differs from that amount, the claim shall be followed by a statement explaining that the claim is based on the reference amount rather than the labeled serving size (e.g., "Diets low in sodium may reduce the risk of high blood pressure, a disease associated with many factors. A serving of _____ ounces of this product conforms to such a diet.").

(B) Where the food that bears the claim is sold in a restaurant or in other establishments in which food that is ready for immediate human consumption is sold, the food can meet the requirements of paragraphs (d)(2)(vi) or (d)(2)(vii) of this section if the firm that sells the food has a reasonable basis on which to believe that the food that bears the claim meets the requirements of paragraphs (d)(2)(vi) or (d)(2)(vii) of this section and provides that basis upon request.

(3) Nutrition labeling shall be provided in the label or labeling of any food for which a health claim is made in accordance with §101.9; for restaurant foods, in accordance with §101.10; or for dietary supplements, in accordance with §101.36.

(e) *Prohibited health claims.* No expressed or implied health claim may be made on the label or in labeling for a food, regardless of whether the food is in conventional food form or dietary supplement form, unless:

(1) The claim is specifically provided for in subpart E of this part; and

(2) The claim conforms to all general provisions of this section as well as to all specific provisions in the appropriate section of subpart E of this part;

(3) None of the disqualifying levels identified in paragraph (a)(5) of this section is exceeded in the food, unless specific alternative levels have been established for the substance in subpart E of this part; or unless FDA has

permitted a claim despite the fact that a disqualifying level of a nutrient is present in the food based on a finding that such a claim will assist consumers in maintaining healthy dietary practices, and, in accordance with the regulation in subpart E of this part that makes such a finding, the label bears a referral statement that complies with §101.13(h), highlighting the nutrient that exceeds the disqualifying level;

(4) Except as provided in paragraph (e)(3) of this section, no substance is present at an inappropriate level as determined in the specific provision authorizing the claim in subpart E of this part;

(5) The label does not represent or purport that the food is for infants and toddlers less than 2 years of age except if the claim is specifically provided for in subpart E of this part; and

(6) Except for dietary supplements or where provided for in other regulations in part 101, subpart E, the food contains 10 percent or more of the Reference Daily Intake or the Daily Reference Value for vitamin A, vitamin C, iron, calcium, protein, or fiber per reference amount customarily consumed prior to any nutrient addition.

(f) The requirements of this section do not apply to:

(1) Infant formulas subject to section 412(h) of the Federal Food, Drug, and Cosmetic Act, and

(2) Medical foods defined by section 5(b) of the Orphan Drug Act.

(g) *Applicability.* The requirements of this section apply to foods intended for human consumption that are offered for sale, regardless of whether the foods are in conventional food form or dietary supplement form.

[58 FR 2533, Jan. 6, 1993; 58 FR 17097, Apr. 1, 1993, as amended at 58 FR 44038, Aug. 18, 1993; 59 FR 425, Jan. 4, 1994; 59 FR 15050, Mar. 31, 1994; 61 FR 40332, Aug. 2, 1996; 62 *FR* 49867, Sept. 23, 1997]

§101.15 Food; prominence of required statements.

(a) A word, statement, or other information required by or under authority of the act to appear on the label may lack that prominence and conspicuousness required by section 403(f) of the act by reason (among other reasons) of:

(1) The failure of such word, statement, or information to appear on the part or panel of the label which is presented or displayed under customary conditions of purchase;

(2) The failure of such word, statement, or information to appear on two or more parts or panels of the label, each of which has sufficient space therefor, and each of which is so designed as to render it likely to be, under customary conditions of purchase, the part or panel displayed;

(3) The failure of the label to extend over the area of the container or package available for such extension, so as to provide sufficient label space for the prominent placing of such word, statement, or information;

(4) Insufficiency of label space (for the prominent placing of such word, statement, or information) resulting from the use of label space for any word, statement, design, or device which is not required by or under authority of the act to appear on the label;

(5) Insufficiency of label space (for the prominent placing of such word, statement, or information) resulting from the use of label space to give materially greater conspicuousness to any other word, statement, or information, or to any design or device; or

(6) Smallness or style of type in which such word, statement, or information appears, insufficient background contrast, obscuring designs or vignettes, or crowding with other written, printed, or graphic matter.

(b) No exemption depending on insufficiency of label space, as prescribed in regulations promulgated under section 403 (e) or (i) of the act, shall apply if such insufficiency is caused by:

(1) The use of label space for any word, statement, design, or device which is not required by or under authority of the act to appear on the label;

(2) The use of label space to give greater conspicuousness to any word, statement, or other information than is required by section 403(f) of the act; or

(3) The use of label space for any representation in a foreign language.

(c)

(1) All words, statements, and other information required by or under authority of the act to appear on the label or labeling shall appear thereon in the English language: *Provided, however,* That in the case of articles distributed solely in the Commonwealth of Puerto Rico or in a Territory where the predominant language is one other than English, the predominant language may be substituted for English.

(2) If the label contains any representation in a foreign language, all words, statements, and other information required by or under authority of the act to appear on the label shall appear thereon in the foreign language:

Provided, however, That individual serving-size packages of foods containing no more than 1 1/2 avoirdupois ounces or no more than 1 1/2 fluid ounces served with meals in restaurants, institutions, and passenger carriers and not intended for sale at retail are exempt from the requirements of this paragraph (c)(2), if the only representation in the foreign language(s) is the name of the food.

(3) If any article of labeling (other than a label) contains any representation in a foreign language, all words, statements, and other information required by or under authority of the act to appear on the label or labeling shall appear on such article of labeling.

§101.17 Food labeling warning and notice statements.

(a) *Self-pressurized containers.*

(1) The label of a food packaged in a self-pressurized container and intended to be expelled from the package under pressure shall bear the following warning:

WARNING-Avoid spraying in eyes. Contents under pressure. Do not puncture or incinerate. Do not store at temperature above 120° F. Keep out of reach of children.

(2) In the case of products intended for use by children, the phrase "except under adult supervision" may be added at the end of the last sentence in the warning required by paragraph (a)(1) of this section.

(3) In the case of products packaged in glass containers, the word "break" may be substituted for the word "puncture" in the warning required by paragraph (a)(1) of this section.

(4) The words "Avoid spraying in eyes" may be deleted from the warning required by paragraph (a)(1) of this section in the case of a product not expelled as a spray.

(b) *Self-pressurized containers with halocarbon or hydrocarbon propellants.*

(1) In addition to the warning required by paragraph (a) of this section, the label of a food packaged in a self-pressurized container in which the propellant consists in whole or in part of a halocarbon or a hydrocarbon shall bear the following warning:

WARNING-Use only as directed. Intentional misuse by deliberately concentrating and inhaling the contents can be harmful or fatal.

(2) The warning required by paragraph (b)(1) of this section is not required for the following products:

(i) Products expelled in the form of a foam or cream, which contain less than 10 percent propellant in the container.

(ii) Products in a container with a physical barrier that prevents escape of the propellant at the time of use.

(iii) Products of a net quantity of contents of less than 2 ounces that are designed to release a measured amount of product with each valve actuation.

(iv) Products of a net quantity of contents of less than one- half ounce.

(c) *Food containing or manufactured with a chlorofluorocarbon or other ozone-depleting substance.* Labeling requirements for foods that contain or are manufactured with a chlorofluorocarbon or other ozone-depleting substance designated by the Environmental Protection Agency (EPA) are set forth in 40 CFR part 82.

(d) *Protein products.*

(1) The label and labeling of any food product in liquid, powdered, tablet, capsule, or similar forms that derives more than 50 percent of its total caloric value from either whole protein, protein hydrolysates, amino acid mixtures, or a combination of these, and that is represented for use in reducing weight shall bear the following warning:

WARNING: Very low calorie protein diets (below 400 Calories per day) may cause serious illness or death. Do Not Use for Weight Reduction in Such Diets Without Medical Supervision. Not for use by infants, children, or pregnant or nursing women.

(2) Products described in paragraph (d)(1) of this section are exempt from the labeling requirements of that paragraph if the protein products are represented as part of a nutritionally balanced diet plan providing 400 or more Calories (kilocalories) per day and the label or labeling of the product specifies the diet plan in detail or provides a brief description of that diet plan and adequate information describing where the detailed diet plan may be obtained and the label and labeling bear the following statement:

NOTICE: For weight reduction, use only as directed in the accompanying diet plan (the name and specific location in labeling of the diet plan may be included in this statement in place of "accompanying diet plan").

21 CFR §101.17

Do not use in diets supplying less than 400 Calories per day without medical supervision.

(3) The label and labeling of food products represented or intended for dietary (food) supplementation that derive more than 50 percent of their total caloric value from either whole protein, protein hydrolysates, amino acid mixtures, or a combination of these, that are represented specifically for purposes other than weight reduction; and that are not covered by the requirements of paragraph (d) (1) and (2) of this section; shall bear the following statement:

NOTICE: Use this product as a food supplement only. Do not use for weight reduction.

(4) The provisions of this paragraph are separate from and in addition to any labeling requirements promulgated by the Federal Trade Commission for protein supplements.

(5) Protein products shipped in bulk form for use solely in the manufacture of other foods and not for distribution to consumers in such container are exempt from the labeling requirements of this paragraph.

(6) The warning and notice statements required by paragraphs (d) (1), (2), and (3) of this section shall appear prominently and conspicuously on the principal display panel of the package label and any other labeling.

(e) *Dietary supplements containing iron or iron salts.*

(1) The labeling of any dietary supplement in solid oral dosage form (e.g., tablets or capsules) that contains iron or iron salts for use as an iron source shall bear the following statement:

WARNING: Accidental overdose of iron-containing products is a leading cause of fatal poisoning in children under 6. Keep this product out of reach of children. In case of accidental overdose, call a doctor or poison control center immediately.

(2)(i) The warning statement required by paragraph (e)(1) of this section shall appear prominently and conspicuously on the information panel of the immediate container label.

(ii) If a product is packaged in unit-dose packaging, and if the immediate container bears labeling but not a label, the warning statement required by paragraph (e)(1) of this section shall appear prominently and conspicuously on the immediate container labeling in a way that maximizes the likelihood that the warning is intact until all of the dosage units to which it applies are used.

(3) Where the immediate container is not the retail package, the warning statement required by paragraph (e)(1) of this section shall also appear prominently and conspicuously on the information panel of the retail package label.

(4) The warning statement shall appear on any labeling that contains warnings.

(5) The warning statement required by paragraph (e)(1) of this section shall be set off in a box by use of hairlines.

(f) *Foods containing psyllium husk.*

(1) Foods containing dry or incompletely hydrated psyllium husk, also known as psyllium seed husk, and bearing a health claim on the association between soluble fiber from psyllium husk and reduced risk of coronary heart disease, shall bear a label statement informing consumers that the appropriate use of such foods requires consumption with adequate amounts of fluids, alerting them of potential consequences of failing to follow usage recommendations, and informing persons with swallowing difficulties to avoid consumption of the product (e.g., "NOTICE: This food should be eaten with at least a full glass of liquid. Eating this product without enough liquid may cause choking. Do not eat this product if you have difficulty in swallowing."). However, a product in conventional food form may be exempt from this requirement if a viscous adhesive mass is not formed when the food is exposed to fluids.

(2) The statement shall appear prominently and conspicuously on the information panel or principal display panel of the package label and any other labeling to render it likely to be read and understood by the ordinary individual under customary conditions of purchase and use. The statement shall be preceded by the word "NOTICE" in capital letters.

[42 FR 14308, Mar. 15, 1977, as amended at 42 FR 22033, Apr. 29, 1977; 49 FR 13690, Apr. 6, 1984; 49 FR 28548, July 13, 1984; 61 FR 20100, May 3, 1996; 62 FR 2249, Jan. 15, 1997; 63 *FR* 8119, Feb. 18, 1998]

§101.18 Misbranding of food.

(a) Among representations in the labeling of a food which render such food misbranded is a false or misleading representation with respect to another food or a drug, device, or cosmetic.

(b) The labeling of a food which contains two or more ingredients may be misleading by reason (among other reasons) of the designation of such food in such labeling by a name which includes or suggests the name of one or more but not all such ingredients, even though the names of all such ingredients are stated elsewhere in the labeling.

(c) Among representations in the labeling of a food which render such food misbranded is any representation that expresses or implies a geographical origin of the food or any ingredient of the food except when such representation is either:

(1) A truthful representation of geographical origin.

(2) A trademark or trade name provided that as applied to the article in question its use is not deceptively misdescriptive. A trademark or trade name composed in whole or in part of geographical words shall not be considered deceptively misdescriptive if it:

(i) Has been so long and exclusively used by a manufacturer or distributor that it is generally understood by the consumer to mean the product of a particular manufacturer or distributor; or

(ii) Is so arbitrary or fanciful that it is not generally understood by the consumer to suggest geographic origin.

(3) A part of the name required by applicable Federal law or regulation.

(4) A name whose market significance is generally understood by the consumer to connote a particular class, kind, type, or style of food rather than to indicate geographical origin.

Subpart B-Specific Food Labeling Requirements

§101.22 Foods; labeling of spices, flavorings, colorings and chemical preservatives.

(a)

(1) The term "artificial flavor" or "artificial flavoring" means any substance, the function of which is to impart flavor, which is not derived from a spice, fruit or fruit juice, vegetable or vegetable juice, edible yeast, herb, bark, bud, root, leaf or similar plant material, meat, fish, poultry, eggs, dairy products, or fermentation products thereof. Artificial flavor includes the substances listed in §§172.515(b) and 182.60 of this chapter except where these are derived from natural sources.

(2) The term "spice" means any aromatic vegetable substance in the whole, broken, or ground form, except for those substances which have been traditionally regarded as foods, such as onions, garlic and celery; whose significant function in food is seasoning rather than nutritional; that is true to name; and from which no portion of any volatile oil or other flavoring principle has been removed. Spices include the spices listed in §182.10 and Part 184 of this chapter, such as the following:

Allspice, Anise, Basil, Bay leaves, Caraway seed, Cardamon, Celery seed, Chervil, Cinnamon, Cloves, Coriander, Cumin seed, Dill seed, Fennel seed, Fenugreek, Ginger, Horseradish, Mace, Marjoram, Mustard flour, Nutmeg, Oregano, Paprika, Parsley, Pepper, black; Pepper, white; Pepper, red; Rosemary, Saffron, Sage, Savory, Star aniseed, Tarragon, Thyme, Turmeric.

Paprika, turmeric, and saffron or other spices which are also colors, shall be declared as "spice and coloring" unless declared by their common or usual name.

(3) The term "natural flavor" or "natural flavoring" means the essential oil, oleoresin, essence or extractive, protein hydrolysate, distillate, or any product of roasting, heating or enzymolysis, which contains the flavoring constituents derived from a spice, fruit or fruit juice, vegetable or vegetable juice, edible yeast, herb, bark, bud, root, leaf or similar plant material, meat, seafood, poultry, eggs, dairy products, or fermentation products thereof, whose significant function in food is flavoring rather than nutritional. Natural flavors include the natural essence or extractives obtained from plants listed in §§182.10, 182.20, 182.40, and 182.50 and part 184 of this chapter, and the substances listed in §172.510 of this chapter.

(4) The term "artificial color" or "artificial coloring" means any "color additive" as defined in §70.3(f) of this chapter.

(5) The term "chemical preservative" means any chemical that, when added to food, tends to prevent or retard deterioration thereof, but does not include common salt, sugars, vinegars, spices, or oils extracted from spices, substances added to food by direct exposure thereof to wood smoke, or chemicals applied for their insecticidal or herbicidal properties.

21 CFR §101.22

(b) A food which is subject to the requirements of section 403(k) of the act shall bear labeling, even though such food is not in package form.

(c) A statement of artificial flavoring, artificial coloring, or chemical preservative shall be placed on the food or on its container or wrapper, or on any two or all three of these, as may be necessary to render such statement likely to be read by the ordinary person under customary conditions of purchase and use of such food. The specific artificial color used in a food shall be identified on the labeling when so required by regulation in part 74 of this chapter to assure safe conditions of use for the color additive.

(d) A food shall be exempt from compliance with the requirements of section 403(k) of the act if it is not in package form and the units thereof are so small that a statement of artificial flavoring, artificial coloring, or chemical preservative, as the case may be, cannot be placed on such units with such conspicuousness as to render it likely to be read by the ordinary individual under customary conditions of purchase and use.

(e) A food shall be exempt while held for sale from the requirements of section 403(k) of the act (requiring label statement of any artificial flavoring, artificial coloring, or chemical preservatives) if said food, having been received in bulk containers at a retail establishment, is displayed to the purchaser with either

   (1) the labeling of the bulk container plainly in view or

   (2) a counter card, sign, or other appropriate device bearing prominently and conspicuously the information required to be stated on the label pursuant to section 403(k).

(f) A fruit or vegetable shall be exempt from compliance with the requirements of section 403(k) of the act with respect to a chemical preservative applied to the fruit or vegetable as a pesticide chemical prior to harvest.

(g) A flavor shall be labeled in the following way when shipped to a food manufacturer or processor (but not a consumer) for use in the manufacture of a fabricated food, unless it is a flavor for which a standard of identity has been promulgated, in which case it shall be labeled as provided in the standard:

   (1) If the flavor consists of one ingredient, it shall be declared by its common or usual name.

   (2) If the flavor consists of two or more ingredients, the label either may declare each ingredient by its common or usual name or may state "All flavor ingredients contained in this product are approved for use in a regulation of the Food and Drug Administration." Any flavor ingredient not contained in one of these regulations, and any nonflavor ingredient, shall be separately listed on the label.

   (3) In cases where the flavor contains a solely natural flavor(s), the flavor shall be so labeled, e.g., "strawberry flavor", "banana flavor", or "natural strawberry flavor". In cases where the flavor contains both a natural flavor and an artificial flavor, the flavor shall be so labeled, e.g., "natural and artificial strawberry flavor". In cases where the flavor contains a solely artificial flavor(s), the flavor shall be so labeled, e.g., "artificial strawberry flavor".

(h) The label of a food to which flavor is added shall declare the flavor in the statement of ingredients in the following way:

   (1) Spice, natural flavor, and artificial flavor may be declared as "spice", "natural flavor", or "artificial flavor", or any combination thereof, as the case may be.

   (2) An incidental additive in a food, originating in a spice or flavor used in the manufacture of the food, need not be declared in the statement of ingredients if it meets the requirements of §101.100(a)(3).

   (3) Substances obtained by cutting, grinding, drying, pulping, or similar processing of tissues derived from fruit, vegetable, meat, fish, or poultry, e.g., powdered or granulated onions, garlic powder, and celery powder, are commonly understood by consumers to be food rather than flavor and shall be declared by their common or usual name.

   (4) Any salt (sodium chloride) used as an ingredient in food shall be declared by its common or usual name "salt."

   (5) Any monosodium glutamate used as an ingredient in food shall be declared by its common or usual name "monosodium glutamate."

   (6) Any pyroligneous acid or other artificial smoke flavors used as an ingredient in a food may be declared as artificial flavor or artificial smoke flavor. No representation may be made, either directly or implied, that a food flavored with pyroligneous acid or other artificial smoke flavor has been smoked or has a true smoked flavor, or that a seasoning sauce or similar product containing pyroligneous acid or other artificial smoke flavor and used to season or flavor other foods will result in a smoked product or one having a true smoked flavor.

   (7) Because protein hydrolysates function in foods as both flavorings and flavor enhancers, no protein hydrolysate used in food for its effects on flavor may be declared simply as "flavor," "natural flavor," or

21 CFR §101.22

"flavoring." The ingredient shall be declared by its specific common or usual name as provided in §102.22 of this chapter.

(i) If the label, labeling, or advertising of a food makes any direct or indirect representations with respect to the primary recognizable flavor(s), by word, vignette, e.g., depiction of a fruit, or other means, or if for any other reason the manufacturer or distributor of a food wishes to designate the type of flavor in the food other than through the statement of ingredients, such flavor shall be considered the characterizing flavor and shall be declared in the following way:

(1) If the food contains no artificial flavor which simulates, resembles or reinforces the characterizing flavor, the name of the food on the principal display panel or panels of the label shall be accompanied by the common or usual name of the characterizing flavor, e.g., "vanilla", in letters not less than one-half the height of the letters used in the name of the food, except that:

(i) If the food is one that is commonly expected to contain a characterizing food ingredient, e.g., strawberries in "strawberry shortcake", and the food contains natural flavor derived from such ingredient and an amount of characterizing ingredient insufficient to independently characterize the food, or the food contains no such ingredient, the name of the characterizing flavor may be immediately preceded by the word "natural" and shall be immediately followed by the word "flavored" in letters not less than one-half the height of the letters in the name of the characterizing flavor, e.g., "natural strawberry flavored shortcake," or "strawberry flavored shortcake".

(ii) If none of the natural flavor used in the food is derived from the product whose flavor is simulated, the food in which the flavor is used shall be labeled either with the flavor of the product from which the flavor is derived or as "artificially flavored."

(iii) If the food contains both a characterizing flavor from the product whose flavor is simulated and other natural flavor which simulates, resembles or reinforces the characterizing flavor, the food shall be labeled in accordance with the introductory text and paragraph (i)(1)(i) of this section and the name of the food shall be immediately followed by the words "with other natural flavor" in letters not less than one-half the height of the letters used in the name of the characterizing flavor.

(2) If the food contains any artificial flavor which simulates, resembles or reinforces the characterizing flavor, the name of the food on the principal display panel or panels of the label shall be accompanied by the common or usual name(s) of the characterizing flavor, in letters not less than one-half the height of the letters used in the name of the food and the name of the characterizing flavor shall be accompanied by the word(s) "artificial" or "artificially flavored", in letters not less than one-half the height of the letters in the name of the characterizing flavor, e.g., "artificial vanilla", "artificially flavored strawberry", or "grape artificially flavored".

(3) Wherever the name of the characterizing flavor appears on the label (other than in the statement of ingredients) so conspicuously as to be easily seen under customary conditions of purchase, the words prescribed by this paragraph shall immediately and conspicuously precede or follow such name, without any intervening written, printed, or graphic matter, except:

(i) Where the characterizing flavor and a trademark or brand are presented together, other written, printed, or graphic matter that is a part of or is associated with the trademark or brand may intervene if the required words are in such relationship with the trademark or brand as to be clearly related to the characterizing flavor; and

(ii) If the finished product contains more than one flavor subject to the requirements of this paragraph, the statements required by this paragraph need appear only once in each statement of characterizing flavors present in such food, e.g., "artificially flavored vanilla and strawberry".

(iii) If the finished product contains three or more distinguishable characterizing flavors, or a blend of flavors with no primary recognizable flavor, the flavor may be declared by an appropriately descriptive generic term in lieu of naming each flavor, e.g., "artificially flavored fruit punch".

(4) A flavor supplier shall certify, in writing, that any flavor he supplies which is designated as containing no artificial flavor does not, to the best of his knowledge and belief, contain any artificial flavor, and that he has added no artificial flavor to it. The requirement for such certification may be satisfied by a guarantee under section 303(c)(2) of the act which contains such a specific statement. A flavor user shall be required to make such a written certification only where he adds to or combines another flavor with a flavor which has been certified by a flavor supplier as containing no artificial flavor, but otherwise such user may rely upon the supplier's certification and need make no separate certification. All such certifications shall be retained by the certifying party throughout the period in which the flavor is supplied and for a minimum of three years

21 CFR §101.22

thereafter, and shall be subject to the following conditions:

(i) The certifying party shall make such certifications available upon request at all reasonable hours to any duly authorized office or employee of the Food and Drug Administration or any other employee acting on behalf of the Secretary of Health and Human Services. Such certifications are regarded by the Food and Drug Administration as reports to the government and as guarantees or other undertakings within the meaning of section 301(h) of the act and subject the certifying party to the penalties for making any false report to the government under 18 U.S.C. 1001 and any false guarantee or undertaking under section 303(a) of the act. The defenses provided under section 303(c)(2) of the act shall be applicable to the certifications provided for in this section.

(ii) Wherever possible, the Food and Drug Administration shall verify the accuracy of a reasonable number of certifications made pursuant to this section, constituting a representative sample of such certifications, and shall not request all such certifications.

(iii) Where no person authorized to provide such information is reasonably available at the time of inspection, the certifying party shall arrange to have such person and the relevant materials and records ready for verification as soon as practicable: *Provided*, That, whenever the Food and Drug Administration has reason to believe that the supplier or user may utilize this period to alter inventories or records, such additional time shall not be permitted. Where such additional time is provided, the Food and Drug Administration may require the certifying party to certify that relevant inventories have not been materially disturbed and relevant records have not been altered or concealed during such period.

(iv) The certifying party shall provide, to an officer or representative duly designated by the Secretary, such qualitative statement of the composition of the flavor or product covered by the certification as may be reasonably expected to enable the Secretary's representatives to determine which relevant raw and finished materials and flavor ingredient records are reasonably necessary to verify the certifications. The examination conducted by the Secretary's representative shall be limited to inspection and review of inventories and ingredient records for those certifications which are to be verified.

(v) Review of flavor ingredient records shall be limited to the qualitative formula and shall not include the quantitative formula. The person verifying the certifications may make only such notes as are necessary to enable him to verify such certification. Only such notes or such flavor ingredient records as are necessary to verify such certification or to show a potential or actual violation may be removed or transmitted from the certifying party's place of business: *Provided*, That, where such removal or transmittal is necessary for such purposes the relevant records and notes shall be retained as separate documents in Food and Drug Administration files, shall not be copied in other reports, and shall not be disclosed publicly other than in a judicial proceeding brought pursuant to the act or 18 U.S.C. 1001.

(j) A food to which a chemical preservative(s) is added shall, except when exempt pursuant to §101.100 bear a label declaration stating both the common or usual name of the ingredient(s) and a separate description of its function, e.g., "preservative", "to retard spoilage", "a mold inhibitor", "to help protect flavor" or "to promote color retention".

(k) The label of a food to which any coloring has been added shall declare the coloring in the statement of ingredients in the manner specified in paragraphs (k)(1) and (k)(2) of this section, except that colorings added to butter, cheese, and ice cream, if declared, may be declared in the manner specified in paragraph (k)(3) of this section, and colorings added to foods subject to §§105.62 and 105.65 of this chapter shall be declared in accordance with the requirements of those sections.

(1) A color additive or the lake of a color additive subject to certification under 721(c) of the act shall be declared by the name of the color additive listed in the applicable regulation in part 74 or part 82 of this chapter, except that it is not necessary to include the "FD&C" prefix or the term "No." in the declaration, but the term "Lake" shall be included in the declaration of the lake of the certified color additive (e.g., Blue 1 Lake). Manufacturers may parenthetically declare an appropriate alternative name of the certified color additive following its common or usual name as specified in part 74 or part 82 of this chapter.

(2) Color additives not subject to certification may be declared as "Artificial Color," "Artificial Color Added," or "Color Added" (or by an equally informative term that makes clear that a color additive has been used in the food). Alternatively, such color additives may be declared as "Colored with ____ " or " ____ color", the blank to be filled with the name of the color additive listed in the applicable regulation in part 73 of this chapter.

(3) When a coloring has been added to butter, cheese, or ice cream, it need not be declared in the ingredient list unless such declaration is required by a regulation in part 73 or part 74 of this chapter to ensure safe conditions

21 CFR §101.30

of use for the color additive. Voluntary declaration of all colorings added to butter, cheese, and ice cream, however, is recommended.

[42 FR 14308, Mar. 15, 1977, as amended at 44 FR 3963, Jan. 19, 1979; 44 FR 37220, June 26, 1979; 54 FR 24891, June 12, 1989; 58 FR 2875, Jan. 6, 1993; 63 FR 14818, Mar. 27, 1998]

§101.29 Labeling of kosher and kosher-style foods.
[Removed August 12, 1997; 62 *FR* 43074]

§101.30 Percentage juice declaration for foods purporting to be beverages that contain fruit or vegetable juice.

(a) This section applies to any food that purports to be a beverage that contains any fruit or vegetable juice (i.e., the product's advertising, label, or labeling bears the name of, or variation on the name of, or makes any other direct or indirect representation with respect to, any fruit or vegetable juice), or the label or labeling bears any vignette (i.e., depiction of a fruit or vegetable) or other pictorial representation of any fruit or vegetable, or the product contains color and flavor that gives the beverage the appearance and taste of containing a fruit or vegetable juice. The beverage may be carbonated or noncarbonated, concentrated, full-strength, diluted, or contain no juice. For example, a soft drink (soda) that does not represent or suggest by its physical characteristics, name, labeling, ingredient statement, or advertising that it contains fruit or vegetable juice does not purport to contain juice and therefore does not require a percent juice declaration.

(b)

(1) If the beverage contains fruit or vegetable juice, the percentage shall be declared by the words "Contains _____ percent (or %) _____ juice" or " _____ percent (or %) juice," or a similar phrase, with the first blank filled in with the percentage expressed as a whole number not greater than the actual percentage of the juice and the second blank (if used) filled in with the name of the particular fruit or vegetable (e.g., "Contains 50 percent apple juice" or "50 percent juice").

(2) If the beverage contains less than 1 percent juice, the total percentage juice shall be declared as "less than 1 percent juice" or "less than 1 percent _____ juice" with the blank filled in with the name of the particular fruit or vegetable.

(3) If the beverage contains 100 percent juice and also contains non-juice ingredients that do not result in a diminution of the juice soluble solids or, in the case of expressed juice, in a change in the volume, when the 100 percent juice declaration appears on a panel of the label that does not also bear the ingredient statement, it must be accompanied by the phrase "with added _____," the blank filled in with a term such as "ingredient(s)," "preservative," or "sweetener," as appropriate (e.g., "100% juice with added sweetener"), except that when the presence of the non-juice ingredient(s) is declared as a part of the statement of identity of the product, this phrase need not accompany the 100 percent juice declaration.

(c) If a beverage contains minor amounts of juice for flavoring and is labeled with a flavor description using terms such as "flavor", "flavored", or "flavoring" with a fruit or vegetable name and does not bear:

(1) The term "juice" on the label other than in the ingredient statement; or

(2) An explicit vignette depicting the fruit or vegetable from which the flavor derives, such as juice exuding from a fruit or vegetable; or

(3) Specific physical resemblance to a juice or distinctive juice characteristic such as pulp then total percentage juice declaration is not required.

(d) If the beverage does not meet the criteria for exemption from total juice percentage declaration as described in paragraph (c) of this section and contains no fruit or vegetable juice, but the labeling or color and flavor of the beverage represents, suggests, or implies that fruit or vegetable juice may be present (e.g., the product advertising or labeling bears the name, a variation of the name, or a pictorial representation of any fruit or vegetable, or the product contains color and flavor that give the beverage the appearance and taste of containing a fruit or vegetable juice), then the label shall declare "contains zero (0) percent (or %) juice". Alternatively, the label may declare "Containing (or contains) no _____ juice", or "no _____ juice", or "does not contain _____ juice", the blank to be filled in with the name of the fruits or vegetables represented, suggested, or implied, but if there is a

21 CFR §101.30

general suggestion that the product contains fruit or vegetable juice, such as the presence of fruit pulp, the blank shall be filled in with the word "fruit" or "vegetable" as applicable (e.g., "contains no fruit juice", or "does not contain fruit juice").

(e) If the beverage is sold in a package with an information panel as defined in §101.2, the declaration of amount of juice shall be prominently placed on the information panel in lines generally parallel to other required information, appearing:

(1) Near the top of the information panel, with no other printed label information appearing above the statement except the brand name, product name, logo, or universal product code; and

(2) In easily legible boldface print or type in distinct contrast to other printed or graphic matter, in a height not less than the largest type found on the information panel except that used for the brand name, product name, logo, universal product code, or the title phrase "Nutrition Facts" appearing in the nutrition information as required by §101.9.

(f) The percentage juice declaration may also be placed on the principal display panel, provided that the declaration is consistent with that presented on the information panel.

(g) If the beverage is sold in a package that does not bear an information panel as defined in §101.2, the percentage juice declaration shall be placed on the principal display panel, in type size not less than that required for the declaration of net quantity of contents statement in §101.105(i), and be placed near the name of the food.

(h)

(1) In enforcing these regulations, the Food and Drug Administration will calculate the labeled percentage of juice from concentrate found in a juice or juice beverage using the minimum Brix levels listed below where single-strength (100 percent) juice has at least the specified minimum Brix listed below:

Juice	100 percent juice[1]
Acerola	6.0
Apple	11.5
Apricot	11.7
Banana	22.0
Blackberry	10.0
Blueberry	10.0
Boysenberry	10.0
Cantaloupe Melon	9.6
Carambola	7.8
Carrot	8.0
Casaba Melon	7.5
Cashew (Caju)	12.0
Celery	3.1
Cherry, dark, sweet	20.0
Cherry, red, sour	14.0
Crabapple	15.4
Cranberry	7.5

21 CFR §101.30

Fruit	Value
Currant (Black)	11.0
Currant (Red)	10.5
Date	18.5
Dewberry	10.0
Elderberry	11.0
Fig	18.2
Gooseberry	8.3
Grape	16.0
Grapefruit	10.0[3]
Guanabana (soursop)	16.0
Guava	7.7
Honeydew melon	9.6
Kiwi	15.4
Lemon	4.5[2]
Lime	4.5[2]
Loganberry	10.5
Mango	13.0
Nectarine	11.8
Orange	11.8[3]
Papaya	11.5
Passion Fruit	14.0
Peach	10.5
Pear	12.0
Pineapple	12.8
Plum	14.3
Pomegranate	16.0
Prune	18.5
Quince	13.3
Raspberry (Black)	11.1
Raspberry (Red)	9.2
Rhubarb	5.7
Strawberry	8.0

21 CFR §101.30

Tangerine	11.8[3]
Tomato	5.0
Watermelon	7.8
Youngberry	10.0

{1} Indicates Brix value unless other value specified.
{2} Indicates anhydrous citric acid percent by weight.
{3} Brix values determined by refractometer for citrus juices may be corrected for citric acid.

(2) If there is no Brix level specified in paragraph (h)(1) of this section, the labeled percentage of that juice from concentrate in a juice or juice beverage will be calculated on the basis of the soluble solids content of the single-strength (unconcentrated) juice used to produce such concentrated juice.

(i) Juices directly expressed from a fruit or vegetable (i.e., not concentrated and reconstituted) shall be considered to be 100 percent juice and shall be declared as "100 percent juice."

(j) Calculations of the percentage of juice in a juice blend or a diluted juice product made directly from expressed juice (i.e., not from concentrate) shall be based on the percentage of the expressed juice in the product computed on a volume/volume basis.

(k) If the product is a beverage that contains a juice whose color, taste, or other organoleptic properties have been modified to the extent that the original juice is no longer recognizable at the time processing is complete, or if its nutrient profile has been diminished to a level below the normal nutrient range for the juice, then that juice to which such a major modification has been made shall not be included in the total percentage juice declaration.

(l) A beverage required to bear a percentage juice declaration on its label, that contains less than 100 percent juice, shall not bear any other percentage declaration that describes the juice content of the beverage in its label or in its labeling (e.g., "100 percent natural" or "100 percent pure"). However, the label or labeling may bear percentage statements clearly unrelated to juice content (e.g., "provides 100 percent of U.S. RDA of vitamin C").

(m) Products purporting to be beverages that contain fruit or vegetable juices are exempted from the provisions of this section until May 8, 1994. All products that are labeled on or after that date shall comply with this section.

[58 FR 2925, Jan. 6, 1993, as amended at 58 FR 44063, Aug. 18, 1993; 58 FR 49192, Sept. 22, 1993]

Subpart C-Specific Nutrition Labeling Requirements and Guidelines
Source: 55 FR 60890, Nov. 27, 1991, unless otherwise noted.

§101.36 Nutrition labeling of dietary supplements.

(a) The label of a dietary supplement that is offered for sale shall bear nutrition labeling in accordance with this regulation unless an exemption is provided for the product in paragraph (h) of this section.

(b) The declaration of nutrition information on the label and in labeling shall contain the following information, using the subheadings and the format specified in paragraph (e) of this section.

(1) *Serving size*–(i) The subheading "Serving Size" shall be placed under the heading "Supplement Facts" and aligned on the left side of the nutrition label. The serving size shall be determined in accordance with §101.9(b) and 101.12(b), Table 2. Serving size for dietary supplements shall be expressed using a term that is appropriate for the form of the supplement, such as "tablets," "capsules," "packets," or "teaspoonfuls."

(ii) The subheading "Servings Per Container" shall be placed under the subheading "Serving Size" and aligned on the left side of the nutrition label, except that this information need not be provided when it is stated in the net quantity of contents declaration.

(2) *Information on dietary ingredients that have a Reference Daily Intake (RDI) or a Daily Reference Value (DRV) as established in §101.9(c) and their subcomponents (hereinafter referred to as "(b)(2)-dietary ingredients")*–(i) The (b)(2)-dietary ingredients to be declared, that is, total calories, calories from fat, total fat, saturated fat, cholesterol, sodium, total carbohydrate, dietary fiber, sugars, protein, vitamin A, vitamin C,

21 CFR §101.36

calcium and iron, shall be declared when they are present in a dietary supplement in quantitative amounts by weight that exceed the amount that can be declared as zero in nutrition labeling of foods in accordance with §101.9(c). Calories from saturated fat and polyunsaturated fat, monounsaturated fat, soluble fiber, insoluble fiber, sugar alcohol, and other carbohydrate may be declared, but they shall be declared when a claim is made about them. Any other vitamins or minerals listed in §101.9(c)(8)(iv) or (c)(9) may be declared, but they shall be declared when they are added to the product for purposes of supplementation, or when a claim is made about them. Any (b)(2)-dietary ingredients that are not present, or that are present in amounts that can be declared as zero in §101.9(c), shall not be declared (e.g., amounts corresponding to less than 2 percent of the RDI for vitamins and minerals). Protein shall not be declared on labels of products that, other than ingredients added solely for technological reasons, contain only individual amino acids.

(A) The names and the quantitative amounts by weight of each (b)(2)-dietary ingredient shall be presented under the heading "Amount Per Serving." When the quantitative amounts by weight are presented in a separate column, the heading may be centered over a column of quantitative amounts, described by paragraph (b)(2)(ii) of this section, if space permits. A heading consistent with the declaration of the serving size, such as "Each Tablet Contains," or "Amount Per 2 Tablets" may be used in place of the heading "Amount Per Serving." Other appropriate terms, such as capsule, packet, or teaspoonful, also may be used in place of the term "Serving."

(B) The names of dietary ingredients that are declared under paragraph (b)(2)(i) of this section shall be presented in a column aligned on the left side of the nutrition label in the order and manner of indentation specified in §101.9(c), except that calcium and iron shall follow pantothenic acid, and sodium and potassium shall follow chloride. This results in the following order for vitamins and minerals: Vitamin A, vitamin C, vitamin D, vitamin E, vitamin K, thiamin, riboflavin, niacin, vitamin B6, folate, vitamin B12, biotin, pantothenic acid, calcium, iron, phosphorus, iodine, magnesium, zinc, selenium, copper, manganese, chromium, molybdenum, chloride, sodium, and potassium. The (b)(2)-dietary ingredients shall be listed according to the nomenclature specified in 101.9 or in paragraph (b)(2)(i)(B)(2) of this section.

(*1*) When "Calories" are declared, they shall be listed first in the column of names, beneath a light bar separating the heading "Amount Per Serving" from the list of names. When "Calories from fat" or "Calories from saturated fat" are declared, they shall be indented beneath "Calories."

(*2*) The following synonyms may be added in parentheses immediately following the name of these (b)(2)-dietary ingredients: Vitamin C (ascorbic acid), thiamin (vitamin B1), riboflavin (vitamin B2), folate (folacin or folic acid), and calories (energy). Alternatively, the term "folic acid" or "folacin" may be listed without parentheses in place of "folate." Energy content per serving may be expressed in kilojoules units, added in parentheses immediately following the statement of caloric content.

(*3*) Beta-carotene may be declared as the percent of vitamin A that is present as beta-carotene, except that the declaration is required when a claim is made about beta-carotene. When declared, the percent shall be declared to the nearest whole percent, immediately adjacent to or beneath the name vitamin A (e.g., "Vitamin A (90% as beta-carotene)"). The amount of beta-carotene in terms of international units (IU) may be included in parentheses following the percent statement (e.g., "Vitamin A (90% (4500 IU) as beta- carotene)").

(ii) The number of calories, if declared, and the quantitative amount by weight per serving of each dietary ingredient required to be listed under paragraph (b)(2)(i) of this section shall be presented either in a separate column aligned to the right of the column of names or immediately following the listing of names within the same column. The quantitative amounts by weight shall represent the weight of the dietary ingredient rather than the weight of the source of the dietary ingredient (e.g., the weight of calcium rather than that of calcium carbonate).

(A) These amounts shall be expressed in the increments specified in §101.9(c)(1) through (c)(7), which includes increments for sodium and potassium.

(B) The amounts of vitamins and minerals, excluding sodium and potassium, shall be the amount of the vitamin or mineral included in one serving of the product, using the units of measurement and the levels of significance given in §101.9(c)(8)(iv), except that zeros following decimal points may be dropped, and additional levels of significance may be used when the number of decimal places indicated is not sufficient to express lower amounts (e.g., the RDI for zinc is given in whole milligrams (mg), but the quantitative amount may be declared in tenths of a mg).

(iii) The percent of the Daily Value of all dietary ingredients declared under paragraph (b)(2)(i) of this section shall be listed, except that the percent for protein may be omitted as provided in §101.9(c)(7); no percent shall be given for subcomponents for which DRV's have not been established (e.g., sugars); and, for labels of dietary supplements of vitamins and minerals that are represented or purported to be for use by infants, children less than 4 years of age, or pregnant or lactating women, no percent shall be given for total fat, saturated fat, cholesterol, total carbohydrate, dietary fiber, vitamin K, selenium, manganese, chromium, molybdenum, chloride, sodium, or potassium.

(A) When information on the percent of Daily Values is listed, this information shall be presented in one column aligned under the heading of "% Daily Value" and to the right of the column of amounts. The headings "% Daily Value (DV)," "% DV," "Percent Daily Value," or "Percent DV" may be substituted for "% Daily Value." The heading "% Daily Value" shall be placed on the same line as the heading "Amount Per Serving." When the acronym "DV" is unexplained in the heading and a footnote is required under (b)(2)(iii)(D), (b)(2)(iii)(F), or (b)(3)(iv) of this section, the footnote shall explain the acronym (e.g. "Daily Value (DV) not established").

(B) The percent of Daily Value shall be calculated by dividing the quantitative amount by weight of each (b)(2)-dietary ingredient by the RDI as established in §101.9(c)(8)(iv) or the DRV as established in §101.9(c)(9) for the specified dietary ingredient and multiplying by 100, except that the percent of Daily Value for protein, when present, shall be calculated as specified in §101.9(c)(7)(ii). The quantitative amount by weight of each dietary ingredient in this calculation shall be the unrounded amount, except that for total fat, saturated fat, cholesterol, sodium, potassium, total carbohydrate, and dietary fiber, the quantitative amount by weight declared on the label (i.e, rounded amount) may be used. The numerical value shall be followed by the symbol for percent (i.e., %).

(C) The percentages based on RDI's and on DRV's shall be expressed to the nearest whole percent, except that for dietary ingredients for which DRV's have been established, "Less than 1%" or "<1%" shall be used to declare the "% Daily Value" when the quantitative amount of the dietary ingredient by weight is great enough to require that the dietary ingredient be listed, but the amount is so small that the "% Daily Value" when rounded to the nearest percent is zero (e.g., a product that contains 1 gram of total carbohydrate would list the percent Daily Value as "Less than 1%" or "<1%").

(D) If the percent of Daily Value is declared for total fat, saturated fat, total carbohydrate, dietary fiber, or protein, a symbol shall follow the value listed for those nutrients that refers to the same symbol that is placed at the bottom of the nutrition label, below the bar required under paragraph (e)(6) of this section and inside the box, that is followed by the statement "Percent Daily Values are based on a 2,000 calorie diet."

(E) The percent of Daily Value shall be based on RDI and DRV values for adults and children 4 or more years of age, unless the product is represented or purported to be for use by infants, children less than 4 years of age, pregnant women, or lactating women, in which case the column heading shall clearly state the intended group. If the product is for persons within more than one group, the percent of Daily Value for each group shall be presented in separate columns as shown in paragraph (e)(10)(ii) of this section.

(F) For declared subcomponents that have no DRV's and, on the labels of dietary supplements of vitamins and minerals that are represented or purported to be for use by infants, children less that 4 years of age, or pregnant or lactating women, for total fat, saturated fat, cholesterol, total carbohydrate, dietary fiber, vitamin K, selenium, manganese, chromium, molybdenum, chloride, sodium, or potassium, a symbol (e.g., an asterisk) shall be placed in the "Percent Daily Value" column that shall refer to the same symbol that is placed at the bottom of the nutrition label, below the last heavy bar and inside the box, and followed by the statement "Daily Value not established."

(G) When calories, calories from fat, or calories from saturated fat are declared, the space under the "% Daily Value" column shall be left blank for these items. When there are no other (b)(2)-dietary ingredients listed for which a value must be declared in the "% Daily Value" column, the column may be omitted as shown in paragraph (e)(10)(vii) of this section. When the "% Daily Value" column is not required, but the dietary ingredients listed are subject to paragraph (b)(2)(iii)(F) of this section, the symbol required in that paragraph shall immediately follow the quantitative amount by weight for each dietary ingredient listed under "Amount Per Serving."

(iv) The quantitative amount by weight and the percent of Daily Value may be presented on a "per unit" basis in addition to on a "per serving" basis, as required in paragraph (b)(2)(ii) of this section. This information shall be presented in additional columns and clearly identified by appropriate headings.

(3) *Information on dietary ingredients for which RDI's and DRV's have not been established*—(i) Dietary

21 *CFR* §101.36

ingredients for which FDA has not established RDI's or DRV's and that are not subject to regulation under paragraph (b)(2) of this section (hereinafter referred to as "other dietary ingredients") shall be declared by their common or usual name when they are present in a dietary supplement, in a column that is under the column of names described in paragraph (b)(2)(i)(B) of this section or, as long as the constituents of an other dietary ingredient are not listed, in a linear display, under the heavy bar described in paragraph (e)(6) of this section, except that if no (b)(2)-dietary ingredients are declared, other dietary ingredients shall be declared directly beneath the heading "Amount Per Serving" described in paragraph (b)(2)(i)(A) of this section.

(ii) The quantitative amount by weight per serving of other dietary ingredients shall be presented in the same manner as the corresponding information required in paragraph (b)(2)(ii) of this section or, when a linear display is used, shall be presented immediately following the name of the other dietary ingredient. The quantitative amount by weight shall be the weight of the other dietary ingredient listed and not the weight of any component, or the source, of that dietary ingredient.

(A) These amounts shall be expressed using metric measures in appropriate units (i.e., 1,000 or more units shall be declared in the next higher set of units, e.g., 1,100 mg shall be declared as 1.1 g).

(B) For any dietary ingredient that is a liquid extract from which the solvent has not been removed, the quantity listed shall be the weight of the total extract with information on the concentration of the dietary ingredient, the solvent used, and the condition of the starting material (i.e., whether it is fresh or dried), e.g., "fresh dandelion root extract, x mg (y:z) in 70% ethanol," where x is the number of mg of the entire extract, y is the weight of the starting material and z is the volume (milliliters) of solvent. Where the solvent has been partially removed (not to dryness), the final concentration shall be stated (e.g., if the original extract was 1:5 and 50 percent of the solvent was removed, then the final concentration shall be stated as 1:2.5).

(C) For a dietary ingredient that is an extract from which the solvent has been removed, the weight of the ingredient shall be the weight of the dried extract. The dried extract shall be described by an appropriately descriptive term that identifies the solvent used, e.g., "dried hexane extract of _____" or "_____, dried hexane extract."

(iii) The constituents of a dietary ingredient described in paragraph (b)(3)(i) of this section may be listed indented under the dietary ingredient and followed by their quantitative amounts by weight, except that dietary ingredients described in paragraph (b)(2) of this section shall be listed in accordance with that section. When the constituents of a dietary ingredient described in paragraph (b)(3)(i) of this section are listed, all other dietary ingredients shall be declared in a column; however, the constituents themselves may be declared in a column or in a linear display.

(iv) Other dietary ingredients shall bear a symbol (e.g., an asterisk) in the column under the heading of "% Daily Value" that refers to the same symbol placed at the bottom of the nutrition label and followed by the statement "Daily Value not established," except that when the heading "% Daily Value" is not used, the symbol shall follow the quantitative amount by weight for each dietary ingredient listed.

(c) A proprietary blend of dietary ingredients shall be included in the list of dietary ingredients described in paragraph (b)(3)(i) of this section and identified by the term "Proprietary Blend" or other appropriately descriptive term or fanciful name and may be highlighted by bold type. Except as specified in this paragraph, all other requirements for the listing of dietary ingredients in dietary supplements are applicable.

(1) Dietary ingredients contained in the proprietary blend that are listed under paragraph (b)(2) of this section shall be declared in accordance with paragraph (b)(2) of this section.

(2) Dietary ingredients contained in the proprietary blend that are listed under paragraph (b)(3) of this section (i.e., "other dietary ingredients") shall be declared in descending order of predominance by weight, in a column or linear fashion, and indented under the term "Proprietary Blend" or other appropriately descriptive term or fanciful name.

(3) The quantitative amount by weight specified for the proprietary blend shall be the total weight of all other dietary ingredients contained in the proprietary blend and shall be placed on the same line to the right of the term "Proprietary Blend" or other appropriately descriptive term or fanciful name underneath the column of amounts described in paragraph (b)(2)(ii) of this section. A symbol (e.g., asterisk), which refers to the same symbol placed at the bottom of the nutrition label that is followed by the statement "Daily Value not established," shall be placed under the heading "% Daily Value," if present, or immediately following the quantitative amount by weight for the proprietary blend.

21 CFR §101.36

(4) The sample label shown in paragraph (e)(10)(v) of this section illustrates one method of nutrition labeling a proprietary blend of dietary ingredients.

(d) The source ingredient that supplies a dietary ingredient may be identified within the nutrition label in parentheses immediately following or indented beneath the name of a dietary ingredient and preceded by the words "as" or "from", e.g., "Calcium (as calcium carbonate)," except that manner of presentation is unnecessary when the name of the dietary ingredient (e.g., Oriental ginseng) or its synonym (e.g., ascorbic acid) is itself the source ingredient. When a source ingredient is identified in parentheses within the nutrition label, or when the name of the dietary ingredient or its synonym is the source ingredient, it shall not be required to be listed again in the ingredient statement that appears outside of the nutrition label. When a source ingredient is not identified within the nutrition label, it shall be listed in an ingredient statement in accordance with 101.4(g), which shall appear outside and immediately below the nutrition label or, if there is insufficient space below the nutrition label, immediately contiguous and to the right of the nutrition label.

(1) Source ingredients shall be identified in accordance with §101.4 (i.e., shall be listed by common or usual name, and the listing of botanicals shall specify the part of the plant from which the ingredient is derived) regardless of whether they are listed in an ingredient statement or in the nutrition label.

(2) When source ingredients are listed within the nutrition label, and two or more are used to provide a single dietary ingredient, all of the sources shall be listed within the parentheses in descending order by weight.

(3) Representations that the source ingredient conforms to an official compendium may be included either in the nutrition label or in the ingredient list (e.g., "Calcium (as calcium carbonate USP)").

(e) Nutrition information specified in this section shall be presented as follows:

(1) The title, "Supplement Facts," shall be set in a type size larger than all other print size in the nutrition label and, unless impractical, shall be set full width of the nutrition label. The title and all headings shall be bolded to distinguish them from other information.

(2) The nutrition information shall be enclosed in a box by using hairlines.

(3) All information within the nutrition label shall utilize:

(i) A single easy-to-read type style,

(ii) All black or one color type, printed on a white or other neutral contrasting background whenever practical,

(iii) Upper- and lowercase letters, except that all uppercase lettering may be utilized for packages that have a total surface area available to bear labeling of less than 12 square inches,

(iv) At least one point leading (i.e., space between lines of text), and

(v) Letters that do not touch.

(4) Except as provided for small and intermediate-sized packages under paragraph (i)(2) of this section, information other than the title, headings, and footnotes shall be in uniform type size no smaller than 8 point. Type size no smaller than 6 point may be used for column headings (e.g., "Amount Per Serving" and "% Daily Value") and for footnotes (e.g., "Percent Daily Values are based on a 2,000 calorie diet").

(5) A hairline rule that is centered between the lines of text shall separate each dietary ingredient required in paragraph (b)(2) and (b)(3) of this section from the dietary ingredient above and beneath it, as shown in paragraph (e)(10) of this section.

(6) A heavy bar shall be placed:

(i) Beneath the subheading "Servings Per Container" except that if "Servings Per Container" is not required and, as a result, not declared, the bar shall be placed beneath the subheading "Serving Size,"

(ii) Beneath the last dietary ingredient to be listed under paragraph (b)(2)(i) of this section, if any, and

(iii) Beneath the last other dietary ingredient to be listed under paragraph (b)(3) of this section, if any.

(7) A light bar shall be placed beneath the headings "Amount Per Serving" and "% Daily Value."

(8) If the product contains two or more separately packaged dietary supplements that differ from each other (e.g., the product has a packet of supplements to be taken in the morning and a different packet to be taken in the afternoon), the quantitative amounts and percent of Daily Value may be presented as specified in this paragraph in individual nutrition labels or in one aggregate nutrition label as illustrated in paragraph (e)(10)(iii) of this section.

(9) In the interest of uniformity of presentation, FDA urges that the information be presented using the graphic specifications set forth in Appendix B to part 101, as applicable.

(10) The following sample labels are presented for the purpose of illustration:

21 CFR §101.36

(i) Multiple vitamins:

Supplement Facts
Serving Size 1 Tablet

	Amount Per Serving	% Daily Value
Vitamin A (as retinyl acetate and 50% as beta-carotene)	5000 IU	100%
Vitamin C (as ascorbic acid)	60 mg	100%
Vitamin D (as cholecalciferol)	400 IU	100%
Vitamin E (as di-alpha tocopheryl acetate)	30 IU	100%
Thiamin (as thiamin mononitrate)	1.5 mg	100%
Riboflavin	1.7 mg	100%
Niacin (as niacinamide)	20 mg	100%
Vitamin B_6 (as pyridoxine hydrochloride)	2.0 mg	100%
Folate (as folic acid)	400 mcg	100%
Vitamin B_{12} (as cyanocobalamin)	6 mcg	100%
Biotin	30 mcg	10%
Pantothenic Acid (as calcium pantothenate)	10 mg	100%

Other ingredients: Gelatin, lactose, magnesium stearate, microcrystalline cellulose, FD&C Yellow No. 6, propylene glycol, propylparaben, and sodium benzoate.

(ii) Multiple vitamins for children and adults:

Supplement Facts
Serving Size 1 Tablet

Amount Per Serving		% Daily Value for Children Under 4 Years of Age	% Daily Value for Adults and Children 4 or more Years of Age
Calories	5		
Total Carbohydrate	1 g	†	< 1%*
Sugars	1 g	†	†
Vitamin A (50% as beta-carotene)	2500 IU	100%	50%
Vitamin C	40 mg	100%	67%
Vitamin D	400 IU	100%	100%
Vitamin E	15 IU	150%	50%
Thiamin	1.1 mg	157%	73%
Riboflavin	1.2 mg	150%	71%
Niacin	14 mg	156%	70%
Vitamin B_6	1.1 mg	157%	55%
Folate	300 mcg	150%	75%
Vitamin B_{12}	5 mcg	167%	83%

* Percent Daily Values are based on a 2,000 calorie diet.
† Daily Value not established.

Other ingredients: Sucrose, sodium ascorbate, stearic acid, gelatin, maltodextrins, artificial flavors, di-alpha tocopheryl acetate, niacinamide, magnesium stearate, Yellow 6, artificial colors, stearic acid, palmitic acid, pyridoxine hydrochloride, thiamin mononitrate, vitamin A acetate, beta-carotene, folic acid, cholecalciferol, and cyanocobalamin.

(iii) Multiple vitamins in packets:

Supplement Facts

Serving Size 1 Packet
Servings Per Container 10

	AM Packet		PM Packet	
Amount Per Serving		% Daily Value		% Daily Value
Vitamin A	2500 IU	50%	2500 IU	50%
Vitamin C	60 mg	100%	60 mg	100%
Vitamin D	400 IU	100%		
Vitamin E	30 IU	100%		
Thiamin	1.5 mg	100%	1.5 mg	100%
Riboflavin	1.7 mg	100%	1.7 mg	100%
Niacin	20 mg	100%	20 mg	100%
Vitamin B$_6$	2.0 mg	100%	2.0 mg	100%
Folic Acid	200 mcg	50%	200 mcg	50%
Vitamin B$_{12}$	3 mcg	50%	3 mcg	50%
Biotin			30 mcg	10%
Pantothenic Acid	5 mg	50%	5 mg	50%

Ingredients: Sodium ascorbate, ascorbic acid, calcium pantothenate, niacinamide, d-alpha tocopheryl acetate, microcrystalline cellulose, artificial flavors, dextrin, starch, mono- and diglycerides, vitamin A acetate, magnesium stearate, gelatin, FD&C Blue #1, FD&C Red #3, artificial colors, thiamin mononitrate, pyridoxine hydrochloride, citric acid, lactose, sorbic acid, tricalcium phosphate, sodium benzoate, sodium caseinate, methylparaben, potassium sorbate, BHA, BHT, ergocalciferol and cyanocobalamin.

(iv) Dietary supplement containing dietary ingredients with and without RDI's and DRV's:

Supplement Facts

Serving Size 1 Capsule

Amount Per Capsule	% Daily Value
Calories 20	
Calories from Fat 20	
Total Fat 2 g	3%*
Saturated Fat 0.5 g	3%*
Polyunsaturated Fat 1 g	†
Monounsaturated Fat 0.5 g	†
Vitamin A 4250 IU	85%
Vitamin D 425 IU	106%
Omega-3 fatty acids 0.5 g	†

* Percent Daily Values are based on a 2,000 calorie diet.
† Daily Value not established.

Ingredients: Cod liver oil, gelatin, water, and glycerin.

21 *CFR* §101.36

(v) A proprietary blend of dietary ingredients:

Supplement Facts
Serving Size 1 tsp (3 g) (makes 8 fl oz prepared)
Servings Per Container 24

	Amount Per Teaspoon	% Daily Value
Calories	10	
Total Carbohydrate	2 g	< 1%*
Sugars	2 g	†
Proprietary blend	0.7 g	
German Chamomile (flower)		†
Hyssop (leaves)		†

* Percent Daily Values are based on a 2,000 calorie diet.
† Daily Value not established.

Other ingredients: Fructose, lactose, starch, and stearic acid.

(vi) Dietary supplement of an herb:

Supplement Facts
Serving Size 1 Capsule

Amount Per Capsule	
Oriental Ginseng, powdered (root)	250 mcg*

* Daily Value not established.

Other ingredients: Gelatin, water, and glycerin.

(vii) Dietary supplement of amino acids:

Supplement Facts
Serving Size 1 Tablet

Amount Per Tablet	
Calories	15
Isoleucine (as L-isoleucine hydrochloride)	450 mg*
Leucine (as L-leucine hydrochloride)	620 mg*
Lysine (as L-lysine hydrochloride)	500 mg*
Methionine (as L-methionine hydrochloride)	350 mg*
Cystine (as L-cystine hydrochloride)	200 mg*
Phenylalanine (as L-phenylalanine hydrochloride)	220 mg*
Tyrosine (as L-tyrosine hydrochloride)	900 mg*
Threonine (as L-threonine hydrochloride)	300 mg*
Valine (as L-valine hydrochloride)	650 mg*

* Daily Value not established.

Other ingredients: Cellulose, lactose, and magnesium stearate.

(11) If space is not adequate to list the required information as shown in the sample labels in paragraph (e)(10) of this section, the list may be split and headings are repeated. The list to the right shall be set off by a line that distinguishes it and sets it apart from Daily Value information given to the left. The following sample label illustrates this display:

Supplement Facts

Serving Size 1 Packet

Amount Per Packet		% Daily Value	Amount Per Packet		% Daily Value
Vitamin A (from cod liver oil)	5,000 IU	100%	Zinc (as zinc oxide)	15 mg	100%
Vitamin C (as ascorbic acid)	250 mg	417%	Selenium (as sodium selenate)	25 mcg	35%
Vitamin D (as ergocalciferol)	400 IU	100%	Copper (as cupric oxide)	1 mg	50%
Vitamin E (as d-alpha tocopherol)	50 IU	500%	Manganese (as manganese sulfate)	5 mg	250%
Thiamin (as thiamin mononitrate)	75 mg	5000%	Chromium (as chromium chloride)	50 mcg	42%
Riboflavin	75 mg	4412%	Molybdenum (as sodium molybdate)	50 mcg	67%
Niacin (as niacinamide)	75 mg	375%	Potassium (as potassium chloride)	10 mg	< 1%
Vitamin B₆ (as pyridoxine hydrochloride)	75 mg	3750%			
Folic Acid	400 mcg	100%	Choline (as choline chloride)	100 mg	*
Vitamin B₁₂ (as cyanocobalamin)	100 mcg	1667%	Betaine (as betaine hydrochloride)	25 mg	*
Biotin	100 mcg	33%	Glutamic Acid (as L-glutamic acid)	25 mg	*
Pantothenic Acid (as calcium pantothenate)	75 mg	750%	Inositol (as inositol monophosphate)	75 mg	*
Calcium (from oystershell)	100 mg	10%	para-Aminobenzoic acid	30 mg	*
Iron (as ferrous fumarate)	10 mg	56%	Deoxyribonucleic acid	50 mg	*
Iodine (from kelp)	150 mcg	100%	Boron	500 mcg	*
Magnesium (as magnesium oxide)	60 mg	15%			

* Daily Value not established

Other Ingredients: Cellulose, stearic acid and silica.

(f)
(1) Compliance with this section will be determined in accordance with §101.9(g)(1) through (g)(8), except that the sample for analysis shall consist of a composite of 12 subsamples (consumer packages) or 10 percent of the number of packages in the same inspection lot, whichever is smaller, randomly selected to be representative of the lot. The criteria on class I and class II nutrients given in §101.9(g)(3) and (g)(4) also are applicable to other dietary ingredients described in paragraph (b)(3)(i) of this section. Reasonable excesses of these other dietary ingredients over labeled amounts are acceptable within current good manufacturing practice.
(2) When it is not technologically feasible, or some other circumstance makes it impracticable, for firms to comply with the requirements of this section, FDA may permit alternative means of compliance or additional exemptions to deal with the situation in accordance with §101.9(g)(9). Firms in need of such special allowances shall make their request in writing to the Office of Food Labeling (HFS–150), Food and Drug Administration, 200 C St. SW., Washington, DC 20204.
(g) Except as provided in paragraphs (i)(2) and (i)(5) of this section, the location of nutrition information on a label shall be in compliance with §101.2.
(h) Dietary supplements are subject to the exemptions specified as follows in:
(1) Section 101.9(j)(1) for foods that are offered for sale by a person who makes direct sales to consumers (i.e., a retailer) who has annual gross sales or business done in sales to consumers that is not more than $500,000 or has annual gross sales made or business done in sales of food to consumers of not more than $50,000, and whose labels, labeling, and advertising do not provide nutrition information or make a nutrient content or health claim;
(2) Section 101.9(j)(18) for foods that are low-volume products (that is, they meet the requirements for units sold in §101.9(j)(18)(i) or (j)(18)(ii)); that, except as provided in §101.9(j)(18)(iv), are the subject of a claim for an exemption that provides the information required under §101.9(j)(18)(iv), that is filed before the beginning of the time period for which the exemption is claimed, and that is filed by a person, whether it is the manufacturer, packer, or distributor, that qualifies to claim the exemption under the requirements for average full-time equivalent employees §101.9(j)(18)(i) or (j)(18)(ii), and whose labels, labeling, and advertising do not provide nutrition information or make a nutrient content or health claim;
(3) Section 101.9(j)(9) for foods shipped in bulk form that are not for distribution to consumers in such form and that are for use solely in the manufacture of other dietary supplements or that are to be processed, labeled, or repacked at a site other than where originally processed or packed.

21 CFR §101.36

(i) Dietary supplements are subject to the special labeling provisions specified in:

(1) Section 101.9(j)(5)(i) for foods, other than infant formula, represented or purported to be specifically for infants and children less than 2 years of age, in that nutrition labels on such foods shall not include calories from fat, calories from saturated fat, saturated fat, polyunsaturated fat, monounsaturated fat, and cholesterol;

(2) Section 101.9(j)(13) for foods in small or intermediate-sized packages, except that:

(i) All information within the nutrition label on small-sized packages, which have a total surface area available to labeling of less than 12 square inches, shall be in type size no smaller than 4.5 point;

(ii) All information within the nutrition label on intermediate-sized packages, which have from 12 to 40 square inches of surface area available to bear labeling, shall be in type size no smaller than 6 point, except that type size no smaller than 4.5 point may be used on packages that have less than 20 square inches available for labeling and more than 8 dietary ingredients to be listed and on packages that have 20 to 40 square inches available for labeling and more than 16 dietary ingredients to be listed.

(iii) When the nutrition information is presented on any panel under §101.9(j)(13)(ii)(D), the ingredient list shall continue to be located immediately below the nutrition label, or, if there is insufficient space below the nutrition label, immediately contiguous and to the right of the nutrition label as specified in §101.4(g).

(iv) When it is not possible for a small or intermediate-sized package that is enclosed in an outer package to comply with these type size requirements, the type size of the nutrition label on the primary (inner) container may be as small as needed to accommodate all of the required label information provided that the primary container is securely enclosed in outer packaging, the nutrition labeling on the outer packaging meets the applicable type size requirements, and such outer packaging is not intended to be separated from the primary container under conditions of retail sale.

(v) Where there is not sufficient space on a small or intermediate-sized package for a nutrition label that meets minimum type size requirements of 4.5 points if hairlines are used in accordance with paragraph (e)(5) of this section, the hairlines may be omitted and replaced by a row of dots connecting the columns containing the name of each dietary ingredient and the quantitative amounts (by weight and as a percent of Daily Value).

(3) Section 101.9(j)(15) for foods in multiunit food containers;

(4) Section 101.9(j)(16) for foods sold in bulk containers; and

(5) Section 101.9(j)(17) for foods in packages that have a total surface area available to bear labeling greater than 40 square inches but whose principal display panel and information panel do not provide sufficient space to accommodate all required label information, except that the ingredient list shall continue to be located immediately below the nutrition label, or, if there is insufficient space below the nutrition label, immediately contiguous and to the right of the nutrition label as specified in §101.4(g).

(j) Dietary supplements shall be subject to the misbranding provisions of §101.9(k).

[62 FR 49849-49858, Sept. 23, 1997]

§101.42 Nutrition labeling of raw fruit, vegetables, and fish.

(a) The Food and Drug Administration (FDA) urges food retailers to provide nutrition information, as provided in §101.9(c), for raw fruit, vegetables, and fish at the point-of-purchase. If retailers choose to provide such information, they should do so in a manner that conforms to the guidelines in §101.45.

(b) In §101.44, FDA has listed the 20 varieties of raw fruit, vegetables, and fish that are most frequently consumed during a year and to which the guidelines apply.

(c) FDA has also defined in §101.43, the circumstances that constitute substantial compliance by food retailers with the guidelines.

(d) By May 8, 1993, FDA will issue a report on actions taken by food retailers to provide consumers with nutrition information for raw fruit, vegetables, and fish under the guidelines established in §101.45.

(1) The report will include a determination of whether there is substantial compliance, as defined in §101.43, with the guidelines.

(2) In evaluating substantial compliance, FDA will consider only the 20 varieties of raw fruit, vegetables, and fish most frequently consumed as identified in §101.44.

(e) If FDA finds that there is substantial compliance with the guidelines for the nutrition labeling of raw fruit and

21 CFR §101.43

vegetables or of fish, the agency will so state in the report, and the guidelines will remain in effect. FDA will reevaluate the market place for substantial compliance every 2 years.

(f) If FDA determines that there is not substantial compliance with the guidelines for raw fruit and vegetables or for raw fish, the agency will at that time issue proposed regulations requiring that any person who offers raw fruit and vegetables or fish to consumers provide, in a manner prescribed by regulations, the nutrition information required by §101.9. Final regulations would have to be issued 6 months after issuance of proposed regulations, and they would become effective 6 months after the date of their promulgation.

§101.43 Substantial compliance of food retailers with the guidelines for the voluntary nutrition labeling of raw fruit, vegetables, and fish.

(a) The Food and Drug Administration (FDA) will judge a food retailer who sells raw agricultural commodities or raw fish to be in compliance with the guidelines in §101.45 with respect to raw agricultural commodities if the retailer displays or provides nutrition labeling for at least 90 percent of the raw agricultural commodities listed in §101.44 that it sells, and with respect to raw fish if the retailer displays or provides nutrition labeling for at least 90 percent of the types of raw fish listed in §101.44 that it sells. To be in compliance, the nutrition labeling shall:
    (1) Be presented in the store or other type of establishment in a manner that is consistent with §101.45(a)(1);
    (2) Be presented in content and format that are consistent with §101.45(a)(2), (a)(3), and (a)(4); and
    (3) Include data that have been provided by FDA in Appendices C and D to part 101 of this chapter, except that the information on potassium is voluntary.
(b) To determine whether there is substantial compliance by food retailers with the guidelines in §101.45 for the voluntary nutrition labeling of raw fruit and vegetables and of raw fish, FDA will select a representative sample of 2,000 stores, allocated by store type and size, for raw fruit and vegetables and for raw fish.
(c) FDA will find that there is substantial compliance with the guidelines in §101.45 if it finds based on paragraph (a) of this section that at least 60 percent of all stores that are evaluated are in compliance.
(d) FDA will evaluate substantial compliance separately for raw agricultural commodities and for raw fish.
[55 FR 60890, Nov. 27, 1991; 61 FR 42760-42761, Aug. 16, 1996]

§101.44 Identification of the 20 most frequently consumed raw fruit, vegetables, and fish in the United States.

(a) The 20 most frequently consumed raw fruit are: Banana, apple, watermelon, orange, cantaloupe, grape, grapefruit, strawberry, peach, pear, nectarine, honeydew melon, plum, avocado, lemon, pineapple, tangerine, sweet cherry, kiwifruit, and lime.
(b) The 20 most frequently consumed raw vegetables are: Potato, iceberg lettuce, tomato, onion, carrot, celery, sweet corn, broccoli, green cabbage, cucumber, bell pepper, cauliflower, leaf lettuce, sweet potato, mushroom, green onion, green (snap) bean, radish, summer squash, and asparagus.
(c) The 20 most frequently consumed raw fish are: Shrimp, cod, pollock, catfish, scallops, salmon (Atlantic/Coho, chum/pink, sockeye), flounder/sole, oysters, orange roughy, Atlantic/Pacific mackerel, ocean perch, rockfish, whiting, clam, haddock, blue crab, rainbow trout, halibut, lobster, and swordfish.
[55 FR 60890, Nov. 27, 1991; 61 FR 42760-42761, Aug. 16, 1996]

§101.45 Guidelines for the voluntary nutrition labeling of raw fruit, vegetables, and fish.

(a) Nutrition labeling for raw fruits, vegetables, and fish listed in § 101.44 should be presented to the public in the following manner:
    (1) Nutrition labeling information should be displayed at the point of purchase by an appropriate means such as by a label affixed to the food or through labeling including shelf labels, signs, posters, brochures, notebooks, or leaflets that are readily available and in close proximity to the foods. The nutrition labeling information may also be supplemented by a video, live demonstration, or other media.
    (2) Serving sizes should be determined, and nutrients declared, in accordance with § 101.9 (b) and (c),

21 CFR §101.45

respectively, except that the nutrition labeling data should be based on the raw edible portion for fruits and vegetables and on the cooked edible portion for fish. The methods used to cook fish should be those that do not add fat, breading, or seasoning (e.g., salt or spices).

(3) When nutrition labeling information is provided for more than one raw fruit, vegetable, or fish on signs, posters, brochures, notebooks, or leaflets, it may be presented in charts with horizontal or vertical columns or as a compilation of individual nutrition labels. Nutrition labeling that is presented in a linear display (see §101.9(j)(13)(ii)(A)(2)) will not be considered to be in compliance. The heading "Nutrition Facts" must be in a type size larger than all other print in the nutrition label. The required information (i.e., headings, serving sizes, list of nutrients, quantitative amounts by weight (except for vitamins and minerals), and percent of Daily Values (DV's) (except for sugars and protein) must be clearly presented and of sufficient type size and color contrast to be plainly legible, with numeric values for percent of DV highlighted in contrast to the quantitative amounts by weight and hairlines between all nutrients.

(i) Declaration of the number of servings per container need not be included in the nutrition labeling of raw fruits, vegetables, and fish.

(ii) Except for the statement "Percent Daily Values are based on a 2,000 calorie diet," the footnote required in §101.9(d)(9) is not required. However, when labeling is provided in brochures, notebooks, leaflets, or similar types of materials, retailers are encouraged to include the footnote.

(iii) When the nutrition labeling information for more than one raw fruit or vegetable is provided on signs, posters, brochures, notebooks, or leaflets, the listings for saturated fat and cholesterol may be omitted from the charts or individual nutrition labels so long as the fact that most fruits and vegetables provide negligible amounts of these nutrients, but that avocados contain 1 gram (g) of fat per ounce, is stated in a footnote (e.g., "Most fruits and vegetables provide negligible amounts of saturated fat and cholesterol; avocados provide 1 g of saturated fat per ounce"). The footnote may also contain information about the polyunsaturated and monounsaturated fat content of avocados. When the nutrition labeling information for raw fish is provided on a chart, the listings for dietary fiber and sugars may be omitted if the following footnote is used, "Fish provide negligible amounts of dietary fiber and sugars."

(4) When nutrition labeling is provided for individual raw fruits, vegetables, or fish on packages or on signs, posters, brochures, notebooks, or leaflets, it should be displayed in accordance with §101.9, except that the declaration of the number of servings per container need not be included. For individual labels provided by retailers on signs and posters, the footnote required in §101.9(d)(9) may be shortened to "Percent Daily Values are based on a 2,000 calorie diet."

(b) Nutrition label values provided by the Food and Drug Administration (FDA) in Appendices C and D to part 101 for the 20 most frequently consumed raw fruits, vegetables, and fish listed in §101.44 shall be used to ensure uniformity in declared values. FDA will publish proposed updates of the 20 most frequently consumed raw fruits, vegetables, and fish and nutrition label data for these foods (or a notice that the data sets have not changed from the previous publication) at least every 4 years in the Federal Register.

(1) The agency encourages the submission of data bases with new or additional nutrient data for any of the most frequently consumed raw fruits, vegetables, and fish to the Office of Food Labeling (HFS-150), Center for Food Safety and Applied Nutrition, Food and Drug Administration, 200 C St. SW., Washington, DC 20204, for review and evaluation. FDA may incorporate these data in the next revision of the nutrition labeling information for the top 20 raw fruits, vegetables, and fish.

(i) Guidance in the development of data bases may be found in the "FDA Nutrition Labeling Manual: A Guide for Developing and Using Data Bases." available from the FDA Office of Food Labeling.

(ii) The submission to FDA should include, but need not be limited to, information on the following: Source of the data (names of investigators, name of organization, place of analyses, dates of analyses), number of samples, sampling design, analytical methods, and statistical treatment of the data. Proposed quantitative label declarations may be included. The proposed values for declaration should be determined in accordance with the "FDA Nutrition Labeling Manual: A Guide for Developing and Using Data Bases."

(2) [Reserved]

(c) Data bases of nutrient values for raw fruits, vegetables, and fish that are not among the 20 most frequently consumed may be used to develop nutrition labeling values for these foods. This includes data bases of nutrient values for specific varieties, species, or cultivars of raw fruits, vegetables, and fish not specifically identified among the 20 most frequently consumed.

(1) The food names and descriptions for the fruits, vegetables, and fish should clearly identify these foods as distinct from foods among the most frequently consumed list for which FDA has provided data.
(2) Guidance in the development of data bases may be found in the "FDA Nutrition Labeling Manual: A Guide for Developing and Using Data Bases."
(3) Nutrition labeling values computed from data bases are subject to the compliance provisions of § 101.9(g).
(i) Compliance with the provisions of § 101.9(g) may be achieved by use of a data base that has been developed following FDA guideline procedures and approved by FDA.
(A) The submission to FDA for approval should include but need not be limited to information on the following: Source of the data (names of investigators, name of organization, place of analyses, dates of analyses), number of samples, sampling design, analytical methods, statistical treatment of the data, and proposed quantitative label declarations. The values for declaration should be determined in accordance with the "FDA Nutrition Labeling Manual: A Guide for Developing and Using Databases."
(B) FDA approval of a data base and nutrition labeling values shall not be considered granted until the Center for Food Safety and Applied Nutrition has agreed to all aspects of the data base in writing. Approvals will be in effect for a limited time, e.g., 10 years, and will be eligible for renewal in the absence of significant changes in agricultural or industry practices (e.g., a change occurs in a predominant variety produced). FDA will take steps to revoke its approval of the data base and nutrition labeling values if FDA monitoring suggests that the data base or nutrition labeling values are no longer representative of the item sold in this country. Approval requests shall be submitted in accordance with the provision of § 101.30 of this chapter.
(ii) [Reserved]
[55 FR 60890, Nov. 27, 1991; 61 FR 14479, April 2, 1996; 61 FR 42760-42761, Aug. 16, 1996]

Subpart D-Specific Requirements for Nutrient Content Claims
Source: 58 FR 2413, Jan. 6, 1993, unless otherwise noted.

§101.54 Nutrient content claims for "good source," "high," "more," and "high potency."

(a) *General requirements.* Except as provided in paragraph (e) of this section, a claim about the level of a nutrient in a food in relation to the Reference Daily Intake (RDI) established for that nutrient in §101.9(c)(8)(iv) or Daily Reference Value (DRV) established for that nutrient in §101.9(c)(9), (excluding total carbohydrates) may only be made on the label or in labeling of the food if:
(1) The claim uses one of the terms defined in this section in accordance with the definition for that term;
(2) The claim is made in accordance with the general requirements for nutrient content claims in §101.13; and
(3) The food for which the claim is made is labeled in accordance with §101.9, §101.10, or §101.36, as applicable.
(b) *"High" claims.*
(1) The terms *"high," "rich in,"* or *"excellent source of"* may be used on the label and in the labeling of foods, except meal products as defined in §101.13(l) and main dish products as defined in §101.13(m), provided that the food contains 20 percent or more of the RDI or the DRV per reference amount customarily consumed.
(2) The terms defined in paragraph (b)(1) of this section may be used on the label and in the labeling of meal products as defined in §101.13(l), main dish products as defined in §101.13(m), and dietary supplements of vitamins or minerals to characterize the level of any substance that is not a vitamin or mineral, provided that the food contains 20 percent or more of the RDI or the DRV per reference amount customarily consumed.
(i) The product contains a food that meets the definition of "high" in paragraph (b)(1) of this section; and
(ii) The label or labeling clearly identifies the food that is the subject of the claim (e.g., the serving of broccoli in this product is high in vitamin C).
(c) *"Good Source" claims.*

21 CFR §101.54

(1) The terms "good source," "contains," or "provides" may be used on the label and in the labeling of foods, except meal products as described in §101.13(l) and main dish products as defined in §101.13(m), provided that the food contains 10 to 19 percent of the RDI or the DRV per reference amount customarily consumed.

(2) The terms defined in paragraph (c)(1) of this section may be used on the label and in the labeling of meal products as defined in §101.13(l) and main dish products as defined in 101.13(m), provided that:

(i) The product contains a food that meets the definition of "good source" in paragraph (c)(1) of this section; and

(ii) The label or labeling clearly identifies the food that is the subject of the claim (e.g., the serving of sweet potatoes in this product is a "good source" of fiber).

(d) *"Fiber" claims.*

(1) If a nutrient content claim is made with respect to the level of dietary fiber, that is, that the product is high in fiber, a good source of fiber, or that the food contains "more" fiber, and the food is not "low" in total fat as defined in §101.62(b)(2) or, in the case of a meal product, as defined in §101.13(l), or main dish product, as defined in §101.13(m), is not "low" in total fat as defined in §101.62(b)(3), then the label shall disclose the level of total fat per labeled serving.

(2) The disclosure shall appear in immediate proximity to such claim, be in a type size no less than one-half the size of the claim and precede the referral statement required in §101.13(g) (e.g., "contains [*x amount*] of total fat per serving. See [*appropriate panel*] for nutrition information").

(e) *"More" claims.*

(1) A relative claim using the terms "more," "fortified," "enriched," "added," "extra," and "plus" may be used on the label or in labeling of foods to describe the level of protein, vitamins, minerals, dietary fiber, or potassium, except as limited by §101.13(j)(1)(i) and except meal products as defined in §101.13(l) and main dish products as defined in §101.13(m), provided that:

(i) The food contains at least 10 percent more of the RDI for vitamins or minerals or of the DRV for protein, dietary fiber, or potassium (expressed as a percent of the Daily Value) per reference amount customarily consumed than an appropriate reference food; and

(ii) Where the claim is based on a nutrient that has been added to the food, that fortification is in accordance with the policy on fortification of foods in §104.20 of this chapter; and

(iii) As required in §101.13(j)(2) for relative claims:

(A) The identity of the reference food and the percentage (or fraction) that the nutrient is greater relative to the RDI or DRV are declared in immediate proximity to the most prominent such claim (e.g., "contains 10 percent more of the Daily Value for fiber than white bread"); and

(B) Quantitative information comparing the level of the nutrient in the product per labeled serving, with that of the reference food that it replaces (e.g., "Fiber content of white bread is 1 gram (g) per serving; (this product) 3.5 g per serving") is declared adjacent to the most prominent claim or to the nutrition label, except that if the nutrition label is on the information panel, the quantitative information may be located elsewhere on the information panel in accordance with §101.2..

(2) A relative claim using the terms "more," "fortified," "enriched," "added," "extra," and "plus" may be used on the label or in labeling to describe the level of protein, vitamins, minerals, dietary fiber or potassium, except as limited in §101.13(j)(1)(i), in meal products as defined in §101.13(l) or main dish products as defined in §101.13(m), provided that:

(i) The food contains at least 10 percent more of the RDI for vitamins or minerals or of the DRV for protein, dietary fiber, or potassium (expressed as a percent of the Daily Value) per 100 g of food than an appropriate reference food.

(ii) Where the claim is based on a nutrient that has been added to the food, that fortification is in accordance with the policy on fortification of foods in §104.20 of this chapter; and

(iii) As required in §101.13(j)(2) for relative claims:

(A) The identity of the reference food and the percentage (or fraction) that the nutrient was increased relative to the RDI or DRV are declared in immediate proximity to the most prominent such claim (e.g., "contains 10 percent more of the Daily Value for fiber per 3 oz than does 'X brand of product'"), and

(B) Quantitative information comparing the level of the nutrient in the product per specified weight, with that of the reference food that it replaces (e.g., "The fiber content of 'X brand of product' is 2 g per 3 oz.

This product contains 4.5 g per 3 oz") is declared adjacent to the most prominent claim or to the nutrition label, except that if the nutrition label is on the information panel, the quantitative information may be located elsewhere on the information panel in accordance with §101.2.

(f) *"High potency" claims.*

(1)(i) The term "high potency" may be used on the label or in the labeling of foods to describe individual vitamins or minerals that are present at 100 percent or more of the RDI per reference amount customarily consumed.

(ii) When the term "high potency" is used to describe individual vitamins or minerals in a product that contains other nutrients or dietary ingredients, the label or labeling shall clearly identify which vitamin or mineral is described by the term "high potency" (e.g., "Botanical 'X' with high potency vitamin E").

(2) The term "high potency" may be used on the label or in the labeling of a multiingredient food product to describe the product if the product contains 100 percent or more of the RDI for at least two-thirds of the vitamins and minerals that are listed in §101.9(c)(8)(iv) and that are present in the product at 2 percent or more of the RDI (e.g., "High potency multivitamin, multimineral dietary supplement tablets").

(3) Where compliance with paragraphs (f)(1)(i), (f)(1)(ii), or (f)(2) of this section is based on a nutrient that has been added to a food (other than a dietary supplement), that fortification shall be in accordance with the policy on fortification of foods in §104.20 of this chapter.

(g) *Nutrient content claims using the term "antioxidant."* A nutrient content claim that characterizes the level of antioxidant nutrients present in a food may be used on the label or in the labeling of that food when:

(1) An RDI has been established for each of the nutrients;

(2) The nutrients that are the subject of the claim have recognized antioxidant activity; that is, when there exists scientific evidence that, following absorption from the gastrointestinal tract, the substance participates in physiological, biochemical, or cellular processes that inactivate free radicals or prevent free radical-initiated chemical reactions;

(3) The level of each nutrient that is the subject of the claim is sufficient to qualify for the §101.54(b), (c), or (e) claim (e.g., to bear the claim "high in antioxidant vitamin C," the product must contain 20 percent or more of the RDI for vitamin C). Beta-carotene may be a subject of the claim when the level of vitamin A present as beta- carotene in the food that bears the claim is sufficient to qualify for the claim. For example, for the claim "good source of antioxidant beta-carotene," 10 percent or more of the RDI for vitamin A must be present as beta- carotene per reference amount customarily consumed; and

(4) The names of the nutrients that are the subject of the claim are included as part of the claim (e.g., "high in antioxidant vitamins C and E"). Alternatively, when used as part of a nutrient content claim, the term "antioxidant" or "antioxidants" (as in "high in antioxidants") may be linked by a symbol (e.g., an asterisk) that refers to the same symbol that appears elsewhere on the same panel of a product label followed by the name or names of the nutrients with recognized antioxidant activity. The list of nutrients shall appear in letters of a type size height no smaller than the larger of one-half of the type size of the largest nutrient content claim or 1/16 inch.

[58 FR 2413, Jan. 6, 1993; 58 FR 17342, Apr. 2, 1993, as amended at 59 FR 394, Jan. 4, 1994; 59 FR 15051, Mar. 31, 1994; 60 FR 17206, Apr. 5, 1995; 61 FR 11731, Mar. 22, 1996; 62 *FR* 31339, June 9, 1997; 62 *FR* 49867, 49880-49881, Sept. 23, 1997]

§101.56 Nutrient content claims for "light" or "lite."

(a) *General requirements.* A claim using the term "light" or "lite" to describe a food may only be made on the label or in labeling of the food if:

(1) The claim uses one of the terms defined in this section in accordance with the definition for that term;

(2) The claim is made in accordance with the general requirements for nutrient content claims in §101.13; and

(3) The food is labeled in accordance with §101.9 or §101.10, where applicable.

(b) *"Light" claims.* The terms "light" or "lite" may be used on the label or in the labeling of foods, except meal products as defined in §101.13(l) and main dish products as defined in §101.13(m), without further qualification, provided that:

21 CFR §101.56

(1) If the food derives 50 percent or more of its calories from fat, its fat content is reduced by 50 percent or more per reference amount customarily consumed compared to an appropriate reference food as specified in §101.13(j)(1); or

(2) If the food derives less than 50 percent of its calories from fat:

(i) The number of calories is reduced by at least one-third (33 1/3 percent) per reference amount customarily consumed compared to an appropriate reference food; or

(ii) Its fat content is reduced by 50 percent or more per reference amount customarily consumed compared to the reference food that it resembles or for which it substitutes as specified in §101.13(j)(1); and

(3) As required in §101.13(j)(2) for relative claims:

(i) The identity of the reference food and the percent (or fraction) that the calories and the fat were reduced are declared in immediate proximity to the most prominent such claim, (e.g., "1/3 fewer calories and 50 percent less fat than our regular cheese cake");

(ii) Quantitative information comparing the level of calories and fat content in the product per labeled serving size, with that of the reference food that it replaces (e.g., "lite cheese cake-200 calories, 4 grams (g) fat; regular cheese cake-300 calories, 8 g fat per serving") is declared adjacent to the most prominent claim or to the nutrition label, except that if the nutrition label is on the information panel, the quantitative information may be located elsewhere on the information panel in accordance with §101.2; and

(iii) If the labeled food contains less than 40 calories or less than 3 g fat per reference amount customarily consumed, the percentage reduction for that nutrient need not be declared.

(4) A "light" claim may not be made on a food for which the reference food meets the definition of "low fat" and "low calorie."

(c)

(1) (i) A product for which the reference food contains 40 calories or less and 3 g fat or less per reference amount customarily consumed may use the term "light" or "lite" without further qualification if it is reduced by 50 percent or more in sodium content compared to the reference food; and

(ii) As required in §101.13(j)(2) for relative claims:

(A) The identity of the reference food and the percent (or fraction) that the sodium was reduced shall be declared in immediate proximity to the most prominent such claim (e.g., 50 percent less sodium than our regular soy sauce); and

(B) Quantitative information comparing the level of sodium per labeled serving size with that of the reference food it replaces (e.g., "lite soy sauce 500 milligrams (mg) sodium per serving, regular soy sauce 1,000 mg per serving") is declared adjacent to the most prominent claim or to the nutrition label, except that if the nutrition label is on the information panel, the quantitative information may be located elsewhere on the information panel in accordance with §101.2.

(2) (i) A product for which the reference food contains more than 40 calories or more than 3 g fat per reference amount customarily consumed may use the term "light in sodium" or "lite in sodium" if it is reduced by 50 percent or more in sodium content compared to the reference food, provided that "light" or "lite" is presented in immediate proximity with "in sodium" and the entire term is presented in uniform type size, style, color, and prominence; and

(ii) As required in §101.13(j)(2) for relative claims:

(A) The identity of the reference food and the percent (or fraction) that the sodium was reduced shall be declared in immediate proximity to the most prominent such claim (e.g., 50 percent less sodium than our regular canned peas); and

(B) Quantitative information comparing the level of sodium per labeled serving size with that of the reference food it replaces (e.g., "light canned peas, 175 milligrams (mg) sodium per serving, regular canned peas 350 mg per serving.") is declared adjacent to the most prominent claim or to the nutrition label, except that if the nutrition label is on the information panel, the quantitative information may be located elsewhere on the information panel in accordance with §101.2.

(iii) Except for meal products as defined in §101.13(l) and main dish products as defined in §101.13(m), a "light in sodium" claim may not be made on a food for which the reference food meets the definition of "low in sodium".

(d)

(1) The terms "light" or "lite" may be used on the label or in the labeling of a meal product as defined in

§101.13(l) and a main dish product as defined in §101.13(m), provided that:

(i) The food meets the definition of:

(A) "Low in calories" as defined in §101.60(b)(3); or

(B) "Low in fat" as defined in §101.62(b)(3); and

(ii) (A) A statement appears on the principal display panel that explains whether "light" is used to mean "low fat," "low calories," or both (e.g., "Light Delight, a low fat meal"); and

(B) The accompanying statement is no less than one-half the type size of the "light" or "lite" claim.

(2) (i) The term "light in sodium" or "lite in sodium" may be used on the label or in the labeling of a meal product as defined in §101.13(l) and a main dish product as defined in §101.13(m), provided that the food meets the definition of "low in sodium" as defined in §101.61(b)(5)(i); and

(ii) "Light" or "lite" and "in sodium" are presented in uniform type size, style, color, and prominence.

(e) Except as provided in paragraphs (b) through (d) of this section, the term "light" or "lite" may not be used to refer to a food that is not reduced in fat by 50 percent, or, if applicable, in calories by 1/3 or, when properly qualified, in sodium by 50 percent unless:

(1) It describes some physical or organoleptic attribute of the food such as texture or color and the information (e.g., "light in color" or "light in texture") so stated, clearly conveys the nature of the product; and

(2) The attribute (e.g., "color" or "texture") is in the same style, color, and at least one-half the type size as the word "light" and in immediate proximity thereto.

(f) If a manufacturer can demonstrate that the word "light" has been associated, through common use, with a particular food to reflect a physical or organoleptic attribute (e.g., light brown sugar, light corn syrup, or light molasses) to the point where it has become part of the statement of identity, such use of the term "light" shall not be considered a nutrient content claim subject to the requirements in this part.

(g) The term "lightly salted" may be used on a product to which has been added 50 percent less sodium than is normally added to the reference food as described in §101.13(j)(1)(i)(B) and (j)(1)(ii)(B), provided that if the product is not "low in sodium" as defined in §101.61(b)(4), the statement "not a low sodium food," shall appear adjacent to the nutrition label of the food bearing the claim, or, if the nutrition label is on the information panel, it may appear elsewhere on the information panel in accordance with §101.2 and the information required to accompany a relative claim shall appear on the label or labeling as specified in §101.13(j)(2).

[58 FR 2413, Jan. 6, 1993; 58 FR 17342, Apr. 2, 1993, as amended at 60 FR 17206, Apr. 5, 1995]

§101.60 Nutrient content claims for the calorie content of foods.

(a) *General requirements.* A claim about the calorie or sugar content of a food may only be made on the label or in the labeling of a food if:

(1) The claim uses one of the terms defined in this section in accordance with the definition for that term;

(2) The claim is made in accordance with the general requirements for nutrient content claims in §101.13;

(3) The food for which the claim is made is labeled in accordance with §101.9, §101.10, or §101.36, as applicable; and

(4) For dietary supplements, claims regarding calories may not be made on products that meet the criteria in §101.60(b)(1) or (b)(2) for "calorie free" or "low calorie" claims except when an equivalent amount of a similar dietary supplement (e.g., another protein supplement) that the labeled food resembles and for which it substitutes, normally exceeds the definition for "low calorie" in §101.60(b)(2).

(b) *"Calorie content claims."*

(1) The terms "calorie free," "free of calories," "no calories," "zero calories," "without calories," "trivial source of calories," "negligible source of calories," or "dietarily insignificant source of calories" may be used on the label or in the labeling of foods, provided that:

(i) The food contains less than 5 calories per reference amount customarily consumed and per labeled serving.

(ii) As required in §101.13(e)(2), if the food meets this condition without the benefit of special processing, alteration, formulation, or reformulation to lower the caloric content, it is labeled to disclose that calories are not usually present in the food (e.g., "cider vinegar, a calorie free food").

21 CFR §101.60

(2) The terms "low calorie," "few calories," "contains a small amount of calories," "low source of calories," or "low in calories" may be used on the label or in labeling of foods, except meal products as defined in §101.13(l) and main dish products as defined in §101.13(m), provided that:

(i) (A) The food has a reference amount customarily consumed greater than 30 grams (g) or greater than 2 tablespoons and does not provide more than 40 calories per reference amount customarily consumed; or

(B) The food has a reference amount customarily consumed of 30 g or less or 2 tablespoons or less and does not provide more than 40 calories per reference amount customarily consumed and, except for sugar substitutes, per 50 g (for dehydrated foods that must be reconstituted before typical consumption with water or a diluent containing an insignificant amount, as defined in §101.9(f)(1), of all nutrients per reference amount customarily consumed, the per 50 g criterion refers to the "as prepared" form).

(ii) If a food meets these conditions without the benefit of special processing, alteration, formulation, or reformulation to vary the caloric content, it is labeled to clearly refer to all foods of its type and not merely to the particular brand to which the label attaches (e.g., "celery, a low calorie food").

(3) The terms defined in paragraph (b)(2) of this section may be used on the label or in labeling of meal products as defined in §101.13(l) or main dish products as defined in §101.13(m), provided that:

(i) The product contains 120 calories or less per 100 g; and

(ii) If the product meets this condition without the benefit of special processing, alteration, formulation, or reformulation to lower the calorie content, it is labeled to clearly refer to all foods of its type and not merely to the particular brand to which it attaches.

(4) The terms "reduced calorie," "reduced in calories," "calorie reduced," "fewer calories," "lower calorie," or "lower in calories" may be used on the label or in the labeling of foods, except as limited by §101.13(j)(1)(i) and except meal products as defined in §101.13(l) and main dish products as defined in §101.13(m), provided that:

(i) The food contains at least 25 percent fewer calories per reference amount customarily consumed than an appropriate reference food as described in §101.13(j)(1); and

(ii) As required in §101.13(j)(2) for relative claims:

(A) The identity of the reference food and the percent (or fraction) that the calories differ between the two foods are declared in immediate proximity to the most prominent such claim (e.g., reduced calorie cupcakes "33 1/3 percent fewer calories than regular cupcakes"); and

(B) Quantitative information comparing the level of the nutrient in the product per labeled serving with that of the reference food that it replaces (e.g., "Calorie content has been reduced from 150 to 100 calories per serving.") is declared adjacent to the most prominent claim or to the nutrition label, except that if the nutrition label is on the information panel, the quantitative information may be located elsewhere on the information panel in accordance with §101.2.

(iii) Claims described in paragraph (b)(4) of this section may not be made on the label or labeling of foods if the reference food meets the definition for "low calorie."

(5) The terms defined in paragraph (b)(4) of this section may be used on the label or in the labeling of meal products as defined in §101.13(l) and main dish products as defined in §101.13(m), provided that:

(i) The food contains at least 25 percent fewer calories per 100 g of food than an appropriate reference food as described in §101.13(j)(1); and

(ii) As required in §101.13(j)(2) for relative claims:

(A) The identity of the reference food and the percent (or fraction) that the calories differ between the two foods are declared in immediate proximity to the most prominent such claim (e.g., Larry's Reduced Calorie Lasagna, "25 percent fewer calories per oz (or 3 oz) than our regular Lasagna"); and

(B) Quantitative information comparing the level of the nutrient in the product per specified weight with that of the reference food that it replaces (e.g., "Calorie content has been reduced from 108 calories per 3 oz to 83 calories per 3 oz.") is declared adjacent to the most prominent claim or to the nutrition label, except that if the nutrition label is on the information panel, the quantitative information may be located elsewhere on the information panel in accordance with §101.2..

(iii) Claims described in paragraph (b)(5) of this section may not be made on the label or labeling of food if the reference food meets the definition for "low calorie."

(c) *Sugar content claims-*

(1) *Use of terms such as "sugar free," "free of sugar," "no sugar," "zero sugar," "without sugar,"*

21 *CFR* §101.60

"sugarless," "trivial source of sugar," "negligible source of sugar," or "dietarily insignificant source of sugar." Consumers may reasonably be expected to regard terms that represent that the food contains no sugars or sweeteners e.g., "sugar free," or "no sugar," as indicating a product which is low in calories or significantly reduced in calories. Consequently, except as provided in paragraph (c)(2) of this section, a food may not be labeled with such terms unless:

(i) The food contains less than 0.5 g of sugars, as defined in §101.9(c)(6)(ii), per reference amount customarily consumed and per labeled serving or, in the case of a meal product or main dish product, less than 0.5 g of sugars per labeled serving; and

(ii) The food contains no ingredient that is a sugar or that is generally understood by consumers to contain sugars unless the listing of the ingredient in the ingredient statement is followed by an asterisk that refers to the statement below the list of ingredients, which states "adds a trivial amount of sugar," "adds a negligible amount of sugar," or "adds a dietarily insignificant amount of sugar;" and

(iii) (A) It is labeled "low calorie" or "reduced calorie" or bears a relative claim of special dietary usefulness labeled in compliance with paragraphs (b)(2), (b)(3), (b)(4), or (b)(5) of this section, or, if a dietary supplement, it meets the definition in paragraph (b)(2) of this section for "low-calorie" but is prohibited by §§101.13(b)(5) and 101.60(a)(4) from bearing the claim; or

(B) Such term is immediately accompanied, each time it is used, by either the statement "not a reduced calorie food," "not a low calorie food," or "not for weight control."

(2) The terms "no added sugar," "without added sugar," or "no sugar added" may be used only if:

(i) No amount of sugars, as defined in §101.9(c)(6)(ii), or any other ingredient that contains sugars that functionally substitute for added sugars is added during processing or packaging; and

(ii) The product does not contain an ingredient containing added sugars such as jam, jelly, or concentrated fruit juice; and

(iii) The sugars content has not been increased above the amount present in the ingredients by some means such as the use of enzymes, except where the intended functional effect of the process is not to increase the sugars content of a food, and a functionally insignificant increase in sugars results; and

(iv) The food that it resembles and for which it substitutes normally contains added sugars; and

(v) The product bears a statement that the food is not "low calorie" or "calorie reduced" (unless the food meets the requirements for a "low" or "reduced calorie" food) and that directs consumers' attention to the nutrition panel for further information on sugar and calorie content.

(3) Paragraph (c)(1) of this section shall not apply to a factual statement that a food, including foods intended specifically for infants and children less than 2 years of age, is unsweetened or contains no added sweeteners in the case of a food that contains apparent substantial inherent sugar content, e.g., juices.

(4) The claims provided for in paragraph (c)(1) and (c)(2) of this section may be used on labels or in labeling of dietary supplements of vitamins or minerals that are intended specifically for use by infants and children less than 2 years of age.

(5) The terms "reduced sugar," "reduced in sugar," "sugar reduced," "less sugar," "lower sugar" or "lower in sugar" may be used on the label or in labeling of foods, except meal products as defined in §101.13(l), main dish products as defined in §101.13(m), and dietary supplements of vitamins or minerals, provided that:

(i) The food contains at least 25 percent less sugar per reference amount customarily consumed than an appropriate reference food as described in §101.13(j)(1); and

(ii) As required in §101.13(j)(2) for relative claims:

(A) The identity of the reference food and the percent (or fraction) that the sugar differs between the two foods are declared in immediate proximity to the most prominent such claim (e.g., "these corn flakes contain 25 percent less sugar than our sugar coated corn flakes"); and

(B) Quantitative information comparing the level of the sugar in the product per labeled serving with that of the reference food that it replaces (e.g., "Sugar content has been lowered from 8 g to 6 g per serving") is declared adjacent to the most prominent claim or to the nutrition label, except that if the nutrition label is on the information panel, the quantitative information may be located elsewhere on the information panel in accordance with §101.2.

(6) The terms defined in paragraph (c)(5) of this section may be used on the label or in the labeling of a meal product as defined in §101.13(l) and a main dish product as defined in §101.13(m), provided that:

21 CFR §101.61

(i) The food contains at least 25 percent less sugars per 100 g of food than an appropriate reference food as described in §101.13(j)(1), and

(ii) As required in §101.13(j)(2) for relative claims:

(A) The identity of the reference food and the percent (or fraction) that the sugars differ between the two foods are declared in immediate proximity to the most prominent such claim (e.g., reduced sweet and sour shrimp dinner, "25 percent less sugar per 3 oz than our regular sweet and sour shrimp dinner"); and

(B) Quantitative information comparing the level of the nutrient in the product per specified weight with that of the reference food that it replaces (e.g., "Sugar content has been reduced from 17 g per 3 oz to 13 g per 3 oz.") is declared adjacent to the most prominent claim or to the nutrition label, except that if the nutrition label is on the information panel, the quantitative information may be located elsewhere on the information panel in accordance with §101.2.

[58 FR 2413, Jan. 6, 1993; 58 FR 17342, Apr. 2, 1993; as amended at 58 FR 44031, Aug. 18, 1993; 59 FR 394, Jan. 4, 1994; 60 FR 17206, Apr. 5, 1995; 62 FR 15342, Mar. 31, 1997; 62 *FR* 49881, Sept. 23, 1997]

§101.61 Nutrient content claims for the sodium content of foods.

(a) *General requirements.* A claim about the level of sodium or salt in a food may only be made on the label or in the labeling of the food if:

(1) The claim uses one of the terms defined in this section in accordance with the definition for that term;

(2) The claim is made in accordance with the general requirements for nutrient content claims in §101.13; and

(3) The food for which the claim is made is labeled in accordance with §101.9, §101.10, or §101.36, as applicable.

(b) *"Sodium content claims."*

(1) The terms "sodium free," "free of sodium," "no sodium," "zero sodium," "without sodium," "trivial source of sodium," "negligible source of sodium," or "dietary insignificant source of sodium" may be used on the label or in the labeling of foods, provided that:

(i) The food contains less than 5 milligrams (mg) of sodium per reference amount customarily consumed and per labeled serving or, in the case of a meal product or a main dish product, less than 5 mg of sodium per labeled serving; and

(ii) The food contains no ingredient that is sodium chloride or is generally understood by consumers to contain sodium, unless the listing of the ingredient in the ingredient statement is followed by an asterisk that refers to the statement below the list of ingredients, which states: "Adds a trivial amount of sodium," "adds a negligible amount of sodium" or "adds a dietarily insignificant amount of sodium;" and

(iii) As required in §101.13(e)(2) if the food meets these conditions without the benefit of special processing, alteration, formulation, or reformulation to lower the sodium content, it is labeled to disclose that sodium is not usually present in the food (e.g., "leaf lettuce, a sodium free food").

(2) The terms "very low sodium," or "very low in sodium," may be used on the label or in labeling of foods, except meal products as defined in §101.13(l) and main dish products as defined in §101.13(m), provided that:

(i) (A) The food has a reference amount customarily consumed greater than 30 grams (g) or greater than 2 tablespoons and contains 35 mg or less sodium per reference amount customarily consumed; or

(B) The food has a reference amount customarily consumed of 30 g or less or 2 tablespoons or less and contains 35 mg or less sodium per reference amount customarily consumed and per 50 g (for dehydrated foods that must be reconstituted before typical consumption with water or a diluent containing an insignificant amount, as defined in §101.9(f)(1), of all nutrients per reference amount customarily consumed, the per 50-g criterion refers to the "as prepared" form);

(ii) If the food meets these conditions without the benefit of special processing, alteration, formulation, or reformulation to vary the sodium content, it is labeled to clearly refer to all foods of its type and not merely to the particular brand to which the label attaches (e.g., "potatoes, a very low-sodium food").

(3) The terms defined in paragraph (b)(2) of this section may be used on the label or in labeling of meal products as defined in §101.13(l) and main dish products as defined in §101.13(m), provided that:

21 *CFR* §101.61

(i) The product contains 35 mg or less of sodium per 100 g of product; and

(ii) If the product meets this condition without the benefit of special processing, alteration, formulation, or reformulation to lower the sodium content, it is labeled to clearly refer to all foods of its type and not merely to the particular brand to which the label attaches.

(4) The terms "low sodium," or "low in sodium," "little sodium," "contains a small amount of sodium," or "low source of sodium" may be used on the label or in the labeling of foods, except meal products as defined in §101.13(l) and main dish products as defined in §101.13(m), provided that:

(i) (A) The food has a reference amount customarily consumed greater than 30 g or greater than 2 tablespoons and contains 140 mg or less sodium per reference amount customarily consumed; or

(B) The food has a reference amount customarily consumed of 30 g or less or 2 tablespoons or less and contains 140 mg or less sodium per reference amount customarily consumed and per 50 g (for dehydrated foods that must be reconstituted before typical consumption with water or a diluent containing an insignificant amount, as defined in §101.9(f)(1), of all nutrients per reference amount customarily consumed, the per 50-g criterion refers to the "as prepared" form); and

(ii) If the food meets these conditions without the benefit of special processing, alteration, formulation, or reformulation to vary the sodium content, it is labeled to clearly refer to all foods of its type and not merely to the particular brand to which the label attaches (e.g., "fresh spinach, a low sodium food"); and

(5) The terms defined in paragraph (b)(4) of this section may be used on the label or in labeling of meal products as defined in §101.13(l) and main dish products as defined in §101.13(m), provided that:

(i) The product contains 140 mg or less sodium per 100 g; and

(ii) If the product meets these conditions without the benefit of special processing, alteration, formulation, or reformulation to lower the sodium content, it is labeled to clearly refer to all foods of its type and not merely to the particular brand to which the label attaches.

(6) The terms "reduced sodium," "reduced in sodium," "sodium reduced," "less sodium," "lower sodium," or "lower in sodium" may be used on the label or in labeling of foods, except meal products as defined in §101.13(l) and main dish products as defined in §101.13(m), provided that:

(i) The food contains at least 25 percent less sodium per reference amount customarily consumed than an appropriate reference food as described in §101.13(j)(1).

(ii) As required for §101.13(j)(2) for relative claims:

(A) The identity of the reference food and the percent (or fraction) that the sodium differs from the labeled food are declared in immediate proximity to the most prominent such claim (e.g., "reduced sodium _____, 50 percent less sodium than regular _____"); and

(B) Quantitative information comparing the level of the sodium in the product per labeled serving with that of the reference food that it replaces (e.g., "Sodium content has been lowered from 300 to 150 mg per serving") is declared adjacent to the most prominent claim or to the nutrition label, except that if the nutrition label is on the information panel, the quantitative information may be located elsewhere on the information panel in accordance with §101.2.

(iii) Claims described in paragraph (b)(6) of this section may not be made on the label or in the labeling of a food if the nutrient content of the reference food meets the definition for "low sodium."

(7) The terms defined in paragraph (b)(6) of this section may be used on the label or in the labeling of meal products as defined in §101.13(l) and main dish products as defined in §101.13(m), provided that:

(i) The food contains at least 25 percent less sodium per 100 g of food than an appropriate reference food as described in §101.13(j)(1), and

(ii) As required in §101.13(j)(2) for relative claims:

(A) The identity of the reference food and the percent (or fraction) that the sodium differs from the reference food are declared in immediate proximity to the most prominent such claim (e.g., reduced sodium eggplant parmigiana dinner "30 percent less sodium per oz (or 3 oz) than our regular eggplant parmigiana dinner").

(B) Quantitative information comparing the level of sodium in the product per specified weight with that of the reference food that it replaces (e.g., "Sodium content has been reduced from 217 mg per 3 oz to 150 mg per 3 oz.") is declared adjacent to the most prominent claim or to the nutrition label, except that if the nutrition label is on the information panel, the quantitative information may be located elsewhere on the information panel in accordance with §101.2.

(iii) Claims described in paragraph (b)(7) of this section may not be made on the label or in the labeling of a food if the nutrient content of the reference food meets the definition for "low sodium."

(c) The term "salt" is not synonymous with "sodium." Salt refers to sodium chloride. However, references to salt content such as "unsalted," "no salt," "no salt added" are potentially misleading.

(1) The term "salt free" may be used on the label or in labeling of foods only if the food is "sodium free" as defined in paragraph (b)(1) of this section.

(2) The terms "unsalted," "without added salt," and "no salt added" may be used on the label or in labeling of foods only if:

(i) No salt is added during processing;

(ii) The food that it resembles and for which it substitutes is normally processed with salt; and

(iii) If the food is not sodium free, the statement, "not a sodium free food" or "not for control of sodium in the diet" appears adjacent to the nutrition label of the food bearing the claim, or, if the nutrition label is on the information panel, it may appear elsewhere on the information panel in accordance with §101.2.

(3) Paragraph (c)(2) of this section shall not apply to a factual statement that a food intended specifically for infants and children less than 2 years of age is unsalted, provided such statement refers to the taste of the food and is not otherwise false and misleading.

[58 FR 2413, Jan. 6, 1993; 58 FR 17342, Apr. 2, 1993; as amended at 58 FR 44032, Aug. 18, 1993; 59 FR 394, Jan. 4, 1994; 60 FR 17206, Apr. 5, 1995]

§101.62 Nutrient content claims for fat, fatty acid, and cholesterol content of foods.

(a) *General requirements.* A claim about the level of fat, fatty acid, and cholesterol in a food may only be made on the label or in the labeling of foods if:

(1) The claim uses one of the terms defined in this section in accordance with the definition for that term;

(2) The claim is made in accordance with the general requirements for nutrient content claims in §101.13;

(3) The food for which the claim is made is labeled in accordance with §101.9, §101.10, or §101.36, as applicable; and

(4) For dietary supplements, claims for fat, saturated fat, and cholesterol may not be made on products that meet the criteria in §101.60(b)(1) or (b)(2) for "calorie free" or "low calorie" claims.

(b) *"Fat content claims."*

(1) The terms "fat free," "free of fat," "no fat," "zero fat," "without fat," "nonfat," "trivial source of fat," "negligible source of fat," or "dietarily insignificant source of fat" or, in the case of milk products, "skim" may be used on the label or in labeling of foods, provided that:

(i) The food contains less than 0.5 gram (g) of fat per reference amount customarily consumed and per labeled serving or, in the case of a meal product or main dish product, less than 0.5 g of fat per labeled serving; and

(ii) The food contains no added ingredient that is a fat or is generally understood by consumers to contain fat unless the listing of the ingredient in the ingredient statement is followed by an asterisk that refers to the statement below the list of ingredients, which states "adds a trivial amount of fat," "adds a negligible amount of fat," or "adds a dietarily insignificant amount of fat;" and

(iii) As required in §101.13(e)(2), if the food meets these conditions without the benefit of special processing, alteration, formulation, or reformulation to lower fat content, it is labeled to disclose that fat is not usually present in the food (e.g., "broccoli, a fat free food").

(2) The terms "low fat," "low in fat," "contains a small amount of fat," "low source of fat," or "little fat" may be used on the label or in labeling of foods, except meal products as defined in §101.13(l) and main dish products as defined in §101.13(m), provided that:

(i)

(A) The food has a reference amount customarily consumed greater than 30 g or greater than 2 tablespoons and contains 3 g or less of fat per reference amount customarily consumed; or

(B) The food has a reference amount customarily consumed of 30 g or less or 2 tablespoons or less and contains 3 g or less of fat per reference amount customarily consumed and per 50 g of food (for dehydrated foods that must be reconstituted before typical consumption with water or a diluent containing an insignificant amount, as defined in §101.9(f)(1), of all nutrients per reference amount customarily consumed, the per 50-g criterion refers to the 'as prepared' form); and

(ii) If the food meets these conditions without the benefit of special processing, alteration, formulation, or reformulation to lower fat content, it is labeled to clearly refer to all foods of its type and not merely to the particular brand to which the label attaches (e.g., "frozen perch, a low fat food").

(3) The terms defined in paragraph (b)(2) of this section may be used on the label or in labeling of meal products as defined in §101.13(l) or main dish products as defined in §101.13(m), provided that:

(i) The product contains 3 g or less of total fat per 100 g and not more than 30 percent of calories from fat; and

(ii) If the product meets these conditions without the benefit of special processing, alteration, formulation, or reformulation to lower fat content, it is labeled to clearly refer to all foods of its type and not merely to the particular brand to which the label attaches.

(4) The terms "reduced fat," "reduced in fat," "fat reduced," "less fat," "lower fat," or "lower in fat" may be used on the label or in the labeling of foods, except meal products as defined in §101.13(l) and main dish products as defined in §101.13(m), provided that:

(i) The food contains at least 25 percent less fat per reference amount customarily consumed than an appropriate reference food as described in §101.13(j)(1); and

(ii) As required in §101.13(j)(2) for relative claims:

(A) The identity of the reference food and the percent (or fraction) that the fat differs between the two foods and are declared in immediate proximity to the most prominent such claim (e.g., "reduced fat-50 percent less fat than our regular brownies"); and

(B) Quantitative information comparing the level of fat in the product per labeled serving with that of the reference food that it replaces (e.g., "Fat content has been reduced from 8 g to 4 g per serving.") is declared adjacent to the most prominent claim or to the nutrition label, except that if the nutrition label is on the information panel, the quantitative information may be located elsewhere on the information panel in accordance with §101.2.

(iii) Claims described in paragraph (b)(4) of this section may not be made on the label or in the labeling of a food if the nutrient content of the reference food meets the definition for "low fat."

(5) The terms defined in paragraph (b)(4) of this section may be used on the label or in the labeling of meal products as defined in §101.13(l) and main dish products as defined in §101.13(m), provided that:

(i) The food contains at least 25 percent less fat per 100 g of food than an appropriate reference food as described in §101.13(j)(1); and

(ii) As required in §101.13(j)(2) for relative claims:

(A) The identity of the reference food and the percent (or fraction) that the fat differs between the two foods are declared in immediate proximity to the most prominent such claim (e.g., reduced fat spinach souffle, "33 percent less fat per 3 oz than our regular spinach souffle"); and

(B) Quantitative information comparing the level of fat in the product per specified weight with that of the reference food that it replaces (e.g., "Fat content has been reduced from 7.5 g per 3 oz to 5 g per 3 oz.") is declared adjacent to the most prominent claim, to the nutrition label, or, if the nutrition label is on the information panel, the quantitative information may be located elsewhere on the information panel in accordance with §101.2.

(iii) Claims described in paragraph (b)(5) of this section may not be made on the label or in the labeling of a food if the nutrient content of the reference food meets the definition for "low fat."

(6) The term "--- percent fat free" may be used on the label or in the labeling of foods, provided that:

(i) The food meets the criteria for "low fat" in paragraph (b)(2) or (b)(3) of this section;

(ii) The percent declared and the words "fat free" are in uniform type size; and

(iii) A "100 percent fat free" claim may be made only on foods that meet the criteria for "fat free" in paragraph (b)(1) of this section, that contain less than 0.5 g of fat per 100 g, and that contain no added fat.

(c) *"Fatty acid content claims."* The label or labeling of foods that bear claims with respect to the level of saturated fat shall disclose the level of total fat and cholesterol in the food in immediate proximity to such claim each time the claim is made and in type that shall be no less than one-half the size of the type used for the claim with respect to the level of saturated fat. Declaration of cholesterol content may be omitted when the food contains less than 2 milligrams (mg) of cholesterol per reference amount customarily consumed or in the case of a meal or main dish

21 CFR §101.62

product less than 2 mg of cholesterol per labeled serving. Declaration of total fat may be omitted with the term defined in paragraph (c)(1) of this section when the food contains less than 0.5 g of total fat per reference amount customarily consumed or, in the case of a meal product or a main dish product, when the product contains less than 0.5 g of total fat per labeled serving. The declaration of total fat may be omitted with the terms defined in paragraphs (c)(2) through (c)(5) of this section when the food contains 3 g or less of total fat per reference amount customarily consumed or in the case of a meal product or a main dish product, when the product contains 3 g or less of total fat per 100 g and not more than 30 percent calories from fat.

(1) The terms "saturated fat free," "free of saturated fat," "no saturated fat," "zero saturated fat," "without saturated fat," "trivial source of saturated fat," "negligible source of saturated fat," or "dietarily insignificant source of saturated fat" may be used on the label or in the labeling of foods, provided that:
(i) The food contains less than 0.5 g of saturated fat and less than 0.5 g trans fatty acid per reference amount customarily consumed and per labeled serving, or in the case of a meal product or main dish product, less than 0.5 g of saturated fat and less than 0.5 g trans fatty acid per labeled serving; and
(ii) The food contains no ingredient that is generally understood by consumers to contain saturated fat unless the listing of the ingredient in the ingredient statement is followed by an asterisk that refers to the statement below the list of ingredients which states, "adds a trivial amount of saturated fat," "adds a negligible amount of saturated fat," or "adds a dietarily insignificant amount of saturated fat;" and
(iii) As required in §101.13(e)(2), if the food meets these conditions without the benefit of special processing, alteration, formulation, or reformulation to lower saturated fat content, it is labeled to disclose that saturated fat is not usually present in the food.
(2) The terms "low in saturated fat," "low saturated fat," "contains a small amount of saturated fat," "low source of saturated fat," or "a little saturated fat" may be used on the label or in the labeling of foods, except meal products as defined in §101.13(l) and main dish products as defined in §101.13(m), provided that:
(i) The food contains 1 g or less of saturated fatty acids per reference amount customarily consumed and not more than 15 percent of calories from saturated fatty acids; and
(ii) If a food meets these conditions without benefit of special processing, alteration, formulation, or reformulation to lower saturated fat content, it is labeled to clearly refer to all foods of its type and not merely to the particular brand to which the label attaches (e.g., "raspberries, a low saturated fat food").
(3) The terms defined in paragraph (c)(2) of this section may be used on the label or in the labeling of meal products as defined in §101.13(l) and main dish products as defined in §101.13(m), provided that:
(i) The product contains 1 g or less of saturated fatty acids per 100 g and less than 10 percent calories from saturated fat; and
(ii) If the product meets these conditions without the benefit of special processing, alteration, formulation, or reformulation to lower saturated fat content, it is labeled to clearly refer to all foods of its type and not merely to the particular brand to which the label attaches.
(4) The terms "reduced saturated fat," "reduced in saturated fat," "saturated fat reduced," "less saturated fat," "lower saturated fat," or "lower in saturated fat" may be used on the label or in the labeling of foods, except as limited by §101.13(j)(1)(i) and except meal products as defined in §101.13(l) and main dish products as defined in §101.13(m), provided that:
(i) The food contains at least 25 percent less saturated fat per reference amount customarily consumed than an appropriate reference food as described in §101.13(j)(1); and
(ii) As required in §101.13(j)(2) for relative claims:
(A) The identity of the reference food and the percent (or fraction) that the saturated fat differs between the two foods are declared in immediate proximity to the most prominent such claim (e.g., "reduced saturated fat. Contains 50 percent less saturated fat than the national average for nondairy creamers"); and
(B) Quantitative information comparing the level of saturated fat in the product per labeled serving with that of the reference food that it replaces (e.g., "Saturated fat reduced from 3 g to 1.5 g per serving") is declared adjacent to the most prominent claim or to the nutrition label, except that if the nutrition label is on the information panel, the quantitative information may be located elsewhere on the information panel in accordance with §101.2.
(iii) Claims described in paragraph (c)(4) of this section may not be made on the label or in the labeling of a food if the nutrient content of the reference food meets the definition for "low saturated fat."

21 CFR §101.62

(5) The terms defined in paragraph (c)(4) of this section may be used on the label or in the labeling of meal products as defined in §101.13(l) and main dish products as defined in §101.13(m), provided that:
(i) The food contains at least 25 percent less saturated fat per 100 g of food than an appropriate reference food as described in §101.13(j)(1), and
(ii) As required in §101.13(j)(2) for relative claims:
(A) The identity of the reference food, and the percent (or fraction) that the fat differs between two foods are declared in immediate proximity to the most prominent such claim (e.g., reduced saturated fat Macaroni and Cheese, "33 percent less saturated fat per 3 oz than our regular Macaroni and Cheese").
(B) Quantitative information comparing the level of saturated fat in the product per specified weight with that of the reference food that it replaces (e.g., "Saturated fat content has been reduced from 2.5 g per 3 oz to 1.7 g per 3 oz.") is declared adjacent to the most prominent claim or to the nutrition label, except that if the nutrition label is on the information panel, the quantitative information may be located elsewhere on the information panel in accordance with §101.2.
(iii) Claims described in paragraph (c)(5) of this section may not be made on the label or in the labeling of a food if the nutrient content of the reference food meets the definition for "low saturated fat."

(d) *"Cholesterol content claims."*

(1) The terms "cholesterol free," "free of cholesterol," "zero cholesterol," "without cholesterol," "no cholesterol," "trivial source of cholesterol," "negligible source of cholesterol," or "dietarily insignificant source of cholesterol" may be used on the label or in the labeling of foods, provided that:
(i) For foods that contain 13 g or less of total fat per reference amount cutomarily consumed, per labeled serving, and per 50 g if the reference amount customarily consumed is 30 g or less or 2 tablespoons or less (for dehydrated foods that must be reconstituted before typical consumption with water or a diluent containing an insignificant amount, as defined in §101.9(f)(1), of all nutrients per reference amount customarily consumed, the per 50-g criterion refers to the 'as prepared' form), or, in the case of meal products, 26.0 g or less total fat per labeled serving, or, in the case of main dish products, 19.5 g or less total fat per labeled serving:
(A) The food contains less than 2 mg of cholesterol per reference amount customarily consumed and per labeling serving or, in the case of a meal product or main dish product, less than 2 mg of cholesterol per labeled serving; and
(B) The food contains no ingredient that is generally understood by consumers to contain cholesterol, unless the listing of the ingredient in the ingredient statement is followed by an asterisk that refers to the statement below the list of ingredients, which states "adds a trivial amount of cholesterol," "adds a negligible amount of cholesterol," or "adds a dietarily insignificant amount of cholesterol;" and
(C) The food contains 2 g or less of saturated fatty acids per reference amount customarily consumed or, in the case of a meal product or main dish product, 2 g or less of saturated fatty acids per labeled serving; and
(D) As required in §101.13(e)(2), if the food contains less than 2 mg of cholesterol per reference amount customarily consumed or in the case of a meal product or main dish product, less than 2 mg of cholesterol per labeled serving without the benefit of special processing, alteration, formulation, or reformulation to lower cholesterol content, it is labeled to disclose that cholesterol is not usually present in the food (e.g., "applesauce, a cholesterol-free food").
(ii) For food that contain more than 13 g of total fat per reference amount customarily consumed, per labeling serving, or per 50 g if the reference amount customarily consumed is 30 g or less or 2 tablespoons or less (for dehydrated foods that must be reconstituted before typical consumption with water or a diluent containing an insignificant amount, as defined in §101.9(f)(1), of all nutrients per reference amount customarily consumed, the per 50-g criterion refers to the 'as prepared' form) or in the case of a meal product, more than 26 g of total fat per labeled serving, or, in the case of a main dish product more than 19.5 g of total fat per labeled serving:
(A) The food contains less than 2 mg of cholesterol per reference amount customarily consumed and per labeling serving or, in the case of a meal product or main dish product, less than 2 mg of cholesterol per labeled serving; and
(B) The food contains no ingredient that is generally understood by consumers to contain cholesterol, unless the listing of the ingredient in the ingredient statement is followed by an asterisk that refers to the statement below the list of ingredients, which states "adds a trivial amount of cholesterol," "adds a

21 CFR §101.62

negligible amount of cholesterol," or "adds a dietarily insignificant amount of cholesterol;" and

(C) The food contains 2 g or less of saturated fatty acids per reference amount customarily consumed or, in the case of a meal product or main dish product less than 2 g of saturated fatty acids per labeled serving; and

(D) The label or labeling discloses the level of total fat in a serving (as declared on the label) of the food. Such disclosure shall appear in immediate proximity to such claim preceding the referral statement required in §101.13(g) in type that shall be no less than one-half the size of the type used for such claim. If the claim appears on more than one panel, the disclosure shall be made on each panel except for the panel that bears nutrition labeling. If the claim appears more than once on a panel, the disclosure shall be made in immediate proximity to the claim that is printed in the largest type; and

(E) As required in §101.13(e)(2), if the food contains less than 2 mg of cholesterol per reference amount customarily consumed or in the case of a meal product or main dish product less than 2 mg of cholesterol per labeled serving without the benefit of special processing, alteration, formulation, or reformulation to lower cholesterol content, it is labeled to disclose that cholesterol is not usually present in the food (e.g., "canola oil, a cholesterol-free food, contains 14 g of fat per serving"); or

(F) If the food contains less than 2 mg of cholesterol per reference amount customarily consumed or in the case of a meal product or main dish product less than 2 mg of cholesterol per labeled serving only as a result of special processing, alteration, formulation, or reformulation, the amount of cholesterol is substantially less (i.e., meets requirements of paragraph (d)(4)(ii)(A) of this section) than the food for which it substitutes as specified in §101.13(d) that has a significant (e.g., 5 percent or more of a national or regional market) market share. As required in §101.13(j)(2) for relative claims:

(*1*) The identity of the reference food and the percent (or fraction) that the cholesterol was reduced are declared in immediate proximity to the most prominent such claim (e.g., "cholesterol-free margarine, contains 100 percent less cholesterol than butter"); and

(*2*) Quantitative information comparing the level of cholesterol in the product per labeled serving with that of the reference food that it replaces (e.g., "Contains no cholesterol compared with 30 mg in one serving of butter. Contains 13 g of fat per serving.") is declared adjacent to the most prominent claim or to the nutrition label, except that if the nutrition label is on the information panel, the quantitative information may be located elsewhere on the information panel in accordance with §101.2.

(2) The terms "low in cholesterol," "low cholesterol," "contains a small amount of cholesterol," "low source of cholesterol," or "little cholesterol" may be used on the label or in the labeling of foods, except meal products as defined in §101.13(l) and main dish products as defined in §101.13(m), provided that:

(i) For foods that have a reference amount customarily consumed greater than 30 g or greater than 2 tablespoons and contain 13 g or less of total fat per reference amount customarily consumed and per labeled serving:

(A) The food contains 20 mg or less of cholesterol per reference amount customarily consumed;

(B) The food contains 2 g or less of saturated fatty acids per reference amount customarily consumed; and

(C) As required in §101.13(e)(2), if the food meets these conditions without the benefit of special processing, alteration, formulation, or reformulation to lower cholesterol content, it is labeled to clearly refer to all foods of that type and not merely to the particular brand to which the label attaches (e.g., "low fat cottage cheese, a low cholesterol food.").

(ii) For foods that have a reference amount customarily consumed of 30 g or less or 2 tablespoons or less and contain 13 g or less of total fat per reference amount customarily consumed, per labeled serving, and per 50 g (for dehydrated foods that must be reconstituted before typical consumption with water or a diluent containing an insignificant amount, as defined in §101.9(f)(1), of all nutrients per reference amount customarily consumed, the per 50-g criterion refers to the 'as prepared' form);

(A) The food contains 20 mg or less of cholesterol per reference amount customarily consumed and per 50 g (for dehydrated foods that must be reconstituted before typical consumption with water or a diluent containing an insignificant amount, as defined in §101.9(f)(1), of all nutrients per reference amount customarily consumed, the per 50-g criterion refers to the 'as prepared' form);

(B) The food contains 2 g or less of saturated fatty acids per reference amount customarily consumed; and

(C) As required in §101.13(e)(2), if the food meets these conditions without the benefit of special processing, alteration, formulation, or reformulation to lower cholesterol content, it is labeled to clearly refer to all foods of that type and not merely to the particular brand to which the label attaches (e.g., "low fat cottage cheese, a low cholesterol food").

(iii) For foods that have a reference amount customarily consumed greater than 30 g or greater than 2 tablespoons and contain more than 13 g of total fat per reference amount customarily consumed or per labeled serving,

(A) The food contains 20 mg or less of cholesterol per reference amount customarily consumed;

(B) The food contains 2 g or less of saturated fatty acids per reference amount customarily consumed;

(C) The label or labeling discloses the level of total fat in a serving (as declared on the label) of the food. Such disclosure shall appear in immediate proximity to such claim preceding the referral statement required in §101.13(g) in type that shall be no less than one-half the size of the type used for such claim. If the claim appears on more than one panel, the disclosure shall be made on each panel except for the panel that bears nutrition labeling. If the claim is made more than once on a panel, the disclosure shall be made in immediate proximity to the claim that is printed in the largest type; and

(D) As required in §101.13(e)(2), if the food meets these conditions without the benefit of special processing, alteration, formulation, or reformulation to lower cholesterol content, it is labeled to clearly refer to all foods of that type and not merely to the particular brand to which the label attaches; or

(E) If the food contains 20 mg or less of cholesterol only as a result of special processing, alteration, formulation, or reformulation, the amount of cholesterol is substantially less (i.e., meets requirements of paragraph (d)(4)(ii)(A) of this section) than the food for which it substitutes as specified in §101.13(d) that has a significant (e.g., 5 percent or more of a national or regional market) market share. As required in §101.13(j)(2) for relative claims:

(*1*) The identity of the reference food and the percent (or fraction) that the cholesterol has been reduced are declared in immediate proximity to the most prominent such claim (e.g., "low-cholesterol peanut butter sandwich crackers, contains 83 percent less cholesterol than our regular peanut butter sandwich crackers"); and

(*2*) Quantitative information comparing the level of cholesterol in the product per labeled serving with that of the reference food that it replaces (e.g., "Cholesterol lowered from 30 mg to 5 mg per serving, contains 13 g of fat per serving.") is declared adjacent to the most prominent claim or to the nutrition label, except that if the nutrition label is on the information panel, the quantitative information may be located elsewhere on the information panel in accordance with §101.2..

(iv) For foods that have a reference amount customarily consumed of 30 g or less or 2 tablespoons or less and contain more than 13 g of total fat per reference amount customarily consumed, per labeled serving, or per 50 g (for dehydrated foods that must be reconstituted before typical consumption with water or a diluent containing an insignificant amount, as defined in §101.9(f)(1), of all nutrients per reference amount customarily consumed, the per 50-g criterion refers to the 'as prepared' form),

(A) The food contains 20 mg or less of cholesterol per reference amount customarily consumed and per 50 g (for dehydrated foods that must be reconstituted before typical consumption with water or a diluent containing an insignificant amount, as defined in §101.9(f)(1), of all nutrients per reference amount customarily consumed, the per 50-g criterion refers to the 'as prepared' form),

(B) The food contains 2 g or less of saturated fatty acids per reference amount customarily consumed;

(C) The label or labeling discloses the level of total fat in a serving (as declared on the label) of the food. Such disclosure shall appear in immediate proximity to such claim preceding the referral statement required in §101.13(g) in type that shall be no less than one-half the size of the type used for such claim. If the claim appears on more than one panel, the disclosure shall be made on each panel except for the panel that bears nutrition labeling. If the claim is made more than once on a panel, the disclosure shall be made in immediate proximity to the claim that is printed in the largest type; and

(D) As required in §101.13(e)(2), if the food meets these conditions without the benefit of special processing, alteration, formulation, or reformulation to lower cholesterol content, it is labeled to clearly refer to all foods of that type and not merely to the particular brand to which the label attaches; or

(E) If the food contains 20 mg or less of cholesterol only as a result of special processing, alteration, formulation, or reformulation, the amount of cholesterol is substantially less (i.e., meets requirements of paragraph (d)(4)(ii)(A) of this section) than the food for which it substitutes as specified in §101.13(d) that has a significant (i.e., 5 percent or more of a national or regional market) market share. As required in

21 CFR §101.62

§101.13(j)(2) for relative claims:

(*1*) The identity of the reference food and the percent (or fraction) that the cholesterol has been reduced are declared in immediate proximity to the most prominent such claim (e.g., "low-cholesterol peanut butter sandwich crackers, contains 83 percent less cholesterol than our regular peanut butter sandwich crackers"); and

(*2*) Quantitative information comparing the level of cholesterol in the product per labeled serving with that of the reference food that it replaces (e.g., "Cholesterol lowered from 30 mg to 5 mg per serving, contains 13 g of fat per serving.") is declared adjacent to the most prominent claim or to the nutrition label, except that if the nutrition label is on the information panel, the quantitative information may be located elsewhere on the information panel in accordance with §101.2.

(3) The terms defined in paragraph (d)(2) of this section may be used on the label and in labeling of meal products as defined in §101.13(l) or a main dish product as defined in §101.13(m) provided that the product meets the requirements of paragraph (d)(2) of this section except that the determination as to whether paragraph (d)(2)(i) or (d)(2)(iii) of this section applies to the product will be made only on the basis of whether the meal product contains 26 g or less of total fat per labeled serving or the main dish product contain 19.5 g or less of total fat per labeled serving, the requirement in paragraphs (d)(2)(i)(A) and (d)(2)(iii)(A) of this section shall be limited to 20 mg of cholesterol per 100 g, and the requirement in paragraphs (d)(2)(i)(B) and (d)(2)(iii)(B) of this section shall be modified to require that the food contain 2 g or less of saturated fat per 100 g rather than per reference amount customarily consumed.

(4) The terms "reduced cholesterol," "reduced in cholesterol," "cholesterol reduced," "less cholesterol," "lower cholesterol," or "lower in cholesterol" except as limited by §101.13(j)(1)(i) may be used on the label or in labeling of foods or foods that substitute for those foods as specified in §101.13(d), excluding meal products as defined in §101.13(l) and main dish products as defined in §101.13(m), provided that:

(i) For foods that contain 13 g or less of total fat per reference amount customarily consumed, per labeled serving, and per 50 g if the reference amount customarily consumed is 30 g or less or 2 tablespoons or less (for dehydrated foods that must be reconstituted before typical consumption with water or a diluent containing an insignificant amount, as defined in §101.9(f)(1), of all nutrients per reference amount customarily consumed, the per 50-g criterion refers to the 'as prepared' form):

(A) The food has been specifically formulated, altered, or processed to reduce its cholesterol by 25 percent or more from the reference food it resembles as defined in §101.13(j)(1) and for which it substitutes as specified in §101.13(d) that has a significant (i.e., 5 percent or more) market share; and

(B) The food contains 2 g or less of saturated fatty acids per reference amount customarily consumed; and

(C) As required in §101.13(j)(2) for relative claims:

(*1*) The identity of the reference food and the percent (or fraction) that the cholesterol has been reduced are declared in immediate proximity to the most prominent such claim; and

(*2*) Quantitative information comparing the level of cholesterol in the product per labeled serving with that of the reference food that it replaces (e.g., "[labeled product] 50 mg cholesterol per serving; [reference product] 30 mg cholesterol per serving") is declared adjacent to the most prominent claim or to the nutrition label, except that if the nutrition label is on the information panel, the quantitative information may be located elsewhere on the information panel in accordance with §101.2.

(ii) For foods that contain more than 13 g of total fat per reference amount customarily consumed, per labeled serving, or per 50 g if the reference amount customarily consumed is 30 g or less or 2 tablespoons or less (for dehydrated foods that must be reconstituted before typical consumption with water or a diluent containing an insignificant amount, as defined in §101.9(f)(1), of all nutrients per reference amount customarily consumed, the per 50-g criterion refers to the 'as prepared' form):

(A) The food has been specifically formulated, altered, or processed to reduce its cholesterol by 25 percent or more from the reference food it resembles as defined in §101.13(j)(1) and for which it substitutes as specified in §101.13(d) that has a significant (i.e., 5 percent or more of a national or regional market) market share;

(B) The food contains 2 g or less of saturated fatty acids per reference amount customarily consumed;

(C) The label or labeling discloses the level of total fat in a serving (as declared on the label) of the food. Such disclosure shall appear in immediate proximity to such claim preceding the referral statement required in §101.13(g) in type that shall be no less than one-half the size of the type used for such claim. If the claim appears on more than one panel, the disclosure shall be made on each panel except for the panel that bears nutrition labeling. If the claim is made more than once on a panel, the disclosure shall be made in immediate proximity to the claim that is printed in the largest type; and

(D) As required in §101.13(j)(2) for relative claims:

(*1*) The identity of the reference food and the percent (or fraction) that the cholesterol has been reduced are declared in immediate proximity to the most prominent such claim (e.g., 25 percent less cholesterol than _____); and

(*2*) Quantitative information comparing the level of cholesterol in the product per labeled serving with that of the reference food that it replaces (e.g., "Cholesterol lowered from 55 mg to 30 mg per serving. Contains 13 g of fat per serving") is declared adjacent to the most prominent claim or to the nutrition label, except that if the nutrition label is on the information panel, the quantitative information may be located elsewhere on the information panel in accordance with §101.2.

(iii) Claims described in paragraph (d)(4) of this section may not be made on the label or in labeling of a food if the nutrient content of the reference food meets the definition for "low cholesterol."

(5) The terms defined in paragraph (d)(4) of this section may be used on the label or in the labeling of meal products as defined in §101.13(l) and main dish products as defined in §101.13(m), provided that:

(i) For meal products that contain 26.0 g or less of total fat per labeled serving or for main dish products that contain 19.5 g or less of total fat per labeled serving;

(A) The food has been specifically formulated, altered, or processed to reduce its cholesterol by 25 percent or more from the reference food it resembles as defined in §101.13(j)(1) and for which it substitutes as specified in §101.13(d) that has a significant (e.g., 5 percent or more of a national or regional market) market share;

(B) The food contains 2 g or less of saturated fatty acids per 100 g; and

(C) As required in §101.13(j)(2) for relative claims:

(*1*) The identity of the reference food, and the percent (or fraction) that the cholesterol has been reduced are declared in immediate proximity to the most prominent such claim (e.g., "25% less cholesterol per 3 oz than _____ ."); and

(*2*) Quantitative information comparing the level of cholesterol in the product per specified weight with that of the reference food that it replaces (e.g., "Cholesterol content has been reduced from 35 mg per 3 oz to 25 mg per 3 oz.") is declared adjacent to the most prominent claim or to the nutrition label, except that if the nutrition label is on the information panel, the quantitative information may be located elsewhere on the information panel in accordance with §101.2.

(ii) For meal products that contain more than 26.0 g of total fat per labeled serving or for main dish products that contain more than 19.5 g of total fat per labeled serving:

(A) The food has been specifically formulated, altered, or processed to reduce its cholesterol by 25 percent or more from the reference food it resembles as defined in §101.13(j)(1) and for which it substitutes as specified in §101.13(d) that has a significant (e.g., 5 percent or more of a national or regional market) market share.

(B) The food contains 2 g or less of saturated fatty acids per 100 g;

(C) The label or labeling discloses the level of total fat in a serving (as declared on the label) of the food. Such disclosure shall appear in immediate proximity to such claim preceding the referral statement required in §101.13(g) in type that shall be no less than one-half the size of the type used for such claim. If the claim appears on more than one panel the disclosure shall be made on each panel except for the panel that bears nutrition labeling. If the claim is made more than once on a panel, the disclosure shall be made in immediate proximity to the claim that is printed in the largest type; and

(D) As required in §101.13(j)(2) for relative claims:

(*1*) The identity of the reference food and the percent (or fraction) that the cholesterol has been reduced are declared in immediate proximity to the most prominent such claim (e.g., 25 percent less cholesterol than _____); and

(*2*) Quantitative information comparing the level of cholesterol in the product per specified weight with that of the reference food that it replaces (e.g., "Cholesterol lowered from 30 mg to 22 mg per 3 oz of product.") is declared adjacent to the most prominent claim or to the nutrition label, except that if the nutrition label is on the information panel, the quantitative information may be located elsewhere on the information panel in accordance with §101.2.

(iii) Claims described in paragraph (d)(5) of this section may not be made on the label or in the labeling of a food if the nutrient content of the reference food meets the definition for "low cholesterol."

(e) *"Lean" and "extra lean" claims.*

(1) The term "lean" may be used on the label or in labeling of foods except meal products as defined in §101.13(l) and main dish products as defined in §101.13(m) provided that the food is a seafood or game meat product and as packaged contains less than 10 g total fat, 4.5 g or less saturated fat, and less than 95 mg cholesterol per reference amount customarily consumed and per 100 g;

(2) The term defined in paragraph (e)(1) of this section may be used on the label or in the labeling of meal products as defined in §101.13(l) or main dish products as defined in §101.13(m) provided that the food contains less than 10 g total fat, 4.5 g or less saturated fat, and less than 95 mg cholesterol per 100 g and per labeled serving;

(3) The term "extra lean" may be used on the label or in the labeling of foods except meal products as defined in §101.13(l) and main dish products as defined in §101.13(m) provided that the food is a discrete seafood or game meat product and as packaged contains less than 5 g total fat, less than 2 g saturated fat, and less than 95 mg cholesterol per reference amount customarily consumed and per 100 g; and

(4) The term defined in paragraph (e)(3) of this section may be used on the label or in labeling of meal products as defined in §101.13(l) and main dish products as defined in §101.13(m) provided that the food contains less than 5 g of fat, less than 2 g of saturated fat, and less than 95 mg of cholesterol per 100 g and per labeled serving.

(f) *Misbranding.* Any label or labeling containing any statement concerning fat, fatty acids, or cholesterol that is not in conformity with this section shall be deemed to be misbranded under sections 201(n), 403(a), and 403(r) of the Federal Food, Drug, and Cosmetic Act.

[58 FR 2413, Jan. 6, 1993; 58 FR 17342, 17343, Apr. 2, 1993, as amended at 58 FR 44032, Aug. 18, 1993; 58 FR 60105, Nov. 15, 1993; 59 FR 394, Jan. 4, 1994; 60 FR 17207, Apr. 5, 1995; 61 FR 59001, Nov. 20, 1996]

§101.65 Implied nutrient content claims and related label statements.

(a) *General requirements.* An implied nutrient content claim can only be made on the label and in labeling of the food if:

(1) The claim uses one of the terms described in this section in accordance with the definition for that term;
(2) The claim is made in accordance with the general requirements for nutrient content claims in §101.13; and
(3) The food is labeled in accordance with §101.9, §101.10, or §101.36, as applicable.

(b) *Label statements that are not implied claims.* Certain label statements about the nature of a product are not nutrient content claims unless such statements are made in a context that would make them an implied claim under §101.13(b)(2). The following types of label statements are generally not implied nutrient content claims and, as such, are not subject to the requirements of §101.13 and this section:

(1) A claim that a specific ingredient or food component is absent from a product, provided that the purpose of such claim is to facilitate avoidance of the substances because of food allergies (see §105.62 of this chapter), food intolerance, religious beliefs, or dietary practices such as vegetarianism or other nonnutrition related reason, e.g., "100 percent milk free;"

(2) A claim about a substance that is nonnutritive or that does not have a nutritive function, e.g., "contains no preservatives," "no artificial colors;"

(3) A claim about the presence of an ingredient that is perceived to add value to the product e.g., "made with real butter," "made with whole fruit," or "contains honey;" except that claims about the presence of ingredients other than vitamins or minerals or that are represented as a source of vitamins or minerals are not allowed on labels or in labeling of dietary supplements of vitamins and minerals that are not in conventional food form.

(4) A statement of identity for a food in which an ingredient constitutes essentially 100 percent of a food (e.g, "corn oil," "oat bran.," "dietary supplement of vitamin C 60 mg tablet");

(5) A statement of identity that names as a characterizing ingredient, an ingredient associated with a

nutrient benefit (e.g., "corn oil margarine," "oat bran muffins," or "whole wheat bagels"), unless such claim is made in a context in which label or labeling statements, symbols, vignettes, or other forms of communication suggest that a nutrient is absent or present in a certain amount; and

(6) A label statement made in compliance with a specific provision of part 105 of this chapter, solely to note that a food has special dietary usefulness relative to a physical, physiological, pathological, or other condition, where the claim identifies the special diet of which the food is intended to be a part.

(c) *Particular implied nutrient content claims.*

(1) Claims about the food or an ingredient therein that suggest that a nutrient or an ingredient is absent or present in a certain amount (e.g., "high in oat bran") are implied nutrient content claims and must comply with paragraph (a) of this section.

(2) The phrases "contains the same amount of [nutrient] as a [food]" and "as much [nutrient] as a [food]" may be used on the label or in the labeling of foods, provided that the amount of the nutrient in the reference food is enough to qualify that food as a "good source" of that nutrient, and the labeled food, on a per serving basis, is an equivalent, good source of that nutrient (e.g., "as much fiber as an apple," "Contains the same amount of Vitamin C as an 8 oz glass of orange juice.").

(3) Claims may be made that a food contains or is made with an ingredient that is known to contain a particular nutrient, or is prepared in a way that affects the content of a particular nutrient in the food, if the finished food is either "low" in or a "good source" of the nutrient that is associated with the ingredient or type of preparation. If a more specific level is claimed (e.g., "high in -----"), that level of the nutrient must be present in the food. For example, a claim that a food contains oat bran is a claim that it is a good source of dietary fiber; that a food is made only with vegetable oil is a claim that it is low in saturated fat; and that a food contains no oil is a claim that it is fat free.

(d) *General nutritional claims.*

(1) Claims about a food that suggest that the food because of its nutrient content may be useful in maintaining healthy dietary practices and that are made in association with an explicit claim or statement about a nutrient (e.g., "healthy, contains 3 grams of fat") are implied nutrient content claims covered by this paragraph.

(2) The term "healthy" or any derivative of the term "healthy," such as "health," "healthful," "healthfully," "healthfulness," "healthier," "healthiest," "healthily," and "healthiness" may be used on the label or in labeling of a food, other than raw, single ingredient seafood or game meat products, main dish products as defined in §101.13(m), and meal products as defined in §101.13(l), as an implied nutrient content claim to denote foods that are useful in constructing a diet that is consistent with dietary recommendations provided that:

(i) The food meets the definition of "low" for fat and saturated fat;

(ii) (A) The food has a reference amount customarily consumed greater than 30 grams (g) or greater than 2 tablespoons and, before January 1, 1998, contains 480 milligrams (mg) sodium or less per reference amount customarily consumed, per labeled serving; or

(B) The food has a reference amount customarily consumed of 30 g or less or 2 tablespoons or less and, before January 1, 1998, contains 480 mg sodium or less per 50 g (for dehydrated foods that must be reconstituted before typical consumption with water or a diluent containing an insignificant amount as defined in §101.9(f)(1), of all nutrients per reference amount customarily consumed, the per 50 g criterion refers to the "as prepared" form);

(C) (1) The food has a reference amount customarily consumed greater than 30 g or greater than 2 tablespoons and, after January 1, 1998, contains 360 mg sodium or less per reference amount customarily consumed, per labeled serving; or

(2) The food has a reference amount customarily consumed of 30 g or less or 2 tablespoons or less and, after January 1, 1998, contains 360 mg sodium or less per 50 g (for dehydrated foods that must be reconstituted before typical consumption with water or a diluent containing an insignificant amount as defined in §101.9(f)(1), of all nutrients per reference amount customarily consumed, the per 50 g criterion refers to the "as prepared" form);

(iii) Cholesterol is not present at a level exceeding the disclosure level as described in §101.13(h);

(iv) The food contains at least 10 percent of the Reference Daily Intake (RDI) or Daily Reference Value (DRV) per reference amount customarily consumed of vitamin A, vitamin C, calcium, iron, protein, or fiber, except for the following:

(A) Raw fruits and vegetables;

(B) Frozen or canned single ingredient fruits and vegetables and mixtures of frozen or canned single ingredient fruits and vegetables, except that ingredients whose addition does not change the nutrient profile of the fruit or vegetable may be added;

(C) Enriched cereal-grain products that conform to a standard of identity in part 136, 137, or 139 of this chapter.

(v) Where compliance with paragraph (d)(2)(iv) of this section is based on a nutrient that has been added to the food, that fortification is in accordance with the policy on fortification of foods in §104.20 of this chapter; and

(vi) The food complies with definitions and declaration requirements established in part 101 of this chapter for any specific nutrient content claim on the label or in labeling.

(3) The term "healthy" or its derivatives may be used on the label or in labeling of raw, single ingredient seafood or game meat as an implied nutrient content claim provided that:

(i) The food contains less than 5 g total fat, less than 2 g saturated fat, and less than 95 mg cholesterol per reference amount customarily consumed and per 100 g;

(ii) (A) The food has a reference amount customarily consumed greater than 30 g or greater than 2 tablespoons and, before January 1, 1998, contains 480 mg sodium or less per reference amount customarily consumed, per labeled serving; or

(B) The food has a reference amount customarily consumed of 30 g or less or 2 tablespoons or less and, before January 1, 1998, contains 480 mg sodium or less per 50 g (for dehydrated foods that must be reconstituted before typical consumption with water or a diluent containing an insignificant amount as defined in §101.9(f)(1), of all nutrients per reference amount customarily consumed, the per 50 g criterion refers to the "as prepared" form);

(C) (1) The food has a reference amount customarily consumed greater than 30 g or greater than 2 tablespoons and, after January 1, 1998, contains 360 mg sodium or less per reference amount customarily consumed, per labeled serving; or

(2) The food has a reference amount customarily consumed of 30 g or less or 2 tablespoons or less and, after January 1, 1998, contains 360 mg sodium or less per 50 g (for dehydrated foods that must be reconstituted before typical consumption with water or a diluent containing an insignificant amount as defined in §101.9(f)(1), of all nutrients per reference amount customarily consumed, the per 50 g criterion refers to the "as prepared" form);

(iii) The food contains at least 10 percent of the RDI or DRV per reference amount customarily consumed, per labeled serving of vitamin A, vitamin C, calcium, iron, protein, or fiber;

(iv) Where compliance with paragraph (d)(3)(iii) of this section is based on a nutrient that has been added to the food, that fortification is in accordance with the policy on fortification of foods in §104.20 of this chapter; and

(v) The food complies with definitions and declaration requirements established in this part for any specific nutrient content claim on the label or in labeling.

(4) The term "healthy" or its derivatives may be used on the label or in labeling of main dish products, as defined in §101.13(m), and meal products, as defined in §101.13(l) as an implied nutrient content claim provided that:

(i) The food meets the definition of "low" for fat and saturated fat;

(ii) (A) Before January 1, 1998, sodium is not present at a level exceeding 600 mg per labeled serving, or

(B) After January 1, 1998, sodium is not present at a level exceeding 480 mg per labeled serving;

(iii) Cholesterol is not present at a level exceeding 90 mg per labeled serving;

(iv) The food contains at least 10 percent of the RDI or DRV per labeled serving of two (for main dish products) or three (for meal products) of the following nutrients-vitamin A, vitamin C, calcium, iron, protein, or fiber;

(v) Where compliance with paragraph (d)(4)(iv) of this section is based on a nutrient that has been added to the food, that fortification is in accordance with the policy on fortification of foods in §104.20 of this chapter; and

(vi) The food complies with definitions and declaration requirements established in this part for any

21 CFR §101.66

specific nutrient content claim on the label or in labeling.

Effective date note: *§§101.65(d)(2)(ii)(C) and 101.65(d)(4)(ii)(B) are stayed until January 1, 2000* [62 FR 15391, Apr. 1, 1997].

[58 FR 2413, Jan. 6, 1993; 58 FR 17343, Apr. 2, 1993, as amended at 59 FR 394, Jan. 4, 1994; 59 FR 24249, May 10, 1994; 59 FR 50828, Oct. 6, 1994; 62 FR 15391, Apr. 1, 1997; 62 *FR* 49858, Sept. 23, 1997; 63 *FR* 14355, March 25, 1998]

§101.67 Use of nutrient content claims for butter.

(a) Claims may be made to characterize the level of nutrients, including fat, in butter if:
    (1) The claim complies with the requirements of §101.13 and with the requirements of the regulations in this subpart that define the particular nutrient content claim that is used and how it is to be presented. In determining whether a claim is appropriate, the calculation of the percent fat reduction in milkfat shall be based on the 80 percent milkfat requirement provided by the statutory standard for butter (21 U.S.C. 321a);
    (2) The product contains cream or milk, including milk constituents (including, but not limited to, whey, casein, modified whey, and salts of casein), or both, with or without added salt, with or without safe and suitable colorings, with or without nutrients added to comply with paragraph (a)(3) of this section, and with or without safe and suitable bacterial cultures. The product may contain safe and suitable ingredients to improve texture, prevent syneresis, add flavor, extend shelf life, improve appearance, and add sweetness. The product may contain water to replace milkfat although the amount of water in the product shall be less than the amount of cream, milk, or milk constituents;
    (3) The product is not nutritionally inferior, as defined in §101.3(e)(4), to butter as produced under 21 U.S.C. 321a; and
    (4) If the product would violate 21 U.S.C. 321a but for the nutrient content claim that characterizes the level of nutrients, that claim shall be an explicit claim that is included as part of the common or usual name of the product.
(b) Deviations from the ingredient provisions of 21 U.S.C. 321a must be the minimum necessary to achieve similar performance characteristics as butter as produced under 21 U.S.C. 321a, or the food will be deemed to be adulterated under section 402(b) of the act. The performance characteristics (e.g., physical properties, organoleptic characteristics, functional properties, shelf life) of the product shall be similar to butter as produced under 21 U.S.C. 321a. If there is a significant difference in performance characteristics (that materially limits the uses of the product compared to butter,) the label shall include a statement informing the consumer of such difference (e.g., if appropriate, "not recommended for baking purposes"). Such statement shall comply with the requirements of §101.13(d). The modified product shall perform at least one of the principal functions of butter substantially as well as butter as produced under 21 U.S.C. 321a.
(c)
    (1) Each of the ingredients used in the food shall be declared on the label as required by the applicable sections of this part.
    (2) Safe and suitable ingredients added to improve texture, prevent syneresis, add flavor, extend shelf life, improve appearance, or add sweetness and water added to replace milkfat shall be identified with an asterisk in the ingredient statement. The statement "*Ingredients not in regular butter" shall immediately follow the ingredient statement in the same type size.

[58 FR 2455, Jan. 6, 1993]

§101.69 Petitions for nutrient content claims.

(a) This section pertains to petitions for claims, expressed or implied, that:
    (1) Characterize the level of any nutrient which is of the type required to be in the label or labeling of food by section 403(q)(1) or (q)(2) of the Federal Food, Drug, and Cosmetic Act (the act); and
    (2) That are not exempted under section 403(r)(5)(A) through (r)(5)(C) of the act from the requirements for

such claims in section 403(r)(2).

(b) Petitions included in this section are:

(1) Petitions for a new (heretofore unauthorized) nutrient content claim;

(2) Petitions for a synonymous term (i.e., one that is consistent with a term defined by regulation) for characterizing the level of a nutrient; and

(3) Petitions for the use of an implied claim in a brand name.

(c) An original and one copy of the petition to be filed under the provisions of section 403(r)(4) of the act shall be submitted, or the petitioner may submit an original and a computer readable disk containing the petition. Contents of the disk should be in a standard format, such as ASCII format. Petitioners interested in submitting a disk should contact FDA's Center for Food Safety and Applied Nutrition for details. If any part of the material submitted is in a foreign language, it shall be accompanied by an accurate and complete English translation. The petition shall state the petitioner's post office address to which published notices as required by section 403 of the act may be sent.

(d) Pertinent information may be incorporated in, and will be considered as part of, a petition on the basis of specific reference to such information submitted to and retained in the files of the Food and Drug Administration. However, any reference to unpublished information furnished by a person other than the applicant will not be considered unless use of such information is authorized (with the understanding that such information may in whole or part be subject to release to the public) in a written statement signed by the person who submitted it. Any reference to published information should be accompanied by reprints or photostatic copies of such references.

(e) If nonclinical laboratory studies are included in a petition submitted under section 403(r)(4) of the act, the petition shall include, with respect to each nonclinical study contained in the petition, either a statement that the study has been, or will be, conducted in compliance with the good laboratory practice regulations as set forth in part 58 of this chapter or, if any such study was not conducted in compliance with such regulations, a brief statement of the reason for the noncompliance.

(f) If clinical investigations are included in a petition submitted under section 403(r)(4) of the act, the petition shall include a statement regarding each such clinical investigation relied upon in the petition that the study either was conducted in compliance with the requirements for institutional review set forth in part 56 of this chapter or was not subject to such requirements in accordance with §56.104 or §56.105 of this chapter, and that it was conducted in compliance with the requirements for informed consent set forth in part 50 of this chapter.

(g) The availability for public disclosure of petitions submitted to the agency under this section will be governed by the rules specified in §10.20(j) of this chapter.

(h) All petitions submitted under this section shall include either a claim for a categorical exclusion under §25.30 or 25.32 of this chapter or an environmental assessment under §25.40 of this chapter.

(i) The data specified under the several lettered headings should be submitted on separate sheets or sets of sheets, suitably identified. If such data have already been submitted with an earlier application from the petitioner, the present petition may incorporate it by specific reference to the earlier petition.

(j) The petition must be signed by the petitioner or by his attorney or agent, or (if a corporation) by an authorized official.

(k) The petition shall include a statement signed by the person responsible for the petition, that to the best of his knowledge, it is a representative and balanced submission that includes unfavorable information, as well as favorable information, known to him pertinent to the evaluation of the petition.

(l) All applicable provisions of part 10-Administrative Practices and Procedures, may be used by the Commissioner of Food and Drugs, the petitioner or any outside party with respect to any agency action on the petition.

(m)

(1) Petitions for a new nutrient content claim shall include the following data and be submitted in the following form.

(Date)

Name of petitioner _____

Post office address _____

Subject of the petition _____

Office of Food Labeling (HFS-150)

Food and Drug Administration,

Department of Health and Human Services,

21 *CFR* §101.69

Washington, DC 20204.

To Whom It May Concern:

The undersigned, _____ submits this petition under section 403(r)(4) of the Federal Food, Drug, and Cosmetic Act (the act) with respect to (statement of the claim and its proposed use).

Attached hereto, and constituting a part of this petition, are the following:

A. A statement identifying the descriptive term and the nutrient that the term is intended to characterize with respect to the level of such nutrient. The statement should address why the use of the term as proposed will not be misleading. The statement should provide examples of the nutrient content claim as it will be used on labels or labeling, as well as the types of foods on which the claim will be used. The statement shall specify the level at which the nutrient must be present or what other conditions concerning the food must be met for the use of the term in labels or labeling to be appropriate, as well as any factors that would make the use of the term inappropriate.

B. A detailed explanation, supported by any necessary data, of why use of the food component characterized by the claim is of importance in human nutrition by virtue of its presence or absence at the levels that such claim would describe. This explanation shall also state what nutritional benefit to the public will derive from use of the claim as proposed, and why such benefit is not available through the use of existing terms defined by regulation under section 403(r)(2)(A)(i) of the act. If the claim is intended for a specific group within the population, the analysis should specifically address nutritional needs of such group, and should include scientific data sufficient for such purpose.

C. Analytical data that shows the amount of the nutrient that is the subject of the claim and that is present in the types of foods for which the claim is intended. The assays should be performed on representative samples using the Association of Official Analytical Chemists International (AOAC International) methods where available. If no AOAC International method is available, the petitioner shall submit the assay method used, and data establishing the validity of the method for assaying the nutrient in the particular food. The validation data should include a statistical analysis of the analytical and product variability.

D. A detailed analysis of the potential effect of the use of the proposed claim on food consumption and of any corresponding changes in nutrient intake. The latter item shall specifically address the intake of nutrients that have beneficial and negative consequences in the total diet. If the claim is intended for a specific group within the population, the above analysis shall specifically address the dietary practices of such group and shall include data sufficient to demonstrate that the dietary analysis is representative of such group.

E. The petitioner is required to submit either a claim for categorical exclusion under §25.30 or 25.32 of this chapter or an environmental assessment under §25.40 of this chapter.

Yours very truly,
Petitioner _____
By _____
(Indicate authority)

(2) Within 15 days of receipt of the petition, the petitioner will be notified by letter of the date on which the petition was received by the agency. Such notice will inform the petitioner:

(i) That the petition is undergoing agency review (in which case a docket number will be assigned to the petition), and the petitioner will subsequently be notified of the agency's decision to file or deny the petition; or

(ii) That the petition is incomplete, e.g., it lacks any of the data required by this part, it presents such data in a manner that is not readily understood, or it has not been submitted in quadruplicate, in which case the petition will be denied, and the petitioner will be notified as to what respect the petition is incomplete.

(3) Within 100 days of the date of receipt of the petition, the Commissioner of Food and Drugs will notify the petitioner by letter that the petition has either been filed or denied. If denied, the notification shall state the reasons therefor. If filed, the date of the notification letter becomes the date of filing for the purposes of section 403(r)(4)(A)(i) of the act. A petition that has been denied shall not be made available to the public. A filed petition shall be available to the public as provided under paragraph (g) of this section.

(4) Within 90 days of the date of filing the Commissioner of Food and Drugs will by letter of notification to the petitioner:

(i) Deny the petition; or

(ii) Inform the petitioner that a proposed regulation to provide for the requested use of the new term will be published in the Federal Register. The Commissioner of Food and Drugs will publish the proposal to

21 CFR §101.69

amend the regulations to provide for the requested use of the nutrient content claim in the Federal Register within 90 days of the date of filing. The proposal will also announce the availability of the petition for public disclosure.

(n) (1) Petitions for a synonymous term shall include the following data and be submitted in the following form.

_____
(Date)
Name of petitioner _____
Post office address _____
Subject of the petition _____
Office of Food Labeling (HFS-150)
Food and Drug Administration,
Department of Health and Human Services,
Washington, DC 20204.

To Whom It May Concern:
The undersigned, _____ submits this petition under section 403(r)(4) of the Federal Food, Drug, and Cosmetic Act (the act) with respect to (statement of the synonymous term and its proposed use in a nutrient content claim that is consistent with an existing term that has been defined under section 403(r)(2) of the act).
Attached hereto, and constituting a part of this petition, are the following:
A. A statement identifying the synonymous descriptive term, the existing term defined by a regulation under section 403(r)(2)(A)(i) of the act with which the synonymous term is claimed to be consistent. The statement should address why the proposed synonymous term is consistent with the term already defined by the agency, and why the use of the synonymous term as proposed will not be misleading. The statement should provide examples of the nutrient content claim as it will be used on labels or labeling, as well as the types of foods on which the claim will be used. The statement shall specify whether any limitations not applicable to the use of the defined term are intended to apply to the use of the synonymous term.
B. A detailed explanation, supported by any necessary data, of why use of the proposed term is requested, including an explanation of whether the existing defined term is inadequate for the purpose of effectively characterizing the level of a nutrient. This item shall also state what nutritional benefit to the public will derive from use of the claim as proposed, and why such benefit is not available through the use of existing term defined by regulation. If the claim is intended for a specific group within the population, the analysis should specifically address nutritional needs of such group, and should include scientific data sufficient for such purpose.
C. The petitioner is required to submit either a claim for categorical exclusion under §25.30 or 25.32 of this chapter or an environmental assessment under §25.40 of this chapter.
Yours very truly,
Petitioner _____
By _____
(Indicate authority)

(2) Within 15 days of receipt of the petition the petitioner will be notified by letter of the date on which the petition was received. Such notice will inform the petitioner:
(i) That the petition is undergoing agency review (in which case a docket number will be assigned to the petition) and the petitioner will subsequently be notified of the agency's decision to grant the petitioner permission to use the proposed term or to deny the petition; or
(ii) That the petition is incomplete, e.g., it lacks any of the data required by this part, it presents such data in a manner that is not readily understood, or it has not been submitted in quadruplicate, in which case the petition will be denied, and the petitioner will be notified as to what respect the petition is incomplete.
(3) Within 90 days of the date of receipt of the petition that is accepted for review (i.e., that has not been found to be incomplete and consequently denied, the Commissioner of Food and Drugs will notify the petitioner by letter of the agency's decision to grant the petitioner permission to use the proposed term, with any conditions or limitations on such use specified, or to deny the petition, in which case the letter shall state the reasons therefor. Failure of the petition to fully address the requirements of this section shall be grounds for denial of the petition.

21 CFR §101.70

(4) As soon as practicable following the agency's decision to either grant or deny the petition, the Commissioner of Food and Drugs will publish a notice in the Federal Register informing the public of his decision. If the petition is granted the Food and Drug Administration will list, the approved synonymous term in the regulations listing terms permitted for use in nutrient content claims.

(o)

(1) Petitions for the use of an implied nutrient content claim in a brand name shall include the following data and be submitted in the following form:

_____
(Date)
Name of petitioner _____
Post office address _____
Subject of the petition _____
Office of Food Labeling (HFS-150)
Food and Drug Administration,
Department of Health and Human Services,
Washington, DC 20204.

To Whom It May Concern:

The undersigned, _____ submits this petition under section 403(r)(4) of the Federal Food, Drug, and Cosmetic Act (the act) with respect to (statement of the implied nutrient content claim and its proposed use in a brand name). Attached hereto, and constituting a part of this petition, are the following:

A. A statement identifying the implied nutrient content claim, the nutrient the claim is intended to characterize, the corresponding term for characterizing the level of such nutrient as defined by a regulation under section 403(r)(2)(A)(i) of the act, and the brand name of which the implied claim is intended to be a part. The statement should address why the use of the brandname as proposed will not be misleading. It should address in particular what information is required to accompany the claim or other ways in which the claim meets the requirements of sections 201(n) and 403(a) of the act. The statement should provide examples of the types of foods on which the brand name will appear. It shall also include data showing that the actual level of the nutrient in the food qualifies the food to bear the corresponding term defined by regulation. Assay methods used to determine the level of a nutrient should meet the requirements stated under petition format item C in paragraph (k)(1) of this section.

B. A detailed explanation, supported by any necessary data, of why use of the proposed brand name is requested. This item shall also state what nutritional benefit to the public will derive from use of the brand name as proposed. If the branded product is intended for a specific group within the population, the analysis should specifically address nutritional needs of such group and should include scientific data sufficient for such purpose.

C. The petitioner is required to submit either a claim for categorical exclusion under §25.30 or 25.32 of this chapter or an environmental assessment under §25.40 of this chapter.

Yours very truly,
Petitioner _____
By _____

(2) Within 15 days of receipt of the petition the petitioner will be notified by letter of the date on which the petition was received. Such notice will inform the petitioner:

(i) That the petition is undergoing agency review (in which case a docket number will be assigned to the petition); or

(ii) That the petition is incomplete, e.g., it lacks any of the data required by this part, it presents such data in a manner that is not readily understood, or it has not been submitted in quadruplicate, in which case the petition will be denied, and the petitioner will be notified as to what respect the petition is incomplete.

(3) The Commissioner of Food and Drugs will publish a notice of the petition in the Federal Register announcing its availability to the public and seeking comment on the petition. The petition shall be available to the public to the extent provided under paragraph (g) of this section. The notice shall allow 30 days for comments.

(4) Within 100 days of the date of receipt of the petition that is accepted for review (i.e., that has not been found to be incomplete and subsequently returned to the petitioner), the Commissioner of Food and Drugs will:

21 CFR §101.70

(i) Notify the petitioner by letter of the agency's decision to grant the petitioner permission to use the proposed brand name if such use is not misleading, with any conditions or limitations on such use specified; or

(ii) Deny the petition, in which case the letter shall state the reasons therefor. Failure of the petition to fully address the requirements of this section shall be grounds for denial of the petition. Should the Commissioner of Food and Drugs not notify the petitioner of his decision on the petition within 100 days, the petition shall be considered to be granted.

(5) As soon as practicable following the granting of a petition, the Commissioner of Food and Drugs will publish a notice in the Federal Register informing the public of such fact.

(Information collection requirements in this section were approved by the Office of Management and Budget (OMB) and assigned OMB control number ------)

[58 FR 2413, Jan. 6, 1993; 58 FR 17343, Apr. 2, 1993, as amended at 58 FR 4033, Aug. 18, 1993; 62 *FR* 40598-40599, July 29, 1997]

Subpart E-Specific Requirements for Health Claims

§101.70 Petitions for health claims.

(a) Any interested person may petition the Food and Drug Administration (FDA) to issue a regulation regarding a health claim. An original and one copy of the petition shall be submitted, or the petitioner may submit an original and a computer readable disk containing the petition. Contents of the disk should be in a standard format, such as ASCII format. (Petitioners interested in submitting a disk should contact the Center for Food Safety and Applied Nutrition for details.) If any part of the material submitted is in a foreign language, it shall be accompanied by an accurate and complete English translation. The petition shall state the petitioner's post office address to which any correspondence required by section 403 of the Federal Food, Drug, and Cosmetic Act may be sent.

(b) Pertinent information may be incorporated in, and will be considered as part of, a petition on the basis of specific reference to such information submitted to and retained in the files of FDA. Such information may include any findings, along with the basis of the findings, of an outside panel with expertise in the subject area. Any reference to published information shall be accompanied by reprints, or easily readable copies of such information.

(c) If nonclinical laboratory studies are included in a petition, the petition shall include, with respect to each nonclinical study contained in the petition, either a statement that the study has been conducted in compliance with the good laboratory practice regulations as set forth in part 58 of this chapter, or, if any such study was not conducted in compliance with such regulations, a brief statement of the reason for the noncompliance.

(d) If clinical or other human investigations are included in a petition, the petition shall include a statement that they were either conducted in compliance with the requirements for institutional review set forth in part 56 of this chapter, or were not subject to such requirements in accordance with §56.104 or §56.105, and a statement that they were conducted in compliance with the requirements for informed consent set forth in part 50 of this chapter.

(e) All data and information in a health claim petition are available for public disclosure after the notice of filing of petition is issued to the petitioner, except that clinical investigation reports, adverse reaction reports, product experience reports, consumer complaints, and other similar data and information shall only be available after deletion of:

(1) Names and any information that would identify the person using the product.

(2) Names and any information that would identify any third party involved with the report, such as a physician or hospital or other institution.

(f) Petitions for a health claim shall include the following data and be submitted in the following form:

```
_____
(Date)
Name of petitioner _____
Post office address _____
Subject of the petition _____
Food and Drug Administration,
Office of Food Labeling (HFS-150),
```

21 CFR §101.70

200 C St. SW.,
Washington, DC 20204,

The undersigned, _____ submits this petition pursuant to section 403(r)(4) or 403(r)(5)(D) of the Federal Food, Drug, and Cosmetic Act with respect to (statement of the substance and its health claim).

Attached hereto, and constituting a part of this petition, are the following:

A. Preliminary requirements. A complete explanation of how the substance conforms to the requirements of §101.14(b) (21 CFR 101.14(b)). For petitions where the subject substance is a food ingredient or a component of a food ingredient, the petitioner should compile a comprehensive list of the specific ingredients that will be added to the food to supply the substance in the food bearing the health claim. For each such ingredient listed, the petitioner should state how the ingredient complies with the requirements of §101.14(b)(3)(ii), e.g., that its use is generally recognized as safe (GRAS), listed as a food additive, or authorized by a prior sanction issued by the agency, and what the basis is for the GRAS claim, the food additive status, or prior sanctioned status.

B. Summary of scientific data. The summary of scientific data provides the basis upon which authorizing a health claim can be justified as providing the health benefit. The summary must establish that, based on the totality of publicly available scientific evidence (including evidence from well-designed studies conducted in a manner which is consistent with generally recognized scientific procedures and principles), there is significant scientific agreement among experts qualified by scientific training and experience to evaluate such claims, that the claim is supported by such evidence.

The summary shall state what public health benefit will derive from use of the claim as proposed. If the claim is intended for a specific group within the population, the summary shall specifically address nutritional needs of such group and shall include scientific data showing how the claim is likely to assist in meeting such needs.

The summary shall concentrate on the findings of appropriate review articles, National Institutes of Health consensus development conferences, and other appropriate resource materials. Issues addressed in the summary shall include answers to such questions as:

1. Is there an optimum level of the particular substance to be consumed beyond which no benefit would be expected?
2. Is there any level at which an adverse effect from the substance or from foods containing the substance occurs for any segment of the population?
3. Are there certain populations that must receive special consideration?
4. What other nutritional or health factors (both positive and negative) are important to consider when consuming the substance?

In addition, the summary of scientific data shall include a detailed analysis of the potential effect of the use of the proposed claim on food consumption, specifically any change due to significant alterations in eating habits and corresponding changes in nutrient intake resulting from such changes in food consumption. The latter item shall specifically address the effect on the intake of nutrients that have beneficial and negative consequences in the total diet. If the claim is intended for a significant subpopulation within the general U.S. population, the analysis shall specifically address the dietary practices of such group, and shall include data sufficient to demonstrate that the dietary analysis is representative of such group (e.g., adolescents or the elderly).

If appropriate, the petition shall explain the prevalence of the disease or health-related condition in the U.S. population and the relevance of the claim in the context of the total daily diet.

Also, the summary shall demonstrate that the substance that is the subject of the proposed claim conforms to the definition of the term "substance" in §101.14(a)(2).

C. Analytical data that show the amount of the substance that is present in representative foods that would be candidates to bear the claim should be obtained from representative samples using methods from the Association of Official Analytical Chemists (AOAC), where available. If no AOAC method is available, the petitioner shall submit the assay method used and data establishing the validity of the method for assaying the substance in food. The validation data should include a statistical analysis of the analytical and product variability.

D. Model health claim. One or more model health claims that represent label statements that may be used on a food label or in labeling for a food to characterize the relationship between the substance in a food to a disease or health-related condition that is justified by the summary of scientific data provided in section C of the petition. The model health claim shall include:

1. A brief capsulized statement of the relevant conclusions of the summary, and
2. A statement of how this substance helps the consumer to attain a total dietary pattern or goal associated with the health benefit that is provided.

E. The petition shall include the following attachments:

1. Copies of any computer literature searches done by the petitioner (e.g., Medline).
2. Copies of articles cited in the literature searches and other information as follows:
a. All information relied upon for the support of the health claim, including copies of publications or other information cited in review articles and used to perform meta-analyses.

21 CFR §101.71

b. All information concerning adverse consequences to any segment of the population (e.g., sensitivity to the substance).

c. All information pertaining to the U.S. population.

F. The petitioner is required to submit either a claim for categorical exclusion under §25.24 of this chapter or an environmental assessment under §25.31 of this chapter.

Yours very truly,
Petitioner _____
By _____
(Indicate authority)

(g) The data specified under the several lettered headings should be submitted on separate pages or sets of pages, suitably identified. If such data have already been submitted with an earlier application from the petitioner or any other final petition, the present petition may incorporate it by specific reference to the earlier petition.

(h) The petition shall include a statement signed by the person responsible for the petition that, to the best of his/her knowledge, it is a representative and balanced submission that includes unfavorable information as well as favorable information, known to him/her to be pertinent to the evaluation of the proposed health claim.

(i) The petition shall be signed by the petitioner or by his/her attorney or agent, or (if a corporation) by an authorized official.

(j) *Agency action on the petition.*

(1) Within 15 days of receipt of the petition, the petitioner will be notified by letter of the date on which the petition was received. Such notice will inform the petitioner that the petition is undergoing agency review and that the petitioner will subsequently be notified of the agency's decision to file for comprehensive review or deny the petition.

(2) Within 100 days of the date of receipt of the petition, FDA will notify the petitioner by letter that the petition has either been filed for comprehensive review or denied. The agency will deny a petition without reviewing the information contained in B. Summary of Scientific Data if the information in A. Preliminary Requirements is inadequate in explaining how the substance conforms to the requirements of §101.14(b). If the petition is denied, the notification will state the reasons therefor, including justification of the rejection of any report from an authoritative scientific body of the U.S. Government. If filed, the date of the notification letter becomes the date of filing for the purposes of this regulation. A petition that has been denied without filing will not be made available to the public. A filed petition will be available to the public to the extent provided under paragraph (e) of this section.

(3) Within 90 days of the date of filing, FDA will by letter of notification to the petitioner:

(i) Deny the petition, or

(ii) Inform the petitioner that a proposed regulation to provide for the requested use of the health claim will be published in the Federal Register. If the petition is denied, the notification will state the reasons therefor, including justification for the rejection of any report from an authoritative scientific body of the U.S. Government. FDA will publish the proposal to amend the regulations to provide for the requested use of the health claim in the Federal Register within 90 days of the date of filing. The proposal will also announce the availability of the petition for public review.

(4)

(i) Within 270 days of the date of publication of the proposal, FDA will publish a final rule that either authorizes use of the health claim or explains why the agency has decided not to authorize one.

(ii) For cause, FDA may extend, no more than twice, the period in which it will publish a final rule; each such extension will be for no more than 90 days. FDA will publish a notice of each extension in the *Federal Register*. The document will state the basis for the extension, the length of the extension, and the date by which the final rule will be published.

[58 FR 2534, Jan. 6, 1993; 58 FR 17097, Apr. 1, 1993, as amended at 59 FR 425, Jan. 4, 1994; 62 FR 28232-28233, May 22, 1997]

21 CFR §101.72

§101.71 Health claims: claims not authorized.

Health claims not authorized for foods in conventional food form or for dietary supplements of vitamins, minerals, herbs, or other similar substances:
(a) Dietary fiber and cancer.
(b) Dietary fiber and cardiovascular disease.
(c) Antioxidant vitamins and cancer.
(d) Zinc and immune function in the elderly.
(e) Omega-3 fatty acids and coronary heart disease.
[58 FR 2534, Jan. 6, 1993, as amended at 58 FR 2548, 2578, 2620, 2639, 2664, 2714, Jan. 6, 1993; 58 FR 17100, Apr. 1, 1993; 59 FR 437, Jan. 4, 1994]

§101.72 Health claims: calcium and osteoporosis.

(a) *Relationship between calcium and osteoporosis.* An inadequate calcium intake contributes to low peak bone mass and has been identified as one of many risk factors in the development of osteoporosis. Peak bone mass is the total quantity of bone present at maturity, and experts believe that it has the greatest bearing on whether a person will be at risk of developing osteoporosis and related bone fractures later in life. Another factor that influences total bone mass and susceptibility to osteoporosis is the rate of bone loss after skeletal maturity. An adequate intake of calcium is thought to exert a positive effect during adolescence and early adulthood in optimizing the amount of bone that is laid down. However, the upper limit of peak bone mass is genetically determined. The mechanism through which an adequate calcium intake and optimal peak bone mass reduce the risk of osteoporosis is thought to be as follows. All persons lose bone with age. Hence, those with higher bone mass at maturity take longer to reach the critically reduced mass at which bones can fracture easily. The rate of bone loss after skeletal maturity also influences the amount of bone present at old age and can influence an individual's risk of developing osteoporosis. Maintenance of an adequate intake of calcium later in life is thought to be important in reducing the rate of bone loss particularly in the elderly and in women during the first decade following menopause.
(b) *Significance of calcium.* Calcium intake is not the only recognized risk factor in the development of osteoporosis, a multifactorial bone disease. Other factors including a person's sex, race, hormonal status, family history, body stature, level of exercise, general diet, and specific life style choices such as smoking and excess alcohol consumption affect the risk of osteoporosis.

(1) Heredity and being female are two key factors identifying those individuals at risk for the development of osteoporosis. Hereditary risk factors include race: Notably, Caucasians and Asians are characterized by low peak bone mass at maturity. Caucasian women, particularly those of northern European ancestry, experience the highest incidence of osteoporosis-related bone fracture. American women of African heritage are characterized by the highest peak bone mass and lowest incidence of osteoporotic fracture, despite the fact that they have low calcium intake.

(2) Maintenance of an adequate intake of calcium throughout life is particularly important for a subpopulation of individuals at greatest risk of developing osteoporosis and for whom adequate dietary calcium intake may have the most important beneficial effects on bone health. This target subpopulation includes adolescent and young adult Caucasian and Asian American women.

(c) *Requirements.*

(1) All requirements set forth in §101.14 shall be met.

(2) Specific requirements.

(i) *Nature of the claim.* A health claim associating calcium with a reduced risk of osteoporosis may be made on the label or lableing of a food describe in paragraph (c)(2)(ii) of this section, provided that:

(A) The claim makes clear that adequate calcium intake throughout life is not the only recognized risk factor in this multifactorial bone disease by listing specific factors, including sex, race, and age that place persons at risk of developing osteoporosis and stating that an adequate level of exercise and a healthful diet are also needed;

(B) The claim does not state or imply that the risk of osteoporosis is equally applicable to the general United States population. The claim shall identify the populations at particular risk for the development of

21 CFR §101.72

osteoporosis. These populations include white (or the term "Caucasian") women and Asian women in their bone forming years (approximately 11 to 35 years of age or the phrase "during teen or early adult years" may be used). The claim may also identify menopausal (or the term "middle-aged") women, persons with a family history of the disease, and elderly (or "older") men and women as being at risk;

(C) The claim states that adequate calcium intake throughout life is linked to reduced risk of osteoporosis through the mechanism of optimizing peak bone mass during adolescence and early adulthood. The phrase "build and maintain good bone health" may be used to convey the concept of optimizing peak bone mass. When reference is made to persons with a family history of the disease, menopausal women, and elderly men and women, the claim may also state that adequate calcium intake is linked to reduced risk of osteoporosis through the mechanism of slowing the rate of bone loss;

(D) The claim does not attribute any degree of reduction in risk of osteoporosis to maintaining an adequate calcium intake throughout life; and

(E) The claim states that a total dietary intake greater than 200 percent of the recommended daily intake (2,000 milligrams (mg) of calcium) has no further known benefit to bone health. This requirement does not apply to foods that contain less than 40 percent of the recommended daily intake of 1,000 mg of calcium per day or 400 mg of calcium per reference amount customarily consumed as defined in §101.12 (b) or per total daily recommended supplement intake.

(ii) *Nature of the food.*

(A) The food shall meet or exceed the requirements for a "high" level of calcium as defined in §101.54(b);

(B) The calcium content of the product shall be assimilable;

(C) Dietary supplements shall meet the United States Pharmacopeia (U.S.P.) standards for disintegration and dissolution applicable to their component calcium salts, except that dietary supplements for which no U.S.P. standards exist shall exhibit appropriate assimilability under the conditions of use stated on the product label;

(D) A food or total daily recommended supplement intake shall not contain more phosphorus than calcium on a weight per weight basis.

(d) *Optional information.*

(1) The claim may include information from paragraphs (a) and (b) of this section.

(2) The claim may include information on the number of people in the United States who have osteoporosis. The sources of this information must be identified, and it must be current information from the National Center for Health Statistics, the National Institutes of Health, or "Dietary Guidelines for Americans."

(e) *Model health claim.* The following model health claims may be used in food labeling to describe the relationship between calcium and osteoporosis:

MODEL HEALTH CLAIM APPROPRIATE FOR MOST CONVENTIONAL FOODS:
Regular exercise and a healthy diet with enough calcium helps teen and young adult white and Asian women maintain good bone health and may reduce their high risk of osteoporosis later in life.

MODEL HEALTH CLAIM APPROPRIATE FOR FOODS EXCEPTIONALLY HIGH IN CALCIUM AND MOST CALCIUM SUPPLEMENTS:
Regular exercise and a healthy diet with enough calcium helps teen and young adult white and Asian women maintain good bone health and may reduce their high risk of osteoporosis later in life. Adequate calcium intake is important, but daily intakes above about 2,000 mg are not likely to provide any additional benefit.

[58 FR 2676, Jan. 6, 1993; 58 FR 17101, Apr. 1, 1993; 62 FR 15342, Mar. 31, 1997]

§101.73 Health claims: dietary lipids and cancer.

(a) *Relationship between fat and cancer.*

(1) Cancer is a constellation of more than 100 different diseases, each characterized by the uncontrolled growth and spread of abnormal cells. Cancer has many causes and stages in its development. Both genetic

21 CFR §101.73

and environmental risk factors may affect the risk of cancer. Risk factors include a family history of a specific type of cancer, cigarette smoking, alcohol consumption, overweight and obesity, ultraviolet or ionizing radiation, exposure to cancer- causing chemicals, and dietary factors.

(2) Among dietary factors, the strongest positive association has been found between total fat intake and risk of some types of cancer. Based on the totality of the publicly available scientific evidence, there is significant scientific agreement among experts, qualified by training and experience to evaluate such evidence, that diets high in total fat are associated with an increased cancer risk. Research to date, although not conclusive, demonstrates that the total amount of fats, rather than any specific type of fat, is positively associated with cancer risk. The mechanism by which total fat affects cancer has not yet been established.

(3) A question that has been the subject of considerable research is whether the effect of fat on cancer is site-specific. Neither human nor animal studies are consistent in the association of fat intake with specific cancer sites.

(4) Another question that has been raised is whether the association of total fat intake to cancer risk is independently associated with energy intakes, or whether the association of fat with cancer risk is the result of the higher energy (caloric) intake normally associated with high fat intake. FDA has concluded that evidence from both animal and human studies indicates that total fat intake alone, independent of energy intake, is associated with cancer risk.

(b) *Significance of the relationship between fat intake and risk of cancer.*

(1) Cancer is ranked as a leading cause of death in the United States. The overall economic costs of cancer, including direct health care costs and losses due to morbidity and mortality, are very high.

(2) U.S. diets tend to be high in fat and high in calories. The average U.S. diet is estimated to contain 36 to 37 percent of calories from total fat. Current dietary guidelines from the Federal Government and other national health professional organizations recommend that dietary fat intake be reduced to a level of 30 percent or less of energy (calories) from total fat. In order to reduce intake of total fat, individuals should choose diets which are high in vegetables, fruits, and grain products (particularly whole grain products), choose lean cuts of meats, fish, and poultry, substitute low-fat dairy products for higher fat products, and use fats and oils sparingly.

(c) *Requirements.*

(1) All requirements set forth in §101.14 shall be met.

(2) Specific requirements.

(i) *Nature of the claim.* A health claim associating diets low in fat with reduced risk of cancer may be made on the label or labeling of a food described in paragraph (c)(2)(ii) of this section, provided that:

(A) The claim states that diets low in fat "may" or "might" reduce the risk of some cancers;

(B) In specifying the disease, the claim uses the following terms: "some types of cancer" or "some cancers";

(C) In specifying the nutrient, the claim uses the term "total fat" or "fat";

(D) The claim does not specify types of fat or fatty acid that may be related to the risk of cancer;

(E) The claim does not attribute any degree of cancer risk reduction to diets low in fat; and

(F) The claim indicates that the development of cancer depends on many factors.

(ii) *Nature of the food.* The food shall meet all of the nutrient content requirements of §101.62 for a "low fat" food; except that fish and game meats (i.e., deer, bison, rabbit, quail, wild turkey, geese, ostrich) may meet the requirements for "extra lean" in §101.62.

(d) *Optional information.*

(1) The claim may identify one or more of the following risk factors for development of cancer: Family history of a specific type of cancer, cigarette smoking, alcohol consumption, overweight and obesity, ultraviolet or ionizing radiation, exposure to cancer-causing chemicals, and dietary factors.

(2) The claim may include information from paragraphs (a) and (b) of this section which summarize the relationship between dietary fat and cancer and the significance of the relationship.

(3) The claim may indicate that it is consistent with "Nutrition and Your Health: Dietary Guidelines for Americans," U.S. Department of Agriculture (USDA) and Department of Health and Human Services (DHHS), Government Printing Office.

(4) The claim may include information on the number of people in the United States who have cancer. The sources of this information must be identified, and it must be current information from the National Center

21 CFR §101.74

for Health Statistics, the National Institutes of Health, or "Nutrition and Your Health: Dietary Guidelines for Americans," USDA and DHHS, Government Printing Office.

(e) *Model health claims.* The following model health claims may be used in food labeling to describe the relationship between dietary fat and cancer:

(1) Development of cancer depends on many factors. A diet low in total fat may reduce the risk of some cancers.

(2) Eating a healthful diet low in fat may help reduce the risk of some types of cancers. Development of cancer is associated with many factors, including a family history of the disease, cigarette smoking, and what you eat.

[58 FR 2801, Jan. 6, 1993; 58 FR 17343, Apr. 2, 1993]

§101.74 Health claims: sodium and hypertension.

(a) *Relationship between sodium and hypertension (high blood pressure).*

(1) Hypertension, or high blood pressure, generally means a systolic blood pressure of greater than 140 millimeters of mercury (mm Hg) or a diastolic blood pressure of greater than 90 mm Hg. Normotension, or normal blood pressure, is a systolic blood pressure below 140 mm Hg and diastolic blood pressure below 90 mm Hg. Sodium is specified here as the chemical entity or electrolyte "sodium" and is distinguished from sodium chloride, or salt, which is 39 percent sodium by weight.

(2) The scientific evidence establishes that diets high in sodium are associated with a high prevalence of hypertension or high blood pressure and with increases in blood pressure with age, and that diets low in sodium are associated with a low prevalence of hypertension or high blood pressure and with a low or no increase of blood pressure with age.

(b) *Significance of sodium in relation to high blood pressure.*

(1) High blood pressure is a public health concern primarily because it is a major risk factor for mortality from coronary heart disease and stroke. Early management of high blood pressure is a major public health goal that can assist in reducing mortality associated with coronary heart disease and stroke. There is a continuum of mortality risk that increases as blood pressures rise. Individuals with high blood pressure are at greatest risk, and individuals with moderately high, high normal, and normal blood pressure are at steadily decreasing risk. The scientific evidence indicates that reducing sodium intake lowers blood pressure and associated risks in many but not all hypertensive individuals. There is also evidence that reducing sodium intake lowers blood pressure and associated risks in many but not all normotensive individuals as well.

(2) The populations at greatest risk for high blood pressure, and those most likely to benefit from sodium reduction, include those with family histories of high blood pressure, the elderly, males because they develop hypertension earlier in life than females, and black males and females. Although some population groups are at greater risk than others, high blood-pressure is a disease of public health concern for all population groups. Sodium intake, alcohol consumption, and obesity are identified risk factors for high blood pressure.

(3) Sodium intakes exceed recommended levels in almost every group in the United States. One of the major public health recommendations relative to high blood pressure is to decrease consumption of salt. On a population-wide basis, reducing the average sodium intake would have a small but significant effect on reducing the average blood pressure, and, consequently, reducing mortality from coronary heart disease and stroke.

(4) Sodium is an essential nutrient, and experts have recommended a safe minimum level of 500 milligrams (mg) sodium per day and an upper level of 2,400 mg sodium per day, the FDA Daily Value for sodium.

(c) *Requirements.*

(1) All requirements set forth in §101.14 shall be met.

(2) Specific requirements.

(i) *Nature of the claim.* A health claim associating diets low in sodium with reduced risk of high blood pressure may be made on the label or labeling of a food described in paragraph (c)(2)(ii) of this section, provided that:

(A) The claim states that diets low in sodium "may" or "might" reduce the risk of high blood pressure;

(B) In specifying the disease, the claim uses the term "high blood pressure";

(C) In specifying the nutrient, the claim uses the term "sodium";

(D) The claim does not attribute any degree of reduction in risk of high blood pressure to diets low in sodium; and

(E) The claim indicates that development of high blood pressure depends on many factors.

(ii) *Nature of the food.* The food shall meet all of the nutrient content requirements of §101.61 for a "low sodium" food.

(d) *Optional information.*

(1) The claim may identify one or more of the following risk factors for development of high blood pressure in addition to dietary sodium consumption: Family history of high blood pressure, growing older, alcohol consumption, and excess weight.

(2) The claim may include information from paragraphs (a) and (b) of this section, which summarizes the relationship between dietary sodium and high blood pressure and the significance of the relationship.

(3) The claim may include information on the number of people in the United States who have high blood pressure. The sources of this information must be identified, and it must be current information from the National Center for Health Statistics, the National Institutes of Health, or "Nutrition and Your Health: Dietary Guidelines for Americans," U.S. Department of Health and Human Services (DHHS) and U.S. Department of Argiculture (USDA), Government Printing Office.

(4) The claim may indicate that it is consistent with "Nutrition and Your Health: U.S. Dietary Guidelines for Americans, DHHS and USDA, Government Printing Office.

(5) In specifying the nutrient, the claim may include the term "salt" in addition to the term "sodium."

(6) In specifying the disease, the claim may include the term "hypertension" in addition to the term "high blood pressure."

(7) The claim may state that individuals with high blood pressure should consult their physicians for medical advice and treatment. If the claim defines high or normal blood pressure, then the health claim must state that individuals with high blood pressure should consult their physicians for medical advice and treatment.

(e) *Model health claims.* The following are model health claims that may be used in food labeling to describe the relationship between dietary sodium and high blood pressure:

(1) Diets low in sodium may reduce the risk of high blood pressure, a disease associated with many factors.

(2) Development of hypertension or high blood pressure depends on many factors. [This product] can be part of a low sodium, low salt diet that might reduce the risk of hypertension or high blood pressure.

[58 FR 2836, Jan. 6, 1993; 58 FR 17100, Apr. 1, 1993]

§101.75 Health claims: dietary saturated fat and cholesterol and risk of coronary heart disease.

(a) *Relationship between dietary saturated fat and cholesterol and risk of coronary heart disease.*

(1) Cardiovascular disease means diseases of the heart and circulatory system. Coronary heart disease is the most common and serious form of cardiovascular disease and refers to diseases of the heart muscle and supporting blood vessels. High blood total- and low density lipoprotein (LDL)- cholesterol levels are major modifiable risk factors in the development of coronary heart disease. High coronary heart disease rates occur among people with high blood cholesterol levels of 240 milligrams/decaliter (mg/dL) (6.21 millimoles per liter (mmol/L)) or above and LDL-cholesterol levels of 160 mg/dL (4.13 mmol/L) or above. Borderline high risk blood cholesterol levels range from 200 to 239 mg/dL (5.17 to 6.18 mmol/L) and 130 to 159 mg/dL (3.36 to 4.11 mmol/L) of LDL-cholesterol. Dietary lipids (fats) include fatty acids and cholesterol. Total fat, commonly referred to as fat, is composed of saturated fat (fatty acids containing no double bonds), and monounsaturated and polyunsaturated fat (fatty acids containing one or more double bonds).

21 CFR §101.75

(2) The scientific evidence establishes that diets high in saturated fat and cholesterol are associated with increased levels of blood total- and LDL-cholesterol and, thus, with increased risk of coronary heart disease. Diets low in saturated fat and cholesterol are associated with decreased levels of blood total- and LDL-cholesterol, and thus, with decreased risk of developing coronary heart disease.

(b) *Significance of the relationship between dietary saturated fat and cholesterol and risk of coronary heart disease.*

(1) Coronary heart disease is a major public health concern in the United States, primarily because it accounts for more deaths than any other disease or group of diseases. Early management of risk factors for coronary heart disease is a major public health goal that can assist in reducing risk of coronary heart disease. There is a continuum of mortality risk from coronary heart disease that increases with increasing levels of blood LDL-cholesterol. Individuals with high blood LDL-cholesterol are at greatest risk. A larger number of individuals with more moderately elevated cholesterol also have increased risk of coronary events; such individuals comprise a substantial proportion of the adult U.S. population. The scientific evidence indicates that reducing saturated fat and cholesterol intakes lowers blood LDL-cholesterol and risk of heart disease in most individuals. There is also evidence that reducing saturated fat and cholesterol intakes in persons with blood cholesterol levels in the normal range also reduces risk of heart disease.

(2) Other risk factors for coronary heart disease include a family history of heart disease, high blood pressure, diabetes, cigarette smoking, obesity (body weight 30 percent greater than ideal body weight), and lack of regular physical exercise.

(3) Intakes of saturated fat exceed recommended levels in many people in the United States. Intakes of cholesterol are, on average, at or above recommended levels. One of the major public health recommendations relative to coronary heart disease risk is to consume less than 10 percent of calories from saturated fat, and an average of 30 percent or less of total calories from all fat. Recommended daily cholesterol intakes are 300 mg or less per day.

(c) *Requirements.*

(1) All requirements set forth in §101.14 shall be met.

(2) Specific requirements.

(i) *Nature of the claim.* A health claim associating diets low in saturated fat and cholesterol with reduced risk of coronary heart disease may be made on the label or labeling of a food described in paragraph (c)(2)(ii) of this section provided that:

(A) The claim states that diets low in saturated fat and cholesterol "may" or "might" reduce the risk of heart disease;

(B) In specifying the disease, the claim uses the terms "heart disease" or "coronary heart disease;"

(C) In specifying the nutrient, the claim uses the terms "saturated fat" and "cholesterol" and lists both;

(D) The claim does not attribute any degree of risk reduction for coronary heart disease to diets low in dietary saturated fat and cholesterol; and

(E) The claim states that coronary heart disease risk depends on many factors.

(ii) *Nature of the food.* The food shall meet all of the nutrient content requirements of §101.62 for a "low saturated fat," "low cholesterol," and "low fat" food; except that fish and game meats (i.e., deer, bison, rabbit, quail, wild turkey, geese, and ostrich) may meet the requirements for "extra lean" in §101.62.

(d) *Optional information.*

(1) The claim may identify one or more of the following risk factors in addition to saturated fat and cholesterol about which there is general scientific agreement that they are major risk factors for this disease: A family history of coronary heart disease, elevated blood total and LDL-cholesterol, excess body weight, high blood pressure, cigarette smoking, diabetes, and physical inactivity.

(2) The claim may indicate that the relationship of saturated fat and cholesterol to heart disease is through the intermediate link of "blood cholesterol" or "blood total- and LDL cholesterol."

(3) The claim may include information from paragraphs (a) and (b) of this section, which summarize the relationship between dietary saturated fat and cholesterol and risk of coronary heart disease, and the significance of the relationship.

(4) In specifying the nutrients, the claim may include the term "total fat" in addition to the terms "saturated fat" and "cholesterol".

(5) The claim may include information on the number of people in the United States who have coronary heart disease. The sources of this information shall be identified, and it shall be current information from

the National Center for Health Statistics, the National Institutes of Health, or "Nutrition and Your Health: Dietary Guidelines for Americans," U.S. Department of Health and Human Services (DHHS) and U.S. Department of Agriculture (USDA), Government Printing Office.

(6) The claim may indicate that it is consistent with "Nutrition and Your Health: Dietary Guidelines for Americans," DHHS and USDA, Government Printing Office.

(7) The claim may state that individuals with elevated blood total- or LDL-cholesterol should consult their physicians for medical advice and treatment. If the claim defines high or normal blood total- or LDL-cholesterol levels, then the claim shall state that individuals with high blood cholesterol should consult their physicians for medical advice and treatment.

(e) *Model health claims.* The following are model health claims that may be used in food labeling to describe the relationship between dietary saturated fat and cholesterol and risk of heart disease:

(1) While many factors affect heart disease, diets low in saturated fat and cholesterol may reduce the risk of this disease;

(2) Development of heart disease depends upon many factors, but its risk may be reduced by diets low in saturated fat and cholesterol and healthy lifestyles;

(3) Development of heart disease depends upon many factors, including a family history of the disease, high blood LDL-cholesterol, diabetes, high blood pressure, being overweight, cigarette smoking, lack of exercise, and the type of dietary pattern. A healthful diet low in saturated fat, total fat, and cholesterol, as part of a healthy lifestyle, may lower blood cholesterol levels and may reduce the risk of heart disease;

(4) Many factors, such as a family history of the disease, increased blood- and LDL-cholesterol levels, high blood pressure, cigarette smoking, diabetes, and being overweight, contribute to developing heart disease. A diet low in saturated fat, cholesterol, and total fat may help reduce the risk of heart disease; and

(5) Diets low in saturated fat, cholesterol, and total fat may reduce the risk of heart disease. Heart disease is dependent upon many factors, including diet, a family history of the disease, elevated blood LDL-cholesterol levels, and physical inactivity.

[58 FR 2757, Jan. 6, 1993]

§101.76 Health claims: fiber-containing grain products, fruits, and vegetables and cancer.

(a) *Relationship between diets low in fat and high in fiber- containing grain products, fruits, and vegetables and cancer risk.*

(1) Cancer is a constellation of more than 100 different diseases, each characterized by the uncontrolled growth and spread of abnormal cells. Cancer has many causes and stages in its development. Both genetic and environmental risk factors may affect the risk of cancer. Risk factors include: A family history of a specific type of cancer, cigarette smoking, overweight and obesity, alcohol consumption, ultraviolet or ionizing radiation, exposure to cancer-causing chemicals, and dietary factors.

(2) The scientific evidence establishes that diets low in fat and high in fiber-containing grain products, fruits, and vegetables are associated with a reduced risk of some types of cancer. Although the specific role of total dietary fiber, fiber components, and the multiple nutrients and other substances contained in these foods are not yet fully understood, many studies have shown that diets low in fat and high in fiber-containing foods are associated with reduced risk of some types of cancer.

(b) *Significance of the relationship between consumption of diets low in fat and high in fiber-containing grain products, fruits, and vegetables and risk of cancer.*

(1) Cancer is ranked as a leading cause of death in the United States. The overall economic costs of cancer, including direct health care costs and losses due to morbidity and mortality, are very high.

(2) U.S. diets tend to be high in fat and low in grain products, fruits, and vegetables. Studies in various parts of the world indicate that populations who habitually consume a diet high in plant foods have lower risks of some cancers. These diets generally are low in fat and rich in many nutrients, including, but not limited to, dietary fiber. Current dietary guidelines from Federal government agencies and nationally recognized health professional organizations recommend decreased consumption of fats (less than 30 percent of calories), maintenance of desirable body weight, and increased consumption of fruits and vegetables (five or more servings daily), and grain products (six or more servings daily).

21 CFR §101.76

(c) *Requirements.*

(1) All requirements set forth in §101.14 shall be met.

(2) Specific requirements.

(i) *Nature of the claim.* A health claim associating diets low in fat and high in fiber-containing grain products, fruits, and vegetables with reduced risk of cancer may be made on the label or labeling of a food described in paragraph (c)(2)(ii) of this section, provided that:

(A) The claim states that diets low in fat and high in fiber- containing grain products, fruits, and vegetables "may" or "might" reduce the risk of some cancers;

(B) In specifying the disease, the claim uses the following terms: "some types of cancer," or "some cancers";

(C) The claim is limited to grain products, fruits, and vegetables that contain dietary fiber;

(D) The claim indicates that development of cancer depends on many factors;

(E) The claim does not attribute any degree of cancer risk reduction to diets low in fat and high in fiber-containing grain products, fruits, and vegetables;

(F) In specifying the dietary fiber component of the labeled food, the claim uses the term "fiber", "dietary fiber" or "total dietary fiber"; and

(G) The claim does not specify types of dietary fiber that may be related to risk of cancer.

(ii) *Nature of the food.*

(A) The food shall be or shall contain a grain product, fruit, or vegetable.

(B) The food shall meet the nutrient content requirements of §101.62 for a "low fat" food.

(C) The food shall meet, without fortification, the nutrient content requirements of §101.54 for a "good source" of dietary fiber.

(d) *Optional information.*

(1) The claim may include information from paragraphs (a) and (b) of this section, which summarize the relationship between diets low in fat and high in fiber- containing grain products, fruits, and vegetables, and some types of cancer and the significance of the relationship.

(2) The claim may identify one or more of the following risk factors for development of cancer: Family history of a specific type of cancer, cigarette smoking, overweight and obesity, alcohol consumption, ultraviolet or ionizing radiation, exposure to cancer causing chemicals, and dietary factors.

(3) The claim may indicate that it is consistent with "Nutrition and Your Health: Dietary Guidelines for Americans," U.S. Department of Agriculture (USDA) and Department of Health and Human Services (DHHS), Government Printing Office.

(4) The claim may include information on the number of people in the United States who have cancer. The sources of this information must be identified, and it must be current information from the National Center for Health Statistics, the National Institutes of Health, or "Nutrition and Your Health: Dietary Guidelines for Americans," USDA and DHHS, Government Printing Office.

(e) *Model health claims.* The following model health claims may be used in food labeling to characterize the relationship between diets low in fat and high in fiber-containing grain products, fruits, and vegetables and cancer risk:

(1) Low fat diets rich in fiber-containing grain products, fruits, and vegetables may reduce the risk of some types of cancer, a disease associated with many factors.

(2) Development of cancer depends on many factors. Eating a diet low in fat and high in grain products, fruits, and vegetables that contain dietary fiber may reduce your risk of some cancers.

[58 FR 2548, Jan. 6, 1993]

§101.77 Health claims: fruits, vegetables, and grain products that contain fiber, particularly soluble fiber, and risk of coronary heart disease.

(a) *Relationship between diets low in saturated fat and cholesterol and high in fruits, vegetables, and grain products that contain fiber, particularly soluble fiber, and risk of coronary heart disease.*

(1) Cardiovascular disease means diseases of the heart and circulatory system. Coronary heart disease is the most common and serious form of cardiovascular disease and refers to diseases of the heart muscle and

21 CFR §101.77

supporting blood vessels. High blood total- and low density lipoprotein (LDL)- cholesterol levels are major modifiable risk factors in the development of coronary heart disease. High coronary heart disease rates occur among people with high blood cholesterol levels of 240 milligrams per deciliter (mg/dL) (6.21 (mmol/L)) or above and LDL-cholesterol levels of 160 mg/dL (4.13 mmol/L) or above. Borderline high risk blood cholesterol levels range from 200 to 239 mg/dL (5.17 to 6.18 mmol/L) and 130 to 159 mg/dL (3.36 to 4.11 mmol/L) of LDL-cholesterol. Dietary lipids (fats) include fatty acids and cholesterol. Total fat, commonly referred to as fat, is composed of saturated fat (fatty acids containing no double bonds), and monounsaturated and polyunsaturated fat (fatty acids containing one or more double bonds).

(2) The scientific evidence establishes that diets high in saturated fat and cholesterol are associated with increased levels of blood total- and LDL-cholesterol and, thus, with increased risk of coronary heart disease. Diets low in saturated fat and cholesterol are associated with decreased levels of blood total- and LDL-cholesterol, and thus, with decreased risk of developing coronary heart disease.

(3) Populations with relatively low blood cholesterol levels tend to have dietary patterns that are not only low in total fat, especially saturated fat and cholesterol, but are also relatively high in fruits, vegetables, and grain products. Although the specific roles of these plant foods are not yet fully understood, many studies have shown that diets high in plant foods are associated with reduced risk of coronary heart disease. These studies correlate diets rich in fruits, vegetables, and grain products and nutrients from these diets, such as some types of fiber, with reduced coronary heart disease risk. Persons consuming these diets frequently have high intakes of dietary fiber, particularly soluble fibers. Currently, there is not scientific agreement as to whether a particular type of soluble fiber is beneficial, or whether the observed protective effects of fruits, vegetables, and grain products against heart disease are due to other components, or a combination of components, in these diets, including, but not necessarily limited to, some types of soluble fiber, other fiber components, other characteristics of the complex carbohydrate content of these foods, other nutrients in these foods, or displacement of saturated fat and cholesterol from the diet.

(b) *Significance of the relationship between diets low in saturated fat and cholesterol, and high in fruits, vegetables, and grain products that contain fiber, particularly soluble fiber, and risk of coronary heart disease.*

(1) Coronary heart disease is a major public health concern in the United States, primarily because it accounts for more deaths than any other disease or group of diseases. Early management of risk factors for coronary heart disease is a major public health goal that can assist in reducing risk of coronary heart disease. There is a continuum of mortality risk from coronary heart disease that increases with increasing levels of blood LDL-cholesterol. Individuals with high blood LDL-cholesterol are at greatest risk. A larger number of individuals with more moderately elevated cholesterol also have increased risk of coronary events; such individuals comprise a substantial proportion of the adult U.S. population. The scientific evidence indicates that reducing saturated fat and cholesterol intakes lowers blood LDL-cholesterol and risk of heart disease in most individuals, including persons with blood cholesterol levels in the normal range. Additionally, consuming diets high in fruits, vegetables, and grain products, foods that contain soluble fiber, may be a useful adjunct to a low saturated fat and low cholesterol diet.

(2) Other risk factors for coronary heart disease include a family history of heart disease, high blood pressure, diabetes, cigarette smoking, obesity (body weight 30 percent greater than ideal body weight), and lack of regular physical exercise.

(3) Intakes of saturated fat exceed recommended levels in many people in the United States. Intakes of cholesterol are, on average, at or above recommended levels. Intakes of fiber- containing fruits, vegetables, and grain products are about half of recommended intake levels. One of the major public health recommendations relative to coronary heart disease risk is to consume less than 10 percent of calories from saturated fat, and an average of 30 percent or less of total calories from all fat. Recommended daily cholesterol intakes are 300 mg or less per day. Recommended total dietary fiber intakes are about 25 grams (g) daily, of which about 25 percent (about 6 g) should be soluble fiber.

(4) Current dietary guidance recommendations encourage decreased consumption of dietary fat, especially saturated fat and cholesterol, and increased consumption of fiber-rich foods to help lower blood LDL-cholesterol levels. Results of numerous studies have shown that fiber-containing fruits, vegetables, and grain products can help lower blood LDL-cholesterol.

(c) *Requirements.*

(1) All requirements set forth in §101.14 shall be met.

(2) Specific requirements.

(i) *Nature of the claim.* A health claim associating diets low in saturated fat and cholesterol and high in fruits, vegetables, and grain products that contain fiber, particularly soluble fiber, with reduced risk of heart disease may be made on the label or labeling of a food described in paragraph (c)(2)(ii) of this section, provided that:

(A) The claim states that diets low in saturated fat and cholesterol and high in fruits, vegetables, and grain products that contain fiber "may" or "might" reduce the risk of heart disease; (B) In specifying the disease, the claim uses the following terms: "heart disease" or "coronary heart disease;"

(C) The claim is limited to those fruits, vegetables, and grains that contain fiber;

(D) In specifying the dietary fiber, the claim uses the term "fiber," "dietary fiber," "some types of dietary fiber," "some dietary fibers," or "some fibers;" the term "soluble fiber" may be used in addition to these terms;

(E) In specifying the fat component, the claims uses the terms "saturated fat" and "cholesterol;" and

(F) The claim indicates that development of heart disease depends on many factors; and

(G) The claim does not attribute any degree of risk reduction for coronary heart disease to diets low in saturated fat and cholesterol and high in fruits, vegetables, and grain products that contain fiber.

(ii) *Nature of the food.*

(A) The food shall be or shall contain a fruit, vegetable, or grain product.

(B) The food shall meet the nutrient content requirements of §101.62 for a "low saturated fat," "low cholesterol," and "low fat" food.

(C) The food contains, without fortification, at least 0.6 g of soluble fiber per reference amount customarily consumed;

(D) The content of soluble fiber shall be declared in the nutrition information panel, consistent with §101.9(c)(6)(i)(A).

(d) *Optional information.*

(1) The claim may identify one or more of the following risk factors for heart disease about which there is general scientific agreement: A family history of coronary heart disease, elevated blood-, total- and LDL-cholesterol, excess body weight, high blood pressure, cigarette smoking, diabetes, and physical inactivity.

(2) The claim may indicate that the relationship of diets low in saturated fat and cholesterol, and high in fruits, vegetables, and grain products that contain fiber to heart disease is through the intermediate link of "blood cholesterol" or "blood total- and LDL-cholesterol."

(3) The claim may include information from paragraphs (a) and (b) of this section, which summarize the relationship between diets low in saturated fat and cholesterol and high in fruits, vegetables, and grain products that contain fiber and coronary heart disease, and the significance of the relationship.

(4) In specifying the nutrients, the claim may include the term "total fat" in addition to the terms "saturated fat" and "cholesterol."

(5) The claim may indicate that it is consistent with "Nutrition and Your Health: Dietary Guidelines for Americans," U.S. Department of Agriculture (USDA) and Department of Health and Human Services (DHHS), Government Printing Office (GPO).

(6) The claim may state that individuals with elevated blood total- and LDL-cholesterol should consult their physicians for medical advice and treatment. If the claim defines high or normal blood total- and LDL-cholesterol levels, then the claim shall state that individuals with high blood cholesterol should consult their physicians for medical advice and treatment.

(7) The claim may include information on the number of people in the United States who have heart disease. The sources of this information shall be identified, and it shall be current information from the National Center for Health Statistics, the National Institutes of Health, or "Nutrition and Your Health: Dietary Guidelines for Americans," USDA and DHHS, GPO.

(e) *Model health claims.* The following model health claims may be used in food labeling to characterize the relationship between diets low in saturated fat and cholesterol and high in fruits, vegetables, and grain products that contain soluble fiber:

(1) Diets low in saturated fat and cholesterol and rich in fruits, vegetables, and grain products that contain some types of dietary fiber, particularly soluble fiber, may reduce the risk of heart disease, a disease

associated with many factors.

(2) Development of heart disease depends on many factors. Eating a diet low in saturated fat and cholesterol and high in fruits, vegetables, and grain products that contain fiber may lower blood cholesterol levels and reduce your risk of heart disease.

[58 FR 2578, Jan. 6, 1993]

§101.78 Health claims: fruits and vegetables and cancer.

(a) *Relationship between substances in diets low in fat and high in fruits and vegetables and cancer risk.*

(1) Cancer is a constellation of more than 100 different diseases, each characterized by the uncontrolled growth and spread of abnormal cells. Cancer has many causes and stages in its development. Both genetic and environmental risk factors may affect the risk of cancer. Risk factors include a family history of a specific type of cancer, cigarette smoking, alcohol consumption, overweight and obesity, ultraviolet or ionizing radiation, exposure to cancer- causing chemicals, and dietary factors.

(2) Although the specific roles of the numerous potentially protective substances in plant foods are not yet understood, many studies have shown that diets high in plant foods are associated with reduced risk of some types of cancers. These studies correlate diets rich in fruits and vegetables and nutrients from these diets, such as vitamin C, vitamin A, and dietary fiber, with reduced cancer risk. Persons consuming these diets frequently have high intakes of these nutrients. Currently, there is not scientific agreement as to whether the observed protective effects of fruits and vegetables against cancer are due to a combination of the nutrient components of diets rich in fruits and vegetables, including but not necessarily limited to dietary fiber, vitamin A (as beta-carotene) and vitamin C, to displacement of fat from such diets, or to intakes of other substances in these foods which are not nutrients but may be protective against cancer risk.

(b) *Significance of the relationship between consumption of diets low in fat and high in fruits and vegetables and risk of cancer.*

(1) Cancer is ranked as a leading cause of death in the United States. The overall economic costs of cancer, including direct health care costs and losses due to morbidity and mortality, are very high.

(2) U.S. diets tend to be high in fat and low in fruits and vegetables. Studies in various parts of the world indicate that populations who habitually consume a diet high in plant foods have lower risks of some cancers. These diets generally are low in fat and rich in many nutrients, including, but not limited to, dietary fiber, vitamin A (as beta-carotene), and vitamin C. Current dietary guidelines from Federal Government agencies and nationally recognized health professional organizations recommend decreased consumption of fats (less than 30 percent of calories), maintenance of desirable body weight, and increased consumption of fruits and vegetables (5 or more servings daily), particularly those fruits and vegetables which contain dietary fiber, vitamin A, and vitamin C.

(c) *Requirements.*

(1) All requirements set forth in §101.14 shall be met.

(2) Specific requirements.

(i) *Nature of the claim.* A health claim associating substances in diets low in fat and high in fruits and vegetables with reduced risk of cancer may be made on the label or labeling of a food described in paragraph (c)(2)(ii) of this section, provided that:

(A) The claim states that diets low in fat and high in fruits and vegetables "may" or "might" reduce the risk of some cancers;

(B) In specifying the disease, the claim uses the following terms: "some types of cancer", or "some cancers";

(C) The claim characterizes fruits and vegetables as foods that are low in fat and may contain vitamin A, vitamin C, and dietary fiber;

(D) The claim characterizes the food bearing the claim as containing one or more of the following, for which the food is a good source under §101.54: dietary fiber, vitamin A, or vitamin C;

(E) The claim does not attribute any degree of cancer risk reduction to diets low in fat and high in fruits and vegetables;

(F) In specifying the fat component of the labeled food, the claim uses the term "total fat" or "fat";

(G) The claim does not specify types of fats or fatty acids that may be related to risk of cancer;

(H) In specifying the dietary fiber component of the labeled food, the claim uses the term "fiber", "dietary fiber", or "total dietary fiber";

(I) The claim does not specify types of dietary fiber that may be related to risk of cancer; and

(J) The claim indicates that development of cancer depends on many factors.

(ii) *Nature of the food.*

(A) The food shall be or shall contain a fruit or vegetable.

(B) The food shall meet the nutrient content requirements of §101.62 for a "low fat" food.

(C) The food shall meet, without fortification, the nutrient content requirements of §101.54 for a "good source" of at least one of the following: vitamin A, vitamin C, or dietary fiber.

(d) *Optional information.*

(1) The claim may include information from paragraphs (a) and (b) of this section, which summarize the relationship between diets low in fat and high in fruits and vegetables and some types of cancer and the significance of the relationship.

(2) The claim may identify one or more of the following risk factors for development of cancer: Family history of a specific type of cancer, cigarette smoking, alcohol consumption, overweight and obesity, ultraviolet or ionizing radiation, exposure to cancer-causing chemicals, and dietary factors.

(3) The claim may use the word "beta-carotene" in parentheses after the term vitamin A, provided that the vitamin A in the food bearing the claim is beta-carotene.

(4) The claim may indicate that it is consistent with "Nutrition and Your Health: Dietary Guidelines for Americans," U.S. Department of Agriculture (USDA) and the Department of Health and Human Services (DHHS), Government Printing Office.

(5) The claim may include information on the number of people in the United States who have cancer. The sources of this information must be identified, and it must be current information from the National Center for Health Statistics, the National Institutes of Health, or "Nutrition and Your Health: Dietary Guidelines for Americans," USDA and DHHS, Government Printing Office.

(e) *Model health claims.* The following model health claims may be used in food labeling to characterize the relationship between substances in diets low in fat and high in fruits and vegetables and cancer:

(1) Low fat diets rich in fruits and vegetables (foods that are low in fat and may contain dietary fiber, vitamin A, and vitamin C) may reduce the risk of some types of cancer, a disease associated with many factors. Broccoli is high in vitamins A and C, and it is a good source of dietary fiber.

(2) Development of cancer depends on many factors. Eating a diet low in fat and high in fruits and vegetables, foods that are low in fat and may contain vitamin A, vitamin C, and dietary fiber, may reduce your risk of some cancers. Oranges, a food low in fat, are a good source of fiber and vitamin C.

[58 FR 2639, Jan. 6, 1993]

§101.79 Health claims: Folate and neural tube defects.

(a) *Relationship between folate and neural tube defects—*

(1) *Definition.* Neural tube defects are serious birth defects of the brain or spinal cord that can result in infant mortality or serious disability. The birth defects anencephaly and spina bifida are the most common forms of neural tube defects and account for about 90 percent of these defects. These defects result from failure of closure of the covering of the brain or spinal cord during early embryonic development. Because the neural tube forms and closes during early pregnancy, the defect may occur before a woman realizes that she is pregnant.

(2) *Relationship.* The available data show that diets adequate in folate may reduce the risk of neural tube defects. The strongest evidence for this relationship comes from an intervention study by the Medical Research Council of the United Kingdom that showed that women at risk of recurrence of a neural tube defect pregnancy who consumed a supplement containing 4 milligrams (mg)(4,000 micrograms (mcg)) folic acid daily before conception and continuing into early pregnancy had a reduced risk of having a child with a neural tube defect. (Products containing this level of folic acid are drugs). In addition, based on its review of a Hungarian intervention trial that reported periconceptional use of a multivitamin and multimineral preparation containing 800 mcg (0.8 mg) of folic acid, and its review of the observational

studies that reported periconceptional use of multivitamins containing 0 to 1,000 mcg of folic acid, the Food and Drug Administration concluded that most of these studies had results consistent with the conclusion that folate, at levels attainable in usual diets, may reduce the risk of neural tube defects.

(b) *Significance of folate*—

(1) *Public health concern.* Neural tube defects occur in approximately 0.6 of 1,000 live births in the United States (i.e., approximately 6 of 10,000 live births; about 2,500 cases among 4 million live births annually). Neural tube defects are believed to be caused by many factors. The single greatest risk factor for a neural tube defect-affected pregnancy is a personal or family history of a pregnancy affected with a such a defect. However, about 90 percent of infants with a neural tube defect are born to women who do not have a family history of these defects. The available evidence shows that diets adequate in folate may reduce the risk of neural tube defects but not of other birth defects.

(2) *Populations at risk.* Prevalence rates for neural tube defects have been reported to vary with a wide range of factors including genetics, geography, socioeconomic status, maternal birth cohort, month of conception, race, nutrition, and maternal health, including maternal age and reproductive history. Women with a close relative (i.e., sibling, niece, nephew) with a neural tube defect, those with insulin-dependent diabetes mellitus, and women with seizure disorders who are being treated with valproic acid or carbamazepine are at significantly increased risk compared with women without these characteristics. Rates for neural tube defects vary within the United States, with lower rates observed on the west coast than on the east coast.

(3) *Those who may benefit.* Based on a synthesis of information from several studies, including those which used multivitamins containing folic acid at a daily dose level of ≥ 400 mcg (≥ 0.4 mg), the Public Health Service has inferred that folate alone at levels of 400 mcg (0.4 mg) per day may reduce the risk of neural tube defects. The protective effect found in studies of lower dose folate measured by the reduction in neural tube defect incidence, ranges from none to substantial; a reasonable estimate of the expected reduction in the United States is 50 percent. It is expected that consumption of adequate folate will avert some, but not all, neural tube defects. The underlying causes of neural tube defects are not known. Thus, it is not known what proportion of neural tube defects will be averted by adequate folate consumption. From the available evidence, the Public Health Service estimates that there is the potential for averting 50 percent of cases that now occur (i.e., about 1,250 cases annually). However, until further research is done, no firm estimate of this proportion will be available.

(c) *Requirements.* The label or labeling of food may contain a folate/neural tube defect health claim provided that:

(1) *General requirements.* The health claim for a food meets all of the general requirements of § 101.14 for health claims, except that a food may qualify to bear the health claim if it meets the definition of the term "good source."

(2) *Specific requirements*—

(i) *Nature of the claim*—

(A) *Relationship.* A health claim that women who are capable of becoming pregnant and who consume adequate amounts of folate daily during their childbearing years may reduce their risk of having a pregnancy affected by spina bifida or other neural tube defects may be made on the label or labeling of food provided that:

(B) *Specifying the nutrient.* In specifying the nutrient, the claim shall use the terms "folate," "folic acid," "folacin," "folate, a B vitamin," "folic acid, a B vitamin," or "folacin, a B vitamin."

(C) *Specifying the condition.* In specifying the health- related condition, the claim shall identify the birth defects as "neural tube defects," "birth defects spina bifida or anencephaly," "birth defects of the brain or spinal cord anencephaly or spina bifida," "spina bifida and anencephaly, birth defects of the brain or spinal cord," "birth defects of the brain or spinal cord;" or "brain or spinal cord birth defects."

(D) *Multifactorial nature.* The claim shall not imply that folate intake is the only recognized risk factor for neural tube defects.

(E) *Reduction in risk.* The claim shall not attribute any specific degree of reduction in risk of neural tube defects from maintaining an adequate folate intake throughout the childbearing years. The claim shall state that some women may reduce their risk of a neural tube defect pregnancy by maintaining adequate intakes of folate during their childbearing years. Optional statements about population-based estimates of risk reduction may be made in accordance with paragraph (c)(3)(vi) of this section.

(F) *Safe upper limit of daily intake.* Claims on foods that contain more than 100 percent of the Daily Value (DV) (400 mcg) when labeled for use by adults and children 4 or more years of age, or 800 mcg when labeled for use by pregnant or lactating women) shall identify the safe upper limit of daily intake with respect to the DV. The safe upper limit of daily intake value of 1,000 mcg (1 mg) may be included in parentheses.

(G) *The claim.* The claim shall not state that a specified amount of folate per serving from one source is more effective in reducing the risk of neural tube defects than a lower amount per serving from another source.

(H) The claim shall state that folate needs to be consumed as part of a healthful diet.

(ii) *Nature of the food—*

(A) *Requirements.* The food shall meet or exceed the requirements for a "good source" of folate as defined in § 101.54;

(B) *Dietary supplements.* Dietary supplements shall meet the United States Pharmacopeia (USP) standards for disintegration and dissolution, except that if there are no applicable USP standards, the folate in the dietary supplement shall be shown to be bioavailable under the conditions of use stated on the product label.

(iii) *Limitation.* The claim shall not be made on foods that contain more than 100 percent of the RDI for vitamin A as retinol or preformed vitamin A or vitamin D per serving or per unit.

(iv) *Nutrition labeling.* The nutrition label shall include information about the amount of folate in the food. This information shall be declared after the declaration for iron if only the levels of vitamin A, vitamin C, calcium, and iron are provided, or in accordance with § 101.9 (c)(8) and (c)(9) if other optional vitamins or minerals are declared.

(3) *Optional information—*

(i) *Risk factors.* The claim may specifically identify risk factors for neural tube defects. Where such information is provided, it may consist of statements from § 101.79(b)(1) or (b)(2) (e.g., Women at increased risk include those with a personal history of a neural tube defect-affected pregnancy, those with a close relative (i.e., sibling, niece, nephew) with a neural tube defect; those with insulin-dependent diabetes mellitus; those with seizure disorders who are being treated with valproic acid or carbamazepine) or from other parts of this paragraph (c)(3)(i).

(ii) *Relationship between folate and neural tube defects.* The claim may include statements from paragraphs (a) and (b) of this section that summarize the relationship between folate and neural tube defects and the significance of the relationship except for information specifically prohibited from the claim.

(iii) *Personal history of a neural tube defect-affected pregnancy.* The claim may state that women with a history of a neural tube defect pregnancy should consult their physicians or health care providers before becoming pregnant. If such a statement is provided, the claim shall also state that all women should consult a health care provider when planning a pregnancy.

(iv) *Daily value.* The claim may identify 100 percent of the DV (100% DV; 400 mcg) for folate as the target intake goal.

(v) *Prevalence.* The claim may provide estimates, expressed on an annual basis, of the number of neural tube defect-affected births among live births in the United States. Current estimates are provided in § 101.79(b)(1), and are approximately 6 of 10,000 live births annually (i.e., about 2,500 cases among 4 million live births annually). Data provided in § 101.79(b)(1) shall be used, unless more current estimates from the U.S. Public Health Service are available, in which case the latter may be cited.

(vi) *Reduction in risk.* An estimate of the reduction in the number of neural tube defect-affected births that might occur in the United States if all women consumed adequate folate throughout their childbearing years may be included in the claim. Information contained in paragraph (b)(3) of this section may be used. If such an estimate (i.e., 50 percent) is provided, the estimate shall be accompanied by additional information that states that the estimate is population-based and that it does not reflect risk reduction that may be experienced by individual women.

(vii) *Diets adequate in folate.* The claim may identify diets adequate in folate by using phrases such as "Sources of folate include fruits, vegetables, whole grain products, fortified cereals, and dietary supplements." or "Adequate amounts of folate can be obtained from diets rich in fruits, dark green leafy vegetables, legumes, whole grain products, fortified cereals, or dietary supplements." or "Adequate amounts of folate can be obtained from diets rich in fruits, including citrus fruits and juices, vegetables, including dark green leafy vegetables, legumes, whole grain products, including breads, rice, and pasta, fortified cereals, or a dietary supplement."

(d) *Model health claims.* The following are examples of model health claims that may be used in food labeling to describe the relationship between folate and neural tube defects:

(1) *Examples 1 and 2.* Model health claims appropriate for foods containing 100 percent or less of the DV for folate per serving or per unit (general population). The examples contain only the required elements:

(i) Healthful diets with adequate folate may reduce a woman's risk of having a child with a brain or spinal cord birth defect.

(ii) Adequate folate in healthful diets may reduce a woman's risk of having a child with a brain or spinal cord birth defect.

(2) *Example 3.* Model health claim appropriate for foods containing 100 percent or less of the DV for folate per serving or per unit. The example contains all required elements plus optional information: Women who consume healthful diets with adequate folate throughout their childbearing years may reduce their risk of having a child with a birth defect of the brain or spinal cord. Sources of folate include fruits, vegetables, whole grain products, fortified cereals, and dietary supplements.

(3) *Example 4.* Model health claim appropriate for foods intended for use by the general population and containing more than 100 percent of the DV of folate per serving or per unit: Women who consume healthful diets with adequate folate may reduce their risk of having a child with birth defects of the brain or spinal cord. Folate intake should not exceed 250% of the DV (1,000 mcg).

[61 FR 8779-8781, Mar. 5, 1996; 61 FR 43119, Aug. 20, 1996; 61 FR 48529, Sept. 13, 1996; 61 FR 49964, Sept. 24, 1996]

§101.80 Health claims: dietary sugar alcohols and dental caries.

(a) *Relationship between dietary carbohydrates and dental caries.*

(1) Dental caries, or tooth decay, is a disease caused by many factors. Both environmental and genetic factors can affect the development of dental caries. Risk factors include tooth enamel crystal structure and mineral content, plaque quantity and quality, saliva quantity and quality, individual immune response, types and physical characteristics of foods consumed, eating behaviors, presence of acid producing oral bacteria, and cultural influences.

(2) The relationship between consumption of fermentable carbohydrates, i.e., dietary sugars and starches, and tooth decay is well established. Sucrose, also known as sugar, is one of the most, but not the only, cariogenic sugars in the diet. Bacteria found in the mouth are able to metabolize most dietary carbohydrates, producing acid and forming dental plaque. The more frequent and longer the exposure of teeth to dietary sugars and starches, the greater the risk for tooth decay.

(3) Dental caries continues to affect a large proportion of Americans. Although there has been a decline in the prevalence of dental caries among children in the United States, the disease remains widespread throughout the population, imposing a substantial burden on Americans. Recent Federal government dietary guidelines recommend that Americans choose diets that are moderate in sugars and avoid excessive snacking. Frequent between-meal snacks that are high in sugars and starches may be more harmful to teeth then brushing.

(4) Sugar alcohols can be used as sweeteners to replace dietary sugars, such as sucrose and corn sweeteners, in foods such as chewing gums and certain confectioneries. Dietary sugar alcohols are significantly less cariogenic than dietary sugars and other fermentable carbohydrates.

(b) *Significance of the relationship between sugar alcohols and dental caries. Sugar alcohols do not promote dental caries.* Sugar alcohols are slowly metabolized by bacteria to form some acid. The rate and amount of acid production is significantly less than that from sucrose and other fermentable carbohydrates and does not cause the loss of important minerals from tooth enamel.

(c) *Requirements.*

(1) All requirements set forth in §101.14 shall be met, except that sugar alcohol-containing foods are exempt from section §101.14(e)(6).

(2) *Specific requirements.*
(i) *Nature of the claim.* A health claim relating sugar alcohols, compared to other carbohydrates, and the nonpromotion of dental caries may be made on the label or labeling of a food described in (c)(2)(ii) of this section, provided that:
(A) The claim shall state that frequent between-meal consumption of foods high in sugars and starches can promote tooth decay.
(B) The claim shall state that the sugar alcohol present in the food "does not promote," "may reduce the risk of," "useful [or is useful] in not promoting," or "expressly [or is expressly] for not promoting" dental caries;
(C) In specifying the nutrient, the claim shall state "sugar alcohol," "sugar alcohols," or the name or names of the sugar alcohols, e.g., "sorbitol."
(D) In specifying the disease, the claim uses the following terms: "dental caries" or "tooth decay."
(E) The claim shall not attribute any degree of the reduction in risk of dental caries to the use of the sugar alcohol-containing food.
(F) The claim shall not imply that consuming sugar alcohol-containing foods is the only recognized means of achieving a reduced risk of dental caries.
(G) Packages with less than 15 square inches of surface area available for labeling are exempt from paragraphs (A) and (C) of this section.
(ii) *Nature of the food.*
(A) The food shall meet the requirement in §101.60(c)(1)(i) with respect to sugars content.
(B) The sugar alcohol in the food shall be xylitol, sorbitol, mannitol, maltitol, isomalt, lactitol, hydrogenated starch hydrolysates, hydrogenated glucose syrups, erythritol, or a combination of these.
(C) When fermentable carbohydrates are present in the sugar alcohol-containing food, the food shall not lower plaque pH below 5.7 by bacterial fermentation either during consumption or up to 30 minutes after consumption, as measured by the indwelling plaque pH test found in "Identification of Low Caries Risk Dietary Components," T. N. Imfeld, Volume 11, Monographs in Oral Science, 1983, which is incorporated by reference in accordance with 5 U.S.C. 552(a) and 1 CFR part 51. Copies may be obtained from Karger AG Publishing Co., P. O. Box, Ch-4009 Basel, Switzerland, or may be examined at the Center for Food Safey and Applied Nutrition's Library, 200 C St. SW., rm. 3321, Washington, DC, or at the Office of the Federal Register, 800 North Capitol St. NW., suite 700, Washington, DC.
(d) *Optional information.*
(1) The claim may include information from paragraphs (a) and (b) of this section, which describe the relationship between diets containing sugar alcohols and dental caries.
(2) The claim may indicate that development of dental caries depends on many factors and may identify one or more of the following risk factors for dental caries: Frequent consumption of fermentable carbohydrates, such as dietary sugars and starches; presence of oral bacteria capable of fermenting carbohydrates; length of time fermentable carbohydrates are in contact with the teeth; lack of exposure to fluoride; individual susceptibility; socioeconomic and cultural factors; and characteristics of tooth enamel, saliva, and plaque.
(3) The claim may indicate that oral hygiene and proper dental care may help to reduce the risk of dental disease.
(4) The claim may indicate that the sugar alcohol serves as a sweetener.
(e) *Model health claim.* The following model health claims may be used in food labeling to describe the relationship between sugar alcohol-containing foods and dental caries.
(1) Example of the full claim:
(i) Frequent eating of foods high in sugars and starches as between-meal snacks can promote tooth decay. The sugar alcohol [name, optional] used to sweeten this food may reduce the risk of dental caries.
(ii) Frequent between-meal consumption of foods high in sugars and starches promotes tooth decay. The sugar alcohols in [name of food] do not promote tooth decay.
(2) Example of the shortened claim for small packages:
(i) Does not promote tooth decay.
(ii) May reduce the risk of tooth decay.
[61 FR 43446-43447, Aug. 23, 1996; 62 *FR* 63655, Dec. 2, 1997]

§101.81 Health claims: Soluble fiber from certain foods and risk of coronary heart disease (CHD).

(a) *Relationship between diets low in saturated fat and cholesterol that include soluble fiber from certain foods and risk of coronary heart disease—*

(1) Cardiovascular disease means diseases of the heart and circulatory system. Coronary heart disease (CHD) is one of the most common and serious forms of cardiovascular disease and refers to diseases of the heart muscle and supporting blood vessels. High blood total cholesterol and low density lipoprotein (LDL)-cholesterol levels are associated with increased risk of developing coronary heart disease. High CHD rates occur among people with high total cholesterol levels of 240 milligrams per deciliter (mg/dL) (6.21 (mmol/L)) or above and LDL-cholesterol levels of 160 mg/dL (4.13 mmol/L) or above. Borderline high risk total cholesterol levels range from 200 to 239 mg/dL (5.17 to 6.18 mmol/L) and 130 to 159 mg/dL (3.36 to 4.11 mmol/L) of LDL-cholesterol. The scientific evidence establishes that diets high in saturated fat and cholesterol are associated with increased levels of blood total- and LDL-cholesterol and, thus, with increased risk of CHD.

(2) Populations with a low incidence of CHD tend to have relatively low blood total cholesterol and LDL-cholesterol levels. These populations also tend to have dietary patterns that are not only low in total fat, especially saturated fat and cholesterol, but are also relatively high in fiber-containing fruits, vegetables, and grain products, such as whole oat products.

(3) Scientific evidence demonstrates that diets low in saturated fat and cholesterol may reduce the risk of CHD. Other evidence demonstrates that the addition of soluble fiber from certain foods to a diet that is low in saturated fat and cholesterol may also help to reduce the risk of CHD.

(b) *Significance of the relationship between diets low in saturated fat and cholesterol that include soluble fiber from certain foods and risk of CHD—*

(1) CHD is a major public health concern in the United States. It accounts for more deaths than any other disease or group of diseases. Early management of risk factors for CHD is a major public health goal that can assist in reducing risk of CHD. High blood total and LDL-cholesterol are major modifiable risk factors in the development of CHD.

(2) Intakes of saturated fat exceed recommended levels in the diets of many people in the United States. One of the major public health recommendations relative to CHD risk is to consume less than 10 percent of calories from saturated fat and an average of 30 percent or less of total calories from all fat. Recommended daily cholesterol intakes are 300 milligrams (mg) or less per day. Scientific evidence demonstrates that diets low in saturated fat and cholesterol are associated with lower blood total and LDL-cholesterol levels. Soluble fiber from certain foods, when included in a low saturated fat and cholesterol diet, also helps to lower blood total and LDL-cholesterol levels.

(c) *Requirements—*

(1) All requirements set forth in § 101.14 shall be met. The label and labeling of foods containing psyllium husk shall be consistent with the provisions of §101.70(f).

(2) *Specific requirements*.

(i) Nature of the Claim. A health claim associating diets low in saturated fat and cholesterol that include soluble fiber from certain foods with reduced risk of heart disease may be made on the label or labeling of a food described in paragraph (c)(2)(iii) of this section, provided that:

(A) The claim states that diets low in saturated fat and cholesterol that include soluble fiber from certain foods "may" or "might" reduce the risk of heart disease;

(B) In specifying the disease, the claim uses the following terms: "heart disease" or "coronary heart disease";

(C) In specifying the substance, the claim uses the term "soluble fiber" qualified by the name of the eligible source of soluble fiber (provided in paragraph (c)(2)(ii) of this section). Additionally, the claim may use the name of the food product that contains the eligible source of soluble fiber;

(D) In specifying the fat component, the claim uses the terms "saturated fat" and "cholesterol";

(E) The claim does not attribute any degree of risk reduction for CHD to diets low in saturated fat and cholesterol and that include soluble fiber from the eligible food sources from paragraph (c)(2)(ii) of this section; and

(F) The claim does not imply that consumption of diets that are low in saturated fat and cholesterol and that include soluble fiber from the eligible food sources from paragraph (c)(2)(ii) of this section is the only recognized means of achieving a reduced risk of CHD.

(G) The claim specifies the daily dietary intake of the soluble fiber source that is necessary to reduce the risk of coronary heart disease and the contribution one serving of the product makes to the specified daily dietary intake level. Daily dietary intake levels for soluble fiber sources listed in paragraph (c)(2)(ii) of this section that have been associated with reduced risk coronary heart disease are:

(*1*) 3 g or more per day of β-glucan soluble fiber from whole oats.

(*2*) 7 g or more per day of soluble fiber from psyllium seed husk.

(ii) Nature of the substance. Eligible sources of soluble fiber.

(A) Beta (β) glucan soluble fiber from the whole oat sources listed below. β-glucan soluble fiber will be determined by method No. 992.28 from the "Official Methods of Analysis of the Association of Official Analytical Chemists International," 16th ed. (1995), which is incorporated by reference in accordance with 5 U.S.C. 552(a) and 1 CFR part 51. Copies may be obtained from the Association of Official Analytical Chemists International, 481 North Frederick Ave., suite 500, Gaithersburg, MD 20877-2504, or may be examined at the Center for Food Safety and Applied Nutrition's Library, 200 C St. SW., rm. 3321, Washington, DC, or at the Office of the Federal Register, 800 North Capitol St. NW., suite 700, Washington, DC;

(*1*) Oat bran. Oat bran is produced by grinding clean oat groats or rolled oats and separating the resulting oat flour by suitable means into fractions such that the oat bran fraction is not more than 50 percent of the original starting material and provides at least 5.5 percent (dry weight basis (dwb)) β-glucan soluble fiber and a total dietary fiber content of 16 percent (dwb), and such that at least one-third of the total dietary fiber is soluble fiber;

(*2*) Rolled oats. Rolled oats, also known as oatmeal, produced from 100 percent dehulled, clean oat groats by steaming, cutting, rolling, and flaking, and provides at least 4 percent (dwb) of β-glucan soluble fiber and a total dietary fiber content of at least 10 percent.

(*3*) Whole oat flour. Whole oat flour is produced from 100 percent dehulled,clean oat groats by steaming and grinding, such that there is no significant loss of oat bran in the final product, and provides at least 4 percent (dwb) of β-glucan soluble fiber and a total dietary fiber content of at least 10 percent (dwb).

(B)(*1*) Psyllium husk from the dried seed coat (epidermis) of the seed of *Plantago* (*P.*) ovata, known as blond psyllium or Indian psyllium, *P. indica*, or *P. psyllium*. To qualify for this claim, psyllium seed husk, also known as psyllium husk, shall have a purity of no less than 95 percent, such that it contains 3 percent or less protein, 4.5 percent or less of light extraneous matter, and 0.5 percent or less of heavy extraneous matter, but in no case may the combined extraneous matter exceed 4.9 percent, as determined by U.S. Pharmacopeia (USP) methods described USP's "The National Formulary," USP 23, NF 18, p. 1341, (1995), which incorporated by reference in accordance with 5 U.S.C. 552(a) and 1 CFR part 51. Copies may be obtained from the U.S. Pharmacopeial Convention, Inc., 12601 Twinbrook Pkwy., Rockville, MD 20852, or may be examined at the Center for Food Safety and Applied Nutrition's Library, 200 C St. SW., rm. 3321, Washington, DC, or at the Office of the Federal Register, 800 North Capitol St. NW., suite 700, Washington, DC;

(*2*) FDA will determine the amount of soluble fiber that is provided by psyllium husk by using a modification of the Association of Official Analytical Chemists' (AOAC's) method for soluble dietary fiber (991.43) described by Lee et al., "Determination of Soluble and Insoluble Dietary Fiber in Psyllium-containing Cereal Products," Journal of the AOAC International, 78 (No. 3):724–729, 1995, which is incorporated by reference in accordance with 5 U.S.C. 552(a) and 1 CFR part 51. Copies may be obtained from the Association of Official Analytical Chemists International, 481 North Frederick Ave., suite 500, Gaithersburg, MD 20877– 2504, or may be examined at the Center for Food Safety and Applied Nutrition's Library, 200 C St. SW., rm. 3321, Washington, DC, or at the Office of the Federal Register, 800 North Capitol St. NW., suite 700, Washington, DC;

(iii) Nature of the Food Eligible to Bear the Claim.

(A) The food product shall include:

21 CFR §101.81

(*1*) One or more of the whole oat foods from paragraph (c)(2)(ii)(A) of this section, and the whole oat foods shall contain at least 0.75 gram (g) per reference amount customarily consumed of the food product; or

(2) Psyllium husk that complies with paragraph (c)(2)(ii)(B) of this section, and the psyllium food shall contain at least 1.7 g of soluble fiber per reference amount customarily consumed of the food product;

(B) The amount of soluble fiber shall be declared in the nutrition label, consistent with § 101.9(c)(6)(i)(A).

(C) The food shall meet the nutrient content requirements in § 101.62 for a "low saturated fat," "low cholesterol," and "low fat" food.

(d) *Optional information*—

(1) The claim may state that the development of heart disease depends on many factors and may identify one or more of the following risk factors for heart disease about which there is general scientific agreement: A family history of CHD; elevated blood total and LDL-cholesterol; excess body weight; high blood pressure; cigarette smoking; diabetes; and physical inactivity. The claim may also provide additional information about the benefits of exercise and management of body weight to help lower the risk of heart disease;

(2) The claim may state that the relationship between intake of diets that are low in saturated fat and cholesterol and that include soluble fiber from the eligible food sources from paragraph (c)(2)(ii) of this section and reduced risk of heart disease is through the intermediate link of "blood cholesterol" or "blood total- and LDL-cholesterol;"

(3) The claim may include information from paragraphs (a) and (b) of this section, which summarize the relationship between diets low in saturated fat and cholesterol that include soluble fiber from certain foods and coronary heart disease and the significance of the relationship;

(4) The claim may specify the name of the eligible soluble fiber;

(5) The claim may state that a diet low in saturated fat and cholesterol that includes soluble fiber from whole oats is consistent with "Nutrition and Your Health: Dietary Guidelines for Americans," U.S. Department of Agriculture (USDA) and Department of Health and Human Services (DHHS), Government Printing Office (GPO);

(6) The claim may state that individuals with elevated blood total-and LDL-cholesterol should consult their physicians for medical advice and treatment. If the claim defines high or normal blood total- and LDL-cholesterol levels, then the claim shall state that individuals with high blood cholesterol should consult their physicians for medical advice and treatment;

(7) The claim may include information on the number of people in the United States who have heart disease. The sources of this information shall be identified, and it shall be current information from the National Center for Health Statistics, the National Institutes of Health, or "Nutrition and Your Health: Dietary Guidelines for Americans," USDA and DHHS, GPO;

(e) *Model health claim.* The following model health claims may be used in food labeling to describe the relationship between diets that are low in saturated fat and cholesterol and that include soluble fiber from certain foods and reduced risk of heart disease:

(1) Soluble fiber from foods such as [name of soluble fiber source from paragraph (c)(2)(ii) of this section and, if desired, the name of food product], as part of a diet low in saturated fat and cholesterol, may reduce the risk of heart disease. A serving of [name of food] supplies ___ grams of the [grams of soluble fiber specified in paragraph (c)(2)(i)(G) of this section] soluble fiber from [name of the soluble fiber source from paragraph (c)(2)(ii) of this section] necessary per day to have this effect.

(2) Diets low in saturated fat and cholesterol that include [___ grams of soluble fiber specified in paragraph (c)(2)(i)(G) of this section] of soluble fiber per day from [name of soluble fiber source from paragraph (c)(2)(ii) of this section and, if desired, the name of food product] may reduce the risk of heart disease. One serving of [name of food] provides ___ grams of this soluble fiber.

[62 FR 3600-3601, Jan. 23, 1997; 62 FR 15344, Mar. 31, 1997; 63 *FR* 8119-8120; Feb. 18, 1998]

Subpart F-Specific Requirements for Descriptive Claims that are Neither Nutrient Content Claims nor Health Claims

21 CFR §101.82

§101.93 Notification procedures for certain types of statements on dietary supplements.

(a)

(1) No later than 30 days after the first marketing of a dietary supplement that bears one of the statements listed in section 403(r)(6) or the Federal Food, Drug, and Cosmetic Act, the manufacturer, packer, or distributor of the dietary supplement shall notify the Office of Special Nutritionals (HFS– 450), Center for Food Safety and Applied Nutrition, Food and Drug Administration, 200 C St. SW., Washington, DC 20204, that it has included such a statement on the label or in the labeling of its product. An original and two copies of this notification shall be submitted.

(2) The notification shall include the following:

(i) The name and address of the manufacturer, packer, or distributor of the dietary supplement that bears the statement;

(ii) The text of the statement that is being made;

(iii) The name of the dietary ingredient or supplement that is the subject of the statement, if not provided in the text of the statement; and (iv) The name of the dietary supplement (including brand name), if not provided in response to paragraph (a)(2)(iii) on whose label, or in whose labeling, the statement appears.

(3) The notice shall be signed by a responsible individual or the person who can certify the accuracy of the information presented and contained in the notice. The individual shall certify that the information contained in the notice is complete and accurate, and that the notifying firm has substantiation that the statement is truthful and not misleading.

(b) *Disclaimer.* The requirements in this section apply to the label or labeling of dietary supplements where the dietary supplement bears a statement that is provided for by section 403(r)(6) of the Federal Food, Drug, and Cosmetic Act (the act), and the manufacturer, packer, or distributor wishes to take advantage of the exemption to section 201(g)(1)(C) of the act that is provided by compliance with section 403(r)(6) of the act.

(c) *Text for disclaimer*.

(1) Where there is one statement, the disclaimer shall be placed in accordance with paragraph (d) of this section and shall state:

This statement has not been evaluated by the Food and Drug Administration. This product is not intended to diagnose, treat, cure, or prevent any disease.

(2) Where there is more than one such statement on the label or in the labeling, each statement shall bear the disclaimer in accordance with paragraph (c)(1) of this section, or a plural disclaimer may be placed in accordance with paragraph (d) of this section and shall state:

These statements have not been evaluated by the Food and Drug Administration. This product is not intended to diagnose, treat, cure, or prevent any disease.

(d) *Placement.* The disclaimer shall be placed adjacent to the statement with no intervening material or linked to the statement with a symbol (e.g., an asterisk) at the end of each such statement that refers to the same symbol placed adjacent to the disclaimer specified in paragraphs (c)(1) or (c)(2) of this section. On product labels and in labeling (e.g., pamphlets, catalogs), the disclaimer shall appear on each panel or page where there such is a statement. The disclaimer shall be set off in a box where it is not adjacent to the statement in question.

(e) *Typesize.* The disclaimer in paragraph (c) of this section shall appear in boldface type in letters of a typesize no smaller than one-sixteenth inch.

[62 *FR* 49867, 49885, Sept. 23, 1997]

§101.95 "Fresh," "freshly frozen," "fresh frozen," "frozen fresh"

The terms defined in this section may be used on the label or in labeling of a food in conformity with the provisions of this section. The requirements of the section pertain to any use of the subject terms as described in paragraphs (a) and (b) of this section that expressly or implicitly refers to the food on labels or labeling, including use in a brand name and use as a sensory modifier. However, the use of the term "fresh" on labels or labeling is not subject to the requirements of paragraph (a) of this section if the term does not suggest or imply that a food is unprocessed or unpreserved. For example, the term "fresh" used to describe pasteurized whole milk is not subject to paragraph (a) of this section because the term does not imply that the food is unprocessed (consumers commonly understand that

21 CFR §101.83

milk is nearly always pasteurized). However, the term "fresh" to describe pasta sauce that has been pasteurized or that contains pasteurized ingredients would be subject to paragraph (a) of this section because the term implies that the food is not processed or preserved. Uses of fresh not subject to this regulation will be governed by the provisions of 403(a) of the Federal Food, Drug, and Cosmetic Act (the act).

(a) The term "fresh," when used on the label or in labeling of a food in a manner that suggests or implies that the food is unprocessed, means that the food is in its raw state and has not been frozen or subjected to any form of thermal processing or any other form of preservation, except as provided in paragraph (c) of this section.

(b) The terms "fresh frozen" and "frozen fresh," when used on the label or in labeling of a food, mean that the food was quickly frozen while still fresh (i.e., the food had been recently harvested when frozen). Blanching of the food before freezing will not preclude use of the term "fresh frozen" to describe the food. "Quickly frozen" means frozen by a freezing system such as blast-freezing (sub-zero Fahrenheit temperature with fast moving air directed at the food) that ensures the food is frozen, even to the center of the food, quickly and that virtually no deterioration has taken place.

(c) *Provisions and restrictions-*

    (1) The following do not preclude the food from use of the term "fresh:"

    (i) The addition of approved waxes or coatings;

    (ii) The post-harvest use of approved pesticides;

    (iii) The application of a mild chlorine wash or mild acid wash on produce; or

    (iv) The treatment of raw foods with ionizing radiation not to exceed the maximum dose of 1 kiloGray in accordance with §179.26 of this chapter.

    (2) A food meeting the definition in paragraph (a) of this section that is refrigerated is not precluded from use of "fresh" as provided by this section.

[58 FR 2426, Jan. 6, 1993]

Subpart G-Exemptions From Food Labeling Requirements

§101.100 Food; exemptions from labeling.

(a) The following foods are exempt from compliance with the requirements of section 403(i)(2) of the act (requiring a declaration on the label of the common or usual name of each ingredient when the food is fabricated from two or more ingredients).

    (1) An assortment of different items of food, when variations in the items that make up different packages packed from such assortment normally occur in good packing practice and when such variations result in variations in the ingredients in different packages, with respect to any ingredient that is not common to all packages. Such exemption, however, shall be on the condition that the label shall bear, in conjunction with the names of such ingredients as are common to all packages, a statement (in terms that are as informative as practicable and that are not misleading) indicating by name other ingredients which may be present.

    (2) A food having been received in bulk containers at a retail establishment, if displayed to the purchaser with either:

    (i) The labeling of the bulk container plainly in view, provided ingredient information appears prominently and conspicuously in lettering of not less than one-fourth of an inch in height; or

    (ii) A counter card, sign, or other appropriate device bearing prominently and conspicuously, but in no case with lettering of less than one-fourth of an inch in height, the information required to be stated on the label pursuant to section 403(i)(2) of the Federal Food, Drug, and Cosmetic Act (the act).

    (3) Incidental additives that are present in a food at insignificant levels and do not have any technical or functional effect in that food. For the purposes of this paragraph (a)(3), incidental additives are:

    (i) Substances that have no technical or functional effect but are present in a food by reason of having been incorporated into the food as an ingredient of another food, in which the substance did have a functional or technical effect.

    (ii) Processing aids, which are as follows:

    (a) Substances that are added to a food during the processing of such food but are removed in some manner from the food before it is packaged in its finished form.

(b) Substances that are added to a food during processing, are converted into constituents normally present in the food, and do not significantly increase the amount of the constituents naturally found in the food.

(c) Substances that are added to a food for their technical or functional effect in the processing but are present in the finished food at insignificant levels and do not have any technical or functional effect in that food.

(iii) Substances migrating to food from equipment or packaging or otherwise affecting food that are not food additives as defined in section 201(s) of the act; or if they are food additives as so defined, they are used in conformity with regulations established pursuant to section 409 of the act.

(4) For the purposes of paragraph (a)(3) of this section, any sulfiting agent (sulfur dioxide, sodium sulfite, sodium bisulfite, potasssium bisulfite, sodium metabisulfite, and potassium metabisulfite) that has been added to any food or to any ingredient in any food and that has no technical effect in that food will be considered to be present in an insignificant amount only if no detectable amount of the agent is present in the finished food. A detectable amount of sulfiting agent is 10 parts per million or more of the sulfite in the finished food. Compliance with this paragraph will be determined using sections 20.123- 20.125, "Total Sulfurous Acid," in "Official Methods of Analysis of the Association of Official Analytical Chemists," 14th Ed. (1984), which is incorporated by reference and the refinements of the "Total Sulfurous Acid" procedure in the "Monier-Williams Procedure (with Modifications) for Sulfites in Foods," which is Appendix A to Part 101. A copy of sections 20.123-20-125 of the Official Methods of Analysis of the Association of Official Analytical Chemists" is available from the Association of Official Analytical Chemists, P.O. Box 540, Benjamin Franklin Station, Washington, DC 20044, or available for inspection at the Office of the Federal Register, 800 North Capitol Street, NW., suite 700, Washington, DC 20001.

(b) A food repackaged in a retail establishment is exempt from the following provisions of the act if the conditions specified are met.

(1) Section 403(e)(1) of the act (requiring a statement on the label of the name and place of business of the manufacturer, packer, or distributor).

(2) Section 403(g)(2) of the act (requiring the label of a food which purports to be or is represented as one for which a definition and standard of identity has been prescribed to bear the name of the food specified in the definition and standard and, insofar as may be required by the regulation establishing the standard the common names of the optional ingredients present in the food), if the food is displayed to the purchaser with its interstate labeling clearly in view, or with a counter card, sign, or other appropriate device bearing prominently and conspicuously the information required by these provisions.

(3) Section 403(i)(1) of the act (requiring the label to bear the common or usual name of the food), if the food is displayed to the purchaser with its interstate labeling clearly in view, or with a counter card, sign, or other appropriate device bearing prominently and conspicuously the common or usual name of the food, or if the common or usual name of the food is clearly revealed by its appearance.

(c) An open container (a container of rigid or semirigid construction, which is not closed by lid, wrapper, or otherwise other than by an uncolored transparent wrapper which does not obscure the contents) of a fresh fruit or fresh vegetable, the quantity of contents of which is not more than 1 dry quart, shall be exempt from the labeling requirements of sections 403(e), (g)(2) (with respect to the name of the food specified in the definition and standard), and (i)(1) of the act; but such exemption shall be on the condition that if two or more such containers are enclosed in a crate or other shipping package, such crate or package shall bear labeling showing the number of such containers enclosed therein and the quantity of the contents of each.

(d) Except as provided by paragraphs (e) and (f) of this section, a shipment or other delivery of a food which is, in accordance with the practice of the trade, to be processed, labeled, or repacked in substantial quantity at an establishment other than that where originally processed or packed, shall be exempt, during the time of introduction into and movement in interstate commerce and the time of holding in such establishment, from compliance with the labeling requirements of section 403 (c), (e), (g), (h), (i), (k), and (q) of the act if:

(1) The person who introduced such shipment or delivery into interstate commerce is the operator of the establishment where such food is to be processed, labeled, or repacked; or

(2) In case such person is not such operator, such shipment or delivery is made to such establishment under a written agreement, signed by and containing the post office addresses of such person and such operator, and containing such specifications for the processing, labeling, or repacking, as the case may be, of such food in such establishment as will ensure, if such specifications are followed, that such food will not be

adulterated or misbranded within the meaning of the act upon completion of such processing, labeling, or repacking. Such person and such operator shall each keep a copy of such agreement until 2 years after the final shipment or delivery of such food from such establishment, and shall make such copies available for inspection at any reasonable hour to any officer or employee of the Department who requests them.

(3) The article is an egg product subject to a standard of identity promulgated in part 160 of this chapter, is to be shipped under the conditions specified in paragraph (d) (1) or (2) of this section and for the purpose of pasteurization or other treatment as required in such standard, and each container of such egg product bears a conspicuous tag or label reading "Caution- This egg product has not been pasteurized or otherwise treated to destroy viable Salmonella microorganisms". In addition to safe and suitable bactericidal processes designed specifically for Salmonella destruction in egg products, the term "other treatment" in the first sentence of this paragraph shall include use in acidic dressings in the processing of which the pH is not above 4.1 and the acidity of the aqueous phase, expressed as acetic acid, is not less than 1.4 percent, subject also to the conditions that:

(i) The agreement required in paragraph (d) (2) of this section shall also state that the operator agrees to utilize such unpasteurized egg products in the processing of acidic dressings according to the specifications for pH and acidity set forth in this paragraph, agrees not to deliver the acidic dressing to a user until at least 72 hours after such egg product is incorporated in such acidic dressing, and agrees to maintain for inspection adequate records covering such processing for 2 years after such processing.

(ii) In addition to the caution statement referred to above, the container of such egg product shall also bear the statement "Unpasteurized --- for use in acidic dressings only", the blank being filled in with the applicable name of the eggs or egg product.

(e) Conditions affecting expiration of exemptions:

(1) An exemption of a shipment or other delivery of a food under paragraph (d) (1) or (3) of this section shall, at the beginning of the act of removing such shipment or delivery, or any part thereof, from such establishment become void ab initio if the food comprising such shipment, delivery, or part is adulterated or misbranded within the meaning of the act when so removed.

(2) An exemption of a shipment or other delivery of a food under paragraph (d) (2) or (3) of this section shall become void ab initio with respect to the person who introduced such shipment or delivery into interstate commerce upon refusal by such person to make available for inspection a copy of the agreement, as required by paragraph (d) (2) or (3) of this section.

(3) An exemption of a shipment or other delivery of a food under paragraph (d) (2) or (3) of this section shall expire:

(i) At the beginning of the act of removing such shipment or delivery, or any part thereof, from such establishment if the food constituting such shipment, delivery, or part is adulterated or misbranded within the meaning of the act when so removed; or

(ii) Upon refusal by the operator of the establishment where such food is to be processed, labeled, or repacked, to make available for inspection a copy of the agreement, as required by such paragraph.

(f) The word "processed" as used in this paragraph shall include the holding of cheese in a suitable warehouse at a temperature of not less than 35° F for the purpose of aging or curing to bring the cheese into compliance with requirements of an applicable definition and standard of identity. The exemptions provided for in paragraph (d) of this section shall apply to cheese which is, in accordance with the practice of the trade, shipped to a warehouse for aging or curing, on condition that the cheese is identified in the manner set forth in one of the applicable following paragraphs, and in such case the provisions of paragraph (e) of this section shall also apply:

(1) In the case of varieties of cheese for which definitions and standards of identity require a period of aging whether or not they are made from pasteurized milk, each such cheese shall bear on the cheese a legible mark showing the date at which the preliminary manufacturing process has been completed and at which date curing commences, and to each cheese, on its wrapper or immediate container, shall be affixed a removable tag bearing the statement "Uncured --- cheese for completion of curing and proper labeling", the blank being filled in with the applicable name of the variety of cheese. In the case of swiss cheese, the date at which the preliminary manufacturing process had been completed and at which date curing commences is the date on which the shaped curd is removed from immersion in saturated salt solution as provided in the definition and standard of identity for swiss cheese, and such cheese shall bear a removable tag reading, "To be cured and labeled as 'swiss cheese,' but if eyes do not form, to be labeled as 'swiss

cheese for manufacturing' ".

(2) In the case of varieties of cheeses which when made from unpasteurized milk are required to be aged for not less than 60 days, each such cheese shall bear a legible mark on the cheese showing the date at which the preliminary manufacturing process has been completed and at which date curing commences, and to each such cheese or its wrapper or immediate container shall be affixed a removable tag reading, "___ cheese made from unpasteurized milk. For completion of curing and proper labeling", the blank being filled in with the applicable name of the variety of cheese.

(3) In the case of cheddar cheese, washed curd cheese, colby cheese, granular cheese, and brick cheese made from unpasteurized milk, each such cheese shall bear a legible mark on the cheese showing the date at which the preliminary manufacturing process has been completed and at which date curing commences, and to each such cheese or its wrapper or immediate container shall be affixed a removable tag reading "___ cheese made from unpasteurized milk. For completion of curing and proper labeling, or for labeling as ___ cheese for manufacturing", the blank being filled in with the applicable name of the variety of cheese.

(g) The label declaration of a harmless marker used to identify a particular manufacturer's product may result in unfair competition through revealing a trade secret. Exemption from the label declaration of such a marker is granted, therefore, provided that the following conditions are met:

(1) The person desiring to use the marker without label declaration of its presence has submitted to the Commissioner of Food and Drugs full information concerning the proposed usage and the reasons why he believes label declaration of the marker should be subject to this exemption; and

(2) The person requesting the exemption has received from the Commissioner of Food and Drugs a finding that the marker is harmless and that the exemption has been granted.

(h) Wrapped fish fillets of nonuniform weight intended to be unpacked and marked with the correct weight at or before the point of retail sale in an establishment other than that where originally packed shall be exempt from the requirement of section 403(e)(2) of the act during introduction and movement in interstate commerce and while held for sale prior to weighing and marking:

(1) Provided, That

(i) The outside container bears a label declaration of the total net weight; and

(ii) The individual packages bear a conspicuous statement "To be weighed at or before time of sale" and a correct statement setting forth the weight of the wrapper;

(2) Provided further, That it is the practice of the retail establishment to weigh and mark the individual packages with a correct net-weight statement prior to or at the point of retail sale. A statement of the weight of the wrapper shall be set forth so as to be readily read and understood, using such term as "wrapper tare-ounce", the blank being filled in with the correct average weight of the wrapper used.

(3) The act of delivering the wrapped fish fillets during the retail sale without the correct net-weight statement shall be deemed an act which results in the product's being misbranded while held for sale. Nothing in this paragraph shall be construed as requiring net-weight statements for wrapped fish fillets delivered into institutional trade provided the outside container bears the required information.

(i) Wrapped clusters (consumer units) of bananas of nonuniform weight intended to be unpacked from a master carton or container and weighed at or before the point of retail sale in an establishment other than that where originally packed shall be exempt from the requirements of section 403(e)(2) of the act during introduction and movement in interstate commerce and while held for sale prior to weighing:

(1) Provided, That

(i) The master carton or container bears a label declaration of the total net weight; and

(ii) The individual packages bear a conspicuous statement "To be weighed at or before the time of sale" and a correct statement setting forth the weight of the wrapper; using such term as "wrapper tare --- ounce", the blank being filled in with the correct average weight of the wrapper used;

(2) Provided further, That it is the practice of the retail establishment to weigh the individual packages either prior to or at the time of retail sale.

(3) The act of delivering the wrapped clusters (consumer units) during the retail sale without an accurate net weight statement or alternatively without weighing at the time of sale shall be deemed an act which results in the product's being misbranded while held for sale. Nothing in this paragraph shall be construed as requiring net-weight statements for clusters (consumer units) delivered into institutional trade, provided that the master container or carton bears the required information.

21 *CFR* §101.105

[42 FR 14308, Mar. 15, 1977, as amended at 51 FR 25017, July 9, 1986; 58 FR 2188, 2876, Jan. 6, 1993]

§101.105 Declaration of net quantity of contents when exempt.

(a) The principal display panel of a food in package form shall bear a declaration of the net quantity of contents. This shall be expressed in the terms of weight, measure, numerical count, or a combination of numerical count and weight or measure. The statement shall be in terms of fluid measure if the food is liquid, or in terms of weight if the food is solid, semisolid, or viscous, or a mixture of solid and liquid; except that such statement may be in terms of dry measure if the food is a fresh fruit, fresh vegetable, or other dry commodity that is customarily sold by dry measure. If there is a firmly established general consumer usage and trade custom of declaring the contents of a liquid by weight, or a solid, semisolid, or viscous product by fluid measure, it may be used. Whenever the Commissioner determines that an existing practice of declaring net quantity of contents by weight, measure, numerical count, or a combination in the case of a specific packaged food does not facilitate value comparisons by consumers and offers opportunity for consumer confusion, he will by regulation designate the appropriate term or terms to be used for such commodity.

(b)

    (1) Statements of weight shall be in terms of avoirdupois pound and ounce.

    (2) Statements of fluid measure shall be in terms of the U.S. gallon of 231 cubic inches and quart, pint, and fluid ounce subdivisions thereof, and shall:

    (i) In the case of frozen food that is sold and consumed in a frozen state, express the volume at the frozen temperature. (ii) In the case of refrigerated food that is sold in the refrigerated state, express the volume at 40° F (4° C).

    (iii) In the case of other foods, express the volume at 68° F (20° C).

    (3) Statements of dry measure shall be in terms of the U.S. bushel of 2,150.42 cubic inches and peck, dry quart, and dry pint subdivisions thereof.

(c) When the declaration of quantity of contents by numerical count does not give adequate information as to the quantity of food in the package, it shall be combined with such statement of weight, measure, or size of the individual units of the foods as will provide such information.

(d) The declaration may contain common or decimal fractions. A common fraction shall be in terms of halves, quarters, eighths, sixteenths, or thirty-seconds; except that if there exists a firmly established general consumer usage and trade custom of employing different common fractions in the net quantity declaration of a particular commodity, they may be employed. A common fraction shall be reduced to its lowest terms; a decimal fraction shall not be carried out to more than two places. A statement that includes small fractions of an ounce shall be deemed to permit smaller variations than one which does not include such fractions.

(e) The declaration shall be located on the principal display panel of the label, and with respect to packages bearing alternate principal panels it shall be duplicated on each principal display panel.

(f) The declaration shall appear as a distinct item on the principal display panel, shall be separated (by at least a space equal to the height of the lettering used in the declaration) from other printed label information appearing above or below the declaration and (by at least a space equal to twice the width of the letter "N" of the style of type used in the quantity of contents statement) from other printed label information appearing to the left or right of the declaration. It shall not include any term qualifying a unit of weight, measure, or count (such as "jumbo quart" and "full gallon") that tends to exaggerate the amount of the food in the container. It shall be placed on the principal display panel within the bottom 30 percent of the area of the label panel in lines generally parallel to the base on which the package rests as it is designed to be displayed: Provided, That on packages having a principal display panel of 5 square inches or less, the requirement for placement within the bottom 30 percent of the area of the label panel shall not apply when the declaration of net quantity of contents meets the other requirements of this part.

(g) The declaration shall accurately reveal the quantity of food in the package exclusive of wrappers and other material packed therewith: Provided, That in the case of foods packed in containers designed to deliver the food under pressure, the declaration shall state the net quantity of the contents that will be expelled when the instructions for use as shown on the container are followed. The propellant is included in the net quantity declaration.

(h) The declaration shall appear in conspicuous and easily legible boldface print or type in distinct contrast (by typography, layout, color, embossing, or molding) to other matter on the package; except that a declaration of net quantity blown, embossed, or molded on a glass or plastic surface is permissible when all label information is so formed on the surface. Requirements of conspicuousness and legibility shall include the specifications that:

(1) The ratio of height to width (of the letter) shall not exceed a differential of 3 units to 1 unit (no more than 3 times as high as it is wide).

(2) Letter heights pertain to upper case or capital letters. When upper and lower case or all lower case letters are used, it is the lower case letter "o" or its equivalent that shall meet the minimum standards.

(3) When fractions are used, each component numeral shall meet one-half the minimum height standards.

(i) The declaration shall be in letters and numerals in a type size established in relationship to the area of the principal display panel of the package and shall be uniform for all packages of substantially the same size by complying with the following type specifications:

(1) Not less than one-sixteenth inch in height on packages the principal display panel of which has an area of 5 square inches or less.

(2) Not less than one-eighth inch in height on packages the principal display panel of which has an area of more than 5 but not more than 25 square inches.

(3) Not less than three-sixteenths inch in height on packages the principal display panel of which has an area of more than 25 but not more than 100 square inches.

(4) Not less than one-fourth inch in height on packages the principal display panel of which has an area of more than 100 square inches, except not less than 1/2 inch in height if the area is more than 400 square inches.

Where the declaration is blown, embossed, or molded on a glass or plastic surface rather than by printing, typing, or coloring, the lettering sizes specified in paragraphs (h)(1) through (4) of this section shall be increased by one-sixteenth of an inch.

(j) On packages containing less than 4 pounds or 1 gallon and labeled in terms of weight or fluid measure:

(1) The declaration shall be expressed both in ounces, with identification by weight or by liquid measure and, if applicable (1 pound or 1 pint or more) followed in parentheses by a declaration in pounds for weight units, with any remainder in terms of ounces or common or decimal fractions of the pound (see examples set forth in paragraphs (m) (1) and (2) of this section), or in the case of liquid measure, in the largest whole units (quarts, quarts and pints, or pints, as appropriate) with any remainder in terms of fluid ounces or common or decimal fractions of the pint or quart (see examples in paragraphs (m) (3) and (4) of this section).

(2) If the net quantity of contents declaration appears on a random package, that is a package which is one of a lot, shipment, or delivery of packages of the same consumer commodity with varying weights and with no fixed weight pattern, it may, when the net weight exceeds 1 pound, be expressed in terms of pounds and decimal fractions of the pound carried out to not more than two decimal places. When the net weight does not exceed 1 pound, the declaration on the random package may be in decimal fractions of the pound in lieu of ounces (see example in paragraph (m)(5) of this section).

(3) The declaration may appear in more than one line. The term "net weight" shall be used when stating the net quantity of contents in terms of weight. Use of the terms "net" or "net contents" in terms of fluid measure or numerical count is optional. It is sufficient to distinguish avoirdupois ounce from fluid ounce through association of terms; for example, "Net wt. 6 oz" or "6 oz Net wt." and "6 fl oz" or "Net contents 6 fl oz".

(k) On packages containing 4 pounds or 1 gallon or more and labeled in terms of weight or fluid measure, the declaration shall be expressed in pounds for weight units with any remainder in terms of ounces or common or decimal fraction of the pound, or in the case of fluid measure, it shall be expressed in the largest whole unit (gallons followed by common or decimal fraction of a gallon or by the next smaller whole unit or units (quarts, or quarts and pints)) with any remainder in terms of fluid ounces or common or decimal fractions of the pint or quart (see paragraph (m)(6) of this section).

(l) [Reserved]

(m) Examples:

(1) A declaration of 1 1/2 pounds weight shall be expressed as "Net Wt. 24 oz (1 lb 8 oz)," "Net Wt. 24 oz (1 1/2 lb)," or "Net Wt. 24 oz (1.5 lb)".

(2) A declaration of three-fourths pound avoirdupois weight shall be expressed as "Net Wt. 12 oz".

(3) A declaration of 1 quart liquid measure shall be expressed as "Net 32 fl oz (1 qt)".

21 CFR §101.105

(4) A declaration of 1 3/4 quarts liquid measure shall be expressed as "Net contents 56 fluid ounces (1 quart 1 1/2 pints)" or as "Net 56 fluid oz (1 qt 1 pt 8 oz)", but not in terms of quart and ounce such as "Net 56 fluid oz (1 quart 24 ounces)".

(5) On a random package, declaration of three-fourths pound avoirdupois may be expressed as "Net Wt. .75 lb".

(6) A declaration of 2 1/2 gallons liquid measure shall be expressed as "Net contents 2 1/2 gallons," "Net contents 2.5 gallons," or "Net contents 2 gallons 2 quarts" and not as "2 gallons 4 pints".

(n) For quantities, the following abbreviations and none other may be employed (periods and plural forms are optional):

weight wt	pint pt
ounce oz	quart qt
pound lb	fluid fl
gallon gal	

(o) Nothing in this section shall prohibit supplemental statements at locations other than the principal display panel(s) describing in nondeceptive terms the net quantity of contents; Provided, that such supplemental statements of net quantity of contents shall not include any term qualifying a unit of weight, measure, or count that tends to exaggerate the amount of the food contained in the package; for example, "jumbo quart" and "full gallon". Dual or combination declarations of net quantity of contents as provided for in paragraphs (a), (c), and (j) of this section (for example, a combination of net weight plus numerical count, net contents plus dilution directions of a concentrate, etc.) are not regarded as supplemental net quantity statements and may be located on the principal display panel.

(p) A separate statement of the net quantity of contents in terms of the metric system is not regarded as a supplemental statement and an accurate statement of the net quantity of contents in terms of the metric system of weight or measure may also appear on the principal display panel or on other panels.

(q) The declaration of net quantity of contents shall express an accurate statement of the quantity of contents of the package. Reasonable variations caused by loss or gain of moisture during the course of good distribution practice or by unavoidable deviations in good manufacturing practice will be recognized. Variations from stated quantity of contents shall not be unreasonably large.

(r) The declaration of net quantity of contents on pickles and pickle products, including relishes but excluding one or two whole pickles in clear plastic bags which may be declared by count, shall be expressed in terms of the U.S. gallon of 231 cubic inches and quart, pint, and fluid ounce subdivisions thereof.

(s) On a multiunit retail package, a statement of the quantity of contents shall appear on the outside of the package and shall include the number of individual units, the quantity of each individual unit, and, in parentheses, the total quantity of contents of the multiunit package in terms of avoirdupois or fluid ounces, except that such declaration of total quantity need not be followed by an additional parenthetical declaration in terms of the largest whole units and subdivisions thereof, as required by paragraph (j)(1) of this section. A multiunit retail package may thus be properly labeled: "6-16 oz bottles- (96 fl oz)" or "3-16 oz cans-(net wt. 48 oz)". For the purposes of this section, "multiunit retail package" means a package containing two or more individually packaged units of the identical commodity and in the same quantity, intended to be sold as part of the multiunit retail package but capable of being individually sold in full compliance with all requirements of the regulations in this part. Open multiunit retail packages that do not obscure the number of units or prevent examination of the labeling on each of the individual units are not subject to this paragraph if the labeling of each individual unit complies with the requirements of paragraphs (f) and (i) of this section. The provisions of this section do not apply to that butter or margarine covered by the exemptions in §1.24(a) (10) and (11) of this chapter.

(t) Where the declaration of net quantity of contents is in terms of net weight and/or drained weight or volume and does not accurately reflect the actual quantity of the contents or the product falls below the applicable standard of fill of container because of equipment malfunction or otherwise unintentional product variation, and the label conforms in all other respects to the requirements of this chapter (except the requirement that food falling below the applicable standard of fill of container shall bear the general statement of substandard fill specified in §130.14(b) of this chapter), the mislabeled food product, including any food product that fails to bear the general statement of substandard fill specified in §130.14(b) of this chapter, may be sold by the manufacturer or processor directly to institutions operated by Federal, State or local governments (schools, prisons, hospitals, etc.): Provided, That:

(1) The purchaser shall sign a statement at the time of sale stating that he is aware that the product is mislabeled to include acknowledgment of the nature and extent of the mislabeling, (e.g., "Actual net weight may be as low as --% below labeled quantity") and that any subsequent distribution by him of said product except for his own institutional use is unlawful. This statement shall be kept on file at the principal place of business of the manufacturer or processor for 2 years subsequent to the date of shipment of the product and shall be available to the Food and Drug Administration upon request.

(2) The product shall be labeled on the outside of its shipping container with the statement(s):

(i) When the variation concerns net weight and/or drained weight or volume, "Product Mislabeled. Actual net weight (drained weight or volume where appropriate) may be as low as --% below labeled quantity. This Product Not for Retail Distribution", the blank to be filled in with the maximum percentage variance between the labeled and actual weight or volume of contents of the individual packages in the shipping container, and

(ii) When the variation is in regard to a fill of container standard, "Product Mislabeled. Actual fill may be as low as -% below standard of fill. This Product Not for Retail Distribution".

(3) The statements required by paragraphs (t)(2) (i) and (ii) of this section, which may be consolidated where appropriate, shall appear prominently and conspicuously as compared to other printed matter on the shipping container and in boldface print or type on a clear, contrasting background in order to render them likely to be read and understood by the purchaser under ordinary conditions of purchase.

[42 FR 14308, Mar. 15, 1977, as amended at 42 FR 15673, Mar. 22, 1977]

§101.108 Temporary exemptions for purposes of conducting authorized food labeling experiments.

(a) The food industry is encouraged to experiment voluntarily, under controlled conditions and in collaboration with the Food and Drug Administration, with graphics and other formats for presenting nutrition and other related food labeling information that is consistent with the current quantitative system in §§101.9 and 105.66 of this chapter.

(b) Any firm that intends to undertake a labeling experiment that requires exemptions from certain requirements of §§101.9 and 105.66 of this chapter should submit a written proposal containing a thorough discussion of each of the following information items that apply to the particular experiment:

(1) A description of the labeling format to be tested;

(2) A statement of the criteria to be used in the experiment for assigning foods to categories, e.g., nutrient or other values defining "low" and "reduced";

(3) A draft of the material to be used in the store, e.g., shelf tags, booklets, posters, etc.;

(4) The dates on which the experiment will begin and end and on which a written report of analysis of the experimental data will be submitted to FDA, together with a commitment not to continue the experiment beyond the proposed ending date without FDA approval;

(5) The geographic area or areas in which the experiment is to be conducted;

(6) The mechanism to measure the effectiveness of the experiment;

(7) The method for conveying to consumers the required nutrition and other labeling information that is exempted from the label during the experiment;

(8) The method that will be or has been used to determine the actual nutritional characteristics of foods for which a claim is made; and

(9) A statement of the sections of the regulations for which an exemption is sought.

(c) The written proposal should be sent to the Dockets Management Branch (HFA-305), Food and Drug Administration, rm. 1-23, 12420 Parklawn Dr., Rockville, MD 20857. The proposal should be clearly identified as a request for a temporary exemption for purposes of conducting authorized food labeling experiments and submitted as a citizen petition under §10.30 of this chapter.

(d) Approval for food labeling experiments will be given by FDA in writing. Foods labeled in violation of existing regulations will be subject to regulatory action unless an FDA-approved exemption to the specific regulation has been granted for that specific product.

(e) Reporting requirements contained in §101.108(b) have been approved by this Office of Management and Budget and assigned number 0910-0151.

[48 FR 15240, Apr. 8, 1983, as amended at 59 FR 14364, Mar. 28, 1994; 62 FR 15342-15343, Mar. 31, 1997]

APPENDIX A TO PART 101

Monier-Williams Procedure (With Modifications) For Sulfites in Food, Center For Food Safety And Applied Nutrition Food And Drug Administration (November 1985)

The AOAC official method for sulfites (*Official Methods of Analysis*, 14th Edition, 20.123-20.125, Association of Official Analytical Chemists) has been modified, in FDA laboratories, to facilitate the determination of sulfites at or near 10 ppm in food. Method instructions, including modifications, are described below.

Apparatus-The apparatus shown diagrammatically (Figure 1) is designed to accomplish the selective transfer of sulfur dioxide from the sample in boiling aqueous hydrochloric acid to a solution of 3% hydrogen peroxide. This apparatus is easier to assemble than the official apparatus and the back pressure inside the apparatus is limited to the unavoidable pressure due to the height of the 3% H_2O_2 solution above the tip of the bubbler (F). Keeping the backpressure as low as possible reduces the likelihood that sulfur dioxide will be lost through leaks.

The apparatus should be assembled as shown in Fig. 1 with a thin film of stopcock grease on the sealing surfaces of all the joints except the joint between the separatory funnel and the flask. Each joint should be clamped together to ensure a complete seal throughout the analysis. The separatory funnel, B, should have a capacity of 100 ml or greater. An inlet adapter, A, with a hose connector (Kontes K-183000 or equivalent) is required to provide a means of applying a head of pressure above the solution. (A pressure equalizing dropping funnel is not recommended because condensate, perhaps with sulfur dioxide, is deposited in the funnel and the side arm.) The round bottom flask, C, is a 1000 ml flask with three 24/40 tapered joints. The gas inlet tube, D, (Kontes K-179000 or equivalent) should be of sufficient length to permit introduction of the nitrogen within 2.5 cm of the bottom of the flask. The Allihn condenser, E, (Kontes K-431000-2430 or equivalent) has a jacket length of 300 mm. The bubbler, F, was fabricated from glass according to the dimensions given in Fig. 2. The 3% hydrogen peroxide solution can be contained in a vessel, G, with an i.d. of ca. 2.5 cm and a depth of 18 cm.

Buret-A 10 ml buret (Fisher Cat. No. 03-848-2A or equivalent) with overflow tube and hose connections for an Ascarite tube or equivalent air scrubbing apparatus. This will permit the maintenance of a carbon dioxide-free atmosphere over the standardized 0.01N sodium hydroxide.

Chilled Water Circulator-The condenser must be chilled with a coolant, such as 20% methanol-water, maintained at 5 °C. A circulating pump equivalent to the Neslab Coolflow 33 is suitable.

Reagents

(a) *Aqueous hydrochloric acid, 4N.*-For each analysis prepare 90 ml of hydrochloric acid by adding 30 ml of concentrated hydrochloric acid (12N) to 60 ml of distilled water.
(b) *Methyl red indicator.*-Dissolve 250 mg of methyl red in 100 ml ethanol.
(c) *Hydrogen peroxide solution, 3%.*-Dilute ACS reagent grade 30% hydrogen peroxide to 3% with distilled water. Just prior to use, add three drops of methyl red indicator and titrate to a yellow end-point using 0.01N sodium hydroxide. If the end-point is exceeded discard the solution and prepare another 3% H_2O_2 solution.
(d) *Standardized titrant, 0.01N NaOH.*-Certified reagent may be used (Fisher SO-5-284). It should be standardized with reference standard potassium hydrogen phthalate.
(e) *Nitrogen.*-A source of high purity nitrogen is required with a flow regulator that will maintain a flow of 200 cc per minute. To guard against the presence of oxygen in the nitrogen, an oxygen scrubbing solution such as an alkaline pyrogallol trap may be used.

Prepare pyrogallol trap as follows:
    1. Add 4.5 g pyrogallol to the trap.
    2. Purge trap with nitrogen for 2 to 3 minutes.
    3. Prepare a KOH solution prepared by adding 65g KOH to 85 ml distilled water (caution: heat).
    4. Add the KOH solution to the trap while maintaining an atmosphere of nitrogen in the trap.

Determination

Assemble the apparatus as shown in Fig. 1. The flask C must be positioned in a heating mantle that is controlled by a power regulating device such as Variac or equivalent. Add 400 ml of distilled water to flask C. Close the stopcock of separatory funnel, B, and add 90 ml of 4N hydrochloric acid to the separatory funnel. Begin the flow of nitrogen at a rate of 200±10 cc/min. The condenser coolant flow must be initiated at this time. Add 30 ml of 3% hydrogen peroxide, which has been titrated to a yellow

APPENDIX A TO PART 101 — continued

end-point with 0.01N NaOH, to container G. After fifteen minutes the apparatus and the distilled water will be thoroughly de-oxygenated and the apparatus is ready for sample introduction. Sample preparation (solids)-Transfer 50 g of food, or a quantity of food with a convenient quantity of SO_2 (500 to 1500 mcg SO2), to a food processor or blender. Add 100 ml of 5% ethanol in water and briefly grind the mixture. Grinding or blending should be continued only until the food is chopped into pieces small enough to pass through the 24/40 point of flask C.

Sample preparation (liquids)-Mix 50 g of the sample, or a quantity with a convenient quantity of SO2 (500 to 1500 mcg SO_2), with 100 ml of 5% ethanol in water.

Sample introduction and distillation-Remove the separatory funnel B, and quantitatively transfer the food sample in aqueous ethanol to flask C. Wipe the tapered joint clean with a laboratory tissue, apply stopcock grease to the outer joint of the separatory funnel, and return the separatory funnel, B, to tapered joint flask C. The nitrogen flow through the 3% hydrogen peroxide solution should resume as soon as the funnel, B, is re-inserted into the appropriate joint in flask C. Examine each joint to ensure that it is sealed.

Apply a head pressure above the hydrochloric acid solution in B with a rubber bulb equipped with a valve. Open the stopcock in B and permit the hydrochloric acid solution to flow into flask C. Continue to maintain sufficient pressure above the acid solution to force the solution into the flask C. The stopcock may be closed, if necessary, to pump up the pressure above the acid and then opened again. Close the stopcock before the last few milliliters drain out of the separatory funnel, B, to guard against the escape of sulfur dioxide into the separatory funnel. Apply the power to the heating mantle. Use a power setting which will cause 80 to 90 drops per minute of condensate to return to the flask from condenser, E. After 1.75 hours of boiling the contents of the 1000 ml flask and remove trap G.

Titration.-Titrate the contents with 0.01N sodium hydroxide. Titrate with 0.01N NaOH to a yellow end-point that persists for at least twenty seconds. Compute the sulfite content, expressed as micrograms sulfur dioxide per gram of food (ppm) as follows:

$$ppm = (32.03 \times V_B \times N \times 1000) \div Wt$$

where
32.03=milliequivalent weight of sulfur dioxide;
V_B=volume of sodium hydroxide titrant of normality, N, required to reach endpoint;
the factor, 1000, converts milliequivalents to microequivalents, and
Wt=weight (g) of food sample introduced into the 1000 ml flask.

APPENDIX A TO PART 101 — continued

**Monier-Williams Procedure (With Modifications) For
Sulfites in Food, Center For Food Safety And Applied Nutrition
Food And Drug Administration (November 1985)**

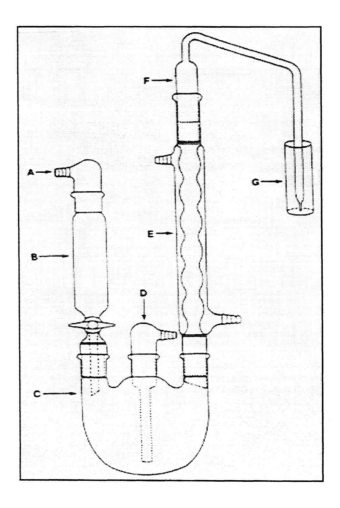

Figure 1. The optimized Monier-Williams apparatus. Component identification is given in text.

APPENDIX A TO PART 101 — continued

**Monier-Williams Procedure (With Modifications) For
Sulfites in Food, Center For Food Safety And Applied Nutrition
Food And Drug Administration (November 1985)**

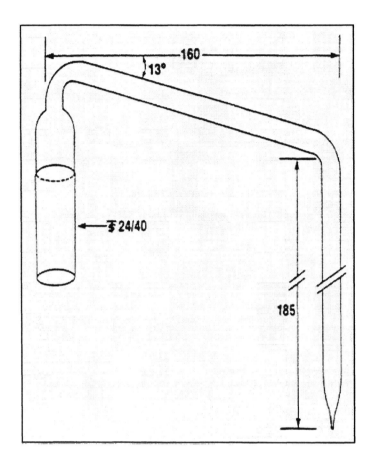

Figure 2. Diagram of bubbler (F in Figure 1). Lengths are given in mm.

[42 FR 14308, Mar. 15, 1977, as amended at 51 FR 25017, July 9, 1986]

APPENDIX B TO PART 101

Graphic Enhancements Used by the FDA

Examples of Graphic Enhancements used by the FDA

A. Overall
1. Nutrition Facts Label is boxed with all black or one color type printed on a white or neutral background.

B. Typeface and size
1. The "Nutrition Facts" label uses 6 point or larger Helvetica Black and/or Helvetica regular type. In order to fit some formats, the typography may be kerned as much as -4 (tighter kerning reduces legibility).

2. Key nutrients and their % Daily Value are set in 8 point Helvetica Black (but "%" is set in Helvetica Regular).

3. "Nutrition Facts" is set in either Franklin Gothic Heavy or Helvetica Black to fit the width of the label flush left and flush right.

4. "Serving Size" and "Servings per container" are set in 8 point Helvetica Regular with 1 point of leading.

5. The table labels (for example: "Amount per Serving") are set 6 point Helvetica Black.

6. Absolute measures of nutrient content (for example: "1g" and nutrient subgroups are set in 8 point Helvetica Regular with 4 points of leading.

7. Vitamins and minerals are set in 8 point Helvetica Regular, with 4 points of leading, separated by 10 point bullets.

8. All type that appears under vitamins and minerals is set in 6 point Helvetica Regular with 1 point of leading.

C. Rules
1. A 7 point rule separates large groupings as shown in example. A 3 point rule separates calorie information from the nutrient information.

2. A hairline rule or 1/4 point rule separates individual nutrients, as shown in the example. Descenders do not touch rule. The top half of the label (nutrient information) has 2 points of leading between the type and the rules; the bottom half of the label (footnotes) has 1 point of leading between the type and the rules.

D. Box
1. All labels are enclosed by 1/2 point box rule within 3 points of text measure.

APPENDIX B TO PART 101 — continued

Graphic Enhancements Used by the FDA

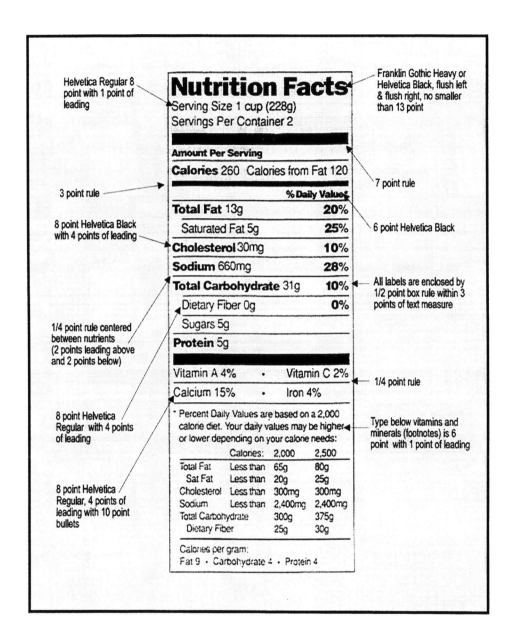

[58 FR 17332, Apr. 2, 1993]

APPENDIX C TO PART 101: Nutrition Facts For Raw Fruits And Vegetables (*Part A*)*

Nutrition facts[1] for raw fruits and vegetables, edible portion	Calories		Total Fat		Saturated Fat		Cholesterol		Sodium		Potassium	
	Total	From fat	g	%	g	%	mg	%	mg	%	mg	%
Banana, 1 medium (126 g/4.5 oz)..................	110	0	0	0	0	0	0	0	0	0	400	11
Apple, 1 medium (154 g/5.5 oz)..................	80	0	0	0	0	0	0	0	0	0	170	5
Watermelon 1/18 medium melon; 2 cups diced pieces (280 g/10.0 oz)..................	80	0	0	0	0	0	0	0	10	0	230	7
Orange, 1 medium (154 g/5.5 oz)..................	70	0	0	0	0	0	0	0	0	0	260	7
Cantaloupe, 1/4 medium (134 g/4.8 oz..................	50	0	0	0	0	0	0	0	25	1	280	8
Grapes, 1 1/2 cups (138 g/4.9 oz)..................	90	10	1	2	0	0	0	0	0	0	270	8
Grapefruit, 1/2 medium (154 g/5.3 oz)..................	60	0	0	0	0	0	0	0	0	0	230	7
Strawberries, 8 medium (147 g/5.3 oz)..................	45	0	0	0	0	0	0	0	0	0	270	8
Peach, 1 medium (98 g/3.5 oz)..................	40	0	0	0	0	0	0	0	0	0	190	5
Pear, 1 medium (166 g/5.9 oz)..................	100	10	1	2	0	0	0	0	0	0	210	6
Nectarine, 1 medium (140 g/5.0 oz)..................	70	0	0.5	1	0	0	0	0	0	0	300	9
Honeydew Melon, 1/10 medium melon (134 g/4.8 oz)..................	50	0	0	0	0	0	0	0	35	1	310	9
Plums, 2 medium (132 g/4.7 oz)..................	80	10	1	2	0	0	0	0	0	0	220	6
Avocado, California, 1/5 medium (30 g/1.1 oz)....	55	45	5	8	1	5	0	0	0	0	170	5
Lemon, 1 medium (58 g/2.1 oz)..................	15	0	0	0	0	0	0	0	5	0	90	3
Pineapple, 2 slices, 3" diameter, 3/4" thick (112 g/4 oz)..................	60	0	0	0	0	0	0	0	10	0	115	3

APPENDIX C TO PART 101: Nutrition Facts For Raw Fruits And Vegetables
(Part A)*

Nutrition facts[1] for raw fruits and vegetables, edible portion	Calories		Total Fat		Saturated Fat		Cholesterol		Sodium		Potassium	
	Total	From fat	g	%	g	%	mg	%	mg	%	mg	%
Tangerine, 1 medium (109 g/3.9 oz).................	50	0	0.5	1	0	0	0	0	0	0	180	5
Sweet cherries, 21 cherries; 1 cup (140 g/5.0 oz)......................	90	0	0.5	1	0	0	0	0	0	0	300	9
Kiwifruit, 2 medium (148 g/5.3 oz).................	100	10	1	2	0	0	0	0	0	0	480	14
Lime, 1 medium (67 g/2.4 oz)........................	20	0	0	0	0	0	0	0	0	0	75	2
Potato, 1 medium (148 g/5.3 oz)........................	100	0	0	0	0	0	0	0	0	0	720	21
Iceberg lettuce, 1/6 medium head (89 g/3.2 oz)...................................	15	0	0	0	0	0	0	0	10	0	120	3
Tomato, 1 medium (148 g/5.3 oz)........................	35	0	0.5	1	0	0	0	0	5	0	360	10
Onion, 1 medium (148 g/5.3 oz)........................	60	0	0	0	0	0	0	0	5	0	240	7
Carrot, 7" long, 1 1/4" diameter (78 g/2.8 oz)...	35	0	0	0	0	0	0	0	40	2	280	8
Celery, 2 medium stalks (110 g/3.9 oz)...............	20	0	0	0	0	0	0	0	100	4	350	10
Sweet corn, kernels from 1 medium ear (90 g/3.2 oz)..................................	80	10	1	2	0	0	0	0	0	0	240	7
Broccoli, 1 medium stalk (148 g/5.3 oz).................	45	0	0.5	1	0	0	0	0	55	2	540	15
Green cabbage, 1/12 medium head (84 g/3.0 oz)..................................	25	0	0	0	0	0	0	0	20	1	190	5
Cucumber, 1/3 medium (99 g/3.5 oz).................	15	0	0	0	0	0	0	0	0	0	170	5
Bell pepper, 1 medium (148 g/5.3 oz).................	30	0	0	0	0	0	0	0	0	0	270	8
Cauliflower, 1/6 medium head (99 g/3.5 oz)...........	25	0	0	0	0	0	0	0	30	1	270	8

APPENDIX C TO PART 101: Nutrition Facts For Raw Fruits And Vegetables (*Part A*)*

Nutrition facts[1] for raw fruits and vegetables, edible portion	Calories		Total Fat		Saturated Fat		Cholesterol		Sodium		Potassium	
	Total	*From fat*	*g*	*%*	*g*	*%*	*mg*	*%*	*mg*	*%*	*mg*	*%*
Leaf lettuce, 1 1/2 cups shredded (85 g/3.0 oz)....	15	0	0	0	0	0	0	0	30	1	230	7
Sweet Potato, medium, 5" long, 2" diameter (130 g/4.6 oz).................	130	0	0	0	0	0	0	0	45	2	350	10
Mushrooms, 5 medium (84 g/3.0 oz)..................	20	0	0	0	0	0	0	0	0	0	300	9
Green onion, 1/4 cup chopped (25 g/0.9 oz).....	10	0	0	0	0	0	0	0	5	0	70	2
Green (snap) beans, 3/4 cup cut (83 g/3.0 oz)......	25	0	0	0	0	0	0	0	0	0	200	6
Radishes, 7 radishes (85 g/3.0 oz)........................	15	0	0	0	0	0	0	0	25	1	230	7
Summer squash, 1/2 medium (98 g/3.5 oz).....	20	0	0	0	0	0	0	0	0	0	260	7
Asparagus, 5 spears (93 g/3.3 oz)........................	25	0	0	0	0	0	0	0	0	0	230	7

[1] Raw, edible weight portion. Percent (%) Daily Values are based on a 2,000 calorie diet.

*(Ed. Note: For legibility, this table has been split into Parts A and B.)

APPENDIX C TO PART 101: Nutrition Facts for Raw Fruits and Vegetables (*Part B*)*

Nutrition facts [1] for raw fruits and vegetables, edible portion	Total Carbohydrate		Dietary Fiber		Sugars	Pro-tein	Vitamin A	Vitamin C	Cal-cium	Iron
	g	%	g	%	g	g	%	%	%	%
Banana, 1 medium (126 g/4.5 oz)............	29	10	4	16	21	1	0	15	0	2
Apple, 1 medium (154 g/5.5 oz)............	22	7	5	20	16	0	2	8	0	2
Watermelon 1/18 medium melon; 2 cups diced pieces (280 g/10.0 oz)......	27	9	2	8	25	1	20	25	2	4
Orange, 1 medium (154 g/5.5 oz)............	21	7	7	28	14	1	2	130	6	2
Cantaloupe, 1/4 medium (134 g/4.8 oz)............	12	4	1	4	11	1	100	80	2	2
Grapes, 1 1/2 cups (138 g/4.9 oz)............	24	8	1	4	23	1	2	25	2	2
Grapefruit, 1/2 medium (154 g/5.3 oz)............	16	5	6	24	10	1	15	110	2	0
Strawberries, 8 medium (147 g/5.3 oz)............	12	4	4	16	8	1	0	160	2	4
Peach, 1 medium (98 g/3.5 oz)............	10	3	2	8	9	1	2	10	0	0
Pear, 1 medium (166 g/5.9 oz)............	25	8	4	16	17	1	0	10	2	0
Nectarine, 1 medium (140 g/5.0 oz)............	16	5	2	8	12	1	4	15	0	2
Honeydew Melon, 1/10 medium melon (134 g/4.8 oz)............	13	4	1	4	12	1	2	45	0	2
Plums, 2 medium (132 g/4.7 oz)............	19	6	2	8	10	1	6	20	0	0
Avocado, California, 1/5 medium (30 g/1.1 oz).......	3	1	3	12	0	1	0	4	0	0
Lemon, 1 medium (58 g/2.1 oz)............	5	2	1	4	1	0	0	40	2	0
Pineapple, 2 slices, 3" diameter, 3/4" thick (112 g/4 oz)............	16	5	1	4	13	1	0	25	2	2
Tangerine, 1 medium (109 g/3.9 oz)............	15	5	3	12	12	1	0	50	4	0

APPENDIX C TO PART 101: Nutrition Facts for Raw Fruits and Vegetables (*Part B*)*

Nutrition facts [1] for raw fruits and vegetables, edible portion	Total Carbohydrate		Dietary Fiber		Sugars	Protein	Vitamin A	Vitamin C	Calcium	Iron
	g	%	g	%	g	g	%	%	%	%
Sweet cherries, 21 cherries; 1 cup (140 g/5.0 oz)...............	22	7	3	12	19	2	2	15	2	2
Kiwifruit, 2 medium (148 g/5.3 oz).........................	24	8	4	16	16	2	2	240	6	4
Lime, 1 medium (67 g/2.4 oz)................................	7	2	2	8	0	0	0	35	0	0
Potato, 1 medium (148 g/5.3 oz).........................	26	9	3	12	3	4	0	45	2	6
Iceberg lettuce, 1/6 medium head (89 g/3.2 oz)................................	3	1	1	4	2	1	4	6	2	2
Tomato, 1 medium (148 g/5.3 oz).........................	7	2	1	4	4	1	20	40	2	2
Onion, 1 medium (148 g/5.3 oz).........................	14	5	3	12	9	2	0	20	4	2
Carrot, 7" long, 1 1/4" diameter (78 g/2.8 oz)......	8	3	2	8	5	1	270	10	2	0
Celery, 2 medium stalks (110 g/3.9 oz)..................	5	2	2	8	0	1	2	15	4	2
Sweet corn, kernels from 1 medium ear (90 g/3.2 oz)................................	18	6	3	12	5	3	2	10	0	2
Broccoli, 1 medium stalk (148 g/5.3 oz)..................	8	3	5	20	3	5	15	220	6	6
Green cabbage, 1/12 medium head (84 g/3.0 oz)................................	5	2	2	8	3	1	0	70	4	2
Cucumber, 1/3 medium (99 g/3.5 oz)..................	3	1	1	4	2	1	4	10	2	2
Bell pepper, 1 medium (148 g/5.3 oz)..................	7	2	2	8	4	1	8	190	2	2
Cauliflower, 1/6 medium head (99 g/3.5 oz).............	5	2	2	8	2	2	0	100	2	2
Leaf lettuce, 1 1/2 cups shredded (85 g/3.0 oz)......	4	1	2	8	2	1	40	6	4	0

APPENDIX C TO PART 101: Nutrition Facts for Raw Fruits and Vegetables (Part B)*

Nutrition facts [1] for raw fruits and vegetables, edible portion	Total Carbohydrate		Dietary Fiber		Sugars	Protein	Vitamin A	Vitamin C	Calcium	Iron
	g	%	g	%	g	g	%	%	%	%
Sweet Potato, medium, 5" long, 2" diameter (130 g/4.6 oz)............	33	11	4	16	7	2	440	30	2	2
Mushrooms, 5 medium (84 g/3.0 oz)..............	3	1	1	4	0	3	0	2	0	2
Green onion, 1/4 cup chopped (25 g/0.9 oz).......	2	1	1	4	1	0	2	8	0	0
Green (snap) beans, 3/4 cup cut (83 g/3.0 oz)........	5	2	3	12	2	1	4	10	4	2
Radishes, 7 radishes (85 g/3.0 oz)................	3	1	0	0	2	1	0	30	2	0
Summer squash, 1/2 medium (98 g/3.5 oz).......	4	1	2	8	2	1	6	30	2	2
Asparagus, 5 spears (93 g/3.3 oz)..................	4	1	2	8	2	2	10	15	2	2

[1] Raw, edible weight portion. Percent (%) Daily Values are based on a 2,000 calorie diet.

*(Ed. Note: For legibility, this table has been split into Parts A and B.)

Source: 61 FR 42761, Aug. 16, 1996

APPENDIX D TO PART 101: Nutrition Facts for Cooked Fish
(Part A)*

Nutrition facts [1] for fish (84g/ 3 oz)	Calories		Total Fat		Saturated Fat		Cholesterol		Sodium		Potassium	
	Total	From fat	g	%	g	%	mg	%	mg	%	mg	%
Shrimp..................	80	10	1	2	0	0	165	55	190	8	140	4
Cod......................	90	0	0.5	1	0	0	45	15	60	3	450	13
Pollock..................	90	10	1	2	0	0	80	27	110	5	360	10
Catfish..................	140	80	9	14	2	10	50	17	40	2	230	7
Scallops, about 6 large or 14 small..................	120	10	1	2	0	0	55	18	260	11	280	8
Salmon, Atlantic/Coho.......	160	60	7	11	1	5	50	17	50	2	490	14
Salmon, Chum/Pink...........	130	35	4	6	1	5	70	23	65	3	410	12
Salmon, Sockeye................	180	80	9	14	1.5	8	75	25	55	2	320	9
Flounder/sole.....................	100	14	1.5	2	0.5	3	60	20	90	4	290	8
Oysters, about 12 medium	100	35	3.5	5	1	5	115	38	190	8	390	11
Orange roughy.................	80	10	1	2	0	0	20	7	70	3	330	9
Mackerel, Atlantic/Pacific..	210	120	13	20	1.5	8	60	20	100	4	400	11
Ocean perch.....................	110	20	2	3	0	0	50	17	95	4	290	8
Rockfish...........................	100	20	2	3	0	0	40	13	70	3	430	12
Whiting.............................	110	25	3	5	0.5	3	70	23	95	4	320	9
Clams, about 12 small........	100	15	1.5	2	0	0	55	18	95	4	530	15
Haddock.............................	100	10	1	2	0	0	80	27	85	4	340	10
Blue crab...........................	100	10	1	2	0	0	90	30	320	13	360	10
Rainbow trout....................	140	50	6	9	2	10	60	20	35	1	370	11
Halibut...............................	110	20	2	3	0	0	35	12	60	3	490	14
Lobster...............................	80	0	0.5	1	0	0	60	20	320	13	300	9
Swordfish...........................	130	35	4.5	7	1	5	40	13	100	4	310	9

[1] Cooked, edible weight portion. Percent (%) Daily Values are based on a 2,000 calorie diet.

*(Ed. Note: For legibility, this table has been split into Parts A and B.)

APPENDIX D TO PART 101: Nutrition Facts for Cooked Fish (Part B)*

Nutrition facts [1] for fish (84 g/ 3 oz)	Total Carbohydrate		Dietary Fiber		Sugars	Protein	Vitamin A	Vitamin C	Calcium	Iron
	g	%	g	%	g	g	%	%	%	%
Shrimp....................	0	0	0	0	0	18	0	0	2	15
Cod.........................	0	0	0	0	0	20	0	0	2	2
Pollock....................	0	0	0	0	0	20	0	0	0	2
Catfish....................	0	0	0	0	0	17	0	0	0	0
Scallops, about 6 large or 14 small......	2	1	0	0	0	22	0	0	2	2
Salmon, Atlantic/Coho.....	0	0	0	0	0	22	0	0	0	4
Salmon, Chum/Pink........	0	0	0	0	0	22	2	0	0	2
Salmon, Sockeye............	0	0	0	0	0	23	4	0	0	2
Flounder/sole.................	0	0	0	0	0	21	0	0	2	2
Oysters, about 12 medium	4	1	0	0	0	10	0	0	6	45
Orange roughy...............	0	0	0	0	0	16	0	0	0	0
Mackerel, Atlantic/Pacific	0	0	0	0	0	21	0	0	0	5
Ocean perch..................	0	0	0	0	0	21	0	0	10	6
Rockfish.......................	0	0	0	0	0	21	4	0	0	2
Whiting........................	0	0	0	0	0	19	2	0	6	0
Clams, about 12 small......	0	0	0	0	0	22	10	0	6	60
Haddock.......................	0	0	0	0	0	21	0	0	2	6
Blue crab......................	0	0	0	0	0	20	0	0	8	4
Rainbow trout................	0	0	0	0	0	21	4	4	6	2
Halibut.........................	0	0	0	0	0	23	2	0	4	4
Lobster.........................	1	0	0	0	0	17	0	0	4	2
Swordfish.....................	0	0	0	0	0	22	2	2	0	4

[1] Cooked, edible weight portion. Percent (%) Daily Values are based on a 2,000 calorie diet.

*(Ed. Note: For legibility, this table has been split into Parts A and B.)

Source: 61 FR 42761–42762, Aug. 16, 1996

Chapter 10—FSIS Labeling Rules for Meat Products

> ## *In This Chapter...*
>
> **Full text of the FSIS meat product labeling rules** (as of May 1, 1998):
>
> - General labeling provisions: 9 *CFR* Part 317, Subpart A — **page 10.3**
>
> - Nutrition labeling provisions: 9 *CFR* Part 317, Subpart B — **page 10.23**

PART 317-LABELING, MARKING DEVICES, AND CONTAINERS
Authority: 21 U.S.C. 601-695; 7 CFR 2.18, 2.53. Source: 35 FR 15580, Oct. 3, 1970, unless otherwise noted.

Subpart A-General

§317.1	Labels required; supervision by Program employee.
§317.2	Labels: definition; required features.
§317.3	Approval of abbreviations of marks of inspection; preparation of marking devices bearing inspection legend without advance approval prohibited; exception.
§317.4	Labeling approval.
§317.5	Generically approved labeling.
§317.6	Approved labels to be used only on products to which they are applicable.
§317.7	Products for foreign commerce; printing labels in foreign language permissible; other deviations.
§317.8	False or misleading labeling or practices generally; specific prohibitions and requirements for labels and containers.
§317.9	Labeling of equine products.
§317.10	Reuse of official inspection marks; reuse of containers bearing official marks, labels, etc.
§317.11	Labeling, filling of containers, handling of labeled products to be only in compliance with regulations.
§317.12	Relabeling products; requirements.
§317.13	Storage and distribution of labels and containers bearing official marks.
§317.14	[Removed and Reserved]
§317.15	[Reserved]
§317.16	Labeling and containers of custom prepared products.
§317.17	Interpretation and statement of labeling policy for cured products; special labeling requirements concerning nitrate and nitrite.
§317.18	Quantity of contents labeling.
§317.19	Definitions and procedures for determining net weight compliance.
§317.20	Scale requirements for accurate weights, repairs, adjustments, and replacement after inspection.
§317.21	Scales: testing of.
§317.22	Handling of failed product.
§317.23	[Reserved]
§317.24	Packaging materials.

9 CFR §317.1

Subpart B-Nutrition Labeling

§317.300	Nutrition labeling of meat products.
§317.301	[Reserved]
§317.302	Location of nutrition information.
§§317.303 - 317.307	[Reserved]
§317.308	Labeling of meat products with number of servings.
§317.309	Nutrition label content.
§317.310	[Reserved]
§317.311	[Reserved]
§317.312	Reference amounts customarily consumed per eating occasion.
§317.313	Nutrient content claims; general principles.
§§317.314 - 317.342	[Reserved]
§317.343	Significant participation for voluntary nutrition labeling.
§317.344	Identification of major cuts of meat products.
§317.345	Guidelines for voluntary nutrition labeling of single-ingredient, raw products.
§317.346 - 317.353	[Reserved]
§317.354	Nutrient content claims for "good source" and "high" and "more".
§317.355	[Reserved]
§317.356	Nutrient content claims for "light" or "lite".
§§317.357 - 317.359	[Reserved]
§317.360	Nutrient content claims for calorie content.
§317.361	Nutrient content claims for sodium content.
§317.362	Nutrient content claims for fat, fatty acids, and cholesterol content.
§317.363	Nutrient content claims for "healthy."
§§317.364 - 317.368	[Reserved]
§317.369	Labeling applications for nutrient content claims.
§§317.370 - 317.379	[Reserved]
§317.380	Label statements relating to usefulness in reducing or maintaining body weight.
§§317.381 - 317.399	[Reserved]
§317.400	Exemption from nutrition labeling.

9 CFR §317.2

Subpart A-General

§317.1 Labels required; supervision by Program employee.

(a) When, in an official establishment, any inspected and passed product is placed in any receptacle or covering constituting an immediate container, there shall be affixed to such container a label as described in §317.2 except that the following do not have to bear such a label.
   (1) Wrappings or dressed carcasses and primal parts ion an unprocessed state, bearing the official inspection legend, if such wrappings are intended solely to protect the product against soiling or excessive drying during transportation or storage, and the wrappings bear no information except company brand names, trade marks, or code numbers which do not include any information required by §317.2;
   (2) Uncolored transparent coverings, such as cellophane, which bear no written, printed, or graphic matter and which enclose any unpackaged or packaged product bearing all markings required by part 316 of this subchapter which are clearly legible through such coverings;
   (3) Animal and transparent artificial casings bearing only the markings required by part 316 of this subchapter;
   (4) Stockinettes used as "operative devices", such as those applied to cured meats in preparation for smoking, whether or not such stockinettes are removed following completion of the operations for which they were applied;
   (5) Containers such as boil-in bags, trays of frozen dinners, and pie pans which bear no information except company brand names, trademarks, code numbers, directions for preparation and serving suggestions, and which are enclosed in a consumer size container that bears a label as described in §317.2;
   (6) Containers of products passed for cooking or refrigeration and moved from an official establishment under §311.1 of this subchapter.
(b) Folders and similar coverings made of paper or similar materials, whether or not they completely enclose the product and which bear any written, printed, or graphic matter, shall bear all features required on a label for an immediate container.
(c) No covering or other container which bears or is to bear a label shall be filled, in whole or in part, except with product which has been inspected and passed in compliance with the regulations in this subchapter, which is not adulterated and which is strictly in accordance with the statements on the label. No such container shall be filled, in whole or in part, and no label shall be affixed thereto, except under supervision of a Program employee.

§317.2 Labels: definition; required features.

(a) A label within the meaning of this part shall mean a display of any printing, lithographing, embossing, stickers, seals, or other written, printed, or graphic matter upon the immediate container (not including package liners) of any product.
(b) Any word, statement, or other information required by this part to appear on the label must be prominently placed thereon with such conspicuousness (as compared with other words, statements, designs, or devices, in the labeling) and in such terms as to render it likely to be read and understood by the ordinary individual under customary conditions of purchase and use. In order to meet this requirement, such information must appear on the principal display panel except as otherwise permitted in this part. Except as provided in §317.7, all words, statements, and other information required by or under authority of the Act to appear on the label or labeling shall appear thereon in the English language: *Provided, however,* That in the case of products distributed solely in Puerto Rico, Spanish may be substituted for English for all printed matter except the USDA inspection legend.
(c) Labels of all products shall show the following information on the principal display panel (except as otherwise permitted in this part), in accordance with the requirements of this part or, if applicable, part 319 of this subchapter:
   (1) The name of the product, which in the case of a product which purports to be or is represented as a product for which a definition and standard of identity or composition is prescribed in part 319 of this subchapter, shall be the name of the food specified in the standard, and in the case of any other product shall be the common or usual name of the food, if any there be, and if there is none, a truthful descriptive designation, as prescribed in paragraph (e) of this section;
   (2) If the product is fabricated from two or more ingredients, the word "ingredients" followed by a list of the ingredients as prescribed in paragraph (f) of this section;

9 CFR §317.2

(3) The name and place of business of the manufacturer, packer, or distributor for whom the product is prepared, as prescribed in paragraph (g) of this section;

(4) An accurate statement of the net quantity of contents, as prescribed in paragraph (h) of this section;

(5) An official inspection legend and, except as otherwise provided in paragraph (i) of this section, the number of the official establishment, in the form required by part 312 of this subchapter;

(6) Any other information required by the regulations in this part or part 319 of this subchapter.

(d) The principal display panel shall be the part of a label that is most likely to be displayed, presented, shown, or examined under customary conditions of display for sale. Where packages bear alternate principal display panels, information required to be placed on the principal display panel shall be duplicated on each principal display panel. The principal display panel shall be large enough to accommodate all the mandatory label information required to be placed thereon by this part and part 319 of this subchapter with clarity and conspicuousness and without obscuring of such information by designs or vignettes or crowding. In determining the area of the principal display panel, exclude tops, bottoms, flanges at tops and bottoms of cans, and shoulders and necks of bottles or jars. The principal display panel shall be:

(1) In the case of a rectangular package, one entire side, the area of which is at least the product of the height times the width of that side.

(2) In the case of a cylindrical or nearly cylindrical container:

(i) An area that is 40 percent of the product of the height of the container times the circumference of the container, or

(ii) A panel, the width of which is one-third of the circumference and the height of which is as high as the container: *Provided, however,* That if there is immediately to the right or left of such principal display panel, a panel which has a width not greater than 20 percent of the circumference and a height as high as the container, and which is reserved for information prescribed in paragraphs (c) (2), (3), and (5), such panel shall be known as the "20 percent panel" and such information may be shown on that panel in lieu of showing it on the principal display panel.

(3) In the case of a container of any other shape, 40 percent of the total surface of the container.

(e) Any descriptive designation used as a product name for a product which has no common or usual name shall clearly and completely identify the product. Product which has been prepared by salting, smoking, drying, cooking, chopping, or otherwise shall be so described on the label unless the name of the product implies, or the manner of packaging shows that the product was subjected to such preparation. The unqualified terms "meat," "meat byproduct," "meat food product," and terms common to the meat industry but not common to consumers such as "picnic," "butt," "cala," "square," "loaf," "spread," "delight," "roll," "plate," "luncheon," and "daisy" shall not be used as names of a product unless accompanied with terms descriptive of the product or with a list of ingredients, as deemed necessary in any specific case by the Administrator in order to assure that the label will not be false or misleading.

(f)

(1) The list of ingredients shall show the common or usual names of the ingredients arranged in the descending order of predominance, except as otherwise provided in this paragraph.

(i) The terms spice, natural flavor, natural flavoring, flavor and flavoring may be used in the following manner:

(A) The term "spice" means any aromatic vegetable substance in the whole, broken, or ground form, with the exceptions of onions, garlic and celery, whose primary function in food is seasoning rather than nutritional and from which no portion of any volatile oil or other flavoring principle has been removed. Spices include the spices listed in 21 CFR 182.10, and 184.

(B) The term "natural flavor," "natural flavoring," "flavor" or "flavoring" means the essential oil, oleoresin, essence or extractive, protein hydrolysate, distillate, or any product or roasting, heating or enzymolysis, which contains the flavoring constituents derived from a spice, fruit or fruit juice, vegetable or vegetable juice, edible yeast, herb, bark, bud, root, leaf or any other edible portion of a plant, meat, seafood, poultry, eggs, dairy products, or fermentation products thereof, whose primary function in food is flavoring rather than nutritional. Natural flavors include the natural essence or extractives obtained from plants listed in 21 CFR 182.10, 182.20, 182.40, 182.50 and 184, and the substances listed in 21 CFR 172.510. The term natural flavor, natural flavoring, flavor or flavoring may also be used to designate spices, powered onion, powdered garlic, and powdered celery.

(ii) The term "corn syrup" may be used to designate either corn syrup or corn syrup solids.

(iii) The term "animal and vegetable fats" or "vegetable and animal fats" may be used to designate the

9 CFR §317.2

ingredients of mixtures of such edible fats in product designated "compound" or "shortening." "Animal fats" as used herein means fat derived from inspected and passed cattle, sheep, swine, or goats.

(iv) When a product is coated with pork fat, gelatin, or other approved substance and a specific declaration of such coating appears contiguous to the name of the product, the ingredient statement need not make reference to the ingredients of such coating.

(v) When two meat ingredients comprise at least 70 percent of the meat and meat byproduct ingredients of a formula and when neither of the two meat ingredients is less than 30 percent by weight of the total meat and meat byproducts used, such meat ingredients may be interchanged in the formula without a change being made in the ingredients statement on labeling materials: *Provided,* That the word "and" in lieu of a comma shall be shown between the declaration of such meat ingredients in the statement of ingredients.

(vi) (A) Product ingredients which are present in individual amounts of 2 percent or less by weight may be listed in the ingredients statement in other than descending order of predominance: *Provided,* That such ingredients are listed by their common or usual names at the end of the ingredients statement and preceded by a quantifying statement, such as "Contains ____ percent of ____," "Less than ____ percent of ____." The percentage of the ingredient(s) shall be filled in with a threshold level of 2 percent, 1.5 percent, 1.0 percent, or 0.5 percent, as appropriate. No ingredient to which the quantifying statement applies may be present in an amount greater than the stated threshold. Such a quantifying statement may also be utilized when an ingredients statement contains a listing of ingredients by individual components. Each component listing may utilize the required quantifying statement at the end of each component ingredients listing.

(B) Such ingredients may be adjusted in the product formulation without a change being made in the ingredients statement on the labeling, provided that the adjusted amount complies with §318.7(c)(4) and part 319 of this subchapter, and does not exceed the amount shown in the quantifying statement. Any such adjustments to the formulation shall be provided to the inspector- in-charge.

(2) On containers of frozen dinners, entrees, pizzas, and similar consumer packaged products in cartons the ingredient statement may be placed on the front riser panel: *Provided,* That the words "see ingredients" followed immediately by an arrow is placed on the principal display panel immediately above the location of such statement without intervening print or designs.

(3) The ingredient statement may be placed on the 20 percent panel adjacent to the principal display panel and reserved for required information, in the case of a cylindrical or nearly cylindrical container.

(4) The ingredients statement may be placed on the information panel, except as otherwise permitted in this subchapter.

(g)

(1) The name or trade name of the person that prepared the product may appear as the name of the manufacturer or packer without qualification on the label. Otherwise the name of the distributor of the product shall be shown with a phrase such as "Prepared for * * *". The place of business of the manufacturer, packer, or distributor shall be shown on the label by city, State, and postal ZIP code when such business is listed in a telephone or city directory, and if not listed in such directory, then the place of business shall be shown by street address, city, State, and postal ZIP code.

(2) The name and place of business of the manufacturer, packer, or distributor may be shown:

(i) On the principal display panel, or

(ii) On the 20 percent panel adjacent to the principal display panel and reserved for required information, in the case of a cylindrical or nearly cylindrical container, or

(iii) On the front riser panel of frozen food cartons, or

(iv) On the information panel.

(h)

(1) The statement of net quantity of contents shall appear on the principal display panel of all containers to be sold at retail intact, in conspicuous and easily legible boldface print or type in distinct contrast to other matter on the container, and shall be declared in accordance with the provisions of this paragraph.

(2) The statement as it is shown on a label shall not be false or misleading and shall express an accurate statement of the quantity of contents of the container. Reasonable variations caused by loss or gain of moisture during the course of good distribution practices or by unavoidable deviations in good manufacturing practices will be recognized. Variations from stated quantity of contents shall be as provided in §317.19. The statement shall not include any term qualifying a unit of weight, measure, or count such as "jumbo quart," "full gallon,"

9 CFR §317.2

"giant quart," "when packed," "minimum," or words of similar importance.

(3) The statement shall be placed on the principal display panel within the bottom 30 percent of the area of the panel in lines generally parallel to the base: Provided, That on packages having a principal display panel of 5 square inches or less, the requirement for placement within the bottom 30 percent of the area of the label panel shall not apply when the statement meets the other requirements of this paragraph (h). In any case, the statement may appear in more than one line. The terms "net weight" or "net wt." shall be used when stating the net quantity of contents in terms of weight, and the term "net contents" or "content" when stating the net quantity of contents in terms of fluid measure.

(4) Except as provided in §317.7, the statement shall be expressed in terms of avoirdupois weight or liquid measure. Where no general consumer usage to the contrary exists, the statement shall be in terms of liquid measure, if the product is liquid, or in terms of weight if the product is solid, semisolid viscous or a mixture of solid and liquid. For example, a declaration of 3/4-pound avoirdupois weight shall be expressed as "Net Wt. 12 oz." except as provided for in paragraph (h)(5) of this section for random weight packages; a declaration of 1 1/2 pounds avoirdupois weight shall be expressed as "Net Wt. 24 oz. (1 lb. 8 oz.)," "Net Wt. 24 oz. (1 1/2 lb.)," or "Net Wt. 24 oz. (1.5 lbs.)."

(5) On packages containing 1 pound or 1 pint and less than 4 pounds or 1 gallon, the statement shall be expressed as a dual declaration both in ounces and (immediately thereafter in parentheses) in pounds, with any remainder in terms of ounces or common or decimal fraction of the pound, or in the case of liquid measure, in the largest whole units with any remainder in terms of fluid ounces or common or decimal fractions of the pint or quart, except that on random weight packages the statement shall be expressed in terms of pounds and decimal fractions of the pound, for packages over 1 pound, and for packages which do not exceed 1 pound the statement may be in decimal fractions of the pound in lieu of ounces. Paragraph (h)(9) of this section permits certain exceptions from the provisions of this paragraph for margarine packages, random weight consumer size packages, and packages of less than 1/2 ounce net weight. Pargraph (h)(12) of this section permits certain exceptions from the provision of this paragraph for multi-unit packages.

(6) The statement shall be in letters and numerals in type size established in relationship to the area of the principal display panel of the package and shall be uniform of all packages of substantially the same size by complying with the following type specifications:

(i) Not less than one-sixteenth inch in height on packages, the principal display panel of which has an area of 5 square inches or less;

(ii) Not less than one-eighth inch in height on packages, the principal display panel of which has an area of more than 5 but not more than 25 square inches;

(iii) Not less than three-sixteenths inch in height on packages, the principal display panel of which has an area of more than 25 but not more than 100 square inches;

(iv) Not less than one-quarter inch in height on packages, the principal display panel of which has an area of more than 100 but not more than 400 square inches.

(v) Not less than one-half inch in height on packages, the principal display panel of which has an area of more than 400 square inches.

(7) The ratio of height to width of letters and numerals shall not exceed a differential of 3 units to 1 unit (no more than 3 times as high as it is wide). Heights pertain to upper case or capital letters. When upper and lower case or all lower case letters are used, it is the lower case letter "o" or its equivalent that shall meet the minimum standards. When fractions are used, each component numeral shall meet one-half the height standards.

(8) The statement shall appear as a distinct item on the principal display panel and shall be separated by a space at least equal to the height of the lettering used in the statement from other printed label information appearing above or below the statement and by a space at least equal to twice the width of the letter "N" of the style of type used in the quantity of contents statement from other printed label information appearing to the left or right of the statement. It shall not include any term qualifying a unit of weight, measure, or count such as, "jumbo quart," "full gallon," "giant quart," "when packed," "Minimum" or words of similar import.

(9) The following exemptions from the requirements contained in this paragraph (h) are hereby established:

(i) Individually wrapped, random weight consumer size packages shipped in bulk containers (as specified in paragraph (h)(11) of this section) and meat products that are subject to shrinkage through moisture loss during good distribution practices and are designated as gray area type of products as defined under §317.19 need not bear a net weight statement when shipped from an official establishment, provided that a net weight shipping

9 CFR §317.2

statement which meets the requirements of paragraph (h)(2) of this section is applied to their shipping container prior to shipping it from the official establishment. Net weight statements so applied to the shipping container are exempt from the type size, dual declaration, and placement requirements of this paragraph, if an accurate statement of net weight is shown conspicuously on the principal display panel of the shipping container. The net weight also shall be applied directly to random weight consumer size packages prior to retail display and sale. The net weight statement on random weight consumer size packages for retail sale shall be exempt from the type size, dual declaration, and placement requirements of this paragraph, if an accurate statement of net weight is shown conspicuously on the principal display panel of the package.

(ii) Individually wrapped and labeled packages of less than 1/2 ounce net weight and random weight consumer size packages shall be exempt from the requirements of this paragraph is they are in a shipping container and the statement of net quantity of contents on the shipping container meets the requirements of paragraph (h)(2) of this section;

(iii) Individually wrapped and labeled packages of less than 1/2 ounce net weight bearing labels declaring net weight, price per pound, and total price, shall be exempt from the type size, dual declaration, and placement requirements of this paragraph, if an accurate statement of net weight is shown conspicuously on the principal display panel of the package.

(iv) Margarine in 1 pound rectangular packages (except packages containing whipped or soft margarine or packages that contain more than four sticks) is exempt from the requirements of paragraphs (h) (3) and (5) of this section regarding the placement of the statement of the net quantity of contents within the bottom 30 percent of the principal display panel and that the statement be expressed both in ounces and in pounds, if the statement appears as "1 pound" or "one pound" in a conspicuous manner on the principal display panel.

(v) Sliced shingle packed bacon in rectangular packages is exempt from the requirements of paragraphs (h)(3) and (h)(5) of this section regarding the placement of the statement of the net quantity of contents within the bottom 30 percent of the principal display panel, and that the statement be expressed both in ounces and in pounds, if the statement appears in a conspicuous manner on the principal display panel.

(10) Labels for containers which bear any representation as to the number of servings contained therein shall bear, contiguous to such representation, and in the same size type as is used for such representation, a statement of the net quantity of each such serving.

(11) As used in this section, a "random weight consumer size package" is one which is one of a lot, shipment or delivery of packages of the same product with varying weights and with no fixed weight pattern.

(12) On a multiunit retail package, a statement of the net quantity of contents shall appear on the outside of the package and shall include the number of individual units, the quantity of each individual unit, and in parentheses, the total net quantity of contents of the multiunit package in terms of avoirdupois or fluid ounces, except that such declaration of total quantity need not be followed by an additional parenthetical declaration in terms of the largest whole units and subdivisions thereof, as required by paragraph (h)(5) of this section. For the purposes of this section, "multiunit retail package" means a package containing two or more individually packaged units of the identical commodity and in the same quantity, with the individual packages intended to be sold as part of the multiunit retail package but capable of being individually sold in full compliance with all requirements of the regulations in this part. Open multiunit retail packages that do not obscure the number of units and the labeling thereon are not subject to this paragraph if the labeling of each individual unit complies with the requirements of paragraphs (h) (2), (3), (6), and (8) of this section.

(i) The official establishment number of the official establishment in which the product was processed under inspection shall be placed as follows:

(1) Within the official inspection legend in the form required by part 312 of this subchapter; or

(2) Outside the official inspection legend elsewhere on the exterior of the container or its labeling, e.g., the lid of a can, if shown in a prominent and legible manner in a size sufficient to insure easy visibility and recognition and accompanied by the prefix "EST"; or

(3) Off the exterior of the container, e.g., on a metal clip used to close casings or bags, or on the back of a paper label of a canned product, or on other packaging or labeling material in the container, e.g., on aluminum pans and trays placed within containers, when a statement of its location is printed contiguous to the official inspection legend, such as "EST. No. on Metal Clip" or "Est. No. on Pan", if shown in a prominent and legible manner in a size sufficient to insure easy visibility and recognition; or

(4) On an insert label placed under a transparent covering if clearly visible and legible and accompanied by the

9 CFR §317.2

prefix "EST".

(j) Labels of any product within any of the following paragraphs shall show the information required by such paragraph for such product:

(1) A label for product which is an imitation of another food shall bear the word "imitation" immediately preceding the name of the food imitated and in the same size and style of lettering as in that name and immediately thereafter the word "ingredients:" and the names of the ingredients arranged in the order of their predominance.

(2) If a product purports to be or is represented for any special dietary use by man, its label shall bear a statement concerning its vitamin, mineral, and other dietary properties upon which the claim for such use is based in whole or in part and shall be in conformity with regulations (21 CFR part 125) established pursuant to sections 403, and 701 of the Federal Food, Drug, and Cosmetic Act (21 U.S.C. 343, 371).

(3) When an artificial smoke flavoring or an smoke flavoring is added as an ingredient in the formula of a meat food product, as permitted in part 318 of this subchapter, there shall appear on the label, in prominent letters and contiguous to the name of the product, a statement such as "Artificial Smoke Flavoring Added" or "Smoke Flavoring Added," as may be applicable, and the ingredient statement shall identify any artificial smoke flavoring or smoke flavoring so added as an ingredient in the formula of the meat food product.

(4) When any other artificial flavoring is permitted under part 318 of this subchapter to be added to a product, the ingredient statement shall identify it as "Artificial Flavoring."

(5) When artificial coloring is added to edible fats as permitted under part 318 of this subchapter such substance shall be declared on the label in a prominent manner and contiguous to the name of the product by the words "Artificially colored" or "Artificial coloring added" or "With added artificial coloring." When natural coloring such as annatto is added to edible fats as permitted under part 318 of this subchapter, such substance shall be declared on the label in the same manner by a phrase such as "Colored with annatto."

(6) When product is placed in a casing to which artificial coloring is applied as permitted under part 318 of this subchapter, there shall appear on the label, in a prominent manner and contiguous to the name of the product, the words, "Artificially colored."

(7) If a casing is removed from product at an official establishment and there is evidence of artificial coloring on the surface of the product, there shall appear on the label, in a prominent manner and contiguous to the name of product, the words "Artificially colored."

(8) When a casing is colored prior to its use as a covering for product and the color is not transferred to the product enclosed in the casing, no reference to color need appear on the label but no such casing may be used if it is misleading or deceptive with respect to color, quality, or kind of product, or otherwise.

(9) Product which bears or contains any other artificial coloring, as permitted under part 318 of this subchapter, shall bear a label stating that fact on the immediate container or if there is none, on the product.

(10) When an antioxidant is added to product as permitted under part 318 of this subchapter, there shall appear on the label in prominent letters and contiguous to the name of the product, a statement identifying the officially approved specific antioxidant by its common name or abbreviation thereof and the purpose for which it is added, such as, "BHA, BHT, and Propylgallate added to help protect flavor."

(11) Containers of meat packed in borax or other preservative for export to a foreign country which permits the use of such preservative shall, at the time of packing, be marked "for export," followed on the next line by the words "packed in preservative," or such equivalent statement as may be approved for this purpose by the Administrator and directly beneath this there shall appear the word "establishment" or abbreviation thereof, followed by the number of the establishment at which the product is packed. The complete statement shall be applied in a conspicuous location and in letters not less than 1 inch in height.

(12) Containers of other product packed in, bearing, or containing any chemical preservative shall bear a label stating that fact.

(13) (i) On the label of any "Mechanically Separated (Species)" described in §319.5(a) of this subchapter, the name of such product shall be followed immediately by the phrase "for processing" unless such product has a protein content of not less than 14 percent and a fat content of not more than 30 percent.

(ii) When any "Mechanically Separated (Species)" described in §319.5 of this subchapter is used as an ingredient in the preparation of a meat food product and such "Mechanically Separated (Species)" contributes 20 mg or more of calcium to a serving of such meat food product, the label of such meat food product shall state the calcium content of such meat food product, determined and expressed as the percentage of the U.S.

9 CFR §317.2

Recommended Daily Allowance (U.S. RDA) in a serving in accordance with 21 CFR 101.9(b)(1), (c)(7) (i) and (iv), and (e), as part of any nutrition information included on such label, or if such meat food product does not bear nutrition labeling information, as part of a prominent statement in immediate conjunction with the list of ingredients, as follows: "A --- serving contains --% of the U.S. RDA of calcium", with the blanks to be filled in, respectively, with the quantity of such product that constitutes a serving and the amount of calcium provided by such serving: *Provided,* That, calcium content need not be stated where

(*a*) the percent of the U.S. RDA of calcium to be declared would not differ from the percent of the U.S. RDA that would be declared if the meat food product contained only hand deboned ingredients or

(*b*) the calcium content of a serving of the meat food product would be 20 percent of the U.S. RDA or more if the meat food product contained only hand deboned ingredients.

(k) Packaged products which require special handling to maintain their wholesome condition shall have prominently displayed on the principal display panel of the label the statement: "Keep Refrigerated," "Keep Frozen," "Perishable Keep Under Refrigeration," or such similar statement as the Administrator may approve in specific cases. Products that are distributed frozen during distribution and thawed prior to or during display for sale at retail shall bear the statement on the shipping container: "Keep Frozen." The consumer-size containers for such products shall bear the statement "Previously Handled Frozen for Your Protection, Refreeze or Keep Refrigerated." For all perishable canned products the statement shall be shown in upper case letters one-fourth inch in height for containers having a net weight of 3 pounds or less, and for containers having a net weight over 3 pounds, the statement shall be in upper case letters at least one-half inch in height.

(l) Safe handling instructions shall be provided for: All meat and meat products of cattle, swine, sheep, goat, horse, or other equine not heat processed in a manner that conforms to the time and temperature combinations in the Table for Time/Temperature Combination For Cooked Beef, Roast Beef, and Cooked Corned Beef in §318.17, or that have not undergone other further processing that would render them ready-to-eat; and all comminuted meat patties not heat processed in a manner that conforms to the time and temperature combinations in the Table for Permitted Heat-Processing Temperature/Time Combinations For Fully-Cooked Patties in §318.23; except as exempted under paragraph (l)(4) of this section.

(1) (i) Safe handling instructions shall accompany every meat or meat product, specified in this paragraph (l) destined for household consumers, hotels, restaurants, or similar institutions and shall appear on the label. The information shall be in lettering no smaller than one-sixteenth of an inch in size and shall be prominently placed with such conspicuousness (as compared with other words, statements, designs or devices in the labeling) as to render it likely to be read and understood by the ordinary individual under customary conditions of purchase and use.

(ii) The safe handling information shall be presented on the label under the heading "Safe Handling Instructions" which shall be set in type size larger than the print size of the rationale statement and handling statements as discussed in paragraphs (l)(2) and (l)(3) of this section. The safe handling information shall be set off by a border and shall be one color type printed on a single color contrasting background whenever practical.

(2) The labels of the meat and meat products specified in this paragraph (l) shall include the following rationale statement as part of the safe handling instructions, "This product was prepared from inspected and passed meat and/or poultry. Some food products may contain bacteria that could cause illness if the product is mishandled or cooked improperly. For your protection, follow these safe handling instructions." This statement shall be placed immediately after the heading and before the safe handling statements.

(3) Meat and meat products, specified in this paragraph (l), shall bear the labeling statements:

(i) Keep refrigerated or frozen. Thaw in refrigerator or microwave. (Any portion of this statement that is in conflict with the product's specific handling instructions, may be omitted, e.g., instructions to cook without thawing.) (A graphic illustration of a refrigerator shall be displayed next to the statement.);

(ii) Keep raw meat and poultry separate from other foods. Wash working surfaces (including cutting boards), utensils, and hands after touching raw meat or poultry. (A graphic illustration of soapy hands under a faucet shall be displayed next to the statement.);

(iii) Cook thoroughly. (A graphic illustration of a skillet shall be displayed next to the statement.); and

(iv) Keep hot foods hot. Refrigerate leftovers immediately or discard. (A graphic illustration of a thermometer shall be displayed next to the statement.)

(4) Meat or meat products intended for further processing at another official establishment are exempt from the requirements prescribed in paragraphs (l)(1) through (l)(3) of this section.

(m)
(1) The information panel is that part of a label that is the first surface to the right of the principal display panel as observed by an individual facing the principal display panel, with the following exceptions:

(i) If the first surface to the right of the principal display panel is too small to accommodate the required information or is otherwise unusable label space, e.g., folded flaps, tear strips, opening flaps, heat-sealed flaps, the next panel to the right of this part of the label may be used.

(ii) If the package has one or more alternate principal display panels, the information panel is to the right of any principal display panel.

(iii) If the top of the container is the principal display panel and the package has no alternate principal display panel, the information panel is any panel adjacent to the principal display panel.

(2) (i) Except as otherwise permitted in this part, all information required to appear on the principal display panel or permitted to appear on the information panel shall appear on the same panel unless there is insufficient space. In determining the sufficiency of the available space, except as otherwise prescribed in this part, any vignettes, designs, and any other nonmandatory information shall not be considered. If there is insufficient space for all required information to appear on a single panel, it may be divided between the principal display panel and the information panel, provided that the information required by any given provision of this part, such as the ingredients statement, is not divided and appears on the same panel.

(ii) All information appearing on the information panel pursuant to this section shall appear in one place without intervening material, such as designs or vignettes.

[35 FR 15580, Oct. 3, 1970, as amended at 36 FR 11903, June 23, 1971; 36 FR 12003, 12004, June 24, 1971; 38 FR 13476, May 22, 1973; 40 FR 2576, Jan. 14, 1975; 40 FR 11347, Mar. 11, 1975; 40 FR 42852, Sept. 17, 1975; 41 FR 48721, Nov. 5, 1976; 47 FR 28256, June 29, 1982; 47 FR 29515, July 7, 1982; 53 FR 28635, July 29, 1988; 55 FR 7294, Mar. 1, 1990; 55 FR 49833, Nov. 30, 1990; 58 FR 38048, July 15, 1993; 58 FR 40468, July 28, 1993; 58 FR 42198, Aug. 9, 1993; 59 FR 14539, Mar. 28, 1994; 59 FR 40212, Aug. 8, 1994; 61 FR 42144, Aug. 14, 1996]

§317.3 Approval of abbreviations of marks of inspection; preparation of marking devices bearing inspection legend without advance approval prohibited; exception.

(a) The Administrator may approve and authorize the use of abbreviations of marks of inspection under the regulations in this subchapter. Such abbreviations shall have the same force and effect as the respective marks for which they are authorized abbreviations.

(b) Except for the purposes of preparing and submitting a sample or samples of the same to the Administrator for approval, no brand manufacturer, printer, or other person shall cast, print, lithograph, or otherwise make any marking device containing any official mark or simulation thereof, or any label bearing any such mark or simulation, without the written authority therefor of the Administrator. However, when any such sample label, or other marking device, is approved by the Administrator, additional supplies of the approved label, or marking device, may be made for use in accordance with the regulations in this subchapter, without further approval by the Administrator. The provisions of this paragraph apply only to labels, or other marking devices, bearing or containing an official inspection legend shown in §312.2(b), §312.3(a) (only the legend appropriate for horse meat food products) or §312.3(b) (only the legend appropriate for other (nonhorse) equine meat food products), or any abbreviations, copy or representation thereof.

(c) No brand manufacturer or other person shall cast or otherwise make, without an official certificate issued in quadruplicate by a Program employee, a brand or other marking device containing an official inspection legend, or simulation thereof, shown in §312.2(a), §312.3(a) (only the legend appropriate for horse carcasses and parts of horse carcasses), §312.3(b) (only the legend appropriate for other equine (nonhorse) carcasses and parts of other (nonhorse) equine carcasses) or §312.7(a).

(1) The certificate is a Food Safety and Inspection Service form for signature by a Program employee and the official establishment ordering the brand or other marking device, bearing a certificate serial number and a letterhead and the seal of the United States Department of Agriculture. The certificate authorizes the making of only the brands or other marking devices of the type and quantity listed on the certificate.

(2) After signing the certificate, the Program employee and the establishment shall each keep a copy, and the remaining two copies shall be given to the brand or other marking device manufacturer.

9 CFR §317.4

(3) The manufacturer of the brands or other marking devices shall engrave or otherwise mark each brand or other marking device with a permanent identifying serial number unique to it. The manufacturer shall list on each of the two copies of the certificate given to the manufacturer the number of each brand or other marking device authorized by the certificate. The manufacturer shall retain one copy of the certificate for the manufacturer's records and return the remaining copy with the brands or other marking devices to the Program employee whose name and address are given on the certificate as the recipient.

(4) In order that all such brands or other marking devices bear identifying numbers, within one year after June 24, 1985, an establishment shall either replace each such brand or other marking device which does not bear an identifying number, or, under the direction of the inspector-in-charge, mark such brand or other marking device with a permanent identifying number.

(Recordkeeping requirements approved by the Office of Management and Budget under control number 0583-0015)
[35 FR 15580, Oct. 3, 1970, as amended at 50 FR 21422, May 24, 1985]

§317.4 Labeling approval.

(a) No final labeling shall be used on any product unless the sketch labeling of such final labeling has been submitted for approval to the Food Labeling Division, Regulatory Programs, Food Safety and Inspection Service, and approved by such division, accompanied by FSIS form, Application for Approval of Labels, Marking, and Devices, except for generically approved labeling authorized for use in §317.5(b). The management of the official establishment or establishment certified under a foreign inspection system, in accordance with part 327 of this subchapter, must maintain a copy of all labeling used, along with the product formulation and processing procedure, in accordance with part 320 of this subchapter. Such records shall be made available to any duly authorized representative of the Secretary upon request.

(b) The Food Labeling Division shall permit submission for approval of only sketch labeling, as defined in §317.4(d), for all products, except as provided in §317.5(b) (2)-(9) and except for temporary use of final labeling as prescribed in paragraph (f) of this section.

(c) All labeling required to be submitted for approval as set forth in §317.4(a) shall be submitted in duplicate to the Food Labeling Division, Regulatory Programs, Food Safety and Inspection Service, U.S. Department of Agriculture, Washington, DC 20250. A parent company for a corporation may submit only one labeling application (in duplicate form) for a product produced in other establishments that are owned by the corporation.

(d) "Sketch" labeling is a printer's proof or equivalent which clearly shows all labeling features, size, location, and indication of final color, as specified in §317.2. FSIS will accept sketches that are hand drawn, computer generated or other reasonable facsimiles that clearly reflect and project the final version of the labeling. Indication of final color may be met by: submission of a color sketch, submission of a sketch which indicates by descriptive language the final colors, or submission with the sketch of previously approved final labeling that indicates the final colors.

(e) Inserts, tags, liners, pasters, and like devices containing printed or graphic matter and for use on, or to be placed within, containers and coverings of product shall be submitted for approval in the same manner as provided for labeling in §317.4(a), except that such devices which contain no reference to product and bear no misleading feature shall be used without submission for approval as prescribed in §317.5(b)(7).

(f)
  (1) Consistent with the requirements of this section, temporary approval for the use of a final label or other final labeling that may otherwise be deemed deficient in some particular may be granted by the Food Labeling Division. Temporary approvals may be granted for a period not to exceed 180 calendar days, under the following conditions:
    (i) The proposed labeling would not misrepresent the product;
    (ii) The use of the labeling would not present any potential health, safety, or dietary problems to the consumer;
    (iii) Denial of the request would create undue economic hardship; and
    (iv) An unfair competitive advantage would not result from the granting of the temporary approval.
  (2) Extensions of temporary approvals may also be granted by the Food Labeling Division provided that the applicant demonstrates that new circumstances, meeting the above criteria, have developed since the original temporary approval was granted.

(g) The inspector-in-charge shall approve meat carcass ink brands and meat food product ink and burning brands,

9 CFR §317.5

which comply with parts 312 and 316 of this subchapter.

[35 FR 15580, Oct. 3, 1970, as amended at 38 FR 29214, Oct. 23, 1973; 48 FR 11418, Mar. 18, 1983; 60 FR 67454, Dec. 29, 1995]

§317.5 Generically approved labeling.

(a)

(1) An official establishment or an establishment certified under a foreign inspection system, in accordance with part 327 of this subchapter, is authorized to use generically approved labeling, as defined in paragraph (b) of this section, without such labeling being submitted for approval to the Food Safety and Inspection Service in Washington or the field, provided the labeling is in accordance with this section and shows all mandatory features in a prominent manner as required in §317.2, and is not otherwise false or misleading in any particular.

(2) The Food Safety and Inspection Service shall select samples of generically approved labeling from the records maintained by official establishments and establishments certified under foreign inspection systems, in accordance with part 327 of this subchapter, as required in §317.4, to determine compliance with labeling requirements. Any finding of false or misleading labeling shall institute the proceedings prescribed in §335.12.

(b) Generically approved labeling is labeling which complies with the following:

(1) Labeling for a product which has a product standard as specified in part 319 of this subchapter or the Standards and Labeling Policy Book and which does not contain any special claims, such as quality claims, nutrient content claims, health claims, negative claims, geographical origin claims, or guarantees, or which is not a domestic product labeled in a foreign language;

(2) Labeling for single-ingredient products (such as beef steak or lamb chops) which does not contain any special claims, such as quality claims, nutrient content claims, health claims, negative claims, geographical origin claims, or guarantees, or which is not a domestic product labeled with a foreign language;

(3) Labeling for containers of products sold under contract specifications to Federal Government agencies, when such product is not offered for sale to the general public, provided that the contract specifications include specific requirements with respect to labeling, and are made available to the inspector-in-charge;

(4) Labeling for shipping containers which contain fully labeled immediate containers, provided such labeling complies with §316.13;

(5) Labeling for products not intended for human food, provided they comply with part 325 of this subchapter;

(6) Meat inspection legends, which comply with parts 312 and 316 of this subchapter;

(7) Inserts, tags, liners, pasters, and like devices containing printed or graphic matter and for use on, or to be placed within containers, and coverings of products, provided such devices contain no reference to product and bear no misleading feature;

(8) Labeling for consumer test products not intended for sale; and

(9) Labeling which was previously approved by the Food Labeling Division as sketch labeling, and the final labeling was prepared without modification or with the following modifications:

(i) All features of the labeling are proportionately enlarged or reduced, provided that all minimum size requirements specified in applicable regulations are met and the labeling is legible;

(ii) The substitution of any unit of measurement with its abbreviation or the substitution of any abbreviation with its unit of measurement, e.g., "lb." for "pound," or "oz." for "ounce," or of the word "pound" for "lb." or "ounce" for "oz.";

(iii) A master or stock label has been approved from which the name and address of the distributor are omitted and such name and address are applied before being used (in such case, the words "prepared for" or similar statement must be shown together with the blank space reserved for the insertion of the name and address when such labels are offered for approval);

(iv) Wrappers or other covers bearing pictorial designs, emblematic designs or illustrations, e.g., floral arrangements, illustrations of animals, fireworks, etc. are used with approved labeling (the use of such designs will not make necessary the application of labeling not otherwise required);

(v) A change in the language or the arrangement of directions pertaining to the opening of containers or the serving of the product;

(vi) The addition, deletion, or amendment of a dated or undated coupon, a cents-off statement, cooking instructions, packer product code information, or UPC product code information;

(vii) Any change in the name or address of the packer, manufacturer or distributor that appears in the signature line;

(viii) Any change in the net weight, provided the size of the net weight statement complies with §317.2;

(ix) The addition, deletion, or amendment of recipe suggestions for the product;

(x) Any change in punctuation;

(xi) Newly assigned or revised establishment numbers for a particular establishment for which use of the labeling has been approved by the Food Labeling Division, Regulatory Programs;

(xii) The addition or deletion of open dating information; (xiii) A change in the type of packaging material on which the label is printed;

(xiv) Brand name changes, provided that there are no design changes, the brand name does not use a term that connotes quality or other product characteristics, the brand name has no geographic significance, and the brand name does not affect the name of the product;

(xv) The deletion of the word "new" on new product labeling;

(xvi) The addition, deletion, or amendment of special handling statements, provided that the change is consistent with §317.2(k);

(xvii) The addition of safe handling instructions as required by §317.2(l);

(xviii) Changes reflecting a change in the quantity of an ingredient shown in the formula without a change in the order of predominance shown on the label, provided that the change in quantity of ingredients complies with any minimum or maximum limits for the use of such ingredients prescribed in parts 318 and 319 of this subchapter;

(xix) Changes in the color of the labeling, provided that sufficient contrast and legibility remain;

(xx) A change in the product vignette, provided that the change does not affect mandatory labeling information or misrepresent the content of the package;

(xxi) A change in the establishment number by a corporation or parent company for an establishment under its ownership;

(xxii) Changes in nutrition labeling that only involve quantitative adjustments to the nutrition labeling information, except for serving sizes, provided the nutrition labeling information maintains its accuracy and consistency;

(xxiii) Deletion of any claim, and the deletion of non-mandatory features or non-mandatory information; and

(xxiv) The addition or deletion of a direct translation of the English language into a foreign language for products marked "for export only."

[48 FR 11419, Mar. 18, 1983, as amended at 58 FR 58919, Nov. 4, 1993; 59 FR 14540, Mar. 28, 1994; 59 FR 40213, Aug. 8, 1994; 60 FR 67455, Dec. 29, 1995]

§317.6 Approved labels to be used only on products to which they are applicable.

Labels shall be used only on products for which they are approved, and only if they have been approved for such products in accordance with §317.3: Provided, That existing stocks of labels approved prior to the effective date of this section and the quantity of which has been identified to the circuit supervisor as being in storage on said date at the official establishment or other identified warehouse for the account of the operator of the official establishment may be used until such stocks are exhausted, but not later than 1 year after the effective date of this section unless such labels conform to all the requirements of this part and part 319 of this subchapter. The Administrator may upon the show of good cause grant individual extension of time as he deems necessary.

§317.7 Products for foreign commerce; printing labels in foreign language permissible; other deviations.

Labels to be affixed to packages of products for foreign commerce may be printed in a foreign language and may show the statement of the quantity of contents in accordance with the usage of the country to which exported and other deviations from the form of labeling required under this part may be approved for such product by the Administrator in specific cases: Provided,

(a) That the proposed labeling accords to the specifications of the foreign purchaser,

9 CFR §317.8

(b) That it is not in conflict with the laws of the country to which the product is intended for export, and

(c) That the outside container is labeled to show that it is intended for export; but if such product is sold or offered for sale in domestic commerce, all the requirements of this subchapter apply. The inspection legend and the establishment number shall in all cases appear in English but in addition, may appear literally translated in a foreign language.

§317.8 False or misleading labeling or practices generally; specific prohibitions and requirements for labels and containers.

(a) No product or any of its wrappers, packaging, or other containers shall bear any false or misleading marking, label, or other labeling and no statement, word, picture, design, or device which conveys any false impression or gives any false indication of origin or quality or is otherwise false or misleading shall appear in any marking or other labeling. No product shall be wholly or partly enclosed in any wrapper, packaging, or other container that is so made, formed, or filled as to be misleading.

(b) The labels and containers of product shall comply with the following provisions, as applicable:

(1) Terms having geographical significance with reference to a locality other than that in which the product is prepared may appear on the label only when qualified by the word "style," "type," or "brand," as the case may be, in the same size and style of lettering as in the geographical term, and accompanied with a prominent qualifying statement identifying the country, State, Territory, or locality in which the product is prepared, using terms appropriate to effect the qualification. When the word "style" or "type" is used, there must be a recognized style or type of product identified with and peculiar to the area represented by the geographical term and the product must possess the characteristics of such style or type, and the word "brand" shall not be used in such a way as to be false or misleading: Provided, That a geographical term which has come into general usage as a trade name and which has been approved by the Administrator as being a generic term may be used without the qualifications provided for in this paragraph. The terms "frankfurter," "vienna," "bologna," "lebanon bologna," "braunschweiger," "thuringer," "genoa," "leona," "berliner," "holstein," "goteborg," "milan," "polish," "italian," and their modifications, as applied to sausages, the terms "brunswick" and "irish" as applied to stews and the term "boston" as applied to pork shoulder butts need not be accompanied with the word "style," "type," or "brand," or a statement identifying the locality in which the product is prepared.

(2) Such terms as "farm" or "country" shall not be used on labels in connection with products unless such products are actually prepared on the farm or in the country: Provided, That if the product is prepared in the same way as on the farm or in the country these terms, if qualified by the word "style" in the same size and style of lettering, may be used: Provided further, That the term "farm" may be used as part of a brand designation when qualified by the word "brand" in the same size and style of lettering, and followed with a statement identifying the locality in which the product is prepared: And Provided further, That the provisions of this paragraph shall not apply to products prepared in accordance with §319.106 of this subchapter. Sausage containing cereal shall not be labeled "farm style" or "country style," and lard not rendered in an open kettle shall not be designated as "farm style" or "country style."

(3) The requirement that the label shall contain the name and place of business of the manufacturer, packer, or distributor shall not relieve any establishment from the requirement that its label shall not be misleading in any particular.

(4) The term "spring lamb" or "genuine spring lamb" is applicable only to carcasses of new-crop lambs slaughtered during the period beginning in March and terminating not beyond the close of the week containing the first Monday in October.

(5) (i) Coverings shall not be of such color, design, or kind as to be misleading with respect to color, quality, or kind of product to which they are applied. For example, transparent or semitransparent coverings for such articles as sliced bacon or fresh (uncooked) meat and meat food products shall not bear lines or other designs of red or other color which give a false impression of leanness of the product. Transparent or semitransparent wrappers, casings, or coverings for use in packaging cured, cured and smoked, or cured and cooked sausage products, and sliced ready-to-eat meat food products may be color tinted or bear red designs on 50 percent of such wrapper or covering: Provided, That the transparent or semitransparent portion of the principal display panel is free of color tinting and red designs: And provided further, That the principal display panel provides at

9 CFR §317.8

least 20 percent unobstructed clear space, consolidated in one area so that the true nature and color of the product is visible to the consumer.

(ii) Packages for sliced bacon that have a transparent opening shall be designed to expose, for viewing, the cut surface of a representative slice. Packages for sliced bacon which meet the following specifications will be accepted as meeting the requirements of this subparagraph provided the enclosed bacon is positioned so that the cut surface of the representative slice can be visually examined:

(*a*) For shingle-packed sliced bacon, the transparent window shall be designed to reveal at least 70 percent of the length (longest dimension) of the representative slice, and this window shall be at least 1 1/2 inches wide. The transparent window shall be located not more than five-eighths inch from the top or bottom edge of a 1-pound or smaller package and not more than three-fourths inch from either the top or bottom edge of a package larger than 1 pound.

(*b*) For stack-packed sliced bacon, the transparent window shall be designed to reveal at least 70 percent of the length (longest dimension) of the representative slice and be at least 1 1/2 inches wide.

(6) The word "fresh" shall not be used on labels to designate product which contains any sodium nitrate, sodium nitrite, potassium nitrate, or potassium nitrite, or which has been salted for preservation.

(7) (i) No ingredient shall be designated on the label as a spice, flavoring, or coloring unless it is a spice, flavoring, or coloring, as the case may be. An ingredient that is both a spice and a coloring, or both a flavoring and a coloring, shall be designated as "spice and coloring", or "flavoring and coloring", as the case may be, unless such ingredient is designated by its common or usual name.

(ii) Any ingredient not designated in §317.2(f)(1)(i) of this part whose function is flavoring, either in whole or in part, must be designated by its common or usual name. Those ingredients which are of livestock and poultry origin must be designated by names that include the species and livestock and poultry tissues from which the ingredients are derived.

(8) As used on labels of product, the term "gelatin" shall mean

(i) the jelly prepared in official establishments by cooking pork skins, tendons, or connective tissue from inspected and passed product, and

(ii) dry commercial gelatin or the jelly resulting from its use.

(9) Product (other than canned product) labeled with the term "loaf" as part of its name:

(i) If distributed from the official establishment in consumer size containers may be in any shape;

(ii) If distributed in a container of a size larger than that sold intact at retail the product shall be prepared in rectangular form, or as in paragraph (b)(9)(iii) of this section;

(iii) If labeled as an "Old Fashioned Loaf" shall be prepared in a traditional form, such as rectangular with rounded top or circular with flat bottom and rounded top.

(10) The term "baked" shall apply only to product which has been cooked by the direct action of dry heat and for a sufficient time to permit the product to assume the characteristics of a baked article, such as the formation of a brown crust on the surface, rendering out of surface fat, and the caramelization of the sugar if applied. Baked loaves shall be heated to a temperature of at least 160° F. and baked pork cuts shall be heated to an internal temperature of at least 170° F.

(11) When products such as loaves are browned by dipping in hot edible oil or by a flame, the label shall state such fact, e.g., by the words "Browned in Hot Cottonseed Oil" or "Browned by a Flame," as the case may be, appearing as part of the product name.

(12) The term "meat" and the names of particular kinds of meat, such as beef, veal, mutton, lamb, and pork, shall not be used in such manner as to be false or misleading.

(13) The word "ham," without any prefix indicating the species of animal from which derived, shall be used in labeling only in connection with the hind legs of swine. Ham shanks as such or ham shank meat as such or the trimmings accruing in the trimming and shaping of hams shall not be labeled "ham" or "ham meat" without qualification. When used in connection with a chopped product the term "ham" or "ham meat" shall not include the skin.

(14) The terms "shankless" and "hockless" shall apply only to hams and pork shoulders from which the shank or hock has been completely removed, thus eliminating the entire tibia and fibula, or radius and ulna, respectively, together with the overlying muscle, skin, and other tissue.

(15) Such terms as "meat extract" or "extract of beef" without qualification shall not be used on labels in connection with products prepared from organs or other parts of the carcass, other than fresh meat. Extracts

prepared from any parts of the carcass other than fresh meat may be properly labeled as extracts with the true name of the parts from which prepared. In the case of extract in fluid form, the word "fluid" shall also appear on the label, as, for example, "fluid extract of beef."

(16) [Reserved]

(17) When any product is enclosed in a container along with a packing substance such as brine, vinegar, or agar jelly, a declaration of the packing substance shall be printed prominently on the label as part of the name of the product, as for example, "frankfurts packed in brine," "lamb tongue packed in vinegar," or "beef tongue packed in agar jelly," as the case may be. The packing substance shall not be used in such a manner as will result in the container being so filled as to be misleading.

(18) "Leaf lard" is lard prepared from fresh leaf fat.

(19) When lard or hardened lard is mixed with rendered pork fat or hardened rendered pork fat, the mixture shall be designated as "rendered pork fat" or "hardened rendered pork fat," as the case may be.

(20) Oil, stearin, or stock obtained from beef or mutton fats rendered at a temperature above 170° F. shall not be designated as "oleo oil," "oleo stearin," or "oleo stock," respectively.

(21) When not more than 20 percent of beef fat, mutton fat, oleo stearin, vegetable stearin, or hardened vegetable fat is mixed with lard or with rendered pork fat, there shall appear on the label, contiguous to and in the same size and style of lettering as the name of the product, the words "beef fat added," "mutton fat added," "oleo stearin added," "vegetable stearin added," or "hardened vegetable fat added," as the case may be. If more than 20 percent is added, the product name shall refer to the particular animal fat or fats used, such as, "Lard and Beef Fat." The designation "vegetable fat" is applicable to vegetable oil, vegetable stearin, or a combination of such oil and stearin, whereas the designations "vegetable oil" and "vegetable stearin" shall be applicable only to the oil and the stearin respectively, when used in meat food products.

(22) Cooked, cured, or pickled pigs feet, pigs knuckles, and similar products, shall be labeled to show that the bones remain in the product, if such is the case. The designation "semi- boneless" shall not be used if less than 50 percent of the total weight of bones has been removed.

(23) When monoglycerides, diglycerides, and/or polyglycerol esters of fatty acids are added to rendered animal fat or a combination of such fat and vegetable fat, there shall appear on the label in a prominent manner and contiguous to the name of the product a statement such as "With Monoglycerides and Diglycerides Added," or "With Diglycerides and Monoglycerides," or "With Polyglycerol Esters of Fatty Acids" as the case may be.

(24) Section 407 of the Federal Food, Drug, and Cosmetic Act contains provisions with respect to colored margarine or colored oleomargarine (21 U.S.C. 347) which are set forth herein as footnote. [1]

---

[1]Sec. 407(a) Colored oleomargarine or colored margarine which is sold in the same State or Territory in which it is produced shall be subject in the same manner and to the same extent to the provisions of this Act as if it had been introduced in interstate commerce.

(b) No person shall sell, or offer for sale, colored oleomargarine or colored margarine unless—

(1) Such oleomargarine or margarine is packaged,

(2) The net weight of the contents of any package sold in a retail establishment is one pound or less,

(3) There appears on the label of the package

(A) The word 'oleomargarine' or 'margarine' in type or lettering at least as large as any other type or lettering on such label, and (B) A full and accurate statement of all the ingredients contained in such oleomargarine, or margarine, and

(4) Each part of the contents of the package is contained in a wrapper which bears the word 'oleomargarine' or 'margarine' in type or lettering not smaller than 20-point type.

The requirements of this subsection shall be in addition to and not in lieu of any of the other requirements of this Act.

(c) No person shall possess in a form ready for serving colored oleomargarine or colored margarine at a public eating place unless a notice that oleomargarine or margarine is served is displayed prominently and conspicuously in such place and in such manner as to render it likely to be read and understood by the ordinary individual being served in such eating place or is printed or is otherwise set forth on the menu in type or lettering not smaller than that normally used to designate the serving of other food items. No person shall serve colored oleomargarine or colored margarine at a public eating place, whether or not any charge is made therefor, unless (1) each separate

9 CFR §317.8

(25) When approved proteolytic enzymes as permitted in part 318 of this subchapter are used on steaks or other raw meat cuts, there shall appear on the label, in a prominent manner, contiguous to the product name, the statement, "Tenderized with [approved enzyme]," to indicate the use of such enzymes. Any other approved substance which may be used in the solution shall also be included in the statement.

When approved inorganic chlorides as permitted in part 318 of this subchapter are used on steaks or other raw meat cuts there shall appear on the label in a prominent manner, contiguous to the product name, the statement, "Tenderized with (names of approved inorganic chloride(s))" to indicate the use of such inorganic chlorides. Any other approved substance which may be in the solution shall also be included in the statement.

(26) When dimethylpolysiloxan is added as an antifoaming agent to rendered fats, its presence shall be declared on the label contiguous to the name of the product. Such declaration shall read "Dimethylpolysiloxan Added."

(27) When pizzas are formulated with crust containing calcium propionate or sodium propionate, there shall appear on the label contiguous to the name of the product the statement "---- added to retard spoilage of crust" preceded by the name of the preservative.

(28) Sausage of the dry varieties treated with potassium sorbate or propylparaben (propyl p-hydroxybenzoate) as permitted by part 318 of this subchapter, shall be marked or labeled with a statement disclosing such treatment and the purpose thereof, such as "dipped in a potassium sorbate solution to retard mold growth."

(29) Meat of goats shall be identified as goat meat or chevon.

(30) The term "Chitterlings" shall apply to the large intestines of swine, or young bovine animals when preceded with the word "Calf" or "Veal." Meat food products that contain chitterlings or calf or veal chitterlings, in accordance with §318.6(b)(8) of this subchapter shall be identified with product names that refer to such ingredients, as for instance, "Chitterling Loaf," "Chitterling Pie," or "Calf Chitterlings and Gravy," and shall be packed in containers having a capacity of 3 pounds or less and of a kind usually sold at retail intact and bearing such other information as is required by this part.

(31) Products that contain blood from livestock as permitted by part 318 of this subchapter shall be labeled with a name that includes the term "blood," and the specific kind of blood shall be declared in the ingredient statement, e.g., "Swine blood," in the manner required by this part.

(32) A calendar date may be shown on labeling when declared in accordance with the provisions of this subparagraph:

(i) The calendar date shall express the month of the year and the day of the month for all products and also the year in the case of products hermetically sealed in metal or glass containers, dried or frozen products, or any other products that the Administrator finds should be labeled with the year because the distribution and marketing practices with respect to such products may cause a label without a year identification to be misleading.

(ii) Immediately adjacent to the calendar date shall be a phrase explaining the meaning of such date, in terms of "packing" date, "sell by" date, or "use before" date, with or without a further qualifying phrase, e.g., "For Maximum Freshness" or "For Best Quality", and such phrases shall be approved by the Administrator as prescribed in §317.4.

(33) [Reserved]

(34) The terms "All," "Pure," "100%," and terms of similar connotation shall not be used on labels for products to identify ingredient content, unless the product is prepared solely from a single ingredient.

(35) When agar-agar is used in canned jellied meat food products, as permitted in part 318 of this subchapter, there shall appear on the label in a prominent manner, contiguous to the product name, a statement to indicate

---

serving bears or is accompanied by labeling identifying it as oleomargarine or margarine, or (2) each separate serving thereof is triangular in shape.

(d) Colored oleomargarine or colored margarine when served with meals at a public eating place shall at the time of such service be exempt from the labeling requirements of section 343 of this Act (except subsection (a) and (f) of section 343 of this title) if it complies with the requirements of subsection (b) of this section.

(e) For the purpose of this section colored oleomargarine or colored margarine is oleomargarine or margarine having a tint or shade containing more than one and six tenths degrees of yellow or of yellow and red collectively, but with an excess of yellow over red, measured in terms of Lovibond tintometer scale or its equivalent" (21 U.S.C. 347).

the use of agar-agar.

(36) When sodium alginate, calcium carbonate, and lactic acid and calcium carbonate (or glucono delta-lactone) are used together in a dry binding matrix in restructured, formed meat food products, as permitted in part 318 of this subchapter, there shall appear on the label contiguous to the product name, a statement to indicate the use of sodium alginate, calcium carbonate and lactic acid and calcium carbonate (or glucono delta-lactone).

(37) [Reserved].

[35 FR 15580, Oct. 3, 1970; 59 FR 12536, Mar. 17, 1994]

Editorial Note: For additional Federal Register citations affecting §317.8, see the List of CFR Sections Affected in the Finding Aids section of this volume.

§317.9 Labeling of equine products.

The immediate containers of any equine products shall be labeled to show the kinds of animals from which derived when the products are sold, transported, offered for sale or transportation or received for transportation in commerce.

§317.10 Reuse of official inspection marks; reuse of containers bearing official marks, labels, etc.

(a) No official inspection legend or other official mark which has been previously used shall be used again for the identification of any product, except as provided for in paragraph (b) of this section.

(b) All stencils, marks, labels, or other labeling on previously used containers, whether relating to any product or otherwise, shall be removed or obliterated before such containers are used for any product, unless such labeling correctly indicates the product to be packed therein and such containers are refilled under the supervision of a Program employee.

§317.11 Labeling, filling of containers, handling of labeled products to be only in compliance with regulations.

(a) No person shall in any official establishment apply or affix, or cause to be applied or affixed, any label to any product prepared or received in such establishment, or to any container thereof, or fill any container at such an establishment, except in compliance with the regulations in this subchapter.

(b) No covering or other container shall be filled, in whole or in part, at any official establishment with any product unless it has been inspected and passed in compliance with the regulations in this subchapter, is not adulterated, and is strictly in accordance with the statements on the label, and such filling is done under the supervision of a Program employee.

(c) No person shall remove, or cause to be removed from an official establishment any product bearing a label unless such label is in compliance with the regulations in this subchapter, or any product not bearing a label required by such regulations.

§317.12 Relabeling products; requirements.

When it is claimed by an official establishment that any of its products which bore labels bearing official marks has been transported to a location other than an official establishment, and it is desired to relabel the product because the labels have become mutilated or otherwise damaged, a request for relabeling the product shall be sent to the Administrator, accompanied with a statement of the reasons therefor. Labeling material intended for relabeling inspected and passed product shall not be transported from an official establishment until permission has been received from the Administrator. The relabeling of inspected and passed product with labels bearing any official marks shall be done under the supervision of a Program inspector. The official establishment shall reimburse the Program, in accordance with the regulations of the Department, for any cost involved in supervising the relabeling of such product.

§317.13 Storage and distribution of labels and containers bearing official marks.

9 CFR §317.17

Labels, wrappers, and containers bearing any official marks, with or without the establishment number, may be transported from one official establishment to any other official establishment provided such shipments are made with the prior authorization of the inspector in charge at point of origin, who will notify the inspector in charge at destination concerning the date of shipment, quantity, and type of labeling material involved. No such material shall be used at the establishment to which it is shipped unless such use conforms with the requirements of this subchapter.

§317.14 [Removed and Reserved]
[Note: This section was removed as of July 1, 1996.]
[47 FR 54287, Dec. 2, 1982; 60 FR 67456, Dec. 29, 1995]

§317.15 [Reserved]

§317.16 Labeling and containers of custom prepared products.

Products that are custom prepared under §303.1(a)(2) of this subchapter must be packaged immediately after preparation and must be labeled (in lieu of information otherwise required by this part 317) with the words "Not For Sale" in lettering not less than three-eighth inch in height. Such exempted custom prepared products or their containers may bear additional labeling provided such labeling is not false or misleading.
[37 FR 4071, Feb. 26, 1972]

§317.17 Interpretation and statement of labeling policy for cured products; special labeling requirements concerning nitrate and nitrite.

(a) With respect to sections 1(n) (7), (9), and (12) of the Act and §317.2, any substance mixed with another substance to cure a product must be identified in the ingredients statement on the label of such product. For example, curing mixtures composed of such ingredients as water, salt, sugar, sodium phosphate, sodium nitrate, and sodium nitrite or other permitted substances which are added to any product, must be identified on the label of the product by listing each such ingredient in accordance with the provisions of §317.2.

(b) Any product, such as bacon and pepperoni, which is required to be labeled by a common or usual name or descriptive name in accordance with §317.2(c)(1) and to which nitrate or nitrite is permitted or required to be added may be prepared without nitrate or nitrite and labeled with such common or usual name or descriptive name when immediately preceded with the term "Uncured" as part of the product name in the same size and style of lettering as the product name, provided that the product is found by the Administrator to be similar in size, flavor, consistency, and general appearance to such product as commonly prepared with nitrate or nitrite, or both.

(c)

(1) Products described in paragraph (b) of this section or §319.2 of this subchapter, which contain no nitrate or nitrite shall bear the statement "No Nitrate or Nitrite Added." This statement shall be adjacent to the product name in lettering of easily readable style and at least one-half the size of the product name.

(2) Products described in paragraph (b) of this section and §319.2 of this subchapter shall bear, adjacent to the product name in lettering of easily readable style and at least one- half the size of the product name, the statement "Not Preserved- Keep Refrigerated Below 40° F. At All Times" unless they have been thermally processed to Fo 3 or more; they have been fermented or pickled to pH of 4.6 or less; or they have been dried to a water activity of 0.92 or less.

(3) Products described in paragraph (b) of this section and §319.2 of this subchapter shall not be subject to the labeling requirements of paragraphs (b) and (c) of this section if they contain an amount of salt sufficient to achieve a brine concentration of 10 percent or more.

[37 FR 16863, Aug. 22, 1972, as amended at 44 FR 48961, Aug. 21, 1979]

§317.18 Quantity of contents labeling.

9 CFR §317.19

Sections 317.18 through 317.22 of this part prescribe the procedures to be followed for determining net weight compliance and prescribe the reasonable variations from the declared net weight on the labels of immediate containers of products in accordance with §317.2(h) of this part.
[55 FR 49834, Nov. 30, 1990]

§317.19 Definitions and procedures for determining net weight compliance.

(a) For the purpose of §§317.18 through 317.22 of this part, the reasonable variations allowed, definitions, and procedures to be used in determining net weight and net weight compliance are described in the National Institute of Standards and Technology (NIST) Handbook 133, "Checking the Net Contents of Packaged Goods," Third Edition, September 1988, and Supplements 1, 2, 3, and 4 dated September 1990, October 1991, October 1992, and October 1994, respectively, which are incorporated by reference, with the exception of the NIST Handbook 133 and Supplements 1, 3, and 4 requirements listed in paragraphs (b) and (c) of this section. Those provisions incorporated by reference herein, are considered mandatory requirements. This incorporation was approved by the Director of the Federal Register in accordance with 5 U.S.C. 552(a) and 1 CFR part 51. (These materials are incorporated as they exist on the date of approval.) A notice of any change in the Handbook cited herein will be published in the Federal Register. Copies may be purchased from the Superintendent of Documents, U.S. Government Printing Office, Washington, DC 20402. It is also available for inspection at the Office of the Federal Register Information Center, room 8401, 1100 L Street NW., Washington, DC 20408.
(b) The following NBS Handbook 133 requirements are not incorporated by reference.

Chapter 2-General Considerations
    2.13.1. Polyethylene Sheeting and Film
    2.13.2. Textiles
    2.13.3. Mulch

Chapter 3-Methods of Test for Packages Labeled by Weight
    3.11 Aerosol Packages
    3.14. Glazed Raw Seafood and Fish
    3.15 Canned Coffee
    3.16 Borax
    3.17 Flour

Chapter 4-Methods of Test for Packages Labeled by Volume
    4.7 Millk
    4.8 Mayonnaise and Salad Dressing
    4.9 Paint, Varnish, and Lacquers–Nonaerosol
    4.11 Peat Moss
    4.12 Bark Mulch
    4.15 Ice Cream Novelties

Chapter 5-Methods of Test for Packages Labeled by Count, Length, Area, Thickness, or Combinations of Quantities
    5.4. Polyethylene Sheeting
    5.5 Paper Plates
    5.6 Sanitary Paper Products
    5.7 Pressed and Blown Glass Tumblers and Stemware

Appendix D: Package Net Contents Regulations
    D.1.1. U.S. Department of Health and Human Services, Food and Drug Administration
    D.1.2. Department of Agriculture, Food Safety and Inspection Service
    D.1.3. Federal Trade Commission

9 CFR §317.19

D.1.4 Environmental Protection Agency
D.1.5. U.S. Department of the Treasury, Bureau of Alcohol, Toboacco, and Firearms

(c) The following requirements of Supplement 1, dated September 1990, Supplement 3, dated October 1992, and Supplement 4, dated 1994, of NIST Handbook 133 are not incorporated by reference.

Supplement 1
Chapter 2-General Considerations
    2.13.1. Polyethylene Sheeting and Film
    2.13.2. Textiles
    2.13.3. Mulch
Chapter 3-Methods of Test for Packages Labeled by Weight
    3.11.4. Exhausting the Aerosol Container
Chapter 4-Methods of Test for Packages Labeled by Volume
    4.6.4. Method D-Determining the Net Contents of Compressed Gas in Cylinders
    4.7. Milk
    4.16. Fresh Oysters Labeled by Volume
Chapter 5-Methods of Test for Packages Labeled by Count, Length, Area, Thickness, or Combinations of Quantities
    5.4. Polyethylene Sheeting

Supplement 3
Chapter 3-Methods of Test for Packages Labeled by Weight
    3.17. Flour and Dry Pet Food
Chapter 5-Methods of Test for Packages Labeled by Count, Length, Area, Thickness, or Combination of Quantities
    5.4. Polyethylene Sheeting
    5.5. Paper Plates
    5.8. Baler Twine
Appendix A. Forms and Worksheets

Supplement 4
    3.11 Aerosol Packages
    3.11.1 Equipment
    3.11.2 Preparation for Test
    3.11.3 The Determination of Net Contents: Part 1
    3.11.4 Exhausting the Aerosol Container
    3.11.5 The Determination of Net Contents: Part 2
Appendix A. Report Forms
[55 FR 49834, Nov. 30, 1990; 60 FR 12884, Mar. 9, 1995]

§317.20 Scale requirements for accurate weights, repairs, adjustments, and replacement after inspection.

(a) All scales used to weigh meat products sold or otherwise distributed in commerce in federally inspected meat establishments shall be installed, maintained and operated to insure accurate weights. Such scales shall meet the applicable requirements contained in National Institute of Standards and Technology Handbook 44, "Specifications, Tolerances and Other Technical Requirements for Weighing and Measuring Devices," 1994 Edition, October 1993, which is incorporated by reference. This incorporation was approved by the Director of the Federal Register in accordance with 5 U.S.C.. 552(a) and 1 CFR part 51. (These materials are incorporated as they exist on the date of approval.) Copies may be purchased from the Superintendent of Documents, U.S. Government Printing Office, Washington, DC 20402. It is also available for inspection at the Office of the Federal Register Information Center, room 8401, 1100 L Street NW., Washington, DC 20408.

(b) All scales used to weigh meat products sold or otherwise distributed in commerce or in States designated under section 301(c) of the Federal Meat Inspection Act, shall be of sufficient capacity to weigh the entire unit and/or package.

(c) No scale shall be used at a federally inspected establishment to weigh meat products unless it has been found upon test and inspection, as specified in NIST Handbook 44, to provide accurate weight. If a scale is reinspected or retested and found to be inaccurate, or if any repairs, adjustments or replacements are made to a scale, it shall not be used until it has been inspected and tested by a USDA official, or a State or local government weights and measures official, or State registered or licensed scale repair firm or person, and it must meet all accuracy requirements as specified in NIST Handbook 44. If a USDA inspector has put a retain tag on a scale it can only be removed by a USDA inspector. As long as the tag is on the scale, it shall not be used.
[55 FR 49834, Nov. 30, 1990; 60 FR 12884, Mar. 9, 1995]

§317.21 Scales: testing of.

(a) The operator of each official establishment that weighs meat food products shall cause such scales to be tested for accuracy, in accordance with the technical requirements of NIST Handbook 44, at least once during the calendar year. In cases where the scales are found not to maintain accuracy between tests, more frequent tests may be required and monitored by an authorized USDA program official.
(b) The operator of each official establishment shall display on or near each scale a valid certification of the scale's accuracy from a State or local government's weights and measures authority or from a State registered or licensed scale repair firm or person, or shall have a net weight program under a Total Quality Control System or Partial Quality Control Program in accordance with §318.4 of this subchapter.
[55 FR 49834, Nov. 30, 1990; 62 *FR* 45024, Aug. 25, 1997]

§317.22 Handling of failed product.

Any lot of product which is found to be out of compliance with net weight requirements upon testing in accordance with §317.19 shall be handled as follows:
(a) A lot tested in an official establishment and found not to comply with net weight requirements may be reprocessed and must be reweighed and remarked to satisfy the net weight requirements of this section and be reinspected, in accordance with the requirements of this Part.
(b) A lot tested outside of an official establishment and found not to comply with net weight requirements must be reweighed and remarked with a proper net weight statement, provided that such reweighing and remarking shall not deface, cover, or destroy any other marking or labeling required under this subchapter and the net quantity of contents is shown with the same prominence as the most conspicuous feature of a label.
[55 FR 49834, Nov. 30, 1990]

§317.23 [Reserved]

§317.24 Packaging materials.

(a) Edible products may not be packaged in a container which is composed in whole or in part of any poisonous or deleterious substances which may render the contents adulterated or injurious to health. All packaging materials must be safe for their intended use within the meaning of section 409 of the Federal Food, Drug, and Cosmetic Act, as amended (FFDCA).
(b) Packaging materials entering the official establishment must be accompanied or covered by a guaranty, or statement of assurance, from the packaging supplier under whose brand name and firm name the material is marketed to the official establishment. The guaranty shall state that the material's intended use complies with the FFDCA and all applicable food additive regulations. The guaranty must identify the material, e.g., by the distinguishing brand name or code designation appearing on the packaging material shipping container; must specify the applicable conditions of use, including temperature limits and any other pertinent limits specified under the FFDCA and food additive regulations; and must be signed by an authorized official of the supplying firm. The guaranty may be limited to a specific shipment of an article, in which case it may be part of or attached to the invoice covering such shipment, or it may be general and continuing, in which case, in its application to any article

9 CFR §317.300

or other shipment of an article, it shall be considered to have been given at the date such article was shipped by the person who gives the guaranty. Guaranties consistent with the Food and Drug Administration's regulations regarding such guaranties (21 CFR 7.12 and 7.13) will be acceptable. The management of the establishment must maintain a file containing guaranties for all food contact packaging materials in the establishment. The file shall be made available to Program inspectors or other Department officials upon request. While in the official establishment, the identity of all packaging materials must be traceable to the applicable guaranty.

(c) The guaranty by the packaging supplier will be accepted by Program inspectors to establish that the use of material complies with the FFDCA and all applicable food additive regulations.

(d) The Department will mmnitor the use of packaging material in official establishments to assure that the requirements of paragraph (a) of this section are met, and may question the basis for any guaranty described under paragraph (b) of this section. Official establishments and packaging suppliers providing written guaranties to those official establishments will be permitted an opportunity to provide information tm designated Department officials as needed to verify the basis for any such guaranty. The required information will include, but is not limited to, manufacturing firm's name, trade name or code designation for the material, complete chemical composition, and use. Selection of a material for review does not in itself affect a material's acceptability. Materials may continue to be used during the review period. However, if information requested from the supplier is not provided within the time indicated in the request-a minimum of 30 days-any applicable guaranty shall cease to be effective, and approval to continue using the specified packaging material in official establishments may be denied. The Administrator may extend this time where reasonable grounds for extension are shown, as, for example, where data must be obtained from suppliers.

(e) The Administrator may disapprove for use in official establishmelts packaging materials whose use cannot be confirmed as complying with FFDCA and applicable food additive regulations. Before approval to use a packaging material is finally denied by the Administrator, the affected official establishment and the supplier of the material shall be given notice and the opportunity to present their views to the Administrator. If the official establishment and the supplier do not accept the Administrator's determination, a hearing in accordance with applicable rules of practice will be held to resolve such dispute. Approval to use the materials pending the outcome of the presentation of views or hearing shall be denied if the Administrator determines that such use may present an imminent hazard to public health.

(f) Periodically, the Administrator will issue to inspectors a listing, by distinguishing brand name or code designation, of packaging materials that have been reviewed and that fail to meet the requirements of paragraph (a) of this section. Listed materials will not be permitted for use in official establishments. If a subsequent review of any material indicates that it meets the requirements of paragraph (a), the material will be deleted from the listing.

(g) Nothing in this section shall affect the authority of Program inspectors to refuse a specific material if he/she determines the material may render products adulterated or injurious to health.

[49 FR 2235, Jan. 19, 1984. Redesignated at 55 FR 49833, Nov. 30, 1990]

Subpart B-Nutrition Labeling
Source: 58 FR 664, Jan. 6, 1993, unless otherwise noted.

§317.300 Nutrition labeling of meat or meat food products.

(a) Nutrition labeling shall be provided for all meat or meat food products intended for human consumption and offered for sale, except single-ingredient, raw products, in accordance with the requirements of §317.309; except as exempted under §317.400 of this Subpart.

(b) Nutrition labeling may be provided for single-ingredient, raw meat or meat food products in accordance with the requirements of §§317.309 and 317.345. Significant participation in voluntary nutrition labeling shall be measured by the Agency in accordance with §§317.343 and 317.344 of this Subpart.

[58 FR 664, as amended at 60 FR 176, Jan. 3, 1995]

§317.301 [Reserved]

§317.302 Location of nutrition information.

9 CFR §317.309

(a) Nutrition information on a label of a packaged meat or meat food product shall appear on the label's principal display panel or on the information panel, except as provided in paragraphs (b) and (c) of this section.
(b) Nutrition information for gift packs may be shown at a location other than on the product label, provided that the labels for these products bear no nutrition claim. In lieu of on the product label, nutrition information may be provided by alternate means such as product label inserts.
(c) Meat or meat food products in packages that have a total surface area available to bear labeling greater than 40 square inches but whose principal display panel and information panel do not provide sufficient space to accommodate all required information may use any alternate panel that can be readily seen by consumers for the nutrition information. In determining the sufficiency of available space for the nutrition information, the space needed for vignettes, designs, and other nonmandatory label information on the principal display panel may be considered.
[58 FR 664, Jan. 6, 1993, as amended at 59 FR 40213, Aug. 8, 1994; 60 FR 176, Jan. 3, 1995]

§§317.303-317.307 [Reserved]

§317.308 Labeling of meat or meat food products with number of servings.

The label of any package of a meat or meat food product that bears a representation as to the number of servings contained in such package shall meet the requirements of §317.2(h)(10).
[60 FR 173, as revised on page 176, Jan. 3, 1995]

§317.309 Nutrition label content.

(a) All nutrient and food component quantities shall be declared in relation to a serving as defined in this section.
(b)
(1) The term "serving" or "serving size" means an amount of food customarily consumed per eating occasion by persons 4 years of age or older, which is expressed in a common household measure that is appropriate to the product. When the product is specially formulated or processed for use by infants or by toddlers, a serving or serving size means an amount of food customarily consumed per eating occasion by infants up to 12 months of age or by children 1 through 3 years of age, respectively.
(2) Except as provided in paragraphs (b)(8), (b)(12), and (b)(14) of this section and for products that are intended for weight control and are available only through a weight-control or weight-maintenance program, serving size declared on a product label shall be determined from the "Reference Amounts Customarily Consumed Per Eating Occasion-- General Food Supply" (Reference Amount(s)) that appear in §317.312(b) using the procedures described in this paragraph (b). For products that are both intended for weight control and available only through a weight-control program, a manufacturer may determine the serving size that is consistent with the meal plan of the program. Such products must bear a statement, "for sale only through the _____ program" (fill in the blank with the name of the appropriate weight- control program, e.g., Smith's Weight Control), on the principal display panel. However, the Reference Amounts in §317.312(b) shall be used for purposes of evaluating whether weight-control products that are available only through a weight-control program qualify for nutrition claims.
(3) The declaration of nutrient and food component content shall be on the basis of the product "as packaged" for all products, except that single-ingredient, raw products may be declared on the basis of the product "as consumed" as set forth in §317.345(a)(1). In addition to the required declaration on the basis of "as packaged" for products other than single-ingredient, raw products, the declaration may also be made on the basis of "as consumed," provided that preparation and cooking instructions are clearly stated.
(4) For products in discrete units (e.g., hot dogs, and individually packaged products within a multi-serving package), and for products which consist of two or more foods packaged and presented to be consumed together where the ingredient represented as the main ingredient is in discrete units (e.g., beef fritters and barbecue sauce), the serving size shall be declared as follows:

9 CFR §317.309

(i) If a unit weighs 50 percent or less of the Reference Amount, the serving size shall be the number of whole units that most closely approximates the Reference Amount for the product category.

(ii) If a unit weighs more than 50 percent but less than 67 percent of the Reference Amount, the manufacturer may declare one unit or two units as the serving size.

(iii) If a unit weighs 67 percent or more but less than 200 percent of the Reference Amount, the serving size shall be one unit.

(iv) If a unit weighs 200 percent or more of the Reference Amount, the manufacturer may declare one unit as the serving size if the whole unit can reasonably be consumed at a single eating occasion.

(v) For products that have Reference Amounts of 100 grams (or milliliter) or larger and are individual units within a multi-serving package, if a unit contains more than 150 percent but less than 200 percent of the Reference Amount, the manufacturer may decide whether to declare the individual unit as 1 or 2 servings.

(vi) For products which consist of two or more foods packaged and presented to be consumed together where the ingredient represented as the main ingredient is in discrete units (e.g., beef fritters and barbecue sauce), the serving size may be the number of discrete units represented as the main ingredient plus proportioned minor ingredients used to make the Reference Amount for the combined product as determined in §317.312(c).

(vii) For packages containing several individual single-serving containers, each of which is labeled with all required information including nutrition labeling as specified in this section (i.e., are labeled appropriately for individual sale as single-serving containers), the serving size shall be 1 unit.

(5) For products in large discrete units that are usually divided for consumption (e.g., pizza), for unprepared products where the entire contents of the package is used to prepare large discrete units that are usually divided for consumption (e.g. pizza kit), and for products which consist of two or more foods packaged and presented to be consumed together where the ingredient represented as the main ingredient is a large discrete unit usually divided for consumption, the serving size shall be the fractional slice of the ready-to-eat product (e.g., 1/8 quiche, 1/4 pizza) that most closely approximates the Reference Amount for the product category. The serving size may be the fraction of the package used to make the Reference Amount for the unprepared product determined in §317.312(d) or the fraction of the large discrete unit represented as the main ingredient plus proportioned minor ingredients used to make the Reference Amount of the combined product determined in §317.312(c). In expressing the fractional slice, manufacturers shall use 1/2, 1/3, 1/4, 1/ 5, 1/6, or smaller fractions that can be generated by further division by 2 or 3.

(6) For nondiscrete bulk products (e.g., whole roast beef, marinated beef tenderloin, large can of chili), and for products which consist of two or more foods packaged and presented to be consumed together where the ingredient represented as the main ingredient is a bulk product (e.g., roast beef and gravy), the serving size shall be the amount in household measure that most closely approximates the Reference Amount for the product category and may be the amount of the bulk product represented as the main ingredient plus proportioned minor ingredients used to make the Reference Amount for the combined product determined in §317.312(c).

(7) For labeling purposes, the term "common household measure" or "common household unit" means cup, tablespoon, teaspoon, piece, slice, fraction (e.g., 1/4 pizza), ounce (oz), or other common household equipment used to package food products (e.g., jar or tray). In expressing serving size in household measures, except as specified in paragraphs (b)(7)(iv), (v), and (vi) of this section, the following rules shall be used:

(i) Cups, tablespoons, or teaspoons shall be used wherever possible and appropriate. Cups shall be expressed in 1/4- or 1/3-cup increments, tablespoons in whole number of tablespoons for quantities less than 1/4 cup but greater than or equal to 2 tablespoons (tbsp), 1, 1 1/3, 1 1/2, or 1 2/3 tbsp for quantities less than 2 tbsp but greater than or equal to 1 tbsp, and teaspoons in whole number of teaspoons for quantities less than 1 tbsp but greater than or equal to 1 teaspoon (tsp), and in 1/4-tsp increments for quantities less than 1 tsp.

(ii) If cups, tablespoons or teaspoons are not applicable, units such as piece, slice, tray, jar, and fraction shall be used.

(iii) If cups, tablespoons and teaspoons, or units such as piece, slice, tray, jar, or fraction are not applicable, ounces may be used. Ounce measurements shall be expressed in 0.5-ounce increments most closely approximating the Reference Amount with rounding indicated by the use of the term "about" (e.g., about 2.5 ounces).

(iv) A description of the individual container or package shall be used for single-serving containers and meal-type products and for individually packaged products within multi-serving containers (e.g., can, box, package, meal, or dinner). A description of the individual unit shall be used for other products in discrete units (e.g., chop, slice, link, or patty).

(v) For unprepared products where the entire contents of the package is used to prepare large discrete units that are usually divided for consumption (e.g., pizza kit), the fraction or portion of the package may be used.

(vi) For products that consist of two or more distinct ingredients or components packaged and presented to be consumed together (e.g., ham with a glaze packet), the nutrition information may be declared for each component or as a composite. The serving size may be provided in accordance with the provisions of paragraphs (b)(4), (b)(5), and (b)(6) of this section.

(vii) For nutrition labeling purposes, a teaspoon means 5 milliliters (mL), a tablespoon means 15 mL, a cup means 240 mL, and 1 oz in weight means 28 grams (g).

(viii) When a serving size, determined from the Reference Amount in §317.312(b) and the procedures described in this section, falls exactly half way between two serving sizes (e.g., 2.5 tbsp), manufacturers shall round the serving size up to the next incremental size.

(8) A product that is packaged and sold individually and that contains less than 200 percent of the applicable Reference Amount shall be considered to be a single-serving container, and the entire content of the product shall be labeled as one serving, except for products that have Reference Amounts of 100 g (or mL) or larger, manufacturers may decide whether a package that contains more than 150 percent but less than 200 percent of the Reference Amount is 1 or 2 servings. Packages sold individually that contain 200 percent or more of the applicable Reference Amount may be labeled as a single-serving if the entire content of the package can reasonably be consumed at a single- eating occasion.

(9) A label statement regarding a serving shall be the serving size expressed in common household measures as set forth in paragraphs (b)(2) through (b)(8) of this section and shall be followed by the equivalent metric quantity in parenthesis (fluids in milliliters and all other foods in grams), except for single-serving containers.

(i) For a single-serving container, the parenthetical metric quantity, which will be presented as part of the net weight statement on the principal display panel, is not required except where nutrition information is required on a drained weight basis according to paragraph (b)(11) of this section. However, if a manufacturer voluntarily provides the metric quantity on products that can be sold as single-servings, then the numerical value provided as part of the serving size declaration must be identical to the metric quantity declaration provided as part of the net quantity of contents statement.

(ii) The gram or milliliter quantity equivalent to the household measure should be rounded to the nearest whole number except for quantities that are less than 5 g (mL). The gram (mL) quantity between 2 and 5 g (mL) should be rounded to the nearest 0.5 g (mL) and the g (mL) quantity less than 2 g (mL) should be expressed in 0.1-g (mL) increments.

(iii) In addition, serving size may be declared in ounce, in parenthesis, following the metric measure separated by a slash where other common household measures are used as the primary unit for serving size, e.g., 1 slice (28 g/1 oz) for sliced bologna. The ounce quantity equivalent to the metric quantity should be expressed in 0.1-oz increments.

(iv) If a manufacturer elects to use abbreviations for units, the following abbreviations shall be used: tbsp for tablespoon, tsp for teaspoon, g for gram, mL for milliliter, and oz for ounce.

(10) Determination of the number of servings per container shall be based on the serving size of the product determined by following the procedures described in this section.

(i) The number of servings shall be rounded to the nearest whole number except for the number of servings between 2 and 5 servings and random weight products. The number of servings between 2 and 5 servings shall be rounded to the nearest 0.5 serving. Rounding should be indicated by the use of the term "about" (e.g., about 2 servings; about 3.5 servings).

(ii) When the serving size is required to be expressed on a drained solids basis and the number of servings varies because of a natural variation in unit size (e.g., pickled pigs feet), the manufacturer may state the typical number of servings per container (e.g., usually 5 servings).

(iii) For random weight products, a manufacturer may declare "varied" for the number of servings per container provided the nutrition information is based on the Reference Amount expressed in ounces. The manufacturer may provide the typical number of servings in parenthesis following the "varied" statement (e.g., varied (approximately 8 servings per pound).

(iv) For packages containing several individual single-serving containers, each of which is labeled with all required information including nutrition labeling as specified in this section (i.e., are labeled appropriately for

9 CFR §317.309

individual sale as single-serving containers), the number of servings shall be the number of individual packages within the total package.

(v) For packages containing several individually packaged multi- serving units, the number of servings shall be determined by multiplying the number of individual multi-serving units in the total package by the number of servings in each individual unit.

(11) The declaration of nutrient and food component content shall be on the basis of product as packaged or purchased with the exception of products that are packed or canned in water, brine, or oil but whose liquid packing medium is not customarily consumed. Declaration of the nutrient and food component content of products that are packed in liquid which is not customarily consumed shall be based on the drained solids.

(12) Serving size for meal-type products as defined in §317.313(l) shall be the entire content (edible portion only) of the package.

(13) Another column of figures may be used to declare the nutrient and food component information in the same format as required by §317.309(e),

(i) Per 100 grams, 100 milliliters, or 1 ounce of the product as packaged or purchased.

(ii) Per one unit if the serving size of a product in discrete units in a multi-serving container is more than one unit.

(14) If a product consists of assortments of meat or meat food products (e.g., variety packs) in the same package, nutrient content shall be expressed on the entire package contents or on each individual product.

(15) If a product is commonly combined with other ingredients or is cooked or otherwise prepared before eating, and directions for such combination or preparations are provided, another column of figures may be used to declare the nutrient contents on the basis of the product as consumed for the product alone (e.g., a cream soup mix may be labeled with one set of Daily Values for the dry mix (per serving), and another set for the serving of the final soup when prepared (e.g., per serving of cream soup mix and 1 cup of vitamin D fortified whole milk)): Provided, That the type and quantity of the other ingredients to be added to the product by the user and the specific method of cooking and other preparation shall be specified prominently on the label.

(c) The declaration of nutrition information on the label or in labeling of a meat or meat food product shall contain information about the level of the following nutrients, except for those nutrients whose inclusion, and the declaration of amounts, is voluntary as set forth in this paragraph. No nutrients or food components other than those listed in this paragraph as either mandatory or voluntary may be included within the nutrition label. Except as provided for in paragraph (f) or (g) of this section, nutrient information shall be presented using the nutrient names specified and in the following order in the formats specified in paragraph (d) or (e) of this section.

(1) "Calories, total," "Total calories," or "Calories": A statement of the caloric content per serving, expressed to the nearest 5-calorie increment up to and including 50 calories, and 10-calorie increment above 50 calories, except that amounts less than 5 calories may be expressed as zero. Energy content per serving may also be expressed in kilojoule units, added in parenthesis immediately following the statement of the caloric content.

(i) Caloric content may be calculated by the following methods. Where either specific or general food factors are used, the factors shall be applied to the actual amount (i.e., before rounding) of food components (e.g., fat, carbohydrate, protein, or ingredients with specific food factors) present per serving.

(A) Using specific Atwater factors (i.e., the Atwater method) given in Table 13, page 25, "Energy Value of Foods--Basis and Derivation," by A. L. Merrill and B. K. Watt, United States Department of Agriculture (USDA), Agriculture Handbook No. 74 (Slightly revised February 1973), which is incorporated by reference. Table 13 of the "Energy Value of Foods--Basis and Derivation," Agriculture Handbook No. 74 is incorporated as it exists on the date of approval. This incorporation by reference was approved by the Director of the Federal Register in accordance with 5 U.S.C. 552(a) and 1 CFR part 51. It is available for inspection at the Office of the Federal Register, suite 700, 800 North Capitol Street, NW., Washington, DC, or at the office of the FSIS Docket Clerk, Room 3171, South Building, 14th and Independence Avenue, SW., Washington, DC. Copies of the incorporation by reference are available from the Product Assessment Division, Regulatory Programs, Food Safety and Inspection Service, U.S. Department of Agriculture, Room 329, West End Court Building, Washington, DC 20250- 3700;

(B) Using the general factors of 4, 4, and 9 calories per gram for protein, total carbohydrate, and total fat, respectively, as described in USDA's Agriculture Handbook No. 74 (Slightly revised February 1973), pages 9-11, which is incorporated by reference. Pages 9-11, Agriculture Handbook No. 74 is incorporated as it exists on the date of approval. This incorporation by reference was approved by the Director of the Federal Register in

9 CFR §317.309

accordance with 5 U.S.C. 552(a) and 1 CFR part 51. (The availability of this incorporation by reference is given in paragraph (c)(1)(i)(A) of this section.);

(C) Using the general factors of 4, 4, and 9 calories per gram for protein, total carbohydrate less the amount of insoluble dietary fiber, and total fat, respectively, as described in USDA's Agriculture Handbook No. 74 (Slightly revised February 1973), pages 9-11, which is incorporated by reference in accordance with 5 U.S.C. 552(a) and 1 CFR part 51. (The availability of this incorporation by reference is given in paragraph (c)(1)(i)(A) of this section.); or

(D) Using data for specific food factors for particular foods or ingredients approved by the Food and Drug Administration (FDA) and provided in parts 172 or 184 of 21 CFR, or by other means, as appropriate.

(ii) "Calories from fat": A statement of the caloric content derived from total fat as defined in paragraph (c)(2) of this section per serving, expressed to the nearest 5-calorie increment, up to and including 50 calories, and the nearest 10-calorie increment above 50 calories, except that label declaration of "calories from fat" is not required on products that contain less than 0.5 gram of fat per serving and amounts less than 5 calories may be expressed as zero. This statement shall be declared as provided in paragraph (d)(5) of this section.

(iii) "Calories from saturated fat" or "Calories from saturated" (VOLUNTARY): A statement of the caloric content derived from saturated fat as defined in paragraph (c)(2)(i) of this section per serving may be declared voluntarily, expressed to the nearest 5- calorie increment, up to and including 50 calories, and the nearest 10-calorie increment above 50 calories, except that amounts less than 5 calories may be expressed as zero. This statement shall be indented under the statement of calories from fat as provided in paragraph (d)(5) of this section.

(2) "Fat, total" or "Total fat": A statement of the number of grams of total fat per serving defined as total lipid fatty acids and expressed as triglycerides. Amounts shall be expressed to the nearest 0.5 (1/2)-gram increment below 5 grams and to the nearest gram increment above 5 grams. If the serving contains less than 0.5 gram, the content shall be expressed as zero.

(i) "Saturated fat" or "Saturated": A statement of the number of grams of saturated fat per serving defined as the sum of all fatty acids containing no double bonds, except that label declaration of saturated fat content information is not required for products that contain less than 0.5 gram of total fat per serving if no claims are made about fat or cholesterol content, and if "calories from saturated fat" is not declared. Saturated fat content shall be indented and expressed as grams per serving to the nearest 0.5 (1/2)-gram increment below 5 grams and to the nearest gram increment above 5 grams. If the serving contains less than 0.5 gram, the content shall be expressed as zero.

(A) "Stearic Acid" (VOLUNTARY): A statement of the number of grams of stearic acid per serving may be declared voluntarily, except that when a claim is made about stearic acid, label declaration shall be required. Stearic acid content shall be indented under saturated fat and expressed to the nearest 0.5 (1/2)-gram increment below 5 grams and the nearest gram increment above 5 grams. If the serving contains less than 0.5 gram, the content shall be expressed as zero.

(B) [Reserved]

(ii) "Polyunsaturated fat" or "Polyunsaturated" (VOLUNTARY): A statement of the number of grams of polyunsaturated fat per serving defined as cis,cis-methylene-interrupted polyunsaturated fatty acids may be declared voluntarily, except that when monounsaturated fat is declared, or when a claim about fatty acids or cholesterol is made on the label or in labeling of a product other than one that meets the criteria in §317.362(b)(1) for a claim for "fat free," label declaration of polyunsaturated fat is required. Polyunsaturated fat content shall be indented and expressed as grams per serving to the nearest 0.5 (1/2)-gram increment below 5 grams and to the nearest gram increment above 5 grams. If the serving contains less than 0.5 gram, the content shall be expressed as zero.

(iii) "Monounsaturated fat" or "Monounsaturated" (VOLUNTARY): A statement of the number of grams of monounsaturated fat per serving defined as cis-monounsaturated fatty acids may be declared voluntarily, except that when polyunsaturated fat is declared, or when a claim about fatty acids or cholesterol is made on the label or in labeling of a product other than one that meets the criteria in §317.362(b)(1) for a claim for "fat free," label declaration of monounsaturated fat is required. Monounsaturated fat content shall be indented and expressed as grams per serving to the nearest 0.5 (1/2)-gram increment below 5 grams and to the nearest gram increment above 5 grams. If the serving contains less than 0.5 gram, the content shall be expressed as zero.

(3) "Cholesterol": A statement of the cholesterol content per serving expressed in milligrams to the nearest

9 CFR §317.309

5-milligram increment, except that label declaration of cholesterol information is not required for products that contain less than 2 milligrams of cholesterol per serving and make no claim about fat, fatty acids, or cholesterol content, or such products may state the cholesterol content as zero. If the product contains 2 to 5 milligrams of cholesterol per serving, the content may be stated as "less than 5 milligrams."

(4) "Sodium": A statement of the number of milligrams of sodium per serving expressed as zero when the serving contains less than 5 milligrams of sodium, to the nearest 5-milligram increment when the serving contains 5 to 140 milligrams of sodium, and to the nearest 10- milligram increment when the serving contains greater than 140 milligrams.

(5) "Potassium" (VOLUNTARY): A statement of the number of milligrams of potassium per serving may be declared voluntarily, except that when a claim is made about potassium content, label declaration shall be required. Potassium content shall be expressed as zero when the serving contains less than 5 milligrams of potassium, to the nearest 5-milligram increment when the serving contains 5 to 140 milligrams of potassium, and to the nearest 10-milligram increment when the serving contains greater than 140 milligrams.

(6) "Carbohydrate, total" or "Total carbohydrate": A statement of the number of grams of total carbohydrate per serving expressed to the nearest gram, except that if a serving contains less than 1 gram, the statement "Contains less than 1 gram" or "less than 1 gram" may be used as an alternative, or, if the serving contains less than 0.5 gram, the content may be expressed as zero. Total carbohydrate content shall be calculated by subtraction of the sum of the crude protein, total fat, moisture, and ash from the total weight of the product. This calculation method is described in USDA's Agriculture Handbook No. 74 (Slightly revised February 1973), pages 2 and 3, which is incorporated by reference. Pages 2 and 3, Agriculture Handbook No. 74 is incorporated as it exists on the date of approval. This incorporation by reference was approved by the Director of the Federal Register in accordance with 5 U.S.C. 552(a) and 1 CFR part 51. (The availability of this incorporation by reference is given in paragraph (c)(1)(i)(A) of this section.)

(i) "Dietary fiber": A statement of the number of grams of total dietary fiber per serving, indented and expressed to the nearest gram, except that if a serving contains less than 1 gram, declaration of dietary fiber is not required, or, alternatively, the statement "Contains less than 1 gram" or "less than 1 gram" may be used, and if the serving contains less than 0.5 gram, the content may be expressed as zero.

(A) "Soluble fiber" (VOLUNTARY): A statement of the number of grams of soluble dietary fiber per serving may be declared voluntarily except when a claim is made on the label or in labeling about soluble fiber, label declaration shall be required. Soluble fiber content shall be indented under dietary fiber and expressed to the nearest gram, except that if a serving contains less than 1 gram, the statement "Contains less than 1 gram" or "less than 1 gram" may be used as an alternative, and if the serving contains less than 0.5 gram, the content may be expressed as zero.

(B) "Insoluble fiber" (VOLUNTARY): A statement of the number of grams of insoluble dietary fiber per serving may be declared voluntarily except when a claim is made on the label or in labeling about insoluble fiber, label declaration shall be required. Insoluble fiber content shall be indented under dietary fiber and expressed to the nearest gram, except that if a serving contains less than 1 gram, the statement "Contains less than 1 gram" or "less than 1 gram" may be used as an alternative, and if the serving contains less than 0.5 gram, the content may be expressed as zero.

(ii) "Sugars": A statement of the number of grams of sugars per serving, except that label declaration of sugars content is not required for products that contain less than 1 gram of sugars per serving if no claims are made about sweeteners, sugars, or sugar alcohol content. Sugars shall be defined as the sum of all free mono- and disaccharides (such as glucose, fructose, lactose, and sucrose). Sugars content shall be indented and expressed to the nearest gram, except that if a serving contains less than 1 gram, the statement "Contains less than 1 gram" or "less than 1 gram" may be used as an alternative, and if the serving contains less than 0.5 gram, the content may be expressed as zero.

(iii) "Sugar alcohol" (VOLUNTARY): A statement of the number of grams of sugar alcohols per serving may be declared voluntarily on the label, except that when a claim is made on the label or in labeling about sugar alcohol or sugars when sugar alcohols are present in the product, sugar alcohol content shall be declared. For nutrition labeling purposes, sugar alcohols are defined as the sum of saccharide derivatives in which a hydroxyl group replaces a ketone or aldehyde group and whose use in the food is listed by FDA (e.g., mannitol or xylitol) or is generally recognized as safe (e.g., sorbitol). In lieu of the term "sugar alcohol," the name of the specific sugar alcohol (e.g., "xylitol") present in the product may be used in the nutrition label, provided [[Page 180]]

9 CFR §317.309

that only one sugar alcohol is present in the product. Sugar alcohol content shall be indented and expressed to the nearest gram, except that if a serving contains less than 1 gram, the statement "Contains less then 1 gram" or "less than 1 gram" may be used as an alternative, and if the serving contains less than 0.5 gram, the content may be expressed as zero.

(iv) "Other carbohydrate" (VOLUNTARY): A statement of the number of grams of other carbohydrate per serving may be declared voluntarily. Other carbohydrate shall be defined as the difference between total carbohydrate and the sum of dietary fiber, sugars, and sugar alcohol, except that if sugar alcohol is not declared (even if present), it shall be defined as the difference between total carbohydrate and the sum of dietary fiber and sugars. Other carbohydrate content shall be indented and expressed to the nearest gram, except that if a serving contains less than 1 gram, the statement "Contains less than 1 gram" or "less than 1 gram" may be used as an alternative, and if the serving contains less than 0.5 gram, the content may be expressed as zero.

(7) "Protein": A statement of the number of grams of protein per serving expressed to the nearest gram, except that if a serving contains less than 1 gram, the statement "Contains less than 1 gram" or "less than 1 gram" may be used as an alternative, and if the serving contains less than 0.5 gram, the content may be expressed as zero. When the protein in products represented or purported to be for adults and children 4 or more years of age has a protein quality value that is a protein digestibility-corrected amino acid score of less than 20 expressed as a percent, or when the protein in a product represented or purported to be for children greater than 1 but less than 4 years of age has a protein quality value that is a protein digestibility- corrected amino acid score of less than 40 expressed as a percent, either of the following shall be placed adjacent to the declaration of protein content by weight: The statement "not a significant source of protein," or a listing aligned under the column headed "Percent Daily Value" of the corrected amount of protein per serving, as determined in paragraph (c)(7)(ii) of this section, calculated as a percentage of the Daily Reference Value (DRV) or Reference Daily Intake (RDI), as appropriate, for protein and expressed as percent of Daily Value. When the protein quality in a product as measured by the Protein Efficiency Ratio (PER) is less than 40 percent of the reference standard (casein) for a product represented or purported to be for infants, the statement "not a significant source of protein" shall be placed adjacent to the declaration of protein content. Protein content may be calculated on the basis of the factor of 6.25 times the nitrogen content of the food as determined by appropriate methods of analysis in accordance with §317.309(h), except when the procedure for a specific food requires another factor.

(i) A statement of the corrected amount of protein per serving, as determined in paragraph (c)(7)(ii) of this section, calculated as a percentage of the RDI or DRV for protein, as appropriate, and expressed as percent of Daily Value, may be placed on the label, except that such a statement shall be given if a protein claim is made for the product, or if the product is represented or purported to be for infants or children under 4 years of age. When such a declaration is provided, it shall be placed on the label adjacent to the statement of grams of protein and aligned under the column headed "Percent Daily Value," and expressed to the nearest whole percent. However, the percentage of the RDI for protein shall not be declared if the product is represented or purported to be for infants and the protein quality value is less than 40 percent of the reference standard.

(ii) The corrected amount of protein (grams) per serving for products represented or purported to be for adults and children 1 or more years of age is equal to the actual amount of protein (grams) per serving multiplied by the amino acid score corrected for protein digestibility. If the corrected score is above 1.00, then it shall be set at 1.00. The protein digestibility-corrected amino acid score shall be determined by methods given in sections 5.4.1, 7.2.1, and 8 in "Protein Quality Evaluation, Report of the Joint FAO/WHO Expert Consultation on Protein Quality Evaluation," Rome, 1990, which is incorporated by reference. Sections 5.4.1, 7.2.1, and 8 of the "Report of the Joint FAO/WHO Expert Consultation on Protein Quality Evaluation," as published by the Food and Agriculture Organization of the United Nations/World Health Organization, is incorporated as it exists on the date of approval. This incorporation by reference was approved by the Director of the Federal Register in accordance with 5 U.S.C. 552(a) and 1 CFR part 51. It is available for inspection at the Office of the Federal Register, suite 700, 800 North Capitol Street, NW., Washington, DC, or at the office of the FSIS Docket Clerk, Room 3171, South Building, 14th and Independence Avenue, SW., Washington, DC. Copies of the incorporation by reference are available from the Product Assessment Division, Regulatory Programs, Food Safety and Inspection Service, U.S. Department of Agriculture, Room 329, West End Court Building, Washington, DC 20250-3700. For products represented or purported to be for infants, the corrected amount of protein (grams) per serving is equal to the actual amount of protein (grams) per serving multiplied by the relative protein quality value. The relative protein quality value shall be determined by dividing the subject product's

9 CFR §317.309

protein PER value by the PER value for casein. If the relative protein value is above 1.00, it shall be set at 1.00.
(iii) For the purpose of labeling with a percent of the DRV or RDI, a value of 50 grams of protein shall be the DRV for adults and children 4 or more years of age, and the RDI for protein for children less than 4 years of age, infants, pregnant women, and lactating women shall be 16 grams, 14 grams, 60 grams, and 65 grams, respectively.

(8) Vitamins and minerals: A statement of the amount per serving of the vitamins and minerals as described in this paragraph, calculated as a percent of the RDI and expressed as percent of Daily Value. (i) For purposes of declaration of percent of Daily Value as provided for in paragraphs (d) through (g) of this section, products represented or purported to be for use by infants, children less than 4 years of age, pregnant women, or lactating women shall use the RDI's that are specified for the intended group. For products represented or purported to be for use by both infants and children under 4 years of age, the percent of Daily Value shall be presented by separate declarations according to paragraph (e) of this section based on the RDI values for infants from birth to 12 months of age and for children under 4 years of age. Similarly, the percent of Daily Value based on both the RDI values for pregnant women and for lactating women shall be declared separately on products represented or purported to be for use by both pregnant and lactating women. When such dual declaration is used on any label, it shall be included in all labeling, and equal prominence shall be given to both values in all such labeling. All other products shall use the RDI for adults and children 4 or more years of age.

(ii) The declaration of vitamins and minerals as a percent of the RDI shall include vitamin A, vitamin C, calcium, and iron, in that order, and shall include any of the other vitamins and minerals listed in paragraph (c)(8)(iv) of this section when they are added, or when a claim is made about them. Other vitamins and minerals need not be declared if neither the nutrient nor the component is otherwise referred to on the label or in labeling or advertising and the vitamins and minerals are:

(A) Required or permitted in a standardized food (e.g., thiamin, riboflavin, and niacin in enriched flour) and that standardized food is included as an ingredient (i.e., component) in another product; or

(B) Included in a product solely for technological purposes and declared only in the ingredients statement. The declaration may also include any of the other vitamins and minerals listed in paragraph (c)(8)(iv) of this section when they are naturally occurring in the food. The additional vitamins and minerals shall be listed in the order established in paragraph (c)(8)(iv) of this section.

(iii) The percentages for vitamins and minerals shall be expressed to the nearest 2-percent increment up to and including the 10-percent level, the nearest 5-percent increment above 10 percent and up to and including the 50-percent level, and the nearest 10-percent increment above the 50-percent level. Amounts of vitamins and minerals present at less than 2 percent of the RDI are not required to be declared in nutrition labeling but may be declared by a zero or by the use of an asterisk (or other symbol) that refers to another asterisk (or symbol) that is placed at the bottom of the table and that is followed by the statement "Contains less than 2 percent of the Daily Value of this (these) nutrient (nutrients)." Alternatively, if vitamin A, vitamin C, calcium, or iron is present in amounts less than 2 percent of the RDI, label declaration of the nutrient(s) is not required if the statement "Not a significant source of _____ (listing the vitamins or minerals omitted)" is placed at the bottom of the table of nutrient values.

(iv) The following RDI's and nomenclature are established for the following vitamins and minerals which are essential in human nutrition:

Vitamin A, 5,000 International Units
Vitamin C, 60 milligrams
Calcium, 1.0 gram
Iron, 18 milligrams
Vitamin D, 400 International Units
Vitamin E, 30 International Units
Thiamin, 1.5 milligrams
Riboflavin, 1.7 milligrams
Niacin, 20 milligrams
Vitamin B_6, 2.0 milligrams
Folate, 0.4 milligram
Vitamin B_{12}, 6 micrograms

9 CFR §317.309

Biotin, 0.3 milligram
Pantothenic acid, 10 milligrams
Phosphorus, 1.0 gram
Iodine, 150 micrograms
Magnesium, 400 milligrams
Zinc, 15 milligrams
Copper, 2.0 milligrams

(v) The following synonyms may be added in parenthesis immediately following the name of the nutrient or dietary component:

Vitamin C--Ascorbic acid
Thiamin—Vitamin B_1
Riboflavin—Vitamin B_2
Folate—Folacin
Calories—Energy

(vi) A statement of the percent of vitamin A that is present as beta-carotene may be declared voluntarily. When the vitamins and minerals are listed in a single column, the statement shall be indented under the information on vitamin A. When vitamins and minerals are arrayed horizontally, the statement of percent shall be presented in parenthesis following the declaration of vitamin A and the percent of Daily Value of vitamin A in the product (e.g., "Percent Daily Value: Vitamin A 50 (90 percent as beta-carotene)"). When declared, the percentages shall be expressed in the same increments as are provided for vitamins and minerals in paragraph (c)(8)(iii) of this section.

(9) For the purpose of labeling with a percent of the DRV, the following DRV's are established for the following food components based on the reference caloric intake of 2,000 calories:

Food component	Unit of measurement	DRV
Fat	gram (g)	65
Saturated fatty acids	do	20
Cholesterol	milligrams (mg)	300
Total carbohydrate	grams (g)	300
Fiber	do	25
Sodium	milligrams (mg)	2,400
Potassium	do	3,500
Protein	grams (g)	50

(d)

(1) Nutrient information specified in paragraph (c) of this section shall be presented on products in the following format, except on products on which dual columns of nutrition information are declared as provided for in paragraph (e) of this section, on those products on which the simplified format is permitted to be used as provided for in paragraph (f) of this section, on products for infants and children less than 4 years of age as provided for in §317.400(c), and on products in packages that have a total surface area available to bear labeling

9 CFR §317.309

of 40 or less square inches as provided for in paragraph (g) of this section.

(i) The nutrition information shall be set off in a box by use of hairlines and shall be all black or one color type, printed on a white or other neutral contrasting background whenever practical.

(ii) All information within the nutrition label shall utilize:

(A) A single easy-to-read type style,

(B) Upper and lower case letters,

(C) At least one point leading (i.e., space between two lines of text) except that at least four points leading shall be utilized for the information required by paragraphs (d)(7) and (d)(8) of this section, and

(D) Letters should never touch.

(iii) Information required in paragraphs (d)(3), (d)(5), (d)(7), and (d)(8) of this section shall be in type size no smaller than 8 point. Except for the heading "Nutrition Facts," the information required in paragraphs (d)(4), (d)(6), and (d)(9) of this section and all other information contained within the nutrition label shall be in type size no smaller than 6 point. When provided, the information described in paragraph (d)(10) of this section shall also be in type no smaller than 6 point.

(iv) The headings required by paragraphs (d)(2), (d)(4), and (d)(6) of this section (i.e., "Nutrition Facts," "Amount Per Serving," and "% Daily Value*"), the names of all nutrients that are not indented according to requirements of paragraph (c) of this section (i.e., Calories, Total fat, Cholesterol, Sodium, Potassium, Total carbohydrate, and Protein), and the percentage amounts required by paragraph (d)(7)(ii) of this section shall be highlighted by bold or extra bold type or other highlighting (reverse printing is not permitted as a form of highlighting) that prominently distinguishes it from other information. No other information shall be highlighted.

(v) A hairline rule that is centered between the lines of text shall separate "Amount Per Serving" from the calorie statements required in paragraph (d)(5) of this section and shall separate each nutrient and its corresponding percent of Daily Value required in paragraphs (d)(7)(i) and (d)(7)(ii) of this section from the nutrient and percent of Daily Value above and below it.

(2) The information shall be presented under the identifying heading of "Nutrition Facts" which shall be set in a type size larger than all other print size in the nutrition label and, except for labels presented according to the format provided for in paragraph (d)(11) of this section, unless impractical, shall be set the full width of the information provided under paragraph (d)(7) of this section.

(3) Information on serving size shall immediately follow the heading. Such information shall include:

(i) "Serving Size": A statement of the serving size as specified in paragraph (b)(9) of this section.

(ii) "Servings Per Container": The number of servings per container, except that this statement is not required on single-serving containers as defined in paragraph (b)(8) of this section.

(4) A subheading "Amount Per Serving" shall be separated from serving size information by a bar.

(5) Information on calories shall immediately follow the heading "Amount Per Serving" and shall be declared in one line, leaving sufficient space between the declaration of "Calories" and "Calories from fat" to allow clear differentiation, or, if "Calories from saturated fat" is declared, in a column with total "Calories" at the top, followed by "Calories from fat" (indented), and "Calories from saturated fat" (indented).

(6) The column heading "% Daily Value," followed by an asterisk (e.g., "% Daily Value*"), shall be separated from information on calories by a bar. The position of this column heading shall allow for a list of nutrient names and amounts as described in paragraph (d)(7) of this section to be to the left of, and below, this column heading. The column headings "Percent Daily Value," "Percent DV," or "% DV" may be substituted for "% Daily Value."

(7) Except as provided for in paragraph (g) of this section, and except as permitted by §317.400(d)(2), nutrient information for both mandatory and any voluntary nutrients listed in paragraph (c) of this section that are to be declared in the nutrition label, except vitamins and minerals, shall be declared as follows:

(i) The name of each nutrient, as specified in paragraph (c) of this section, shall be given in a column and followed immediately by the quantitative amount by weight for that nutrient appended with a "g" for grams or "mg" for milligrams.

(ii) A listing of the percent of the DRV as established in paragraphs (c)(7)(iii) and (c)(9) of this section shall be given in a column aligned under the heading "% Daily Value" established in paragraph (d)(6) of this section with the percent expressed to the nearest whole percent for each nutrient declared in the column described in paragraph (d)(7)(i) of this section for which a DRV has been established, except that the percent for protein may be omitted as provided in paragraph (c)(7) of this section. The percent shall be calculated by dividing either the amount declared on the label for each nutrient or the actual amount of each nutrient (i.e., before rounding) by the DRV for the nutrient, except that the percent for protein shall be calculated as specified in paragraph (c)(7)(ii) of this section. The numerical value shall be followed by the symbol for percent (i.e., %).

(8) Nutrient information for vitamins and minerals shall be separated from information on other nutrients by a bar and shall be arrayed horizontally (e.g., Vitamin A 4%, Vitamin C 2%, Calcium 15%, Iron 4%) or may be listed in two columns, except that when more than four vitamins and minerals are declared, they may be declared vertically with percentages listed under the column headed "% Daily Value."

(9) A footnote, preceded by an asterisk, shall be placed beneath the list of vitamins and minerals and shall be separated from that list by a hairline.

(i) The footnote shall state: Percent Daily Values are based on a 2,000 calorie diet. Your daily values may be higher or lower depending on your calorie needs.

	Calories:	2,000	2,500
Total fat	Less than	65 g	80 g
Saturated fat	Less than	20 g	25 g
Cholesterol	Less than	300 mg	300 mg
Sodium	Less than	2,400 mg	2,400 mg
Total carbohydrate		300 g	375 g
Dietary fiber		25 g	30 g

(ii) If the percent of Daily Value is given for protein in the Percent of Daily Value column as provided in paragraph (d)(7)(ii) of this section, protein shall be listed under dietary fiber, and a value of 50 g shall be inserted on the same line in the column headed "2,000" and value of 65 g in the column headed "2,500." (iii) If potassium is declared in the column described in paragraph (d)(7)(i) of this section, potassium shall be listed under sodium and the DRV established in paragraph (c)(9) of this section shall be inserted on the same line in the numeric columns.

(iv) The abbreviations established in paragraph (g)(2) of this section may be used within the footnote.

(10) Caloric conversion information on a per-gram basis for fat, carbohydrate, and protein may be presented beneath the information required in paragraph (d)(9), separated from that information by a hairline. This information may be presented horizontally (i.e., "Calories per gram: Fat 9, Carbohydrate 4, Protein 4") or vertically in columns.

(11) (i) If the space beneath the information on vitamins and minerals is not adequate to accommodate the information required in paragraph (d)(9) of this section, the information required in paragraph (d)(9) may be moved to the right of the column required in paragraph (d)(7)(ii) of this section and set off by a line that distinguishes it and sets it apart from the percent of Daily Value information. The caloric conversion information provided for in paragraph (d)(10) of this section may be presented beneath either side or along the full length of the nutrition label.

(ii) If the space beneath the mandatory declaration of iron is not adequate to accommodate any remaining vitamins and minerals to be declared or the information required in paragraph (d)(9) of this section, the remaining information may be moved to the right and set off by a line that distinguishes it and sets it apart from the percent of Daily Value information given to the left. The caloric conversion information provided for in paragraph (d)(10) of this section may be presented beneath either side or along the full length of the nutrition

9 CFR §317.309

label.

(iii) If there is not sufficient continuous vertical space (i.e., approximately 3 inches) to accommodate the required components of the nutrition label up to and including the mandatory declaration of iron, the nutrition label may be presented in a tabular display in which the footnote required by paragraph (d)(9) of the section is given to the far right of the label, and additional vitamins and minerals beyond the four that are required (i.e., vitamin A, vitamin C, calcium, and iron) are arrayed horizontally following declarations of the required vitamins and minerals.

(12) The following sample label illustrates the provisions of paragraph (d) of this section:

Nutrition Facts
Serving Size 1 cup (228g)
Servings Per Container 2

Amount Per Serving

Calories 260   Calories from Fat 120

	% Daily Value*
Total Fat 13g	20%
Saturated Fat 5g	25%
Cholesterol 30mg	10%
Sodium 660mg	28%
Total Carbohydrate 31g	10%
Dietary Fiber 0g	0%
Sugars 5g	
Protein 5g	

Vitamin A 4%   •   Vitamin C 2%
Calcium 15%   •   Iron 4%

* Percent Daily Values are based on a 2,000 calorie diet. Your daily values may be higher or lower depending on your calorie needs:

	Calories:	2,000	2,500
Total Fat	Less than	65g	80g
Sat Fat	Less than	20g	25g
Cholesterol	Less than	300mg	300mg
Sodium	Less than	2,400mg	2,400mg
Total Carbohydrate		300g	375g
Dietary Fiber		25g	30g

Calories per gram:
Fat 9 • Carbohydrate 4 • Protein 4

(13) (i) Nutrition labeling on the outer label of packages of meat or meat food products that contain two or more products in the same packages (e.g., variety packs) or of packages that are used interchangeably for the same

type of food (e.g., meat salad containers) may use an aggregate display.

(ii) Aggregate displays shall comply with format requirements of paragraph (d) of this section to the maximum extent possible, except that the identity of each food shall be specified to the right of the "Nutrition Facts" title, and both the quantitative amount by weight (i.e., g/mg amounts) and the percent Daily Value for each nutrient shall be listed in separate columns under the name of each food.

(14) When nutrition labeling appears in a second language, the nutrition information may be presented in a separate nutrition label for each language or in one nutrition label with the information in the second language following that in English. Numeric characters that are identical in both languages need not be repeated (e.g., "Protein/ Proteinas 2 g"). All required information must be included in both languages.

(e) Nutrition information may be presented for two or more forms of the same product (e.g., both "raw" and "cooked") or for common combinations of foods as provided for in paragraph (b) of this section, or for different units (e.g., per 100 grams) as provided for in paragraph (b) of this section, or for two or more groups for which RDI's are established (e.g., both infants and children less than 4 years of age) as provided for in paragraph (c)(8)(i) of this section. When such dual labeling is provided, equal prominence shall be given to both sets of values. Information shall be presented in a format consistent with paragraph (d) of this section, except that:

(1) Following the subheading of "Amount Per Serving," there shall be two or more column headings accurately describing the forms of the same product (e.g., "raw" and "roasted"), the combinations of foods, the units, or the RDI groups that are being declared. The column representing the product as packaged and according to the label serving size based on the Reference Amount in §317.312(b) shall be to the left of the numeric columns.

(2) When the dual labeling is presented for two or more forms of the same product, for combinations of foods, or for different units, total calories and calories from fat (and calories from saturated fat, when declared) shall be listed in a column and indented as specified in paragraph (d)(5) of this section with quantitative amounts declared in columns aligned under the column headings set forth in paragraph (e)(1) of this section.

(3) Quantitative information by weight required in paragraph (d)(7)(i) of this section shall be specified for the form of the product as packaged and according to the label serving size based on the Reference Amount in §317.312(b).

(i) Quantitative information by weight may be included for other forms of the product represented by the additional column(s) either immediately adjacent to the required quantitative information by weight for the product as packaged and according to the label serving size based on the Reference Amount in §317.312(b) or as a footnote. (A) If such additional quantitative information is given immediately adjacent to the required quantitative information, it shall be declared for all nutrients listed and placed immediately following and differentiated from the required quantitative information (e.g., separated by a comma). Such information shall not be put in a separate column.

(B) If such additional quantitative information is given in a footnote, it shall be declared in the same order as the nutrients are listed in the nutrition label. The additional quantitative information may state the total nutrient content of the product identified in the second column or the nutrient amounts added to the product as packaged for only those nutrients that are present in different amounts than the amounts declared in the required quantitative information. The footnote shall clearly identify which amounts are declared. Any subcomponents declared shall be listed parenthetically after principal components (e.g., 1/2 cup skim milk contributes an additional 40 calories, 65 mg sodium, 6 g total carbohydrate (6 g sugars), and 4 g protein). (ii) Total fat and its quantitative amount by weight shall be followed by an asterisk (or other symbol) (e.g., "Total fat (2 g)*") referring to another asterisk (or symbol) at the bottom of the nutrition label identifying the form(s) of the product for which quantitative information is presented.

(4) Information required in paragraphs (d)(7)(ii) and (d)(8) of this section shall be presented under the subheading "% DAILY VALUE" and in columns directly under the column headings set forth in paragraph (e)(1) of this section.

9 CFR §317.309

(5) The following sample label illustrates the provisions of paragraph (e) of this section:

Nutrition Facts

Serving Size 1/12 package
(44g, about 1/4 cup dry mix)
Servings Per Container 12

Amount Per Serving	Mix	Baked
Calories	190	280
Calories from Fat	45	140
	% Daily Value**	
Total Fat 5g*	8%	24%
Saturated Fat 2g	10%	13%
Cholesterol 0mg	0%	23%
Sodium 300mg	13%	13%
Total Carbohydrate 34g	11%	11%
Dietary Fiber 0g	0%	0%
Sugars 18g		
Protein 2g		
Vitamin A	0%	0%
Vitamin C	0%	0%
Calcium	6%	8%
Iron	2%	4%

*Amount in Mix
**Percent Daily Values are based on a 2,000 calorie diet. Your daily values may be higher or lower depending on your calorie needs:

	Calories:	2,000	2,500
Total Fat	Less than	65g	80g
Sat Fat	Less than	20g	25g
Cholesterol	Less than	300mg	300mg
Sodium	Less than	2,400mg	2,400mg
Total Carbohydrate		300g	375g
Dietary Fiber		25g	30g

Calories per gram:
Fat 9 • Carbohydrate 4 • Protein 4

(f)
(1) Nutrition information may be presented in a simplified format as set forth herein when any required nutrients, other than the core nutrients (i.e., calories, total fat, sodium, total carbohydrate, and protein), are present in insignificant amounts. An insignificant amount shall be defined as that amount that may be rounded to zero in nutrition labeling, except that for total carbohydrate, dietary fiber, sugars and protein, it shall be an amount less than 1 gram.
(2) The simplified format shall include information on the following nutrients:

9 CFR §317.309

(i) Total calories, total fat, total carbohydrate, sodium, and protein;

(ii) Any of the following that are present in more than insignificant amounts: Calories from fat, saturated fat, cholesterol, dietary fiber, sugars, vitamin A, vitamin C, calcium, and iron; and

(iii) Any vitamins and minerals listed in paragraph (c)(8)(iv) of this section when they are added in fortified or fabricated foods.

(3) Other nutrients that are naturally present in the product in more than insignificant amounts may be voluntarily declared as part of the simplified format.

(4) Any required nutrient, other than a core nutrient, that is present in an insignificant amount may be omitted from the tabular listing, provided that the following statement is included at the bottom of the nutrition label, "Not a significant source of _____." The blank shall be filled in with the appropriate nutrient or food component. Alternatively, amounts of vitamins and minerals present in insignificant amounts may be declared by the use of an asterisk (or symbol) that is placed at the bottom of the table of nutrient values and that is followed by the statement "Contains less than 2 percent of the Daily Value of this (these) nutrient (nutrients)."

(5) Except as provided for in paragraph (g) of this section and in §317.400(c) and (d), nutrient information declared in the simplified format shall be presented in the same manner as specified in paragraphs (d) or (e) of this section, except that the footnote required in paragraph (d)(9) of this section is not required. When the footnote is omitted, an asterisk shall be placed at the bottom of the label followed by the statement "Percent Daily Values are based on a 2,000 calorie diet" and, if the term "Daily Value" is not spelled out in the heading, a statement that "DV" represents "Daily Value."

(g) Foods in packages that have a total surface area available to bear labeling of 40 or less square inches may modify the requirements of paragraphs (c) through (f) of this section and §317.302(a) by one or more of the following means:

(1) (i) Presenting the required nutrition information in a tabular or linear (i.e., string) fashion, rather than in vertical columns if the product has a total surface area available to bear labeling of less than 12 square inches, or if the product has a total surface area available to bear labeling of 40 or less square inches and the package shape or size cannot accommodate a standard vertical column or tabular display on any label panel. Nutrition information may be given in a linear fashion only if the package shape or size will not accommodate a tabular display.

(ii) When nutrition information is given in a linear display, the nutrition information shall be set off in a box by the use of a hairline. The percent Daily Value is separated from the quantitative amount declaration by the use of parenthesis, and all nutrients, both principal components and subcomponents, are treated similarly. Bolding is required only on the title "Nutrition Facts" and is allowed for nutrient names for "Calories," "Total fat," "Cholesterol," "Sodium," "Total carbohydrate," and "Protein."

(2) Using any of the following abbreviations:

Serving size—Serv size
Servings per container—Servings
Calories from fat—Fat cal
Calories from saturated fat—Sat fat cal
Saturated fat—Sat fat
Monounsaturated fat—Monounsat fat
Polyunsaturated fat—Polyunsat fat
Cholesterol—Cholest
Total carbohydrate—Total carb
Dietary fiber—Fiber
Soluble fiber—Sol fiber
Insoluble fiber—Insol fiber
Sugar alcohol—Sugar alc
Other carbohydrate—Other carb

(3) Omitting the footnote required in paragraph (d)(9) of this section and placing another asterisk at the bottom of the label followed by the statement "Percent Daily Values are based on a 2,000 calorie diet" and, if the term "Daily Value" is not spelled out in the heading, a statement that "DV" represents "Daily Value."

9 CFR §317.309

(4) Presenting the required nutrition information on any other label panel.

(h) Compliance with this section shall be determined as follows:

(1) A production lot is a set of food production consumer units that are from one production shift. Alternatively, a collection of consumer units of the same size, type, and style produced under conditions as nearly uniform as possible, designated by a common container code or marking, constitutes a production lot.

(2) The sample for nutrient analysis shall consist of a composite of a minimum of six consumer units, each from a production lot. Alternatively, the sample for nutrient analysis shall consist of a composite of a minimum of six consumer units, each randomly chosen to be representative of a production lot. In each case, the units may be individually analyzed and the results of the analyses averaged, or the units would be composited and the composite analyzed. In both cases, the results, whether an average or a single result from a composite, will be considered by the Agency to be the nutrient content of a composite. All analyses shall be performed by appropriate methods and procedures used by the Department for each nutrient in accordance with the "Chemistry Laboratory Guidebook," or, if no USDA method is available and appropriate for the nutrient, by appropriate methods for the nutrient in accordance with the 1990 edition of the "Official Methods of Analysis" of the AOAC International, formerly Association of Official Analytical Chemists, 15th ed., which is incorporated by reference, unless a particular method of analysis is specified in §317.309(c), or, if no USDA, AOAC, or specified method is available and appropriate, by other reliable and appropriate analytical procedures as so determined by the Agency. The "Official Methods of Analysis" is incorporated as it exists on the date of approval. This incorporation by reference was approved by the Director of the Federal Register in accordance with 5 U.S.C. 552(a) and 1 CFR part 51. Copies may be purchased from the AOAC International, 2200 Wilson Blvd., suite 400, Arlington, VA 22201. It is also available for inspection at the Office of the Federal Register Information Center, suite 700, 800 North Capitol Street, NW., Washington, DC.

(3) Two classes of nutrients are defined for purposes of compliance:

(i) Class I. Added nutrients in fortified or fabricated foods; and

(ii) Class II. Naturally occurring (indigenous) nutrients. If any ingredient which contains a naturally occurring (indigenous) nutrient is added to a food, the total amount of such nutrient in the final food product is subject to Class II requirements unless the same nutrient is also added, which would make the total amount of such nutrient subject to Class I requirements.

(4) A product with a label declaration of a vitamin, mineral, protein, total carbohydrate, dietary fiber, other carbohydrate, polyunsaturated or monounsaturated fat, or potassium shall be deemed to be misbranded under section 1(n) of the Federal Meat Inspection Act (21 U.S.C. 601(n)(1)) unless it meets the following requirements:

(i) Class I vitamin, mineral, protein, dietary fiber, or potassium. The nutrient content of the composite is at least equal to the value for that nutrient declared on the label.

(ii) Class II vitamin, mineral, protein, total carbohydrate, dietary fiber, other carbohydrate, polyunsaturated or monounsaturated fat, or potassium. The nutrient content of the composite is at least equal to 80 percent of the value for that nutrient declared on the label; *Provided,* That no regulatory action will be based on a determination of a nutrient value which falls below this level by an amount less than the variability generally recognized for the analytical method used in that product at the level involved, and inherent nutrient variation in a product.

(5) A product with a label declaration of calories, sugars, total fat, saturated fat, cholesterol, or sodium shall be deemed to be misbranded under section 1(n) of the Federal Meat Inspection Act (21 U.S.C. 601(n)(1)) if the nutrient content of the composite is greater than 20 percent in excess of the value for that nutrient declared on the label; Provided, That no regulatory action will be based on a determination of a nutrient value which falls above this level by an amount less than the variability generally recognized for the analytical method used in that product at the level involved, and inherent nutrient variation in a product.

(6) The amount of a vitamin, mineral, protein, total carbohydrate, dietary fiber, other carbohydrate, polyunsaturated or monounsaturated fat, or potassium may vary over labeled amounts within good manufacturing practice. The amount of calories, sugars, total fat, saturated fat, cholesterol, or sodium may vary under labeled amounts within good manufacturing practice.

(7) Compliance will be based on the metric measure specified in the label statement of serving size.

(8) The management of the establishment must maintain records to support the validity of nutrient declarations contained on product labels. Such records shall be made available to the inspector or any duly authorized

9 CFR §317.310

representative of the Agency upon request.

(9) The compliance provisions set forth in paragraph (h)(1) through (8) of this section shall not apply to single-ingredient, raw meat (including ground beef) products, including those that have been previously frozen, when nutrition labeling is based on the most current representative data base values contained in USDA's National Nutrient Data Bank or its published form, the Agriculture Handbook No. 8 series available from the Government Printing Office.

(Paperwork requirements were approved by the Office of Management and Budget under control number 0583-0088)

[58 FR 664, Jan. 6, 1993; 58 FR 43788, Aug. 18, 1993; 58 FR 47627, Sept. 10, 1993; 59 FR 45194, Sept. 1, 1994; 60 FR 176, Jan. 3, 1995]

§317.310 [Reserved]

§317.311 [Reserved]

§317.312 Reference amounts customarily consumed per eating occasion.

(a) The general principles followed in arriving at the reference amounts customarily consumed per eating occasion (Reference Amount(s)), as set forth in paragraph (b) of this section, are:

(1) The Reference Amounts are calculated for persons 4 years of age or older to reflect the amount of food customarily consumed per eating occasion by persons in this population group. These Reference Amounts are based on data set forth in appropriate national food consumption surveys.

(2) The Reference Amounts are calculated for an infant or child under 4 years of age to reflect the amount of food customarily consumed per eating occasion by infants up to 12 months of age or by children 1 through 3 years of age, respectively. These Reference Amounts are based on data set forth in appropriate national food consumption surveys. Such Reference Amounts are to be used only when the product is specially formulated or processed for use by an infant or by a child under 4 years of age.

(3) An appropriate national food consumption survey includes a large sample size representative of the demographic and socioeconomic characteristics of the relevant population group and must be based on consumption data under actual conditions of use.

(4) To determine the amount of food customarily consumed per eating occasion, the mean, median, and mode of the consumed amount per eating occasion were considered.

(5) When survey data were insufficient, FSIS took various other sources of information on serving sizes of food into consideration. These other sources of information included:

(i) Serving sizes used in dietary guidance recommendations or recommended by other authoritative systems or organizations;

(ii) Serving sizes recommended in comments;

(iii) Serving sizes used by manufacturers and grocers; and

(iv) Serving sizes used by other countries.

(6) Because they reflect the amount customarily consumed, the Reference Amount and, in turn, the serving size declared on the product label are based on only the edible portion of food, and not bone, seed, shell, or other inedible components.

(7) The Reference Amount is based on the major intended use of the product (e.g., a mixed dish measurable with a cup as a main dish and not as a side dish).

(8) The Reference Amounts for products that are consumed as an ingredient of other products, but that may also be consumed in the form in which they are purchased (e.g., ground beef), are based on use in the form purchased.

(9) FSIS sought to ensure that foods that have similar dietary usage, product characteristics, and customarily consumed amounts have a uniform Reference Amount.

(b) The following Product Categories and Reference Amounts shall be used as the basis for determining serving sizes for specific products:

9 CFR §317.312

TABLE 1 Reference Amounts Customarily Consumed per Eating Occasion — Infant and Toddler Foods [1, 2, 3]

Product category	Reference amount
Infant & Toddler Foods:	
Dinner Dry Mix	15 g
Dinner, ready-to-serve, strained type	60 g
Dinner, soups, ready-to-serve junior type	110 g
Dinner, stew or soup ready-to-serve toddlers	170 g
Plain meats and meat sticks, ready-to-serve	55 g

{1} These values represent the amount of food customarily consumed per eating occasion and were primarily derived from the 1977-1978 and the 1987-1988 Nationwide Food Consumption Surveys conducted by the U.S. Department of Agriculture.

{2} Unless otherwise noted in the Reference Amount column, the Reference Amounts are for the ready-to-serve or almost ready-to-serve form of the product (i.e., heat and serve). If not listed separately, the Reference Amount for the unprepared form (e.g., dehydrated cereal) is the amount required to make one Reference Amount of the prepared form.

{3} Manufacturers are required to convert the Reference Amount to the label serving size in a household measure most appropriate to their specific product using the procedures established by regulation.

TABLE 2 Reference Amounts Customarily Consumed Per Eating Occasion — General Food Supply [1, 2, 3, 4, 5]

Product category	Reference amount Ready-to-serve	Reference amount Ready-to-cook
Egg mixtures, (western style omelet, souffle, egg foo young	110 g	n/a.
Lard, margarine, shortening	1 tbsp	n/a.
Salad and potato toppers; e.g., bacon bits	7 g	n/a.
Bacon (bacon, beef breakfast strips, pork breakfast strips, pork rinds)	15 g	54 g = bacon. 30 g = breakfast strips.
Dried; e.g., jerky, dried beef, Parma ham sausage products with a moisture/protein ratio of less than 2:1; e.g., pepperoni	30 g	n/a.
Snacks; e.g., meat snack food sticks	30 g	n/a.
Luncheon meat, bologna, Canadian style bacon, pork pattie crumbles, beef pattie crumbles, blood pudding, luncheon loaf, old fashioned loaf, berlinger, bangers, minced luncheon roll, thuringer, liver sausage, mortadella, uncured sausage (franks), ham and cheese loaf, P&P loaf, scrapple souse, head cheese, pizza loaf, olive loaf, pate, deviled ham, sandwich spread, teawurst, cervelet, Lebanon bologna, potted meat food product, taco fillings, meat pie fillings	55 g	n/a.
Linked meat sausage products, Vienna sausage, frankfurters, pork sausage, imitation frankfurters, bratwurst, kielbasa, Polish sausage, summer sausage, mettwurst, smoked country sausage, smoked sausage, smoked or pickled meat, pickled pigs feet	55 g	n/a. 75 g = uncooked sausage.

9 CFR §317.312

Product category	Reference amount Ready-to-serve	Reference amount Ready-to-cook
Entrees without sauce, cuts of meat including marinated, tenderized, injected cuts of meat, beef patty, corn dog, croquettes, fritters, cured ham, dry cured ham, dry cured cappicola, corned beef, pastrami, country ham, pork shoulder picnic, meatballs, pureed adult foods..	85 g	114 g.
Canned meats, canned beef, canned pork.[4]................................	55 g	n/a.
Entrees with sauce, barbecued meats in sauce......................................	140 g	n/a.
Mixed dishes NOT measurable with a cup;[5] e.g., burrito, egg roll, enchilada, pizza, pizza roll, quiche, all types of sandwiches, cracker and meat lunch type packages, gyro, stromboli, burger on a bun, frank on a bun, calzone, taco, pockets stuffed with meat, foldovers, stuffed vegetables with meat, shish kabobs, empanada.......................	140 g (plus 55 g for products with sauce toppings)	n/a.
Mixed dishes measurable with a cup; e.g., meat casserole, macaroni and cheese with meat, pot pie, spaghetti with sauce, meat chili, chili with beans, meat hash, creamed chipped beef, beef ravioli in sauce, beef stroganoff, Brunswick stew, goulash, meat stew, ragout, meat lasagna, meat filled pasta..	1 cup	n/a.
Salads-pasta or potato, potato salad with bacon, macaroni and meat salad...	140 g	n/a.
Salads-all other meat, salads, ham salad...	100 g	n/a.
Soups-all varieties..	245 g	n/a.
Major main entree type sauce; e.g., spaghetti sauce with meat, spaghetti sauce with meatballs...	125 g	n/a.
Minor main entree sauce; e.g., pizza sauce with meat, gravy..............	1/4 cup	n/a.

Seasoning mixes dry, bases, extracts, dried broths and stock/juice, freeze dry trail mix products with meat.
 	
As reconstituted:
   Amount to make one Reference Amount of the final dish; e.g.,
     Gravy...
     Major main entree type sauce...
     Soup...
     Entree measurable with a cup..
	
	1/4 cup
	125 g
	245 g
	1 cup
		
		n/a.
		n/a.
		n/a.
		n/a.

{1} These values represent the amount of food customarily consumed per eating occasion and were primarily derived from the 1977-78 and the 1987-88 Nationwide Food Consumption Surveys conducted by the U.S. Department of Agriculture.

{2} Manufacturers are required to convert the Reference Amounts to the label serving size in a household measure most appropriate to their specific product using the procedures established by regulation.

{3} Examples listed under Product Category are not all inclusive or exclusive. Examples are provided to assist manufacturers in identifying appropriate product Reference Amount.

{4} If packed or canned in liquid, the Reference Amount is for the drained solids, except for products in which both the solids and liquids are customarily consumed.

{5} Pizza sauce is part of the pizza and is not considered to be sauce topping.

9 CFR §317.312

(c) For products that have no Reference Amount listed in paragraph (b) of this section for the unprepared or the prepared form of the product and that consist of two or more foods packaged and presented to be consumed together (e.g., lunch meat with cheese and crackers), the Reference Amount for the combined product shall be determined using the following rules:

(1) For bulk products, the Reference Amount for the combined product shall be the Reference Amount, as established in paragraph (b) of this section, for the ingredient that is represented as the main ingredient plus proportioned amounts of all minor ingredients.

(2) For products where the ingredient represented as the main ingredient is one or more discrete units, the Reference Amount for the combined product shall be either the number of small discrete units or the fraction of the large discrete unit that is represented as the main ingredient that is closest to the Reference Amount for that ingredient as established in paragraph (b) of this section plus proportioned amounts of all minor ingredients.

(3) If the Reference Amounts are in compatible units, they shall be summed (e.g., ingredients in equal volumes such as tablespoons). If the Reference Amounts are in incompatible units, the weights of the appropriate volumes should be used (e.g., grams of one ingredient plus gram weight of tablespoons of a second ingredient).

(d) If a product requires further preparation, e.g., cooking or the addition of water or other ingredients, and if paragraph (b) of this section provides a Reference Amount for the product in the prepared form, then the Reference Amount for the unprepared product shall be determined using the following rules:

(1) Except as provided for in paragraph (d)(2) of this section, the Reference Amount for the unprepared product shall be the amount of the unprepared product required to make the Reference Amount for the prepared product as established in paragraph (b) of this section.

(2) For products where the entire contents of the package is used to prepare one large discrete unit usually divided for consumption, the Reference Amount for the unprepared product shall be the amount of the unprepared product required to make the fraction of the large discrete unit closest to the Reference Amount for the prepared product as established in paragraph (b) of this section.

(e) The Reference Amount for an imitation or substitute product or altered product as defined in §317.313(d), such as a "low calorie" version, shall be the same as for the product for which it is offered as a substitute.

(f) The Reference Amounts set forth in paragraphs (b) through (e) of this section shall be used in determining whether a product meets the criteria for nutritional claims. If the serving size declared on the product label differs from the Reference Amount, and the product meets the criteria for the claim only on the basis of the Reference Amount, the claim shall be followed by a statement that sets forth the basis on which the claim is made. That statement shall include the Reference Amount as it appears in paragraph (b) of this section followed, in parentheses, by the amount in common household measure if the Reference Amount is expressed in measures other than common household measures.

(g) The Administrator, on his or her own initiative or on behalf of any interested person who has submitted a labeling application, may issue a proposal to establish or amend a Product Category or Reference Amount identified in paragraph (b) of this section.

(1) Labeling applications and supporting documentation to be filed under this section shall be submitted in quadruplicate, except that the supporting documentation may be submitted on a computer disc copy. If any part of the material submitted is in a foreign language, it shall be accompanied by an accurate and complete English translation. The labeling application shall state the applicant's post office address.

(2) Pertinent information will be considered as part of an application on the basis of specific reference to such information submitted to and retained in the files of the Food Safety and Inspection Service. However, any reference to unpublished information furnished by a person other than the applicant will not be considered unless use of such information is authorized (with the understanding that such information may in whole or part be subject to release to the public) in a written statement signed by the person who submitted it. Any reference to published information should be accompanied by reprints or photostatic copies of such references.

(3) The availability for public disclosure of labeling applications, along with supporting documentation, submitted to the Agency under this section will be governed by the rules specified in subchapter D, title 9.

(4) Data accompanying the labeling application, such as food consumption data, shall be submitted on separate sheets, suitably identified. If such data has already been submitted with an earlier labeling application from the applicant, the present labeling application must provide the data.

(5) The labeling application must be signed by the applicant or by his or her attorney or agent, or (if a

9 CFR §317.312

corporation) by an authorized official.

(6) The labeling application shall include a statement signed by the person responsible for the labeling application, that to the best of his or her knowledge, it is a representative and balanced submission that includes unfavorable information, as well as favorable information, known to him or her pertinent to the evaluation of the labeling application.

(7) Labeling applications for a new Reference Amount and/or Product Category shall be accompanied by the following data which shall be submitted in the following form to the Director, Food Labeling Division, Regulatory Programs, Food Safety and Inspection Service, Washington, DC 20250:

(Date)

The undersigned, ------ submits this labeling application pursuant to 9 CFR 317.312 with respect to Reference Amount and/or Product Category.

Attached hereto, in quadruplicate, or on a computer disc copy, and constituting a part of this labeling application, are the following:

(i) A statement of the objective of the labeling application;
(ii) A description of the product;
(iii) A complete sample product label including nutrition label, using the format established by regulation;
(iv) A description of the form in which the product will be marketed;
(v) The intended dietary uses of the product with the major use identified (e.g., ham as a luncheon meat);
(vi) If the intended use is primarily as an ingredient in other foods, list of foods or food categories in which the product will be used as an ingredient with information on the prioritization of the use;
(vii) The population group for which the product will be offered for use (e.g., infants, children under 4 years of age);
(viii) The names of the most closely-related products (or in the case of foods for special dietary use and imitation or substitute foods, the names of the products for which they are offered as substitutes);
(ix) The suggested Reference Amount (the amount of edible portion of food as consumed, excluding bone, skin or other inedible components) for the population group for which the product is intended with full description of the methodology and procedures that were used to determine the suggested Reference Amount. In determining the Reference Amount, general principles and factors in paragraph (a) of this section should be followed.
(x) The suggested Reference Amount shall be expressed in metric units. Reference Amounts for foods shall be expressed in grams except when common household units such as cups, tablespoons, and teaspoons are more appropriate or are more likely to promote uniformity in serving sizes declared on product labels. For example, common household measures would be more appropriate if products within the same category differ substantially in density such as mixed dishes measurable with a cup.
(A) In expressing the Reference Amount in grams, the following general rules shall be followed:
(1) For quantities greater than 10 grams, the quantity shall be expressed in nearest 5 grams increment.
(2) For quantities less than 10 grams, exact gram weights shall be used.
(B) [Reserved]
(xi) A labeling application for a new subcategory of food with its own Reference Amount shall include the following additional information:
(A) Data that demonstrate that the new subcategory of food will be consumed in amounts that differ enough from the Reference Amount for the parent category to warrant a separate Reference Amount. Data must include sample size, and the mean, standard deviation, median, and modal consumed amount per eating occasion for the product identified in the labeling application and for other products in the category. All data must be derived from the same survey data.
(B) Documentation supporting the difference in dietary usage and product characteristics that affect the consumption size that distinguishes the product identified in the labeling application from the rest of the products in the category.
(xii) In conducting research to collect or process food consumption data in support of the labeling application, the following general guidelines should be followed.
(A) Sampled population selected should be representative of the demographic and socioeconomic characteristics of the target population group for which the food is intended.
(B) Sample size (i.e., number of eaters) should be large enough to give reliable estimates for customarily consumed amounts.
(C) The study protocol should identify potential biases and describe how potential biases are controlled for or, if not possible to control, how they affect interpretation of results.
(D) The methodology used to collect or process data including study design, sampling procedures, materials used (e.g.,

9 CFR §317.312

questionnaire, interviewer's manual), procedures used to collect or process data, methods or procedures used to control for unbiased estimates, and procedures used to correct for nonresponse, should be fully documented.

(xiii) A statement concerning the feasibility of convening associations, corporations, consumers, and other interested parties to engage in negotiated rulemaking to develop a proposed rule.

Yours very truly,

Applicant

By _____
(Indicate authority)

(8) Upon receipt of the labeling application and supporting documentation, the applicant shall be notified, in writing, of the date on which the labeling application was received. Such notice shall inform the applicant that the labeling application is undergoing Agency review and that the applicant shall subsequently be notified of the Agency's decision to consider for further review or deny the labeling application.

(9) Upon review of the labeling application and supporting documentation, the Agency shall notify the applicant, in writing, that the labeling application is either being considered for further review or that it has been summarily denied by the Administrator.

(10) If the labeling application is summarily denied by the Administrator, the written notification shall state the reasons therefor, including why the Agency has determined that the proposed Reference Amount and/or Product Category is false or misleading. The notification letter shall inform the applicant that the applicant may submit a written statement by way of answer to the notification, and that the applicant shall have the right to request a hearing with respect to the merits or validity of the Administrator's decision to deny the use of the proposed Reference Amount and/or Product Category.

(i) If the applicant fails to accept the determination of the Administrator and files an answer and requests a hearing, and the Administrator, after review of the answer, determines the initial determination to be correct, the Administrator shall file with the Hearing Clerk of the Department the notification, answer, and the request for a hearing, which shall constitute the complaint and answer in the proceeding, which shall thereafter be conducted in accordance with the Department's Uniform Rules of Practice.

(ii) The hearing shall be conducted before an administrative law judge with the opportunity for appeal to the Department's Judicial Officer, who shall make the final determination for the Secretary. Any such determination by the Secretary shall be conclusive unless, within 30 days after receipt of notice of such final determination, the applicant appeals to the United States Court of Appeals for the circuit in which the applicant has its principal place of business or to the United States Court of Appeals for the District of Columbia Circuit.

(11) If the labeling application is not summarily denied by the Administrator, the Administrator shall publish in the Federal Register a proposed rule to amend the regulations to authorize the use of the Reference Amount and/or Product Category. The proposal shall also summarize the labeling application, including where the supporting documentation can be reviewed. The Administrator's proposed rule shall seek comment from consumers, the industry, consumer and industry groups, and other interested persons on the labeling application and the use of the proposed Reference Amount and/or Product Category. After public comment has been received and reviewed by the Agency, the Administrator shall make a determination on whether the proposed Reference Amount and/or Product Category shall be approved for use on the labeling of meat food products.

(i) If the Reference Amount and/or Product Category is denied by the Administrator, the Agency shall notify the applicant, in writing, of the basis for the denial, including the reason why the Reference Amount and/or Product Category on the labeling was determined by the Agency to be false or misleading. The notification letter shall also inform the applicant that the applicant may submit a written statement by way of answer to the notification, and that the applicant shall have the right to request a hearing with respect to the merits or validity of the Administrator's decision to deny the use of the proposed Reference Amount and/or Product Category.

(A) If the applicant fails to accept the determination of the Administrator and files an answer and requests a hearing, and the Administrator, after review of an answer, determines the initial determination to be correct, the Administrator shall file with the Hearing Clerk of the Department the notification, answer, and the request for a hearing, which shall constitute the complaint and answer in the proceeding, which shall thereafter be conducted in accordance with the Department's Uniform Rules of Practice.

(B) The hearing shall be conducted before an administrative law judge with the opportunity for appeal to the Department's Judicial Officer, who shall make the final determination for the Secretary. Any such determination by the Secretary shall be conclusive unless, within 30 days after receipt of the notice of such final determination, the applicant appeals to the United States Court of Appeals for the circuit in which the applicant has its principal place of business or to the United States Court of Appeals for the District of Columbia Circuit.

(ii) If the Reference Amount and/or Product Category is approved, the Agency shall notify the applicant, in writing, and shall also publish in the Federal Register a final rule amending the regulations to authorize the use of the Reference Amount and/or Product Category.

(Paperwork requirements were approved by the Office of Management and Budget under control number 0583-0088)

[58 FR 664, Jan. 6, 1993; 58 FR 43788, Aug. 18, 1993 as amended at 58 FR 47627, Sept. 10, 1993; 59 FR 45196, Sept. 1, 1994; 60 FR 186, Jan. 3, 1995]

§317.313 Nutrient content claims; general principles.

(a) This section applies to meat or meat food products that are intended for human consumption and that are offered for sale.

(b) A claim which, expressly or by implication, characterizes the level of a nutrient (nutrient content claim) of the type required in nutrition labeling pursuant to §317.309, may not be made on a label or in labeling of that product unless the claim is made in accordance with the applicable provisions in this subpart.

(1) An expressed nutrient content claim is any direct statement about the level (or range) of a nutrient in the product, e.g., "low sodium" or "contains 100 calories."

(2) An implied nutrient content claim is any claim that:

(i) Describes the product or an ingredient therein in a manner that suggests that a nutrient is absent or present in a certain amount (e.g., "high in oat bran"); or

(ii) Suggests that the product, because of its nutrient content, may be useful in maintaining healthy dietary practices and is made in association with an explicit claim or statement about a nutrient (e.g., "healthy, contains 3 grams (g) of fat").

(3) Except for claims regarding vitamins and minerals described in paragraph (q)(3) of this section, no nutrient content claims may be made on products intended specifically for use by infants and children less than 2 years of age unless the claim is specifically provided for in subpart B of this part.

(4) Reasonable variations in the spelling of the terms defined in applicable provisions in this subpart and their synonyms are permitted provided these variations are not misleading (e.g., "hi" or "lo").

(c) Information that is required or permitted by §317.309 to be declared in nutrition labeling, and that appears as part of the nutrition label, is not a nutrient content claim and is not subject to the requirements of this section. If such information is declared elsewhere on the label or in labeling, it is a nutrient content claim and is subject to the requirements for nutrient content claims.

(d) A "substitute" product is one that may be used interchangeably with another product that it resembles, i.e., that it is organoleptically, physically, and functionally (including shelf life) similar to, and that it is not nutritionally inferior to unless it is labeled as an "imitation."

(1) If there is a difference in performance characteristics that materially limits the use of the product, the product may still be considered a substitute if the label includes a disclaimer adjacent to the most prominent claim as defined in paragraph (j)(2)(iii) of this section, informing the consumer of such difference (e.g., "not recommended for frying").

(2) This disclaimer shall be in easily legible print or type and in a size no less than that required by §317.2(h) for the net quantity of contents statement, except where the size of the claim is less than two times the required size of the net quantity of contents statement, in which case the disclaimer statement shall be no less than one-half the size of the claim but no smaller than 1/16-inch minimum height, except as permitted by §317.400(d)(2).

(e)

(1) Because the use of a "free" or "low" claim before the name of a product implies that the product differs from other products of the same type by virtue of its having a lower amount of the nutrient, only products that have been specially processed, altered, formulated, or reformulated so as to lower the amount of the nutrient in the

9 CFR §317.313

product, remove the nutrient from the product, or not include the nutrient in the product, may bear such a claim (e.g., "low sodium beef noodle soup").

(2) Any claim for the absence of a nutrient in a product, or that a product is low in a nutrient when the product has not been specially processed, altered, formulated, or reformulated to qualify for that claim shall indicate that the product inherently meets the criteria and shall clearly refer to all products of that type and not merely to the particular brand to which the labeling attaches (e.g., "lard, a sodium free food").

(f) A nutrient content claim shall be in type size and style no larger than two times that of the statement of identity and shall not be unduly prominent in type style compared to the statement of identity.

(g) Labeling information required in Secs. 317.313, 317.354, 317.356, 317.360, 317.361, 317.362, and 317.380, whose type size is not otherwise specified, is required to be in letters and/or numbers no less than 1/16 inch in height, except as permitted by §317.400(d)(2).

(h) [Reserved]

(i) Except as provided in §317.309 or in paragraph (q)(3) of this section, the label or labeling of a product may contain a statement about the amount or percentage of a nutrient if:

(1) The use of the statement on the product implicitly characterizes the level of the nutrient in the product and is consistent with a definition for a claim, as provided in subpart B of this part, for the nutrient that the label addresses. Such a claim might be, "less than 10 g of fat per serving;"

(2) The use of the statement on the product implicitly characterizes the level of the nutrient in the product and is not consistent with such a definition, but the label carries a disclaimer adjacent to the statement that the product is not "low" in or a "good source" of the nutrient, such as "only 200 milligrams (mg) sodium per serving, not a low sodium product." The disclaimer must be in easily legible print or type and in a size no less than required by §317.2(h) for the net quantity of contents, except where the size of the claim is less than two times the required size of the net quantity of contents statement, in which case the disclaimer statement shall be no less than one-half the size of the claim but no smaller than 1/16-inch minimum height, except as permitted by §317.400(d)(2);

(3) The statement does not in any way implicitly characterize the level of the nutrient in the product and it is not false or misleading in any respect (e.g., "100 calories" or "5 grams of fat"), in which case no disclaimer is required.

(4) "Percent fat free" claims are not authorized by this paragraph. Such claims shall comply with §317.362(b)(6).

(j) A product may bear a statement that compares the level of a nutrient in the product with the level of a nutrient in a reference product. These statements shall be known as "relative claims" and include "light," "reduced," "less" (or "fewer"), and "more" claims.

(1) To bear a relative claim about the level of a nutrient, the amount of that nutrient in the product must be compared to an amount of nutrient in an appropriate reference product as specified in this paragraph (j).

(i) (A) For "less" (or "fewer") and "more" claims, the reference product may be a dissimilar product within a product category that can generally be substituted for one another in the diet or a similar product.

(B) For "light," "reduced," and "added" claims, the reference product shall be a similar product, and

(ii) (A) For "light" claims, the reference product shall be representative of the type of product that includes the product that bears the claim. The nutrient value for the reference product shall be representative of a broad base of products of that type; e.g., a value in a representative, valid data base; an average value determined from the top three national (or regional) brands, a market basket norm; or, where its nutrient value is representative of the product type, a market leader. Firms using such a reference nutrient value as a basis for a claim, are required to provide specific information upon which the nutrient value was derived, on request, to consumers and appropriate regulatory officials.

(B) For relative claims other than "light," including "less" and "more" claims, the reference product may be the same as that provided for "light" in paragraph (j)(1)(ii)(A) of this section or it may be the manufacturer's regular product, or that of another manufacturer, that has been offered for sale to the public on a regular basis for a substantial period of time in the same geographic area by the same business entity or by one entitled to use its trade name, provided the name of the competitor is not used on the labeling of the product. The nutrient values used to determine the claim when comparing a single manufacturer's product to the labeled product shall be either the values declared in nutrition labeling or the actual nutrient values, provided that the resulting labeling is internally consistent (i.e., that the values stated in the nutrition information, the nutrient values in the

9 CFR §317.313

accompanying information, and the declaration of the percentage of nutrient by which the product has been modified are consistent and will not cause consumer confusion when compared), and that the actual modification is at least equal to the percentage specified in the definition of the claim.

(2) For products bearing relative claims:

(i) The label or labeling must state the identity of the reference product and the percent (or fraction) of the amount of the nutrient in the reference product by which the nutrient has been modified, (e.g., "50 percent less fat than 'reference product'" or "1/3 fewer calories than 'reference product'"); and

(ii) This information shall be immediately adjacent to the most prominent claim in easily legible boldface print or type, in distinct contrast to other printed or graphic matter, that is no less than that required by §317.2(h) for net quantity of contents, except where the size of the claim is less than two times the required size of the net quantity of contents statement, in which case the referral statement shall be no less than one-half the size of the claim, but no smaller than 1/16-inch minimum height, except as permitted by §317.400(d)(2).

(iii) The determination of which use of the claim is in the most prominent location on the label or labeling will be made based on the following factors, considered in order:

(A) A claim on the principal display panel adjacent to the statement of identity;

(B) A claim elsewhere on the principal display panel;

(C) A claim on the information panel; or

(D) A claim elsewhere on the label or labeling.

(iv) The label or labeling must also bear:

(A) Clear and concise quantitative information comparing the amount of the subject nutrient in the product per labeled serving size with that in the reference product; and

(B) This statement shall appear adjacent to the most prominent claim or to the nutrition information.

(3) A relative claim for decreased levels of a nutrient may not be made on the label or in labeling of a product if the nutrient content of the reference product meets the requirement for a "low" claim for that nutrient.

(k) The term "modified" may be used in the statement of identity of a product that bears a relative claim that complies with the requirements of this part, followed immediately by the name of the nutrient whose content has been altered (e.g., "modified fat product"). This statement of identity must be immediately followed by the comparative statement such as "contains 35 percent less fat than 'reference product'." The label or labeling must also bear the information required by paragraph (j)(2) of this section in the manner prescribed.

(l) For purposes of making a claim, a "meal-type product" shall be defined as a product that:

(1) Makes a significant contribution to the diet by weighing at least 6 ounces, but no more than 12 ounces per serving (container), and

(2) Contains ingredients from two or more of the following four food groups:

(i) Bread, cereal, rice and pasta group,

(ii) Fruits and vegetables group,

(iii) Milk, yogurt, and cheese group, and

(iv) Meat, poultry, fish, dry beans, eggs, and nuts group, and

(3) Is represented as, or is in a form commonly understood to be a breakfast, lunch, dinner, meal, main dish, entree, or pizza. Such representations may be made either by statements, photographs, or vignettes.

(m) [Reserved]

(n) Nutrition labeling in accordance with §317.309, shall be provided for any food for which a nutrient content claim is made.

(o) Compliance with requirements for nutrient content claims shall be in accordance with §317.309(h).

(p)

(1) Unless otherwise specified, the reference amount customarily consumed set forth in §317.312(b) through (e) shall be used in determining whether a product meets the criteria for a nutrient content claim. If the serving size declared on the product label differs from the reference amount customarily consumed, and the amount of the nutrient contained in the labeled serving does not meet the maximum or minimum amount criterion in the definition for the descriptor for that nutrient, the claim shall be followed by the criteria for the claim as required by §317.312(f) (e.g., "very low sodium, 35 mg or less per 55 grams").

(2) The criteria for the claim shall be immediately adjacent to the most prominent claim in easily legible print or type and in a size that is no less than that required by §317.2(h) for net quantity of contents, except where the size of the claim is less than two times the required size of the net quantity of contents statement, in which case

9 CFR §317.343

the criteria statement shall be no less than one-half the size of the claim but no smaller than 1/16-inch minimum height, except as permitted by §317.400(d)(2).

(q) The following exemptions apply:

(1) Nutrient content claims that have not been defined by regulation and that appear as part of a brand name that was in use prior to November 27, 1991, may continue to be used as part of that brand name, provided they are not false or misleading under section 1(n) of the Act (21 U.S.C. 601(n)(1)).

(2) [Reserved]

(3) A statement that describes the percentage of a vitamin or mineral in the food, including foods intended specifically for use by infants and children less than 2 years of age, in relation to a Reference Daily Intake (RDI) as defined in §317.309 may be made on the label or in the labeling of a food without a regulation authorizing such a claim for a specific vitamin or mineral.

(4) The requirements of this section do not apply to infant formulas and medical foods, as described in 21 CFR 101.13(q)(4).

(5) [Reserved]

(6) Nutrient content claims that were part of the name of a product that was subject to a standard of identity as of November 27, 1991, are not subject to the requirements of paragraph (b) of this section whether or not they meet the definition of the descriptive term.

(7) Implied nutrient content claims may be used as part of a brand name, provided that the use of the claim has been authorized by FSIS. Labeling applications requesting approval of such a claim may be submitted pursuant to §317.369.

[58 FR 664, Jan. 6, 1993; 58 FR 43788, Aug. 18, 1993, as amended at 58 FR 47627, Sept. 10, 1993; 59 FR 40213, Aug. 8, 1994; 59 FR 45196, Sept. 1, 1994; 60 FR 187, Jan. 3, 1995]

§§317.314-317.342 [Reserved]

§317.343 Significant participation for voluntary nutrition labeling.

(a) In evaluating significant participation for voluntary nutrition labeling, FSIS will consider only the major cuts of single-ingredient, raw meat products, as identified in §317.344, including those that have been previously frozen.

(b) FSIS will judge a food retailer to be participating at a significant level if the retailer provides nutrition labeling information for at least 90 percent of the major cuts of single- ingredient, raw meat products, listed in §317.344, that it sells, and if the nutrition label is consistent in content and format with the mandatory program, or nutrition information is displayed at point-of-purchase in an approriate manner.

(c) To determine whether there is significant participation by retailers under the voluntary nutrition labeling guidelines, FSIS will select a representative sample of companies allocated by type and size.

(d) FSIS will find that significant participation by food retailers exists if at least 60 percent of all companies that are evaluated are participating in accordance with the guidelines.

(e) FSIS will evaluate significant participation of the voluntary program every 2 years beginning in May 1995.

(1) If significant participation is found, the voluntary nutrition labeling guidelines shall remain in effect.

(2) If significant participation is not found, FSIS shall initiate rulemaking to require nutrition labeling on those products under the voluntary program.

§317.344 Identification of major cuts of meat products.

The major cuts of single-ingredient, raw meat products are: Beef chuck blade roast, beef loin top loin steak, beef rib roast large end, beef round eye round steak, beef round top round steak, beef round tip roast, beef chuck arm pot roast, beef loin sirloin steak, beef round bottom round steak, beef brisket (whole, flat half, or point half), beef rib steak small end, beef loin tenderloin steak, ground beef regular without added seasonings, ground beef about 17% fat, pork loin chop, pork loin country style ribs, pork loin top loin chop boneless, pork loin rib chop, pork spareribs, pork loin tenderloin, pork loin sirloin roast, pork shoulder blade steak, pork loin top roast boneless, ground pork, lamb shank, lamb shoulder arm chop, lamb shoulder blade chop, lamb rib roast, lamb loin chop, lamb leg (whole, sirloin half, or shank half), veal shoulder arm steak, veal shoulder blade steak, veal rib roast, veal loin chop, and veal cutlets.

9 CFR §317.345

[58 FR 664, Jan. 6, 1993, as amended at 59 FR 45196, Sept. 1, 1994]

§317.345 Guidelines for voluntary nutrition labeling of single- ingredient, raw products.

(a) Nutrition information on the cuts of single-ingredient, raw meat products, including those that have been previously frozen, shall be provided in the following manner:

(1) If a retailer or manufacturer chooses to provide nutrition information on the label of these products, these products shall be subject to all requirements of the mandatory nutrition labeling program, except that nutrition labeling may be declared on the basis of either "as consumed" or "as packaged." In addition, the declaration of the number of servings per container need not be included in nutrition labeling of single-ingredient, raw meat products (including ground beef), including those that have been previously frozen.

(2) A retailer may choose to provide nutrition information at the point-of-purchase, such as by posting a sign, or by making the information readily available in brochures, notebooks, or leaflet form in close proximity to the food. The nutrition labeling information may also be supplemented by a video, live demonstration, or other media. If a nutrition claim is made on point-of-purchase materials all of the requirements of the mandatory nutrition labeling program apply. However, if only nutrition information--and not a nutrition claim--is supplied on point-of-purchase materials:

(i) The requirements of the mandatory nutrition labeling program apply, but the nutrition information may be supplied on an "as packaged" or "as consumed," basis;

(ii) The listing of percent of Daily Value for the nutrients (except vitamins and minerals specified in §317.309(c)(8)) and footnote required by §317.309(d)(9) may be omitted; and

(iii) The point-of-purchase materials are not subject to any of the format requirements.

(b) [Reserved]

(c) The declaration of nutrition information may be presented in a simplified format as specified in §317.309(f) for the mandatory nutrition labeling program.

(d) The nutrition label data should be based on either the raw or cooked edible portions of meat cuts with external cover fat at trim levels reflecting current marketing practices. If data are based on cooked portions, the methods used to cook the products must be specified and should be those which do not add nutrients from other ingredients such as flour, breading, and salt. Additional nutritional data may be presented on an optional basis for the raw or cooked edible portions of the separable lean of meat cuts.

(e) Nutrient data that are the most current representative data base values contained in USDA's National Nutrient Data Bank or its published form, the Agriculture Handbook No. 8 series, may be used for nutrition labeling of single-ingredient, raw meat products (including ground beef), including those that have been previously frozen. These data may be composite data that reflect different quality grades of beef or other variables affecting nutrient content. Alternatively, data that reflect specific grades or other variables may be used, except that if data are used on labels attached to a product which is labeled as to grade of meat or other variables, the data must represent the product in the package when such data are contained in the representative data base. When data are used on labels attached to a product, the data must represent the edible meat tissues present in the package.

(f) If the nutrition information is in accordance with paragraph (e) of this section, a nutrition label or labeling will not be subject to the Agency compliance review under §317.309(h), unless a nutrition claim is made on the basis of the representative data base values.

(g) Retailers may use data bases that they believe reflect the nutrient content of single-ingredient, raw meat products (including ground beef), including those that have been previously frozen; however, such labeling shall be subject to the compliance procedures of paragraph (e) of this section and the requirements specified in this subpart for the mandatory nutrition labeling program.

[58 FR 664, Jan. 6, 1993, as amended at 58 FR 47627, Sept. 10, 1993; 60 FR 189, Jan. 3, 1995]

§317.346-317.353 [Reserved]

§317.354 Nutrient content claims for "good source," "high," and "more."

(a) *General requirements.* Except as provided in paragraph (e) of this section, a claim about the level of a nutrient in

9 CFR §317.354

a product in relation to the Reference Daily Intake (RDI) or Daily Reference Value (DRV) established for that nutrient (excluding total carbohydrate) in §317.309(c), may only be made on the label or in labeling of the product if:

(1) The claim uses one of the terms defined in this section in accordance with the definition for that term;

(2) The claim is made in accordance with the general requirements for nutrient content claims in §317.313; and

(3) The product for which the claim is made is labeled in accordance with §317.309.

(b) *"High" claims.*

(1) The terms "high," "rich in," or "excellent source of" may be used on the label or in labeling of products, except meal-type products as defined in §317.313(l), provided that the product contains 20 percent or more of the RDI or the DRV per reference amount customarily consumed.

(2) The terms defined in paragraph (b)(1) of this section may be used on the label or in labeling of a meal-type product as defined in §317.313(l), provided that:

(i) The product contains a food that meets the definition of "high" in paragraph (b)(1) of this section; and

(ii) The label or labeling clearly identifies the food that is the subject of the claim (e.g., "the serving of broccoli in this meal is high in vitamin C").

(c) *"Good Source" claims.*

(1) The terms "good source," "contains," or "provides" may be used on the label or in labeling of products, except meal-type products as described in §317.313(l), provided that the product contains 10 to 19 percent of the RDI or the DRV per reference amount customarily consumed.

(2) The terms defined in paragraph (c)(1) of this section may be used on the label or in labeling of a meal-type product as defined in §317.313(l), provided that:

(i) The product contains a food that meets the definition of "good source" in paragraph (c)(1) of this section; and

(ii) The label or labeling clearly identifies the food that is the subject of the claim (e.g., "the serving of sweet potatoes in this meal is a good source of fiber").

(d) *Fiber claims.*

(1) If a nutrient content claim is made with respect to the level of dietary fiber, i.e., that the product is high in fiber, a good source of fiber, or that the product contains "more" fiber, and the product is not "low" in total fat as defined in §317.362(b)(2) or, in the case of a meal-type product, is not "low" in total fat as defined in §317.362(b)(3), then the labeling shall disclose the level of total fat per labeled serving size (e.g., "contains 12 grams (g) of fat per serving"); and

(2) The disclosure shall appear in immediate proximity to such claim and be in a type size no less than one-half the size of the claim.

(e) *"More" claims.*

(1) A relative claim using the terms "more" and "added" may be used on the label or in labeling to describe the level of protein, vitamins, minerals, dietary fiber, or potassium in a product, except meal-type products as defined in §317.313(l), provided that:

(i) The product contains at least 10 percent more of the RDI or the DRV for protein, vitamins, minerals, dietary fiber, or potassium (expressed as a percent of the Daily Value) per reference amount customarily consumed than an appropriate reference product as described in §317.313(j)(1); and

(ii) As required in §317.313(j)(2) for relative claims:

(A) The identity of the reference product and the percent (or fraction) that the nutrient is greater relative to the RDI or DRV are declared in immediate proximity to the most prominent such claim (e.g., "contains 10 percent more of the Daily Value for fiber than 'reference product'"); and

(B) Quantitative information comparing the level of the nutrient in the product per labeled serving size with that of the reference product that it replaces is declared adjacent to the most prominent claim or to the nutrition information (e.g., "fiber content of 'reference product' is 1 g per serving; 'this product' contains 4 g per serving").

(2) A relative claim using the terms "more" and "added" may be used on the label or in labeling to describe the level of protein, vitamins, minerals, dietary fiber, or potassium in meal-type products as defined in §317.313(l), provided that:

(i) The product contains at least 10 percent more of the RDI or the DRV for protein, vitamins, minerals, dietary fiber, or potassium (expressed as a percent of the Daily Value) per 100 g of product than an appropriate reference product as described in §317.313(j)(1); and

(ii) As required in §317.313(j)(2) for relative claims:

(A) The identity of the reference product and the percent (or fraction) that the nutrient is greater relative to the RDI or DRV are declared in immediate proximity to the most prominent such claim (e.g., "contains 10 percent more of the Daily Value for fiber per 3 ounces (oz) than does 'reference product'"), and

(B) Quantitative information comparing the level of the nutrient in the meal-type product per specified weight with that of the reference product that it replaces is declared adjacent to the most prominent claim or to the nutrition information (e.g., "fiber content of 'reference product' is 2 g per 3 oz; 'this product' contains 5 g per 3 oz").

[59 FR 40213, as amended at 60 FR 189, Jan. 3, 1995]

§317.355 [Reserved]

§317.356 Nutrient content claims for "light" or "lite".

(a) *General requirements.* A claim using the terms "light" or "lite" to describe a product may only be made on the label or in labeling of the product if:

(1) The claim uses one of the terms defined in this section in accordance with the definition for that term;

(2) The claim is made in accordance with the general requirements for nutrient content claims in §317.313; and

(3) The product for which the claim is made is labeled in accordance with §317.309.

(b) *"Light" claims.* The terms "light" or "lite" may be used on the label or in labeling of products, except meal-type products as defined in §317.313(l), without further qualification, provided that:

(1) If the product derives 50 percent or more of its calories from fat, its fat content is reduced by 50 percent or more per reference amount customarily consumed compared to an appropriate reference product as described in §317.313(j)(1); or

(2) If the product derives less than 50 percent of its calories from fat:

(i) The number of calories is reduced by at least one-third (33 1/3 percent) per reference amount customarily consumed compared to an appropriate reference product as described in §317.313(j)(1); or

(ii) Its fat content is reduced by 50 percent or more per reference amount customarily consumed compared to the appropriate reference product as described in §317.313(j)(1); and

(3) As required in §317.313(j)(2) for relative claims:

(i) The identity of the reference product and the percent (or fraction) that the calories and the fat were reduced are declared in immediate proximity to the most prominent such claim (e.g., "1/3 fewer calories and 50 percent less fat than the market leader"); and

(ii) Quantitative information comparing the level of calories and fat content in the product per labeled serving size with that of the reference product that it replaces is declared adjacent to the most prominent claim or to the nutrition information (e.g., "lite 'this product'--200 calories, 4 grams (g) fat; regular 'reference product'-- 300 calories, 8 g fat per serving"); and

(iii) If the labeled product contains less than 40 calories or less than 3 g fat per reference amount customarily consumed, the percentage reduction for that nutrient need not be declared.

(4) A "light" claim may not be made on a product for which the reference product meets the definition of "low fat" and "low calorie."

(c)

(1) (i) A product for which the reference product contains 40 calories or less and 3 g fat or less per reference amount customarily consumed may use the terms "light" or "lite" without further qualification if it is reduced by 50 percent or more in sodium content compared to the reference product; and

(ii) As required in §317.313(j)(2) for relative claims:

(A) The identity of the reference product and the percent (or fraction) that the sodium was reduced are declared in immediate proximity to the most prominent such claim (e.g., "50 percent less sodium than the market leader"); and

(B) Quantitative information comparing the level of sodium per labeled serving size with that of the reference product it replaces is declared adjacent to the most prominent claim or to the nutrition information (e.g., "lite 'this product'--500 milligrams (mg) sodium per serving; regular 'reference product'--1,000 mg sodium per serving").

9 CFR §317.356

(2) (i) A product for which the reference product contains more than 40 calories or more than 3 g fat per reference amount customarily consumed may use the terms "light in sodium" or "lite in sodium" if it is reduced by 50 percent or more in sodium content compared to the reference product, provided that "light" or "lite" is presented in immediate proximity with "in sodium" and the entire term is presented in uniform type size, style, color, and prominence; and

(ii) As required in §317.313(j)(2) for relative claims:

(A) The identity of the reference product and the percent (or fraction) that the sodium was reduced are declared in immediate proximity to the most prominent such claim (e.g., "50 percent less sodium than the market leader"); and

(B) Quantitative information comparing the level of sodium per labeled serving size with that of the reference product it replaces is declared adjacent to the most prominent claim or to the nutrition information (e.g., or "lite 'this product'--170 mg sodium per serving; regular 'reference product'--350 mg per serving").

(3) Except for meal-type products as defined in §317.313(l), a "light in sodium" claim may not be made on a product for which the reference product meets the definition of "low in sodium."

(d)

(1) The terms "light" or "lite" may be used on the label or in labeling of a meal-type product as defined in §317.313(l), provided that:

(i) The product meets the definition of:

(A) "Low in calories" as defined in §317.360(b)(3); or

(B) "Low in fat" as defined in §317.362(b)(3); and (ii)(A) A statement appears on the principal display panel that explains whether "light" is used to mean "low fat," "low calories," or both (e.g., "Light Delight, a low fat meal"); and (B) The accompanying statement is no less than one-half the type size of the "light" or "lite" claim.

(2) (i) The terms "light in sodium" or "lite in sodium" may be used on the label or in labeling of a meal-type product as defined in §317.313(l), provided that the product meets the definition of "low in sodium" as defined in §317.361(b)(5)(i); and

(ii) "Light" or "lite" and "in sodium" are presented in uniform type size, style, color, and prominence.

(3) The term "light" or "lite" may be used in the brand name of a product to describe the sodium content, provided that:

(i) The product is reduced by 50 percent or more in sodium content compared to the reference product;

(ii) A statement specifically stating that the product is "light in sodium" or "lite in sodium" appears:

(A) Contiguous to the brand name; and

(B) In uniform type size, style, color, and prominence as the product name; and

(iii) As required in §317.313(j)(2) for relative claims: (A) The identity of the reference product and the percent (or fraction) that the sodium was reduced are declared in immediate proximity to the most prominent such claim; and

(B) Quantitative information comparing the level of sodium per labeled serving size with that of the reference product it replaces is declared adjacent to the most prominent claim or to the nutrition information.

(e) Except as provided in paragraphs (b) through (d) of this section, the terms "light" or "lite" may not be used to refer to a product that is not reduced in fat by 50 percent, or, if applicable, in calories by 1/3 or, when properly qualified, in sodium by 50 percent unless:

(1) It describes some physical or organoleptic attribute of the product such as texture or color and the information (e.g., "light in color" or "light in texture") so stated, clearly conveys the nature of the product; and

(2) The attribute (e.g., "color" or "texture") is in the same style, color, and at least one-half the type size as the word "light" and in immediate proximity thereto.

(f) If a manufacturer can demonstrate that the word "light" has been associated, through common use, with a particular product to reflect a physical or organoleptic attribute to the point where it has become part of the statement of identity, such use of the term "light" shall not be considered a nutrient content claim subject to the requirements in this part.

(g) The term "lightly salted" may be used on a product to which has been added 50 percent less sodium than is normally added to the reference product as described in §317.313(j)(1)(i)(B) and (j)(1)(ii)(B), provided that if the product is not "low in sodium" as defined in §317.361(b)(4), the statement "not a low sodium food," shall appear adjacent to the nutrition information and the information required to accompany a relative claim shall appear on the label or labeling as specified in §317.313(j)(2).

9 CFR §317.360

[58 FR 664, Jan. 6, 1993; 58 FR 43788, Aug. 18, 1993, as amended at 59 FR 40213, Aug. 8, 1994 and 60 FR 189, Jan. 3, 1995]

§§317.357-317.359 [Reserved]

§317.360 Nutrient content claims for calorie content.

(a) *General requirements.* A claim about the calorie or sugar content of a product may only be made on the label or in labeling of the product if:
    (1) The claim uses one of the terms defined in this section in accordance with the definition for that term;
    (2) The claim is made in accordance with the general requirements for nutrient content claims in §317.313; and
    (3) The product for which the claim is made is labeled in accordance with §317.309.

(b) *Calorie content claims.*
    (1) The terms "calorie free," "free of calories," "no calories," "zero calories," "without calories," "trivial source of calories," "negligible source of calories," or "dietarily insignificant source of calories" may be used on the label or in labeling of products, provided that:
    (i) The product contains less than 5 calories per reference amount customarily consumed and per labeled serving size; and
    (ii) If the product meets this condition without the benefit of special processing, alteration, formulation, or reformulation to lower the caloric content, it is labeled to clearly refer to all products of its type and not merely to the particular brand to which the label attaches.

    (2) The terms "low calorie," "few calories," "contains a small amount of calories," "low source of calories," or "low in calories" may be used on the label or in labeling of products, except meal-type products as defined in §317.313(l), provided that:
    (i) (A) The product has a reference amount customarily consumed greater than 30 grams (g) or greater than 2 tablespoons (tbsp) and does not provide more than 40 calories per reference amount customarily consumed; or
    (B) The product has a reference amount customarily consumed of 30 g or less or 2 tbsp or less and does not provide more than 40 calories per reference amount customarily consumed and per 50 g (for dehydrated products that must be reconstituted before typical consumption with water or a diluent containing an insignificant amount, as defined in §317.309(f)(1), of all nutrients per reference amount customarily consumed, the per-50-g criterion refers to the "as prepared" form).
    (ii) If the product meets these conditions without the benefit of special processing, alteration, formulation, or reformulation to lower the caloric content, it is labeled to clearly refer to all products of its type and not merely to the particular brand to which the label attaches.

    (3) The terms defined in paragraph (b)(2) of this section may be used on the label or in labeling of a meal-type product as defined in §317.313(l), provided that:
    (i) The product contains 120 calories or less per 100 g of product; and
    (ii) If the product meets this condition without the benefit of special processing, alteration, formulation, or reformulation to lower the calorie content, it is labeled to clearly refer to all products of its type and not merely to the particular brand to which it attaches.

    (4) The terms "reduced calorie," "reduced in calories," "calorie reduced," "fewer calories," "lower calorie," or "lower in calories" may be used on the label or in labeling of products, except meal-type products as defined in §317.313(l), provided that:
    (i) The product contains at least 25 percent fewer calories per reference amount customarily consumed than an appropriate reference product as described in §317.313(j)(1); and
    (ii) As required in §317.313(j)(2) for relative claims:
    (A) The identity of the reference product and the percent (or fraction) that the calories differ between the two products are declared in immediate proximity to the most prominent such claim (e.g., lower calorie 'product'--"33 1/3 percent fewer calories than our regular 'product'"); and
    (B) Quantitative information comparing the level of calories in the product per labeled serving size with that of the reference product that it replaces is declared adjacent to the most prominent claim or to the nutrition information (e.g., "calorie content has been reduced from 150 to 100 calories per serving").
    (iii) Claims described in paragraph (b)(4) of this section may not be made on the label or in labeling of products

9 CFR §317.360

if the reference product meets the definition for "low calorie."

(5) The terms defined in paragraph (b)(4) of this section may be used on the label or in labeling of a meal-type product as defined in §317.313(l), provided that:

(i) The product contains at least 25 percent fewer calories per 100 g of product than an appropriate reference product as described in §317.313(j)(1); and

(ii) As required in §317.313(j)(2) for relative claims:

(A) The identity of the reference product and the percent (or fraction) that the calories differ between the two products are declared in immediate proximity to the most prominent such claim (e.g., "calorie reduced 'product', 25% less calories per ounce (oz) (or 3 oz) than our regular 'product'"); and

(B) Quantitative information comparing the level of calories in the product per specified weight with that of the reference product that it replaces is declared adjacent to the most prominent claim or to the nutrition information (e.g., "calorie content has been reduced from 110 calories per 3 oz to 80 calories per 3 oz").

(iii) Claims described in paragraph (b)(5) of this section may not be made on the label or in labeling of products if the reference product meets the definition for "low calorie."

(c) *Sugar content claims.*

(1) Terms such as "sugar free," "free of sugar," "no sugar," "zero sugar," "without sugar," "sugarless," "trivial source of sugar," "negligible source of sugar," or "dietarily insignificant source of sugar" may reasonably be expected to be regarded by consumers as terms that represent that the product contains no sugars or sweeteners, e.g., "sugar free," or "no sugar," as indicating a product which is low in calories or significantly reduced in calories. Consequently, except as provided in paragraph (c)(2) of this section, a product may not be labeled with such terms unless:

(i) The product contains less than 0.5 g of sugars, as defined in §317.309(c)(6)(ii), per reference amount customarily consumed and per labeled serving size or, in the case of a meal-type product, less than 0.5 g of sugars per labeled serving size;

(ii) The product contains no ingredient that is a sugar or that is generally understood by consumers to contain sugars unless the listing of the ingredient in the ingredients statement is followed by an asterisk that refers to the statement below the list of ingredients, which states: "Adds a trivial amount of sugar," "adds a negligible amount of sugar," or "adds a dietarily insignificant amount of sugar;" and

(iii) (A) It is labeled "low calorie" or "reduced calorie" or bears a relative claim of special dietary usefulness labeled in compliance with paragraphs (b)(2), (b)(3), (b)(4), or (b)(5) of this section; or

(B) Such term is immediately accompanied, each time it is used, by either the statement "not a reduced calorie product," "not a low calorie product," or "not for weight control."

(2) The terms "no added sugar," "without added sugar," or "no sugar added" may be used only if:

(i) No amount of sugars, as defined in §317.309(c)(6)(ii), or any other ingredient that contains sugars that functionally substitute for added sugars is added during processing or packaging;

(ii) The product does not contain an ingredient containing added sugars such as jam, jelly, or concentrated fruit juice;

(iii) The sugars content has not been increased above the amount present in the ingredients by some means such as the use of enzymes, except where the intended functional effect of the process is not to increase the sugars content of a product, and a functionally insignificant increase in sugars results;

(iv) The product that it resembles and for which it substitutes normally contains added sugars; and

(v) The product bears a statement that the product is not "low calorie" or "calorie reduced" (unless the product meets the requirements for a "low" or "reduced calorie" product) and that directs consumers' attention to the nutrition panel for further information on sugar and calorie content.

(3) Paragraph (c)(1) of this section shall not apply to a factual statement that a product, including products intended specifically for infants and children less than 2 years of age, is unsweetened or contains no added sweeteners in the case of a product that contains apparent substantial inherent sugar content, e.g., juices.

(4) The terms "reduced sugar," "reduced in sugar," "sugar reduced," "less sugar," "lower sugar," or "lower in sugar" may be used on the label or in labeling of products, except meal-type products as defined in §317.313(l), provided that:

(i) The product contains at least 25 percent less sugars per reference amount customarily consumed than an appropriate reference product as described in §317.313(j)(1); and

(ii) As required in §317.313(j)(2) for relative claims:

(A) The identity of the reference product and the percent (or fraction) that the sugars differ between the two products are declared in immediate proximity to the most prominent such claim (e.g., "this product contains 25 percent less sugar than our regular product"); and

(B) Quantitative information comparing the level of the sugar in the product per labeled serving size with that of the reference product that it replaces is declared adjacent to the most prominent claim or to the nutrition information (e.g., "sugar content has been lowered from 8 g to 6 g per serving").

(5) The terms defined in paragraph (c)(4) of this section may be used on the label or in labeling of a meal-type product as defined in §317.313(l), provided that:

(i) The product contains at least 25 percent less sugars per 100 g of product than an appropriate reference product as described in §317.313(j)(1); and

(ii) As required in §317.313(j)(2) for relative claims:

(A) The identity of the reference product and the percent (or fraction) that the sugars differ between the two products are declared in immediate proximity to the most prominent such claim (e.g., reduced sugar 'product'--25% less sugar than our regular 'product'"); and

(B) Quantitative information comparing the level of the nutrient in the product per specified weight with that of the reference product that it replaces is declared adjacent to the most prominent claim or to the nutrition information (e.g., "sugar content has been reduced from 17 g per 3 oz to 13 g per 3 oz").

[59 FR 40213, Aug. 8, 1994, as amended at 60 FR 191, Jan. 3, 1995]

§317.361 Nutrient content claims for the sodium content.

(a) *General requirements.* A claim about the level of sodium in a product may only be made on the label or in labeling of the product if:

(1) The claim uses one of the terms defined in this section in accordance with the definition for that term;

(2) The claim is made in accordance with the general requirements for nutrient content claims in §317.313; and

(3) The product for which the claim is made is labeled in accordance with §317.309.

(b) *Sodium content claims.*

(1) The terms "sodium free," "free of sodium," "no sodium," "zero sodium," "without sodium," "trivial source of sodium," "negligible source of sodium," or "dietarily insignificant source of sodium" may be used on the label or in labeling of products, provided that:

(i) The product contains less than 5 milligrams (mg) of sodium per reference amount customarily consumed and per labeled serving size or, in the case of a meal-type product, less than 5 mg of sodium per labeled serving size;

(ii) The product contains no ingredient that is sodium chloride or is generally understood by consumers to contain sodium unless the listing of the ingredient in the ingredients statement is followed by an asterisk that refers to the statement below the list of ingredients, which states: "Adds a trivial amount of sodium," "adds a negligible amount of sodium" or "adds a dietarily insignificant amount of sodium"; and

(iii) If the product meets these conditions without the benefit of special processing, alteration, formulation, or reformulation to lower the sodium content, it is labeled to clearly refer to all products of its type and not merely to the particular brand to which the label attaches.

(2) The terms "very low sodium" or "very low in sodium" may be used on the label or in labeling of products, except meal-type products as defined in §317.313(l), provided that:

(i) (A) The product has a reference amount customarily consumed greater than 30 grams (g) or greater than 2 tablespoons (tbsp) and contains 35 mg or less sodium per reference amount customarily consumed; or

(B) The product has a reference amount customarily consumed of 30 g or less or 2 tbsp or less and contains 35 mg or less sodium per reference amount customarily consumed and per 50 g (for dehydrated products that must be reconstituted before typical consumption with water or a diluent containing an insignificant amount, as defined in §317.309(f)(1), of all nutrients per reference amount customarily consumed, the per-50-g criterion refers to the "as prepared" form); and

(ii) If the product meets these conditions without the benefit of special processing, alteration, formulation, or reformulation to lower the sodium content, it is labeled to clearly refer to all products of its type and not merely to the particular brand to which the label attaches.

(3) The terms defined in paragraph (b)(2) of this section may be used on the label or in labeling of a meal-type

9 CFR §317.361

product as defined in §317.313(l), provided that:

(i) The product contains 35 mg or less of sodium per 100 g of product; and

(ii) If the product meets this condition without the benefit of special processing, alteration, formulation, or reformulation to lower the sodium content, it is labeled to clearly refer to all products of its type and not merely to the particular brand to which the label attaches.

(4) The terms "low sodium," "low in sodium," "little sodium," "contains a small amount of sodium," or "low source of sodium" may be used on the label and in labeling of products, except meal-type products as defined in §317.313(l), provided that:

(i) (A) The product has a reference amount customarily consumed greater than 30 g or greater than 2 tbsp and contains 140 mg or less sodium per reference amount customarily consumed; or

(B) The product has a reference amount customarily consumed of 30 g or less or 2 tbsp or less and contains 140 mg or less sodium per reference amount customarily consumed and per 50 g (for dehydrated products that must be reconstituted before typical consumption with water or a diluent containing an insignificant amount, as defined in §317.309(f)(1), of all nutrients per reference amount customarily consumed, the per-50-g criterion refers to the "as prepared" form); and

(ii) If the product meets these conditions without the benefit of special processing, alteration, formulation, or reformulation to lower the sodium content, it is labeled to clearly refer to all products of its type and not merely to the particular brand to which the label attaches.

(5) The terms defined in paragraph (b)(4) of this section may be used on the label or in labeling of a meal-type product as defined in §317.313(l), provided that:

(i) The product contains 140 mg or less sodium per 100 g of product; and

(ii) If the product meets these conditions without the benefit of special processing, alteration, formulation, or reformulation to lower the sodium content, it is labeled to clearly refer to all products of its type and not merely to the particular brand to which the label attaches.

(6) The terms "reduced sodium," "reduced in sodium," "sodium reduced," "less sodium," "lower sodium," or "lower in sodium" may be used on the label or in labeling of products, except meal-type products as defined in §317.313(l), provided that:

(i) The product contains at least 25 percent less sodium per reference amount customarily consumed than an appropriate reference product as described in §317.313(j)(1); and

(ii) As required in §317.313(j)(2) for relative claims:

(A) The identity of the reference product and the percent (or fraction) that the sodium differs between the two products are declared in immediate proximity to the most prominent such claim (e.g., "reduced sodium 'product', 50 percent less sodium than regular 'product'"); and

(B) Quantitative information comparing the level of sodium in the product per labeled serving size with that of the reference product that it replaces is declared adjacent to the most prominent claim or to the nutrition information (e.g., "sodium content has been lowered from 300 to 150 mg per serving").

(iii) Claims described in paragraph (b)(6) of this section may not be made on the label or in labeling of a product if the nutrient content of the reference product meets the definition for "low sodium."

(7) The terms defined in paragraph (b)(6) of this section may be used on the label or in labeling of a meal-type product as defined in §317.313(l), provided that:

(i) The product contains at least 25 percent less sodium per 100 g of product than an appropriate reference product as described in §317.313(j)(1); and

(ii) As required in §317.313(j)(2) for relative claims:

(A) The identity of the reference product and the percent (or fraction) that the sodium differs between the two products are declared in immediate proximity to the most prominent such claim (e.g., "reduced sodium 'product'--30% less sodium per 3 oz than our 'regular product'"); and

(B) Quantitative information comparing the level of sodium in the product per specified weight with that of the reference product that it replaces is declared adjacent to the most prominent claim or to the nutrition information (e.g., "sodium content has been reduced from 220 mg per 3 oz to 150 mg per 3 oz").

(iii) Claims described in paragraph (b)(7) of this section may not be made on the label or in labeling of products if the nutrient content of the reference product meets the definition for "low sodium."

(c) The term "salt" is not synonymous with "sodium." Salt refers to sodium chloride. However, references to salt content such as "unsalted," "no salt," "no salt added" are potentially misleading.

(1) The term "salt free" may be used on the label or in labeling of products only if the product is "sodium free" as defined in paragraph (b)(1) of this section.

(2) The terms "unsalted," "without added salt," and "no salt added" may be used on the label or in labeling of products only if:

(i) No salt is added during processing;

(ii) The product that it resembles and for which it substitutes is normally processed with salt; and

(iii) If the product is not sodium free, the statement, "not a sodium free product" or "not for control of sodium in the diet" appears adjacent to the nutrition information of the product bearing the claim.

(3) Paragraph (c)(2) of this section shall not apply to a factual statement that a product intended specifically for infants and children less than 2 years of age is unsalted, provided such statement refers to the taste of the product and is not false or otherwise misleading.

[59 FR 40214, Aug. 8, 1994, as amended at 60 FR 192, Jan. 3, 1995]

§317.362 Nutrient content claims for fat, fatty acids, and cholesterol content.

(a) *General requirements*. A claim about the level of fat, fatty acid, and cholesterol in a product may only be made on the label or in labeling of products if:

(1) The claim uses one of the terms defined in this section in accordance with the definition for that term;

(2) The claim is made in accordance with the general requirements for nutrient content claims in §317.313; and

(3) The product for which the claim is made is labeled in accordance with §317.309.

(b) *Fat content claims*.

(1) The terms "fat free," "free of fat," "no fat," "zero fat," "without fat," "nonfat," "trivial source of fat," "negligible source of fat," or "dietarily insignificant source of fat" may be used on the label or in labeling of products, provided that:

(i) The product contains less than 0.5 gram (g) of fat per reference amount customarily consumed and per labeled serving size or, in the case of a meal-type product, less than 0.5 g of fat per labeled serving size;

(ii) The product contains no added ingredient that is a fat or is generally understood by consumers to contain fat unless the listing of the ingredient in the ingredients statement is followed by an asterisk that refers to the statement below the list of ingredients, which states: "Adds a trivial amount of fat," "adds a negligible amount of fat," or "adds a dietarily insignificant amount of fat"; and (iii) If the product meets these conditions without the benefit of special processing, alteration, formulation, or reformulation to lower the fat content, it is labeled to clearly refer to all products of its type and not merely to the particular brand to which the label attaches.

(2) The terms "low fat," "low in fat," "contains a small amount of fat," "low source of fat," or "little fat" may be used on the label and in labeling of products, except meal-type products as defined in §317.313(l), provided that:

(i) (A) The product has a reference amount customarily consumed greater than 30 g or greater than 2 tablespoons (tbsp) and contains 3 g or less of fat per reference amount customarily consumed; or

(B) The product has a reference amount customarily consumed of 30 g or less or 2 tbsp or less and contains 3 g or less of fat per reference amount customarily consumed and per 50 g (for dehydrated products that must be reconstituted before typical consumption with water or a diluent containing an insignificant amount, as defined in §317.309(f)(1), of all nutrients per reference amount customarily consumed, the per-50-g criterion refers to the "as prepared" form).

(ii) If the product meets these conditions without the benefit of special processing, alteration, formulation, or reformulation to lower the fat content, it is labeled to clearly refer to all products of its type and not merely to the particular brand to which the label attaches.

(3) The terms defined in paragraph (b)(2) of this section may be used on the label or in labeling of a meal-type product as defined in §317.313(l), provided that:

(i) The product contains 3 g or less of total fat per 100 g of product and not more than 30 percent of calories from fat; and

(ii) If the product meets these conditions without the benefit of special processing, alteration, formulation, or reformulation to lower the fat content, it is labeled to clearly refer to all products of its type and not merely to the particular brand to which the label attaches.

(4) The terms "reduced fat," "reduced in fat," "fat reduced," "less fat," "lower fat," or "lower in fat" may be

9 CFR §317.362

used on the label or in labeling of products, except meal-type products as defined in §317.313(l), provided that:

(i) The product contains at least 25 percent less fat per reference amount customarily consumed than an appropriate reference product as described in §317.313(j)(1); and

(ii) As required in §317.313(j)(2) for relative claims:

(A) The identity of the reference product and the percent (or fraction) that the fat differs between the two products are declared in immediate proximity to the most prominent such claim (e.g., "reduced fat--50 percent less fat than our regular 'product'"); and

(B) Quantitative information comparing the level of fat in the product per labeled serving size with that of the reference product that it replaces is declared adjacent to the most prominent claim or to the nutrition information (e.g., "fat content has been reduced from 8 g to 4 g per serving").

(iii) Claims described in paragraph (b)(4) of this section may not be made on the label or in labeling of a product if the nutrient content of the reference product meets the definition for "low fat."

(5) The terms defined in paragraph (b)(4) of this section may be used on the label or in labeling of a meal-type product as defined in §317.313(l), provided that:

(i) The product contains at least 25 percent less fat per 100 g of product than an appropriate reference product as described in §317.313(j)(1); and

(ii) As required in §317.313(j)(2) for relative claims:

(A) The identity of the reference product and the percent (or fraction) that the fat differs between the two products are declared in immediate proximity to the most prominent such claim (e.g., "reduced fat 'product', 33 percent less fat per 3 oz than our regular 'product'"); and

(B) Quantitative information comparing the level of fat in the product per specified weight with that of the reference product that it replaces is declared adjacent to the most prominent such claim or to the nutrition information (e.g., "fat content has been reduced from 8 g per 3 oz to 5 g per 3 oz").

(iii) Claims described in paragraph (b)(5) of this section may not be made on the label or in labeling of a product if the nutrient content of the reference product meets the definition for "low fat."

(6) The term "_____ percent fat free" may be used on the label or in labeling of products, provided that:

(i) The product meets the criteria for "low fat" in paragraph (b)(2) or (b)(3) of this section;

(ii) The percent declared and the words "fat free" are in uniform type size; and

(iii) A "100 percent fat free" claim may be made only on products that meet the criteria for "fat free" in paragraph (b)(1) of this section, that contain less than 0.5 g of fat per 100 g, and that contain no added fat.

(iv) A synonym for "_____ percent fat free" is "_____ percent lean."

(c) *Fatty acid content claims.*

(1) The terms "saturated fat free," "free of saturated fat," "no saturated fat," "zero saturated fat," "without saturated fat," "trivial source of saturated fat," "negligible source of saturated fat," or "dietarily insignificant source of saturated fat" may be used on the label or in labeling of products, provided that:

(i) The product contains less than 0.5 g of saturated fat and less than 0.5 g trans fatty acids per reference amount customarily consumed and per labeled serving size or, in the case of a meal-type product, less than 0.5 g of saturated fat and less than 0.5 g trans fatty acids per labeled serving size;

(ii) The product contains no ingredient that is generally understood by consumers to contain saturated fat unless the listing of the ingredient in the ingredients statement is followed by an asterisk that refers to the statement below the list of ingredients, which states: "Adds a trivial amount of saturated fat," "adds a negligible amount of saturated fat," or "adds a dietarily insignificant amount of saturated fat;" and

(iii) If the product meets these conditions without the benefit of special processing, alteration, formulation, or reformulation to lower saturated fat content, it is labeled to clearly refer to all products of its type and not merely to the particular brand to which the label attaches.

(2) The terms "low in saturated fat," "low saturated fat," "contains a small amount of saturated fat," "low source of saturated fat," or "a little saturated fat" may be used on the label or in labeling of products, except meal-type products as defined in §317.313(l), provided that:

(i) The product contains 1 g or less of saturated fat per reference amount customarily consumed and not more than 15 percent of calories from saturated fat; and

(ii) If the product meets these conditions without benefit of special processing, alteration, formulation, or reformulation to lower saturated fat content, it is labeled to clearly refer to all products of its type and not merely to the particular brand to which the label attaches.

(3) The terms defined in paragraph (c)(2) of this section may be used on the label or in labeling of a meal-type product as defined in §317.313(l), provided that:

(i) The product contains 1 g or less of saturated fat per 100 g and less than 10 percent calories from saturated fat; and

(ii) If the product meets these conditions without the benefit of special processing, alteration, formulation, or reformulation to lower saturated fat content, it is labeled to clearly refer to all products of its type and not merely to the particular brand to which the label attaches.

(4) The terms "reduced saturated fat," "reduced in saturated fat," "saturated fat reduced," "less saturated fat," "lower saturated fat," or "lower in saturated fat" may be used on the label or in labeling of products, except meal-type products as defined in §317.313(l), provided that:

(i) The product contains at least 25 percent less saturated fat per reference amount customarily consumed than an appropriate reference product as described in §317.313(j)(1); and

(ii) As required in §317.313(j)(2) for relative claims:

(A) The identity of the reference product and the percent (or fraction) that the saturated fat differs between the two products are declared in immediate proximity to the most prominent such claim (e.g., "reduced saturated fat 'product', contains 50 percent less saturated fat than the national average for 'product'"); and

(B) Quantitative information comparing the level of saturated fat in the product per labeled serving size with that of the reference product that it replaces is declared adjacent to the most prominent claim or to the nutrition information (e.g., "saturated fat reduced from 3 g to 1.5 g per serving").

(iii) Claims described in paragraph (c)(4) of this section may not be made on the label or in labeling of a product if the nutrient content of the reference product meets the definition for "low saturated fat."

(5) The terms defined in paragraph (c)(4) of this section may be used on the label or in labeling of a meal-type product as defined in §317.313(l), provided that:

(i) The product contains at least 25 percent less saturated fat per 100 g of product than an appropriate reference product as described in §317.313(j)(1); and

(ii) As required in §317.313(j)(2) for relative claims:

(A) The identity of the reference product and the percent (or fraction) that the saturated fat differs between the two products are declared in immediate proximity to the most prominent such claim (e.g., "reduced saturated fat 'product'," "50 percent less saturated fat than our regular 'product'"); and

(B) Quantitative information comparing the level of saturated fat in the product per specified weight with that of the reference product that it replaces is declared adjacent to the most prominent claim or to the nutrition information (e.g., "saturated fat content has been reduced from 2.5 g per 3 oz to 1.5 g per 3 oz").

(iii) Claims described in paragraph (c)(5) of this section may not be made on the label or in labeling of a product if the nutrient content of the reference product meets the definition for "low saturated fat."

(d) *Cholesterol content claims.*

(1) The terms "cholesterol free," "free of cholesterol," "zero cholesterol," "without cholesterol," "no cholesterol," "trivial source of cholesterol," "negligible source of cholesterol," or "dietarily insignificant source of cholesterol" may be used on the label or in labeling of products, provided that:

(i) The product contains less than 2 milligrams (mg) of cholesterol per reference amount customarily consumed and per labeled serving size or, in the case of a meal-type product as defined in §317.313(l), less than 2 mg of cholesterol per labeled serving size;

(ii) The product contains no ingredient that is generally understood by consumers to contain cholesterol, unless the listing of the ingredient in the ingredients statement is followed by an asterisk that refers to the statement below the list of ingredients, which states: "Adds a trivial amount of cholesterol," "adds a negligible amount of cholesterol," or "adds a dietarily insignificant amount of cholesterol";

(iii) The product contains 2 g or less of saturated fat per reference amount customarily consumed or, in the case of a meal-type product as defined in §317.313(l), 2 g or less of saturated fat per labeled serving size; and

(iv) If the product meets these conditions without the benefit of special processing, alteration, formulation, or reformulation to lower cholesterol content, it is labeled to clearly refer to all products of its type and not merely to the particular brand to which it attaches; or

(v) If the product meets these conditions only as a result of special processing, alteration, formulation, or reformulation, the amount of cholesterol is reduced by 25 percent or more from the reference product it replaces as described in §317.313(j)(1) and for which it substitutes as described in §317.313(d) that has a significant

9 CFR §317.362

(e.g., 5 percent or more of a national or regional market) market share. As required in §317.313(j)(2) for relative claims:

(A) The identity of the reference product and the percent (or fraction) that the cholesterol was reduced are declared in immediate proximity to the most prominent such claim (e.g., "cholesterol free 'product', contains 100 percent less cholesterol than 'reference product'"); and

(B) Quantitative information comparing the level of cholesterol in the product per labeled serving size with that of the reference product that it replaces is declared adjacent to the most prominent claim or to the nutrition information (e.g., "contains no cholesterol compared with 30 mg in one serving of 'reference product'").

(2) The terms "low in cholesterol," "low cholesterol," "contains a small amount of cholesterol," "low source of cholesterol," or "little cholesterol" may be used on the label or in labeling of products, except meal-type products as defined in §317.313(l), provided that:

(i) (A) If the product has a reference amount customarily consumed greater than 30 g or greater than 2 tbsp:

(1) The product contains 20 mg or less of cholesterol per reference amount customarily consumed; and

(2) The product contains 2 g or less of saturated fat per reference amount customarily consumed; or

(B) If the product has a reference amount customarily consumed of 30 g or less or 2 tbsp or less:

(1) The product contains 20 mg or less of cholesterol per reference amount customarily consumed and per 50 g (for dehydrated products that must be reconstituted before typical consumption with water or a diluent containing an insignificant amount, as defined in §317.309(f)(1), of all nutrients per reference amount customarily consumed, the per-50-g criterion refers to the "as prepared" form); and

(2) The product contains 2 g or less of saturated fat per reference amount customarily consumed.

(ii) If the product meets these conditions without the benefit of special processing, alteration, formulation, or reformulation to lower cholesterol content, it is labeled to clearly refer to all products of its type and not merely to the particular brand to which the label attaches; or

(iii) If the product contains 20 mg or less of cholesterol only as a result of special processing, alteration, formulation, or reformulation, the amount of cholesterol is reduced by 25 percent or more from the reference product it replaces as described in §317.313(j)(1) and for which it substitutes as described in §317.313(d) that has a significant (e.g., 5 percent or more of a national or regional market) market share. As required in §317.313(j)(2) for relative claims:

(A) The identity of the reference product and the percent (or fraction) that the cholesterol has been reduced are declared in immediate proximity to the most prominent such claim (e.g., "low cholesterol 'product', contains 85 percent less cholesterol than our regular 'product'"); and

(B) Quantitative information comparing the level of cholesterol in the product per labeled serving size with that of the reference product that it replaces is declared adjacent to the most prominent claim or to the nutrition information (e.g., "cholesterol lowered from 30 mg to 5 mg per serving").

(3) The terms defined in paragraph (d)(2) of this section may be used on the label or in labeling of a meal-type product as defined in §317.313(l), provided that:

(i) The product contains 20 mg or less of cholesterol per 100 g of product;

(ii) The product contains 2 g or less of saturated fat per 100 g of product; and

(iii) If the product meets these conditions without the benefit of special processing, alteration, formulation, or reformulation to lower cholesterol content, it is labeled to clearly refer to all products of its type and not merely to the particular brand to which the label attaches.

(4) The terms "reduced cholesterol," "reduced in cholesterol," "cholesterol reduced," "less cholesterol," "lower cholesterol," or "lower in cholesterol" may be used on the label or in labeling of products or products that substitute for those products as specified in §317.313(d), excluding meal-type products as defined in §317.313(l), provided that:

(i) The product has been specifically formulated, altered, or processed to reduce its cholesterol by 25 percent or more from the reference product it replaces as described in §317.313(j)(1) and for which it substitutes as described in §317.313(d) that has a significant (e.g., 5 percent or more of a national or regional market) market share;

(ii) The product contains 2 g or less of saturated fat per reference amount customarily consumed; and

(iii) As required in §317.313(j)(2) for relative claims: (A) The identity of the reference product and the percent (or fraction) that the cholesterol has been reduced are declared in immediate proximity to the most prominent such claim (e.g., "25 percent less cholesterol than 'reference product'"); and

9 CFR §317.362

(B) Quantitative information comparing the level of cholesterol in the product per labeled serving size with that of the reference product that it replaces is declared adjacent to the most prominent claim or to the nutrition information (e.g., "cholesterol lowered from 55 mg to 30 mg per serving").

(iv) Claims described in paragraph (d)(4) of this section may not be made on the label or in labeling of a product if the nutrient content of the reference product meets the definition for "low cholesterol."

(5) The terms defined in paragraph (d)(4) of this section may be used on the label or in labeling of a meal-type product as defined in §317.313(l), provided that:

(i) The product has been specifically formulated, altered, or processed to reduce its cholesterol by 25 percent or more from the reference product it replaces as described in §317.313(j)(1) and for which it substitutes as described in §317.313(d) that has a significant (e.g., 5 percent or more of a national or regional market) market share;

(ii) The product contains 2 g or less of saturated fat per 100 g of product; and

(iii) As required in §317.313(j)(2) for relative claims:

(A) The identity of the reference product and the percent (or fraction) that the cholesterol has been reduced are declared in immediate proximity to the most prominent such claim (e.g., "25% less cholesterol than 'reference product'"); and

(B) Quantitative information comparing the level of cholesterol in the product per specified weight with that of the reference product that it replaces is declared adjacent to the most prominent claim or to the nutrition information (e.g., "cholesterol content has been reduced from 35 mg per 3 oz to 25 mg per 3 oz).

(iv) Claims described in paragraph (d)(5) of this section may not be made on the label or in labeling of a product if the nutrient content of the reference product meets the definition for "low cholesterol."

(e) *"Lean" and "Extra Lean" claims.*

(1) The term "lean" may be used on the label or in labeling of a product, provided that the product contains less than 10 g of fat, 4.5 g or less of saturated fat, and less than 95 mg of cholesterol per 100 g of product and per reference amount customarily consumed for individual foods, and per 100 g of product and per labeled serving size for meal-type products as defined in §317.313(l).

(2) The term "extra lean" may be used on the label or in labeling of a product, provided that the product contains less than 5 g of fat, less than 2 g of saturated fat, and less than 95 mg of cholesterol per 100 g of product and per reference amount customarily consumed for individual foods, and per 100 g of product and per labeled serving size for meal-type products as defined in §317.313(l).

[58 FR 664, Jan. 6, 1993; 58 FR 43788, Aug. 18, 1993, as amended at 58 FR 47627, Sept. 10, 1993; 59 FR 40214, Aug. 8, 1994, as amended at 60 FR 193, Jan. 3, 1995]

§317.363 Nutrient content claims for "healthy."

(a) The term "healthy," or any other derivative of the term "health," may be used on the labeling of any meat or meat food product, provided that the product is labeled in accordance with §317.309 and §317.313.

(b)

(1) The product shall meet the requirements for "low fat" and "low saturated fat," as defined in §317.362, except that single-ingredient, raw products may meet the total fat and saturated fat criteria for "extra lean" in §317.362.

(2) The product shall not contain more than 60 milligrams (mg) of cholesterol per reference amount customarily consumed, per labeled serving size, and, only for foods with reference amounts customarily consumed of 30 grams (g) or less or 2 tablespoons (tbsp) or less, per 50 g, and, for dehydrated products that must be reconstituted with water or a diluent containing an insignificant amount, as defined in §317.309(f)(1), of all nutrients, the per-50-g criterion refers to the prepared form, except that:

(i) A meal-type product, as defined in §317.313(l), and including meal-type products that weigh more than 12 ounces (oz) per serving (container), shall not contain more than 90 mg of cholesterol per labeled serving size; and

(ii) Single-ingredient, raw products may meet the cholesterol criterion for "extra lean" in §317.362.

(3) The product shall not contain more than 360 mg of sodium, except that it shall not contain more than 480 mg of sodium effective through January 1, 2000 per reference amount customarily consumed, per labeled serving size, and, only for foods with reference amounts customarily consumed of 30 g or less or 2 tbsp or less, per 50 g, and, for dehydrated products that must be reconstituted with water or a diluent containing an insignificant

9 CFR §317.369

amount, as defined in §317.309(f)(1), of all nutrients, the per-50-g criterion refers to the prepared form, except that:

(i) A meal-type product, as defined in §317.313(l), and including meal-type products that weigh more than 12 oz per serving (container), shall not contain more than 480 mg of sodium, except that it shall not contain more than 600 mg of sodium effective through January 1, 2000, per labeled serving size; and

(ii) The requirements of this paragraph (b)(3) do not apply to single-ingredient, raw products.

(4) The product shall contain 10 percent or more of the Reference Daily Intake or Daily Reference Value as defined in §317.309 for vitamin A, vitamin C, iron, calcium, protein, or fiber per reference amount customarily consumed prior to any nutrient addition, except that:

(i) A meal-type product, as defined in §317.313(l), and including meal-type products that weigh at least 6 oz but less than 10 oz per serving (container), shall meet the level for two of the nutrients per labeled serving size; and

(ii) A meal-type product, as defined in §317.313(l), and including meal-type products that weigh 10 oz or more per serving (container), shall meet the level for three of the nutrients per labeled serving size.

[59 FR 24228, May 10, 1994; 60 FR 196, Jan. 3, 1995; 63 *FR* 7281, Feb. 13, 1998]

§§317.364 - 317.368 [Reserved]

§317.369 Labeling applications for nutrient content claims.

(a) This section pertains to labeling applications for claims, express or implied, that characterize the level of any nutrient required to be on the label or in labeling of product by this subpart.

(b) Labeling applications included in this section are:

   (1) Labeling applications for a new (heretofore unauthorized) nutrient content claim,

   (2) Labeling applications for a synonymous term (i.e., one that is consistent with a term defined by regulation) for characterizing the level of a nutrient, and

   (3) Labeling applications for the use of an implied claim in a brand name.

(c) Labeling applications and supporting documentation to be filed under this section shall be submitted in quadruplicate, except that the supporting documentation may be submitted on a computer disc copy. If any part of the material submitted is in a foreign language, it shall be accompanied by an accurate and complete English translation. The labeling application shall state the applicant's post office address.

(d) Pertinent information will be considered as part of an application on the basis of specific reference to such information submitted to and retained in the files of the Food Safety and Inspection Service. However, any reference to unpublished information furnished by a person other than the applicant will not be considered unless use of such information is authorized (with the understanding that such information may in whole or part be subject to release to the public) in a written statement signed by the person who submitted it. Any reference to published information should be accompanied by reprints or photostatic copies of such references.

(e) If nonclinical laboratory studies accompany a labeling application, the applicant shall include, with respect to each nonclinical study included with the application, either a statement that the study has been, or will be, conducted in compliance with the good laboratory practice regulations as set forth in part 58 of chapter 1, title 21, or, if any such study was not conducted in compliance with such regulations, a brief statement of the reason for the noncompliance.

(f) If clinical investigations accompany a labeling application, the applicant shall include, with respect to each clinical investigation included with the application, either a statement that the investigation was conducted in compliance with the requirements for institutional review set forth in part 56 of chapter 1, title 21, or was not subject to such requirements in accordance with §56.194 or §56.105, and that it was conducted in compliance with the requirements for informed consents set forth in part 50 of chapter 1, title 21.

(g) The availability for public disclosure of labeling applications, along with supporting documentation, submitted to the Agency under this section will be governed by the rules specified in subchapter D, title 9.

(h) The data specified under this section to accompany a labeling application shall be submitted on separate sheets, suitably identified. If such data has already been submitted with an earlier labeling application from the applicant, the present labeling application must provide the data.

(i) The labeling application must be signed by the applicant or by his or her attorney or agent, or (if a corporation) by an authorized official.

9 CFR §317.369

(j) The labeling application shall include a statement signed by the person responsible for the labeling application, that to the best of his or her knowledge, it is a representative and balanced submission that includes unfavorable information, as well as favorable information, known to him or her pertinent to the evaluation of the labeling application.

(k)

(1) Labeling applications for a new nutrient content claim shall be accompanied by the following data which shall be submitted in the following form to the Director, Food Labeling Division, Regulatory Programs, Food Safety and Inspection Service, Washington, DC 20250.

(Date)

The undersigned, _____ , submits this labeling application pursuant to 9 CFR 317.369 with respect to (statement of the claim and its proposed use).

Attached hereto, in quadruplicate, or on a computer disc copy, and constituting a part of this labeling application, are the following:

(i) A statement identifying the nutrient content claim and the nutrient that the term is intended to characterize with respect to the level of such nutrient. The statement shall address why the use of the term as proposed will not be misleading. The statement shall provide examples of the nutrient content claim as it will be used on labels or labeling, as well as the types of products on which the claim will be used. The statement shall also specify the level at which the nutrient must be present or what other conditions concerning the product must be met for the appropriate use of the term in labels or labeling, as well as any factors that would make the use of the term inappropriate.

(ii) A detailed explanation supported by any necessary data of why use of the food component characterized by the claim is of importance in human nutrition by virtue of its presence or absence at the levels that such claim would describe. This explanation shall also state what nutritional benefit to the public will derive from use of the claim as proposed and why such benefit is not available through the use of existing terms defined by regulation. If the claim is intended for a specific group within the population, the analysis shall specifically address nutritional needs of such group, and scientific data sufficient for such purpose, and data and information to the extent necessary to demonstrate that consumers can be expected to understand the meaning of the term under the proposed conditions of use.

(iii) Analytical data that demonstrates the amount of the nutrient that is present in the products for which the claim is intended. The assays should be performed on representative samples in accordance with 317.309(h). If no USDA or AOAC methods are available, the applicant shall submit the assay method used, and data establishing the validity of the method for assaying the nutrient in the particular food. The validation data shall include a statistical analysis of the analytical and product variability.

(iv) A detailed analysis of the potential effect of the use of the proposed claim on food consumption, and any corresponding changes in nutrient intake. The analysis shall specifically address the intake of nutrients that have beneficial and negative consequences in the total diet. If the claim is intended for a specific group within the population, the analysis shall specifically address the dietary practices of such group, and shall include data sufficient to demonstrate that the dietary analysis is representative of such group.

Yours very truly,

Applicant
By _____
(Indicate authority)

(2) Upon receipt of the labeling application and supporting documentation, the applicant shall be notified, in writing, of the date on which the labeling application was received. Such notice shall inform the applicant that the labeling application is undergoing Agency review and that the applicant shall subsequently be notified of the Agency's decision to consider for further review or deny the labeling application.

(3) Upon review of the labeling application and supporting documentation, the Agency shall notify the applicant, in writing, that the labeling application is either being considered for further review or that it has been summarily denied by the Administrator.

(4) If the labeling application is summarily denied by the Administrator, the written notification shall state the reasons therefor, including why the Agency has determined that the proposed nutrient content claim is false or misleading. The notification letter shall inform the applicant that the applicant may submit a written statement

by way of answer to the notification, and that the applicant shall have the right to request a hearing with respect to the merits or validity of the Administrator's decision to deny the use of the proposed nutrient content claim.
(i) If the applicant fails to accept the determination of the Administrator and files an answer and requests a hearing, and the Administrator, after review of the answer, determines the initial determination to be correct, the Administrator shall file with the Hearing Clerk of the Department the notification, answer, and the request for a hearing, which shall constitute the complaint and answer in the proceeding, which shall thereafter be conducted in accordance with the Department's Uniform Rules of Practice.
(ii) The hearing shall be conducted before an administrative law judge with the opportunity for appeal to the Department's Judicial Officer, who shall make the final determination for the Secretary. Any such determination by the Secretary shall be conclusive unless, within 30 days after receipt of notice of such final determination, the applicant appeals to the United States Court of Appeals for the circuit in which the applicant has its principal place of business or to the United States Court of Appeals for the District of Columbia Circuit.
(5) If the labeling application is not summarily denied by the Administrator, the Administrator shall publish in the Federal Register a proposed rule to amend the regulations to authorize the use of the nutrient content claim. The proposal shall also summarize the labeling application, including where the supporting documentation can be reviewed. The Administrator's proposed rule shall seek comment from consumers, the industry, consumer and industry groups, and other interested persons on the labeling application and the use of the proposed nutrient content claim. After public comment has been received and reviewed by the Agency, the Administrator shall make a determination on whether the proposed nutrient content claim shall be approved for use on the labeling of meat and meat food products.
(i) If the claim is denied by the Administrator, the Agency shall notify the applicant, in writing, of the basis for the denial, including the reason why the claim on the labeling was determined by the Agency to be false or misleading. The notification letter shall also inform the applicant that the applicant may submit a written statement by way of answer to the notification, and that the applicant shall have the right to request a hearing with respect to the merits or validity of the Administrator's decision to deny the use of the proposed nutrient content claim.
(A) If the applicant fails to accept the determination of the Administrator and files an answer and requests a hearing, and the Administrator, after review of the answer, determines the initial determination to be correct, the Administrator shall file with the Hearing Clerk of the Department the notification, answer, and the request for a hearing, which shall constitute the complaint and answer in the proceeding, which shall thereafter be conducted in accordance with the Department's Uniform Rules of Practice.
(B) The hearing shall be conducted before an administrative law judge with the opportunity for appeal to the Department's Judicial Officer, who shall make final determination for the Secretary. Any such determination by the Secretary shall be conclusive unless, within 30 days after receipt of the notice of such final determination, the applicant appeals to the United States Court of Appeals for the circuit in which the applicant has its principal place of business or to the United States Court of Appeals for the District of Columbia Circuit. (ii) If the claim is approved, the Agency shall notify the applicant, in writing, and shall also publish in the Federal Register a final rule amending the regulations to authorize the use of the claim.
(l)
(1) Labeling applications for a synonymous term shall be accompanied by the following data which shall be submitted in the following form to the Director, Food Labeling Division, Regulatory Programs, Food Safety and Inspection Service, Washington, DC 20250:

(Date)

The undersigned, _____ submits this labeling application pursuant to 9 CFR 317.369 with respect to (statement of the synonymous term and its proposed use in a nutrient content claim that is consistent with an existing term that has been defined under subpart B of part 317).
Attached hereto, in quadruplicate, or on a computer disc copy, and constituting a part of this labeling application, are the following:
(i) A statement identifying the synonymous term, the existing term defined by a regulation with which the synonymous term

9 CFR §317.369

is claimed to be consistent, and the nutrient that the term is intended to characterize the level of. The statement shall address why the use of the synonymous term as proposed will not be misleading. The statement shall provide examples of the nutrient content claim as it will be used on labels or labeling, as well as the types of products on which the claim will be used. The statement shall also specify whether any limitations not applicable to the use of the defined term are intended to apply to the use of the synonymous term.

(ii) A detailed explanation supported by any necessary data of why use of the proposed term is requested, including whether the existing defined term is inadequate for the purpose of effectively characterizing the level of a nutrient. This explanation shall also state what nutritional benefit to the public will derive from use of the claim as proposed, and why such benefit is not available through the use of existing terms defined by regulation. If the claim is intended for a specific group within the population, the analysis shall specifically address nutritional needs of such group, scientific data sufficient for such purpose, and data and information to the extent necessary to demonstrate that consumers can be expected to understand the meaning of the term under the proposed conditions of use.

Yours very truly,

Applicant

By _____
(Indicate authority)

(2) Upon receipt of the labeling application and supporting documentation, the applicant shall be notified, in writing, of the date on which the labeling application was received. Such notice shall inform the applicant that the labeling application is undergoing Agency review and that the applicant shall subsequently be notified of the Agency's decision to consider for further review or deny the labeling application.

(3) Upon review of the labeling application and supporting documentation, the Agency shall notify the applicant, in writing, that the labeling application is either being considered for further review or that it has been summarily denied by the Administrator.

(4) If the labeling application is summarily denied by the Administrator, the written notification shall state the reasons therefor, including why the Agency has determined that the proposed synonymous term is false or misleading. The notification letter shall inform the applicant that the applicant may submit a written statement by way of answer to the notification, and that the applicant shall have the right to request a hearing with respect to the merits or validity of the Administrator's decision to deny the use of the proposed synonymous term.

(i) If the applicant fails to accept the determination of the Administrator and files an answer and requests a hearing, and the Administrator, after review of the answer, determines the initial determination to be correct, the Administrator shall file with the Hearing Clerk of the Department the notification, answer, and the request for a hearing, which shall constitute the complaint and answer in the proceeding, which shall thereafter be conducted in accordance with the Department's Uniform Rules of Practice.

(ii) The hearing shall be conducted before an administrative law judge with the opportunity for appeal to the Department's Judicial Officer, who shall make the final determination for the Secretary. Any such determination by the Secretary shall be conclusive unless, within 30 days after receipt of notice of such final determination, the applicant appeals to the United States Court of Appeals for the circuit in which the applicant has its principal place of business or to the United States Court of Appeals for the District of Columbia Circuit.

(5) If the claim is approved, the Agency shall notify the applicant, in writing, and shall publish in the Federal Register a notice informing the public that the synonymous term has been approved for use.

(m)

(1) Labeling applications for the use of an implied nutrient content claim in a brand name shall be accompanied by the following data which shall be submitted in the following form to the Director, Food Labeling Division, Regulatory Programs, Food Safety and Inspection Service, Washington, DC 20250:

(Date)

The undersigned, ------ submits this labeling application pursuant to 9 CFR 317.369 with respect to (statement of the implied nutrient content claim and its proposed use in a brand name).

Attached hereto, in quadruplicate, or on a computer disc copy, and constituting a part of this labeling application, are the

9 CFR §317.369

following:
(i) A statement identifying the implied nutrient content claim, the nutrient the claim is intended to characterize, the corresponding term for characterizing the level of such nutrient as defined by a regulation, and the brand name of which the implied claim is intended to be a part. The statement shall address why the use of the brand-name as proposed will not be misleading. The statement shall provide examples of the types of products on which the brand name will appear. It shall also include data showing that the actual level of the nutrient in the food would qualify the label of the product to bear the corresponding term defined by regulation. Assay methods used to determine the level of a nutrient shall meet the requirements stated under labeling application format in paragraph (k)(1)(iii) of this section.
(ii) A detailed explanation supported by any necessary data of why use of the proposed brand name is requested. This explanation shall also state what nutritional benefit to the public will derive from use of the brand name as proposed. If the branded product is intended for a specific group within the population, the analysis shall specifically address nutritional needs of such group and scientific data sufficient for such purpose.

Yours very truly,

Applicant

By _____

(2) Upon receipt of the labeling application and supporting documentation, the applicant shall be notified, in writing, of the date on which the labeling application was received. Such notice shall inform the applicant that the labeling application is undergoing Agency review and that the applicant shall subsequently be notified of the Agency's decision to consider for further review or deny the labeling application.
(3) Upon review of the labeling application and supporting documentation, the Agency shall notify the applicant, in writing, that the labeling application is either being considered for further review or that it has been summarily denied by the Administrator.
(4) If the labeling application is summarily denied by the Administrator, the written notification shall state the reasons therefor, including why the Agency has determined that the proposed implied nutrient content claim is false or misleading. The notification letter shall inform the applicant that the applicant may submit a written statement by way of answer to the notification, and that the applicant shall have the right to request a hearing with respect to the merits or validity of the Administrator's decision to deny the use of the proposed implied nutrient content claim.
(i) If the applicant fails to accept the determination of the Administrator and files an answer and requests a hearing, and the Administrator, after review of the answer, determines the initial determination to be correct, the Administrator shall file with the Hearing Clerk of the Department the notification, answer, and the request for a hearing, which shall constitute the complaint and answer in the proceeding, which shall thereafter be conducted in accordance with the Department's Uniform Rules of Practice.
(ii) The hearing shall be conducted before an administrative law judge with the opportunity for appeal to the Department's Judicial Officer, who shall make the final determination for the Secretary. Any such determination by the Secretary shall be conclusive unless, within 30 days after receipt of notice of such final determination, the applicant appeals to the United States Court of Appeals for the circuit in which the applicant has its principal place of business or to the United States Court of Appeals for the District of Columbia Circuit.
(5) If the labeling application is not summarily denied by the Administrator, the Administrator shall publish a notice of the labeling application in the Federal Register seeking comment on the use of the implied nutrient content claim. The notice shall also summarize the labeling application, including where the supporting documentation can be reviewed. The Administrator's notice shall seek comment from consumers, the industry, consumer and industry groups, and other interested persons on the labeling application and the use of the implied nutrient content claim. After public comment has been received and reviewed by the Agency, the Administrator shall make a determination on whether the implied nutrient content claim shall be approved for use on the labeling of meat food products.
(i) If the claim is denied by the Administrator, the Agency shall notify the applicant, in writing, of the basis for the denial, including the reason why the claim on the labeling was determined by the Agency to be false or misleading. The notification letter shall also inform the applicant that the applicant may submit a written statement by way of answer to the notification, and that the applicant shall have the right to request a hearing

with respect to the merits or validity of the Administrator's decision to deny the use of the proposed implied nutrient content claim.

(A) If the applicant fails to accept the determination of the Administrator and files an answer and requests a hearing, and the Administrator, after review of the answer, determines the initial determination to be correct, the Administrator shall file with the Hearing Clerk of the Department the notification, answer, and the request for a hearing, which shall thereafter be conducted in accordance with the Department's Uniform Rules of Practice.

(B) The hearing shall be conducted before an administrative law judge with the opportunity for appeal to the Department's Judicial Officer, who shall make the final determination for the Secretary. Any such determination by the Secretary shall be conclusive unless, within 30 days after receipt of the notice of such final determination, the applicant appeals to the United States Court of Appeals for the circuit in which the applicant has its principal place of business or to the United States Court of Appeals for the District of Columbia Circuit.

(ii) If the claim is approved, the Agency shall notify the applicant, in writing, and shall also publish in the Federal Register a notice informing the public that the implied nutrient content claim has been approved for use.

(Paperwork requirements were approved by the Office of Management and Budget under control number 0583-0088)

[58 FR 664, Jan. 6, 1993, as amended at 59 FR 45196, Sept. 1, 1994; 60 FR 196, Jan. 3, 1995]

§§317.370-317.379 [Reserved]

§317.380 Label statements relating to usefulness in reducing or maintaining body weight.

(a) *General requirements.* Any product that purports to be or is represented for special dietary use because of usefulness in reducing body weight shall bear:

(1) Nutrition labeling in conformity with §317.309 of this subpart, unless exempt under that section, and

(2) A conspicuous statement of the basis upon which the product claims to be of special dietary usefulness.

(b) *Nonnutritive ingredients.*

(1) Any product subject to paragraph (a) of this section that achieves its special dietary usefulness by use of a nonnutritive ingredient (i.e., one not utilized in normal metabolism) shall bear on its label a statement that it contains a nonnutritive ingredient and the percentage by weight of the nonnutritive ingredient.

(2) A special dietary product may contain a nonnutritive sweetener or other ingredient only if the ingredient is safe for use in the product under the applicable law and regulations of this chapter. Any product that achieves its special dietary usefulness in reducing or maintaining body weight through the use of a nonnutritive sweetener shall bear on its label the statement required by paragraph (b)(1) of this section, but need not state the percentage by weight of the nonnutritive sweetener. If a nutritive sweetener(s) as well as nonnutritive sweetener(s) is added, the statement shall indicate the presence of both types of sweetener; e.g., "Sweetened with nutritive sweetener(s) and nonnutritive sweetener(s)."

(c) *"Low calorie" foods.* A product purporting to be "low calorie" must comply with the criteria set forth for such foods in §317.360.

(d) *"Reduced calorie" foods and other comparative claims.* A product purporting to be "reduced calorie" or otherwise containing fewer calories than a reference food must comply with the criteria set forth for such foods in §317.360(b) (4) and (5).

(e) *"Label terms suggesting usefulness as low calorie or reduced calorie foods".*

(1) Except as provided in paragraphs (e)(2) and (e)(3) of this section, a product may be labeled with terms such as "diet," "dietetic," "artificially sweetened," or "sweetened with nonnutritive sweetener" only if the claim is not false or misleading, and the product is labeled "low calorie" or "reduced calorie" or bears another comparative calorie claim in compliance with the applicable provisions in this subpart.

(2) Paragraph (e)(1) of this section shall not apply to any use of such terms that is specifically authorized by regulation governing a particular food, or, unless otherwise restricted by regulation, to any use of the term "diet" that clearly shows that the product is offered solely for a dietary use other than regulating body weight, e.g., "for low sodium diets."

(3) Paragraph (e)(1) of this section shall not apply to any use of such terms on a formulated meal replacement or other product that is represented to be of special dietary use as a whole meal, pending the issuance of a regulation governing the use of such terms on foods.

9 CFR §317.400

(f) *"Sugar free" and "no added sugar"*. Criteria for the use of the terms "sugar free" and "no added sugar" are provided for in §317.360(c).

[58 FR 664, Jan. 6, 1993; 58 FR 43788, Aug. 18, 1993, as amended at 58 FR 47627, Sept. 10, 1993; 58 FR 66075, Dec. 17, 1993; 60 FR 196, Jan. 3, 1995]

§§317.381-317.399 [Reserved]

§317.400 Exemption from nutrition labeling.

(a) The following meat or meat food products are exempt from nutrition labeling:

(1) Food products produced by small businesses provided that the labels for these products bear no nutrition claims or nutrition information,

(i) A food product, for the purposes of the small business exemption, is defined as a formulation, not including distinct flavors which do not significantly alter the nutritional profile, sold in any size package in commerce.

(ii) For purposes of this paragraph, a small business is any single-plant facility or multi-plant company/firm that employs 500 or fewer people and produces no more than the following amounts of pounds of the product qualifying the firm for exemption from this subpart:

(iii) For purposes of this paragraph, calculation of the amount of pounds shall be based on the most recent 2-year average of business activity. Where firms have been in business less than 2 years or where products have been produced for less than 2 years, reasonable estimates must indicate that the annual pounds produced will not exceed the amounts specified.

(A) During the first year of implementation of nutrition labeling, from July 1994 to July 1995, 250,000 pounds or less,

(B) During the second year of implementation of nutrition labeling, from July 1995 to July 1996, 175,000 pounds or less, and

(C) During the third year of implementation and subsequent years thereafter, 100,000 pounds or less.

(2) Products intended for further processing, provided that the labels for these products bear no nutrition claim or nutrition information,

(3) Products that are not for sale to consumers, provided that the labels for these products bear no nutrition claims or nutrition information,

(4) Products in small packages that are individually wrapped packages of less than 1/2 ounce net weight, provided that the labels for these products bear no nutrition claims or nutrition information,

(5) Products custom slaughtered or prepared,

(6) Products intended for export, and

(7) The following products prepared and served or sold at retail provided that the labels or the labeling of these products bear no nutrition claims or nutrition information:

(i) Ready-to-eat products that are packaged or portioned at a retail store or similar retail-type establishment; and

(ii) Multi-ingredient products (e.g., sausage) processed at a retail store or similar retail-type establishment.

(b) Restaurant menus generally do not constitute labeling or fall within the scope of these regulations.

(c)

(1) Foods represented to be specifically for infants and children less than 2 years of age shall bear nutrition labeling as provided in paragraph (c)(2) of this section, except such labeling shall not include calories from fat, calories from saturated fat, saturated fat, stearic acid, polyunsaturated fat, monounsaturated fat, and cholesterol.

(2) Foods represented or purported to be specifically for infants and children less than 4 years of age shall bear nutrition labeling except that:

(i) Such labeling shall not include declarations of percent of Daily Value for total fat, saturated fat, cholesterol, sodium, potassium, total carbohydrate, and dietary fiber;

(ii) Nutrient names and quantitative amounts by weight shall be presented in two separate columns;

(iii) The heading "Percent Daily Value" required in § 317.309(d)(6) shall be placed immediately below the quantitative information by weight for protein;

(iv) The percent of the Daily Value for protein, vitamins, and minerals shall be listed immediately below the heading "Percent Daily Value"; and

(v) Such labeling shall not include the footnote specified at 21 CFR 317.309(d)(9).

9 CFR §317.400

(d)

(1) Products in packages that have a total surface area available to bear labeling of less than 12 square inches are exempt from nutrition labeling, provided that the labeling for these products bear no nutrition claims or other nutrition information. The manufacturer, packer, or distributor shall provide, on the label of packages that qualify for and use this exemption, an address or telephone number that a consumer can use to obtain the required nutrition information (e.g., "For nutrition information call 1-800-123-4567").

(2) When such products bear nutrition labeling, either voluntarily or because nutrition claims or other nutrition information is provided, all required information shall be in a type size no smaller than 6 point or all upper case type of 1/16-inch minimum height, except that individual serving-size packages of meat products that have a total area available to bear labeling of 3 square inches or less may provide all required information in a type size no smaller than 1/32-inch minimum height.

[58 FR 664, Jan. 6, 1993, as amended at 58 FR 47627, Sept. 10, 1993; 59 FR 45196, Sept. 1, 1994; 60 FR 197, Jan. 3, 1995]

Chapter 11—FSIS Labeling Rules for Poultry Products

> ### In This Chapter...
>
> **Full text of the FSIS poultry labeling regulations** (as of May 1, 1998):
>
> - General labeling provisions: 9 *CFR* Part 381, Subpart N — **page 11.3**
>
> - Nutrition labeling provisions: 9 *CFR* Part 381, Subpart Y — **page 11.23**

PART 381-POULTRY PRODUCTS INSPECTION REGULATIONS
Authority: 7 U.S.C. 138F; 7 U.S.C. 450; 21 U.S.C. 451-470; 7 CFR 2.18, 2.53. Source: 37 FR 9706, May 16, 1972, unless otherwise noted.

Subpart N - Labeling and Containers

§381.115	Containers of inspected and passed poultry products required to be labeled.
§381.116	Wording on labels of immediate containers.
§381.117	Name of product.
§381.118	Ingredients statement.
§381.119	Declaration of artificial flavoring or coloring.
§381.120	Antioxidants; chemical preservatives; and other additives.
§381.121	Quantity of contents.
§381.121a	Quantity of contents labeling.
§381.121b	Definitions and procedures for determining net weight compliance.
§381.121c	Scale requirements for accurate weights, repairs, adjustments, and replacement after inspection.
§381.121d	Scales; testing of.
§381.121e	Handling of failed product.
§381.122	Identification of manufacturer, packer or distributor.
§381.123	Official inspection mark; official establishment number.
§381.124	Dietary food claims.
§381.125	Special handling label requirements.
§381.126	Date of packing and date of processing; contents of cans.
§381.127	Wording on labels of shipping containers.
§381.128	Labels in foreign languages.
§381.129	False or misleading labeling or containers.
§381.130	False or misleading labeling or containers; orders to withhold from use.
§381.131	Preparation of labeling or other devices bearing official inspection marks without advance approval prohibited; exceptions.
§381.132	Labeling approval.
§381.133	Generically approved labeling.
§381.134	Requirement of formulas.
§381.135	Irradiated poultry product.
§381.136	Affixing of official identification.
§381.137	Evidence of labeling and devices approval.

§381.138 Unauthorized use or disposition of approved labeling or devices.
§381.139 Removal of official identifications.
§381.140 Relabeling poultry products.
§381.141 [Removed and Reserved]
§381.143 [Reserved]
§381.144 Packaging materials.

Subpart Y - Nutrition Labeling

§381.400 Nutrition labeling of poultry products.
§381.401 [Reserved]
§381.402 Location of nutrition information.
§381.408 Labeling of poultry products with number of servings.
§381.409 Nutrition label content.
§381.410 [Reserved]
§381.411 [Reserved]
§381.412 Reference amounts customarily consumed per eating occasion.
§381.413 Nutrient content claims; general principles.
§381.414 -
 381.442 [Reserved]
§381.443 Significant participation for voluntary nutrition labeling.
§381.444 Identification of major cuts of poultry products.
§381.445 Guidelines for voluntary nutrition labeling of single-ingredient, raw products.
§§381.446 -
 381.453 [Reserved]
§381.454 Nutrient content claims for "good source" and "high", and "more".
§381.455 [Reserved]
§381.456 Nutrient content claims for "light" or "lite".
§§381.457 -
 381.459 [Reserved]
§381.460 Nutrient content claims for calorie content.
§381.461 Nutrient content claims for sodium content.
§381.462 Nutrient content claims for fat, fatty acids, and cholesterol content.
§381.463 Nutrient content claims for "healthy."
§§381.464 -
 381.468 [Reserved]
§381.469 Labeling applications for nutrient content claims.
§381.470 -
 381.479 [Reserved]
§381.480 Label statements relating to usefulness in reducing or maintaining body weight.
§§381.481 -
 381.499 [Reserved]
§381.500 Exemption from nutrition labeling.

9 CFR §381.116

Subpart N-Labeling and Containers

§381.115 Containers of inspected and passed poultry products required to be labeled.

Except as may be authorized in specific cases by the Administrator with respect to shipment of poultry products between official establishments, each shipping container and each immediate container of any inspected and passed poultry product shall at the time it leaves the official establishment bear a label which contains information, and has been approved, in accordance with this subpart.

§381.116 Wording on labels of immediate containers.

(a) Each label for use on immediate containers for inspected and passed poultry products shall bear on the principal display panel (except as otherwise permitted in the regulations), the items of information required by this subpart. Such items of information shall be in distinctly legible form. Except as provided in §381.128, all words, statements and other information required by or under authority of the Act to appear on the label or labeling shall appear thereon in the English language: *Provided, however*, That in the case of products distributed solely in Puerto Rico, Spanish may be substituted for English for all printed matter except the USDA inspection legend.

(b) The principal display panel shall be the part of a label that is most likely to be displayed, presented, shown, or examined under customary conditions of display for sale. The principal display panel shall be large enough to accommodate all the mandatory label information required to be placed thereon by the regulations with clarity and conspicuousness and without being obscured by design or vignettes, or crowding. Where packages bear alternate principal display panels, information required to be placed on the principal display panel shall be duplicated on each principal display panel. The area that is to bear the principal display panel shall be:

   (1) In the case of a rectangular package, one entire side, the area of which is the product of the height times the width of that side.

   (2) In the case of a cylindrical or nearly cylindrical container:

   (i) An area on the side of the container that is 40 percent of the product of the height of the container times the circumference, or

   (ii) A panel, the width of which is one-third of the circumference and the height of which is as high as the container: *Provided, however,* That there is, immediately to the right or left of such principal display panel, a panel which has a width not greater than 20 percent of the circumference and a height as high as the container, and which is reserved for information prescribed in §§381.118, 381.122, and 381.123. Such panel shall be known as the "20 percent panel" and such information may be shown on that panel in lieu of showing it on the principal display panel as provided in this §381.116.

   (3) In the case of a container of any other shape, 40 percent of the total surface of the container.

In determining the area of the principal display panel, exclude tops, bottoms, flanges at tops and bottoms of cans, and shoulders and necks of bottles or jars.

(c)

   (1) The information panel is that part of a label that is the first surface to the right of the principal display panel as observed by an individual facing the principal display panel, with the following exceptions:

   (i) If the first surface to the right of the principal display panel is too small to accommodate the required information or is otherwise unusable label space, e.g., folded flaps, tear strips, opening flaps, heat-sealed flaps, the next panel to the right of this part of the label may be used.

   (ii) If the package has one or more alternate principal display panels, the information panel is to the right of any principal display panel.

   (iii) If the top of the container is the principal display panel and the package has no alternate principal display panel, the information panel is any panel adjacent to the principal display panel.

   (2) (i) Except as otherwise permitted in this part, all information required to appear on the principal display panel or permitted to appear on the information panel shall appear on the same panel unless there is insufficient

9 CFR §381.117

space. In determining the sufficiency of the available space, except as otherwise prescribed in this part, any vignettes, designs, and any other nonmandatory information shall not be considered. If there is insufficient space for all required information to appear on a single panel, it may be divided between the principal display panel and the information panel, provided that the information required by any given provision of this part, such as the ingredients statement, is not divided and appears on the same panel.

(ii) All information appearing on the information panel pursuant to this section shall appear in one place without intervening material, such as designs or vignettes.

[37 FR 9706, May 16, 1972, as amended at 40 FR 11347, Mar. 11, 1975; 59 FR 40214, Aug. 8, 1994]

§381.117 Name of product and other labeling..

(a) The label shall show the name of the product, which, in the case of a poultry product which purports to be or is represented as a product for which a definition and standard of identity or composition is prescribed in subpart P, shall be the name of the food specified in the standard, and in the case of any other poultry product shall be the common or usual name of the food, if any there be, and if there is none, a truthful descriptive designation.

(b) The name of the product required to be shown on labels for fresh or frozen raw whole carcasses of poultry shall be in either of the following forms: The name of the kind (such as chicken, turkey, or duck) preceded by the qualifying term "young" or "mature" or "old", whichever is appropriate; or the appropriate class name as described in §381.170(a). The name of the kind may be used in addition to the class name, but the name of the kind alone without the qualifying age or class term is not acceptable as the name of the product, except that the name "chicken" may be used without such qualification with respect to a ready-to-cook pack of fresh or frozen cut- up young chickens, or a half of a young chicken, and the name "duckling" may be used without such qualification with respect to a ready-to-cook pack of fresh or frozen young ducks. The class name may be appropriately modified by changing the word form, such as using the term "roasting chicken", rather than "roaster." The appropriate names for cut-up parts are set forth in §381.170(b). When naming parts cut from young poultry, the identity of both the kind of poultry and the name of the part shall be included in the product name. The product name for parts or portions cut from mature poultry shall include, along with the part or portion name, the class name or the qualifying term "mature." The name of the product for cooked or heat processed poultry products shall include the kind name of the poultry from which the product was prepared but need not include the class name or the qualifying term "mature."

(c) Poultry products containing light and dark chicken or turkey meat in quantities other than the natural proportions, as indicated in Table 1 in this paragraph, must have a qualifying statement in conjunction with the name of the product indicating, as shown in Table 1, the types of meat actually used, except that when the product contains less than 10 percent cooked deboned poultry meat or is processed in such a manner that the character of the light and dark meat is not distinguishable, the qualifying statement will not be required, unless the product bears a label referring to the light or dark meat content. In the latter case, the qualifying statement is required if the light and dark meat are not present in natural proportions. The qualifying statement must be in type at least one-half the size and of equal boldness as the name of the product; e.g., Boned Turkey (Dark Meat).

TABLE 1

Label terminology	Percent light meat	Percent dark meat
Natural proportions..............................	50-65	50-35
Light or white meat.............................	100	0
Dark meat..	0	100
Light and dark meat............................	51-65	49-35

9 CFR §381.118

Dark and light meat............................	35-49	65-51
Mostly white meat..............................	66 or more	34 or less
Mostly dark meat...............................	34 or less	66 or more

(d) Boneless poultry products shall be labeled in a manner that accurately describes their actual form and composition. The product name shall specify the form of the product (e.g., emulsified, finely chopped, etc.), and the kind name of the poultry, and if the product does not consist of natural proportions of skin and fat, as they occur in the whole carcass, shall also include terminology that describes the actual composition. If the product is cooked, it shall be so labeled. For the purpose of this paragraph, natural proportions of skin, as found on a whole chicken or turkey carcass, will be considered to be as follows:

	Percent	
	Raw	Cooked
Chicken................................	20	25
Turkey...................................	15	20

Boneless poultry product shall not have a bone solids content of more than 1 percent, calculated on a weight basis.
(e) On the label of any "Mechanically Separated (Kind of Poultry) " described in §381.173, the name of such product shall be followed immediately by the phrase: "with excess skin" unless such product is made from poultry product that does not include skin in excess of the natural proportion of skin present on the whole carcass, as specified in paragraph (d) of this section. Appropriate terminology on the label shall indicate if heat treatment has been used in the preparation of the product. The labeling information described in this paragraph shall be identified on the label before the product leaves the establishment at which it is manufactured.
[37 FR 9706, May 16, 1972, as amended at 55 FR 7294, Mar. 1, 1990; 55 FR 26422, June 28, 1990; 58 FR 38049, July 15, 1993; 59 FR 40215, Aug. 8, 1994; 60 FR 55983, Nov. 3, 1995]

§381.118 Ingredients statement.

(a)
(1) The label shall show a statement of the ingredients in the poultry product if the product is fabricated from two or more ingredients. Such ingredients shall be listed by their common or usual names in the order of their descending proportions, except as prescribed in paragraph (a)(2) of this section.
(2) (i) Product ingredients which are present in individual amounts of 2 percent or less by weight may be listed in the ingredients statement in other than descending order of predominance: *Provided,* That such ingredients are listed by their common or usual names at the end of the ingredients statement and preceded by a quantifying statement, such as "Contains _____ percent or less of _____," or "Less than _____ percent of _____." The percentage of the ingredient(s) shall be filled in with a threshold level of 2 percent, 1.5 percent, 1.0 percent, or 0.5 percent, as appropriate. No ingredient to which the quantifying statement applies may be present in an amount greater than the stated threshold. Such a quantifying statement may also be utilized when an ingredients statement contains a listing of ingredients by individual components. Each component listing may utilize the required quantifying statement at the end of each component ingredients listing.
(ii) Such ingredients may be adjusted in the product formulation without a change being made in the ingredients statement on the labeling, provided that the adjusted amount complies with §381.147(f)(4) and subpart P of this

9 CFR §381.119

part, and does not exceed the amount shown in the quantifying statement. Any such adjustments to the formulation shall be provided to the inspector-in-charge.

(b) For the purpose of this paragraph, the term "chicken meat," unless modified by an appropriate adjective, is construed to mean deboned white and dark meat; whereas the term "chicken" may include other edible parts such as skin and fat not in excess of their natural proportions, in addition to the chicken meat. If the term "chicken meat:" is listed and the product also contains skin, giblets, and fat, it is necessary to list each such ingredient. Similar principles shall be followed in listing ingredients of poultry products processed from other kinds of poultry.

(c) The terms spice, natural flavor, natural flavoring, flavor or flavoring may be used in the following manner:

(1) The term "spice" means any aromatic vegetable substance in the whole, broken, or ground form, with the exceptions of onions, garlic, and celery, whose primary function in food is seasoning rather than nutritional and from which no portion of any volatile oil or other flavoring principle has been removed. Spices include the spices listed in 21 CFR 182.10 and 184.

(2) The term "natural flavor," "natural flavoring," "flavor," or "flavoring" means the essential oil, oleoresin, essence or extractive, protein hydrolysate, distillate, or any product of roasting, heating or enzymolysis, which contains the flavoring constituents derived from a spice, fruit, or fruit juice, vegetable or vegetable juice, edible yeast, herb, bark, bud, root, leaf, or any other edible portions of a plant, meat, seafood, poultry, eggs, dairy products, or fermentation products thereof, whose primary function in food is flavoring rather than nutritional. Natural flavors include the natural essence or extractives obtained from plants listed in 21 CFR 182.10, 182.20, 182.40, 182.50, and 184, and the substances listed in 21 CFR 172.510, The term natural flavor, natural flavoring, flavor, or flavoring may also be used to designate spices, powdered onion, powdered garlic, and powdered celery.

(i) Natural flavor, natural flavoring, flavor or flavoring as described in paragraphs (c)(1) and (2) of this section, which are also colors shall be designated as "natural flavor and coloring," "natural flavoring and coloring," "flavor and coloring," or "flavoring and coloring" unless designated by their common or usual name.

(ii) Any ingredient not designated in paragraphs (c)(1) and (2) whose function is flavoring, either in whole or in part, must be designated by its common or usual name. Those ingredients which are of livestock or poultry origin must be designated by names that include the species and livestock and poultry tissues from which the ingredients are derived.

(d) On containers of frozen dinners, entrees, and pizzas, and similarly packaged products in cartons, the ingredient statement may be placed on the front riser panel: *Provided*, That the words "see ingredients," followed immediately by an arrow pointing to the front riser panel, are placed on the principal display panel immediately above the location of such statement, without intervening printing or designs.

(e) The ingredients statement may be placed on the information panel, except as otherwise permitted in this subchapter.

(f) Establishments may interchange the identity of two kinds of poultry (e.g., chicken and turkey, chicken meat and turkey meat) used in a product formulation without changing the product's ingredient statement or product name under the following conditions:

(1)(i) The two kinds of poultry used must comprise at least 70 percent by weight of the poultry and the poultry ingredients [e.g. giblets, skin or fat in excess of natural proportions, or mechanically separated (kind)] used; and,

(ii) Neither of the two kinds of poultry used can be less than 30 percent by weight of the total poultry and poultry ingredients used;

(2) The word "and" in lieu of a comma must be shown between the declaration of the two kinds of poultry in the ingredients statement and in the product name.

[37 FR 9706, May 16, 1972, as amended at 55 FR 7294, Mar. 1, 1990; 55 FR 26422, June 28, 1990; 58 FR 38049, July 15, 1993; 59 FR 40215, Aug. 8, 1994; 63 FR 11360, Mar. 9, 1998]

§381.119 Declaration of artificial flavoring or coloring.

(a) When an artificial smoke flavoring or a smoke flavoring is added as an ingredient in the formula of any poultry product, there shall appear on the label, in prominent letters and contiguous to the name of the product, a statement such as "Artificial Smoke Flavoring Added" or "Smoke Flavoring Added," as applicable, and the ingredient statement shall identify any artificial smoke flavoring or smoke flavoring added as an ingredient in the formula of

9 CFR §381.121

the poultry product.

(b) Any poultry product which bears or contains any artificial flavoring other than an artificial smoke flavoring or a smoke flavoring, or bears or contains any artificial coloring shall bear a statement stating that fact on the immediate container or, if there is none, on the product.

§381.120 Antioxidants; chemical preservatives; and other additives.

When an antioxidant is added to a poultry product, there shall appear on the label in prominent letters and contiguous to the name of the product, a statement showing the name of the antioxidant and the purpose for which it is added, such as "BHA added to help protect the flavor." Immediate containers of poultry products packed in, bearing, or containing any chemical preservative shall bear a label stating that fact and naming the additive and the purpose of its use. Immediate containers of poultry products packed in, bearing or containing any other chemical additive shall bear a label naming the additive and the purpose of its use when required by the Administrator in specific cases. When approved proteolytic enzymes as permitted in §381.147 of this subchapter are used in mature poultry muscle tissue, there shall appear on the label, in a prominent manner, contiguous to the product name, the statement "Tenderized with [approved enzyme]," to indicate the use of such enzymes. Any other approved substance which may be used in the solution shall also be included in the statement. When approved inorganic chlorides as permitted in §381.147 of this subchapter are used in mature poultry muscle tissue, there shall appear on the label, in a prominent manner, contiguous to the product name, the statement, "Tenderized with (name of approved inorganic chloride(s))" to indicate the use of such inorganic chlorides. Any other approved substance which may be used in the solution shall also be included in the statement.
[37 FR 9706, May 16, 1972, as amended at 45 FR 58820, Sept. 5, 1980; 49 FR 18999, May 4, 1984]

§381.121 Quantity of contents.

(a) The label shall bear a statement of the quantity of contents in terms of weight or measures as provided in paragraph (c)(5) of this section. However, the Administrator may approve the use of labels for certain types of consumer packages which do not bear a statement of the net weight that would otherwise be required under this subparagraph: *Provided,* That the shipping container bears a statement "Net weight to be marked on consumer packages prior to display and sale": *And provided further,* That the total net weight of the contents of the shipping container is marked on such container: *And provided further,* That the shipping container bears a statement "Tare weight of consumer package" and in close proximity thereto, the actual tare weight (weight of packaging material), weighed to the nearest one-eighth ounce or less, of the individual consumer package in the shipping container. The above-specified statements may be added to approved shipping container labels upon approval by the inspector in charge.

(b) When a poultry product and a nonpoultry product are separately wrapped and are placed in a single immediate container bearing the same name of both products, the net weight on such immediate container may be the total net weight of the products, or such immediate container may show the net weights of the poultry product and the nonpoultry product separately. Notwithstanding the other provisions of this paragraph, the label on consumer size retail packages of stuffed poultry and other stuffed poultry products must show the total net weight of the poultry product, and in close proximity thereto, a statement specifying the minimum weight of the poultry in the product.

(c)

(1) The statement of net quantity of contents shall appear (except as otherwise permitted under this paragraph (c)), on the principal display panel of all containers to be sold at retail intact, in conspicuous and easily legible boldface print or type, in distinct contrast to other matter on the container, and shall be declared in accordance with the provisions of this paragraph (c). An unused tare weight, as defined in section 381.121b of this subchapter, may be printed adjacent to the statement of net quantity of contents when the product is packaged totally with impervious packaging material and is packed with a usable medium.

(2) The statement shall be placed on the principal display panel within the bottom 30 percent of the area of the

9 CFR §381.121

panel, in lines generally parallel to the base: *Provided,* That on packages having a principal display panel of 5 square inches or less, the requirement for placement within the bottom 30 percent of the area of the label panel shall not apply when the statement meets the other requirements of this paragraph. The declaration may appear in more than one line.

(3) The statement shall be in letters and numerals in type size established in relationship to the area of the principal display panel of the package and shall be uniform for all packages of substantially the same size by complying with the following type specifications:

(i) Not less than one-sixteenth inch in height on containers, the principal display panel of which has an area of 5 square inches or less;

(ii) Not less than one-eighth inch in height on containers, the principal display panel of which has an area of more than 5 but not more than 25 square inches;

(iii) Not less than three-sixteenth inch in height on containers, the principal display panel of which has an area of more than 25 but not more than 100 square inches;

(iv) Not less than one-quarter inch in height on containers, the principal display panel of which has an area of more than 100 but not more than 400 square inches;

(v) Not less than one-half inch in height on containers, the principal display panel of which has an area of more than 400 square inches.

(vi) The ratio of height to width of letters and numerals shall not exceed a differential of 3 units to 1 unit (no more than 3 times as high as it is wide). This height standard pertains to upper case or capital letters. When upper and lower case or all lower case letters are used, it is the lower case letter "o" or its equivalent that shall meet the minimum standards. When fractions are used, each component numeral shall meet one- half the height standards.

(4) The statement shall appear as a distinct item on the principal display panel and shall be separated, from other label information appearing to the left or right of the statement, by a space at least equal in width to twice the width of the letter "N" of the style of type used in the quantity of contents statement and shall be separated from other label information appearing above or below the statement by a space at least equal in height to the height of the lettering used in the statement.

(5) The terms "net weight" or "net wt." shall be used when stating the net quantity of contents in terms of weight, and the term "net contents" or "contents" when stating the net quantity of contents in terms of fluid measure. Except as provided in §381.128, the statement shall be expressed in terms of avoirdupois weight or liquid measure. Where no general consumer usage to the contrary exists, the statement shall be in terms of liquid measure, if the product is liquid, or in terms of weight if the product is solid, semisolid, viscous or a mixture of solid and liquid. On packages containing less than 1 pound or 1 pint, the statement shall be expressed in ounces or fractions of a pint, respectively. On packages containing 1 pound or 1 pint or more, and less than 4 pounds or 1 gallon, the statement shall be expressed as a dual declaration both in ounces and (immediately thereafter in parenthesis) in pounds, with any remainder in terms of ounces or common or decimal fraction of the pound, or in the case of liquid measure, in the largest whole units with any remainder in terms of fluid ounces or common or decimal fraction of the pint or quart. For example, a declaration of three-fourths pound avoirdupois weight shall be expressed as "Net Wt. 12 oz."; a declaration of 1 1/2 pounds avoirdupois weight shall be expressed as "Net Wt. 24 oz. (1 lb. 8 oz.)," "Net Wt. 24 oz. (1 1/2 lb.)," or "Net Wt. 24 oz. (1.5 lbs.)." However, on random weight packages the statement shall be expressed in terms of pounds and decimal fractions of the pound, for packages over 1 pound, and for packages which do not exceed 1 pound the statement may be in decimal fractions of the pound in lieu of ounces. The numbers may be written in provided the unit designation is printed. Paragraphs (c) (8) and (9) of this section permit certain exceptions to this paragraph for multi-unit packages, and random weight consumer size and small packages (less than 1/2 ounce), respectively.

(6) The statement as it is shown on a label shall not be false or misleading and shall express an accurate statement of the quantity of contents of the container. Reasonable variations caused by loss or gain of moisture during the course of good distribution practices or by unavoidable deviations in good manufacturing practices will be recognized. Variations from stated quantity of contents shall be as provided in section 381.121b of this subchapter. The statement shall not include any term qualifying a unit of weight, measure, or count such as "jumbo quart," "full gallon," "giant quart," "when packed," "minimum," or words of similar importance except as provided in paragraph (b) of this section.

(7) Labels for containers which bear any representation as to the number of servings contained therein shall bear,

contiguous to such representation, and in the same size type as is used for such representation, a statement of the net quantity of each such serving.

(8) On a multiunit retail package, a statement of the quantity of contents shall appear on the outside of the package and shall include the number of individual units, the quantity of each individual unit, and, in parentheses, the total quantity of contents of the multiunit package in terms of avoirdupois or fluid ounces, except that such declaration of total quantity need not be followed by an additional parenthetical declaration in terms of the largest whole units and subdivisions thereof, as otherwise required by this paragraph (c). "A multiunit retail package" is a package containing two or more individually packaged units of the identical commodity and in the same quantity, with the individual packages intended to be sold as part of the multiunit retail package but capable of being sold individually. Open multiunit retail packages that do not obscure the number of units and the labeling thereon are not subject to this paragraph (c) (8) if the labeling of each individual unit complies with the requirements of this paragraph (c).

(9) The following exemptions from the requirements contained in this section are hereby established:

(i) Individually wrapped, random weight consumer size packages of poultry products (as specified in paragraph (c)(10) of this section) and poultry products that are subject to shrinkage through moisture loss during good distribution practices and are designated as gray area type of products as defined in NBS handbook 133, section 3.18.2, need not bear a net weight statement when shipped from an official establishment provided a net weight shipping statement which meets the requirements of paragraph (c)(6) of this section is applied to the shipping container prior to shipping it from the official establishment. Net weight statements so applied to the shipping container are exempt from the type size, dual declaration, and placement requirements of this paragraph if an accurate statement of net weight is shown conspicuously on the principal display panel of the shipping container. The net weight also shall be applied directly to random weight consumer size packages prior to retail display and sale. The net weight statement of random weight consumer size packages for retail sale shall be exempt from the type size, dual declaration, and placement requirements of this paragraph if an accurate statement of net weight is shown conspicuously on the principal display panel of the package.

(ii) Individually wrapped and labeled packages of less than 1/2 ounce net weight and random weight consumer size packages shall be exempt from the requirements of this paragraph if they are in a shipping container and the statement of net quantity of contents on the shipping container meets the requirements of paragraph (c)(6) of this section;

(iii) Individually wrapped and labeled packages of less than 1/2 ounce net weight bearing labels declaring net weight, price per pound, and total price, shall be exempt from the type size, dual declaration, and placement requirements of this paragraph if an accurate statement of net weight is shown conspicuously on the principal display panel of the package.

(10) As used in this section a "random weight consumer size package" is one of a lot, shipment or delivery of packages of the same product, with varying weights and with no fixed weight pattern.

[37 FR 9706, May 16, 1972, as amended at 39 FR 4569, Feb. 5, 1974; 53 FR 28635, July 29, 1988; 55 FR 49835, Nov. 30, 1990]

§381.121a Quantity of contents labeling.

Sections 381.121a through 381.121e of this part prescribe the procedures to be followed for determining net weight compliance and prescribe the reasonable variations from the declared net weight on the labels of immediate containers of products in accordance with §381.121 of this part.

[55 FR 49835, Nov. 30, 1990]

§381.121b Definitions and procedures for determining net weight compliance.

(a) For the purpose of §381.121b of this part, the reasonable variations allowed, definitions, and procedures to be used in determining net weight and net weight compliance are described in the National Institute of Standards and

9 CFR §381.121

Technology (NIST) Handbook 133, "Checking the Net Contents of Packaged Goods," Third Edition, September 1988, and Supplements 1, 2, 3, and 4 dated September 1990, October 1991, October 1992, and October 1994, respectively, which are incorporated by reference, with the exception of the NIST Handbook 133 and Supplements 1 and 3 requirements listed in paragraphs (b) and (c) of this section. Those provisions, incorporated by reference herein, are considered mandatory requirements. This incorporation was approved by the Director of the Federal Register in accordance with 5 U.S.C. 552(a) and 1 CFR part 51. (These materials are incorporated as they exist on the date of approval.) Copies may be purchased from the Superintendent of Documents, U.S. Government Printing Office, Washington, DC 20402. It is also available for inspection at the Office of the Federal Register Information Center, room 8401, 1100 L Street NW., Washington, DC 20408.

(b) The following NBS Handbook 133 requirements are not incorporated by reference.

Chapter 2-General Considerations
- 2.13.1 Polyethylene Sheeting and Film
- 2.13.2 Textiles
- 2.13.3 Mulch

Chapter 3-Methods of Test for Packages Labeled by Weight
- 3.11. Aerosol Packages
- 3.14. Glazed Raw Seafood and Fish
- 3.15. Canned Coffee
- 3.16. Borax
- 3.17. Flour

Chapter 4-Methods of Test for Packages Labeled by Volume
- 4.7. Milk
- 4.8. Mayonnaise and Salad Dressing
- 4.9. Paint, Varnish, and Lacquers-Nonaerosol
- 4.11. Peat Moss
- 4.12. Bark Mulch
- 4.15. Ice Cream Novelties

Chapter 5-Methods of Test for Packages Labeled by Count, Length, Area, Thickness, or Combinations of Quantities
- 5.4. Polyethlene Sheeting
- 5.5. Paper Plates
- 5.6. Sanitary Paper Products
- 5.7. Pressed and Blown Glass Tumblers and Stemware

Appendix D: Package Net Contents Regulations
- D.1.1 U.S. Department of Health and Human Servcies, Food and Drug Administration
- D.1.2 U.S. Department of Agriculture, Food Safety and Inspection Servcie
- D.1.3 Federal Trade Commission
- D.1.4 Environmental Protection Agency
- D.1.5 U.S. Department of the Treasury, Bureau of Alcohol, Tobacco, and Firearms

(c) The following requirements of Supplement 1 dated September 1990, Supplement 3 dated October 1992, and Supplement 4 dated October 1994, of NIST Handbook 133 are not incorporated by reference.

Supplement 1
Chapter 2: General Considerations
- 2.13.1. Polyethylene Sheeting and Film
- 2.13.2. Textiles
- 2.13.3. Mulch

9 CFR §381.121

Chapter 3: Methods of Test for Packages Labeled by Weight
    3.11.4. Exhausting the Aerosol Container
Chapter 4: Methods of Test for Packages Labeled by Volume
    4.6.4. Method D: Determining the Net Contents of Compressed Gas in Cylinders
    4.7. Milk
    4.16. Fresh Oysters Labeled by Volume
Chapter 5: Methods of Test for Packages Labeled by Count, Length, Area, Thickness, or Combinations of Quantities
    5.4. Polyethylene Sheeting

Supplement 3
Chapter 3: Methods of Test for Packages Labeled by Weight
    3.17. Flour and Dry Pet Food
Chapter 5: Methods of Test for Packages Labeled by Count, Length, Area, Thickness, or Combinations of Quantities
    5.4. Polyethylene Sheeting
    5.5. Paper Plates
    5.8. Baler Twine
Appendix A. Forms and Worksheets

Supplement 4
    3.11 Aerosol Packages
    3.11.1 Equipment
    3.11.2 Preparation for Test
    3.11.3 The Determination of Net Contents: Part 1
    3.11.4 Exhausting the Aerosol Container
    3.11.5 The Determination of Net Contents: Part 2
Appendix A. Report Forms

[55 FR 49835, Nov. 30, 1990; 60 FR 12884, Mar. 9, 1995]

§381.121c Scale requirements for accurate weights, repairs, adjustments, and replacement after inspection.

(a) All scales used to weight poultry products sold or otherwise distributed in commerce in federally inspected poultry plants shall be installed, maintained, and operated to insure accurate weights. Such scales shall meet the applicable requirements contained in National Institute of Standards and Technology (NIST) Handbook 44, "Specifications, Tolerances and Other Technical Requirements for Weighing and Measuring Devices," 1994 Edition, October 1993, which is incorporated by reference. This incorporation was approved by the Director of the Federal Register in accordance with 5 U.S.C. 552(a) and 1 CFR part 51. (These materials are incorporated as they exist on the date of approval.) A notice of any change in the Handbook cited herein will be published in the Federal Register. Copies may be purchased from the Superintendent of Documents, U.S. Government Printing Office, Washington, DC 20402. It is also available for inspection at the Office of the Federal Register Information Center, room 8401, 1100 L Street NW., Washington, DC 20408.

(b) All scales used to weigh poultry products sold or otherwise distributed in commerce or in State designated under section 5(c) of the Poultry Products Inspection Act, shall be of sufficient capacity to weigh the entire unit and/or package.

(c) No scale shall be used at a federally inspected establishment to weigh poultry products unless it has been found upon test and inspection as specified in NIST Handbook 44 to provide accurate weight. If a scale is inspected or tested and found to be inaccurate, or if any repairs, adjustments or replacements are made to a scale, it shall not be used until it has been reinspected and retested by a USDA official, or a State or local government weights and

9 CFR §381.122

measures official, or a State registered or licensed scale repair firm or person, and it must meet all accuracy requirements as specified in NIST Handbook 44. If a USDA inspector has put a "Retain" tag on a scale it can only be removed by a USDA inspector. As long as the tag is on the scale, it shall not be used.
[55 FR 49836, Nov. 30, 1990; 60 FR 12884, Mar. 9, 1995]

§381.121d Scales; testing of.

(a) The operator of each official establishment that weighs poultry food products shall cause such scales to be tested for accuracy in accordance with the technical requirements of NIST Handbook 44, at least once during the calendar year. In cases where the scales are found not to maintain accuracy between tests, more fequent tests may be required and monitored by an authorized USDA program official.
(b) The operator of each official establishment shall display on or near each scale a valid certification of the scale's accuracy from a State or local government's weights and measures authority or from a State registered or licensed scale repair firm or person, or shall have a net weight program under a Total Quality Control System or Partial Quality Control Program in accordance with §381.145 of this subchapter.
[55 FR 49836, Nov. 30, 1990; 62 FR 45026, Aug. 25, 1997]

§381.121e Handling of failed product.

Any lot of product which is found to be out of compliance with net weight requirements upon testing in accordance with §381.121b of this subchapter shall be handled as follows:
(a) A lot tested in an official establishment and found not to comply with net weight requirements may be reprocessed and must be reweighed and remarked to satisfy the net weight requirements of this section, and be reinspected in accordance with the requirements of this part.
(b) A lot tested outside of an official establishment and found not to comply with net weight requirements must be reweighed and remarked with a proper net weight statement, provided that such reweighing and remarking shall not deface, cover, or destroy any other marking or labeling required under this subchapter and the net quantity of contents is shown with the same prominence as the most conspicuous feature of a label.
[55 FR 49836, Nov. 30, 1990]

§381.122 Identification of manufacturer, packer or distributor.

The name and address, including zip code, of the manufacturer, packer, or distributor shall be shown on the label and if only the name and address of the distributor is shown, it shall be qualified by such term as "packed for," "distributed by," or "distributors." The name and place of business of the manufacturer, packer, or distributor may be shown on the principal display panel, on the 20-percent panel of the principal display panel reserved for required information, on the front riser panel of frozen food cartons, or on the information panel.
[37 FR 9706, May 16, 1972, as amended at 59 FR 40215, Aug. 8, 1994]

§381.123 Official inspection mark; official establishment number.

The immediate container of every inspected and passed poultry product shall bear:
(a) The official inspection legend; and
(b) The official establishment number of the official establishment in which the product was processed under inspection and placed as follows:
    (1) Within the official inspection legend in the form required by Subpart M of this part; or
    (2) Outside the official inspection legend elsewhere on the exterior of the container or its labeling, e.g., the lid of a can, if shown in a prominent and legible manner in a size sufficient to insure easy visibility and recognition and accompanied by the prefix "P"; or

(3) Off the exterior of the container, e.g., on a metal clip used to close casings or bags, or on the back of a paper label of a canned product, or on other packaging or labeling in the container, e.g., on aluminum pans and trays placed within containers, when a statement of its location is printed contiguous to the official inspection legend, such as "Plant No. on Package Closure" or "Plant No. on Pan", if shown in a prominent and legible manner in a size sufficient to ensure easy visibility and recognition; or

(4) On an insert label placed under a transparent covering if clearly visible and legible and accompanied by the prefix "P".

[47 FR 29515, July 7, 1982]

§381.124 Dietary food claims.

If a product purports to be or is represented for any special dietary use by man, its label shall bear a statement concerning its vitamin, mineral, and other dietary properties upon which the claim for such use is based in whole or in part and shall be in conformity with regulations (21 CFR part 125) established pursuant to sections 403 and 701 of the Federal Food, Drug, and Cosmetic Act (21 U.S.C. 343, 371).

§381.125 Special handling label requirements.

(a) Packaged products which require special handling to maintain their wholesome condition shall have prominently displayed on the principal display panel of the label the statement: "Keep Refrigerated," "Keep Frozen," "Keep Refrigerated or Frozen," "Perishable-Keep Under Refrigeration," or such similar statement as the Administrator may approve in specific cases. The immediate containers for products that are frozen during distribution and intended to be thawed prior to or during display for sale shall bear the statement "Shipped/Stored and Handled Frozen for Your Protection, Keep Refrigerated or Freeze." For all canned perishable products, the statement shall be shown in upper case letters one-fourth inch in height for containers having a net weight of 3 pounds or less, and for containers having a net weight over 3 pounds, the statement shall be shown in letters one-half inch in height.

(b) Safe handling instructions shall be provided for all poultry products not heat processed in accordance with the provisions of §381.150(b) or that have not undergone other further processing that would render them ready-to-eat, except as exempted under paragraph (b)(4) of this section.

(1) (i) Safe handling instructions shall accompany the poultry products, specified in this paragraph (b), destined for household consumers, hotels, restaurants, or similar institutions and shall appear on the label. The information shall be in lettering no smaller than one-sixteenth of an inch in size and shall be prominently placed with such conspicuousness (as compared with other words, statements, designs or devices in the labeling) as to render it likely to be read and understood by the ordinary individual under customary conditions of purchase and use.

(ii) The safe handling information shall be presented on the label under the heading "Safe Handling Instructions" which shall be set in type size larger than the print size of the rationale statement and handling statements as discussed in paragraphs (b)(2) and (b)(3) of this section. The safe handling information shall be set off by a border and shall be one color type printed on a single color contrasting background whenever practical.

(2) (i) The labels of the poultry products, specified in this paragraph (b) and prepared from inspected and passed poultry, shall include the following rationale statement as part of the safe handling instructions, "This product was prepared from inspected and passed meat and/or poultry. Some food products may contain bacteria that could cause illness if the product is mishandled or cooked improperly. For your protection, follow these safe handling instructions." This statement shall be placed immediately after the heading and before the safe handling statements.

(ii) The labels of the poultry products, specified in this paragraph (b) and prepared pursuant to § 381.10(a) (2), (5), (6), and (7), shall include the following rationale statement as part of the safe handling instructions, "Some food products may contain bacteria that could cause illness if the product is mishandled or cooked improperly. For your protection, follow these safe handling instructions." This statement shall be placed immediately after the heading and before the safe handling statements.

(3) Poultry products, specified in this paragraph (b), shall bear the labeling statements.

(i) Keep refrigerated or frozen. Thaw in refrigerator or microwave. (Any portion of this statement that is in conflict with the product's specific handling instructions may be omitted, e.g., instructions to cook without thawing.) (A graphic illustration of a refrigerator shall be displayed next to the statement.);

(ii) Keep raw meat and poultry separate from other foods. Wash working surfaces (including cutting boards), utensils, and hands after touching raw meat or poultry. (A graphic illustration of soapy hands under a faucet shall be displayed next to the statement.);

(iii) Cook thoroughly. (A graphic illustration of a skillet shall be displayed next to the statement.); and

(iv) Keep hot foods hot. Refrigerate leftovers immediately or discard. (A graphic illustration of a thermometer shall be displayed next to the statement.)

(4) Poultry products intended for further processing at another official establishment are exempt from the requirements prescribed in paragraphs (b)(1) through (b)(3) of this section.

[37 FR 9706, May 16, 1972, as amended at 39 FR 4569, Feb. 5, 1974; 59 FR 14540, Mar. 28, 1994]

§381.126 Date of packing and date of processing; contents of cans.

(a) Either the immediate container or the shipping container of all poultry food products shall be plainly and permanently marked by code or otherwise with the date of packing. If calendar dating is used, it must be accompanied by an explanatory statement, as provided in §381.129(c)(2).

(b) The immediate container for dressed poultry shall be marked with a lot number which shall be the number of the day of the year on which the poultry was slaughtered or a coded number.

(c) All canned products shall be plainly and permanently marked, by code or otherwise, on the containers, with the identity of the contents and date of canning, except that canned products packed in glass containers are not required to be marked with the date of canning if such information appears on the shipping container. If calendar dating is used, it must be accompanied by an explanatory statement, as provided in §381.129(c)(2).

(d) If any marking is by code, the inspector in charge shall be informed as to its meaning.

[37 FR 9706, May 16, 1972, as amended at 39 FR 28516, Aug. 8, 1974; 39 FR 35784, Oct. 4, 1974]

§381.127 Wording on labels of shipping containers.

(a) Each label for use on a shipping container for inspected and passed poultry products shall bear, in distinctly legible form, the following information:

(1) The official inspection legend.

(2) The official establishment number of the official establishment in which the poultry product was inspected, either within the official inspection mark, or elsewhere on the container clearly visible and in proximity to the official inspection mark.

§381.128 Labels in foreign languages.

Any label to be affixed to a container of any dressed poultry or other poultry product for foreign commerce may be printed in a foreign language. However, the official inspection legend and establishment number shall appear on the label in English, but in addition, may be literally translated into such foreign language. Each such label shall be subject to the applicable provisions of §§381.115 to 381.141, inclusive. Deviations from the form of labeling required under the regulations may be approved by the Administrator in specific cases and such modified labeling may be used for poultry products to be exported: *Provided,*

(a) That the proposed labeling accords to the specifications of the foreign purchaser,

(b) that it is not in conflict with the Act or the laws of the country to which it is intended for export, and

(c) that the outside of the shipping container is labeled to show that it is intended for export; but if such product is sold or offered for sale in domestic commerce, all the requirements of the regulations shall apply.

§381.129 False or misleading labeling or containers.

(a) No poultry product subject to the Act shall have any false or misleading labeling or any container that is so made, formed, or filled as to be misleading. However, established trade names and other labeling and containers which are not false or misleading and which are approved by the Administrator in the regulations or in specific cases are permitted.

(b) No statement, word, picture, design, or device which is false or misleading in any particular or conveys any false impression or gives any false indication of origin, identity, or quality, shall appear on any label. For example:

(1) Official grade designations such as the letter grades A, B, and C may be used in labeling individual carcasses of poultry or containers of poultry products only if such articles have been graded by a licensed grader of the Federal or Federal- State poultry grading service and found to qualify for the indicated grade.

(2) Terms having geographical significance with reference to a particular locality may be used only when the product was produced in that locality.

(3) "Fresh frozen", "quick frozen", "frozen fresh", and terms of similar import apply only to ready-to-cook poultry processed in accordance with §381.66(f)(1). Ready-to-cook poultry handled in any other manner and dressed poultry may be labeled "frozen" only if it is frozen in accordance with §381.66(f)(2) under Department supervision and is in fact in a frozen state. "Individually quick frozen (Kind)" and terms of similar import are applicable only to poultry products that are frozen as stated on the label and whose component parts can be easily separated at time of packing.

(4) Poultry products labeled with a term quoted in any paragraph of §381.170(b) shall comply with the specifications in the applicable paragraph. However, parts of poultry may be cut in any manner the processor desires as long as the labeling appropriately reflects the contents of the container of such poultry.

(5) The terms "All," "Pure," "100%," and terms of similar connotation shall not be used on labels for products to identify ingredient content, unless the product is prepared solely from a single ingredient.

(6)(i) Raw poultry product whose internal temperature has ever been below 26°F may not bear a label declaration of "fresh." Raw poultry product bearing a label declaration of "fresh" but whose internal temperature has ever been below 26°F is mislabeled. The "fresh" designation may be deleted from such product in accordance with §381.133(b)(9)(xxiv). The temperature of individual packages of raw poultry product within an official establishment may deviate below the 26°F standard by 1° (i.e., have a temperature of 25°F) and still be labeled "fresh." The temperature of individual packages of raw poultry product outside an official establishment may deviate below the 26°F standard by 2° (i.e., have a temperature of 24°F) and still be labeled "fresh." The average temperature of poultry product lots of each specific product type must be 26°F. Product described in this paragraph is not subject to the freezing procedures required in §381.66(f)(2) of this subchapter.

(ii) Raw poultry product whose internal temperature has ever been at or below 0°F must be labeled with the descriptive term "frozen," except when such labeling duplicates or conflicts with the labeling requirements in §381.125 of this subchapter. The word "previously" may be placed next to the term "frozen" on an optional basis. The descriptive term must be prominently displayed on the principal display panel of the label. If additional labeling containing the descriptive term is affixed to the label, it must be prominently affixed to the label. The additional labeling must be so conspicuous (as compared with other words, statements, designs, or devices in the labeling) that it is likely to be read and understood by the ordinary individual under customary conditions of purchase and use. Product described in this paragraph is subject to the freezing procedures required in §381.66(f)(2) of this subchapter.

(iii) Raw poultry product whose internal temperature has ever been below 26°F, but is above 0°F, is not required to bear any specific descriptive term. Raw poultry product whose internal temperature has ever been below 26°F, but is above 0°F, may bear labeling with an optional, descriptive term, provided the optional, descriptive term does not cause the raw poultry product to become misbranded. If used, an optional, descriptive term must be prominently displayed on the principal display panel of the label. If additional labeling containing the optional, descriptive term is affixed to the label, it must be prominently affixed on the label. The additional labeling must be so conspicuous (as compared with other words, statements, designs, or devices in the labeling)

9 CFR §381.129

that it is likely to be read and understood by the ordinary individual under customary conditions of purchase and use.

(iv) Handling and relabeling of products.

(A) Except as provided under paragraph (b)(6)(iii)(C) of this section, when any inspected and passed product has become misbranded under this subpart after it has been transported from an official establishment, such product may be transported in commerce to an official establishment after oral permission is obtained from the Area Supervisor of the area in which that official establishment is located. The transportation of the product may be to the official establishment from which it had been transported or to another official establishment designated by the person desiring to handle the product. The transportation shall be authorized only for the purpose of the relabeling of the product. The Area Supervisor shall record the authorization and other information necessary to identify the product and shall provide a copy of the record to the inspector at the establishment receiving the product. The shipper shall be furnished a copy of the authorization record upon request.

(B) Upon the arrival of the shipment at the official establishment, a careful inspection shall be made of the product by the inspector, and if it is found that the product is not adulterated, it may be received into the establishment; but if the product is found to be adulterated, it shall at once be condemned and disposed of in accordance with §381.95 of this subchapter. Wholesome product will be relabeled in accordance with paragraph (b)(6) (i) or (ii) of this section, as appropriate.

(C) When any inspected and passed product has become misbranded under this subpart after it has been transported from an official establishment, the owner may transport the product in commerce to a retail entity for relabeling in accordance with paragraph (b)(6) (i) or (ii) of this section, as appropriate, or to other end users, such as hotels, restaurants or similar institutions; or, relabel the product in accordance with paragraph (b)(6) (i) or (ii) of this section, as appropriate if the product is already at a retail entity. A hotel, restaurant or similar institution is not required to relabel product misbranded under this subpart; *Provided,* That the product is prepared in meals or as entrees only for sale or service directly to individual consumers at such institutions, and that the mark of inspection is removed or obliterated. Oral permission shall be obtained from the Area Officer-in-Charge of the Compliance Program for the area in which the product is located prior to such transportation or relabeling. The Area Officer-in-Charge shall record the authorization and other information necessary to identify the product, and shall furnish a copy of the authorization record upon request. Before being offered for sale at a retail entity, such product shall be relabeled.

(c) A calendar date may be shown on labeling when declared in accordance with the provisions of this paragraph:

(1) The calendar date shall express the month of the year and the day of the month for all products and also the year in the case of products hermetically sealed in metal or glass containers, dried or frozen products, or any other products that the Administrator finds should be labeled with the year because the distribution and marketing practices with respect to such products may cause a label without a year identification to be misleading.

(2) Immediately adjacent to the calendar date shall be a phrase explaining the meaning of such date in terms of "packing" date, "sell by" date, or "use before" date, with or without a further qualifying phrase, e.g., "For Maximum Freshness" or "For Best Quality", and such phrases shall be approved by the Administrator as prescribed in §381.132.

(d) When sodium alginate, calcium carbonate, lactic acid, and calcium lactate are used together in a dry binding matrix in ground and formed poultry products, as permitted in §381.147 of this subchapter, there shall appear on the label contiguous to the product name, a statement to indicate the use of sodium alginate, calcium carbonate, lactic acid, and calcium lactate.

[37 FR 9706, May 16, 1972, as amended at 39 FR 28516, Aug. 8, 1974; 39 FR 42339, Dec. 5, 1974; 55 FR 5977, Feb. 21, 1990; 60 FR 44412, Aug. 25, 1995; 61 FR 66200-66201, Dec. 17, 1996; 61 FR 68821, Dec. 30, 1996]

§381.130 False or misleading labeling or containers; orders to withhold from use.

If the Administrator has reason to believe that any marking or other labeling or the size or form of any container in use or proposed for use with respect to any article subject to the Act is false or misleading in any particular, he may direct that the use of the article be withheld unless it is modified in such manner as the Administrator may prescribe so that it will not be false or misleading. If the person using or proposing to use the labeling or container does not accept the determination of the Administrator, he may request a hearing, but the use of the labeling or container shall, if the Administrator so directs, be withheld pending hearing and final determination by the Secretary in accordance with applicable rules of practice. Any such determination with respect to the matter by the Secretary shall be conclusive unless, within 30 days after the receipt of notice of such final determination, the person adversely affected thereby appeals to the U.S. Court of Appeals for the Circuit in which he has his principal place of business, or to the U.S. Court of Appeals for the District of Columbia Circuit. The provisions of section 204 of the Packers and Stockyards Act of 1921, as amended, shall be applicable to appeals taken under this section.

§381.131 Preparation of labeling or other devices bearing official inspection marks without advance approval prohibited; exceptions.

(a) Except for the purposes of preparing and submitting a sample or samples of the same to the Administrator for approval, no brand manufacturer, printer, or other person shall cast, print, lithograph, or otherwise make any marking device containing any official mark or simulation thereof, or any label bearing any such mark or simulation, without the written authority therefor of the Administrator. However, when any such sample label, or other marking device, is approved by the Administrator, additional supplies of the approved label, or marking device, may be made for use in accordance with the regulations in this subchapter, without further approval by the Administrator. The provisions of this paragraph do not apply to marking devices containing the official inspection legend shown in Figure 5 of §381.102.

(b) No brand manufacturer or other person shall cast or otherwise make, without an official certificate issued in quadruplicate by a Program employee, a marking device containing the official inspection legend shown in Figure 5 of §381.102 or any simulation of that legend.

    (1) The certificate is a Food Safety and Inspection Service form for signature by a Program employee and the official establishment ordering the marking device, bearing a certificate serial number and a letterhead and the seal of the United States Department of Agriculture. The certificate authorizes the making of only the devices of the type and quantity listed on the certificate.

    (2) After signing the certificate, the Program employee and the establishment shall each keep a copy, and the remaining two copies shall be given to the marking device manufacturer.

    (3) The manufacturer of the marking devices shall engrave or otherwise mark each marking device with a permanent identifying serial number unique to it. The manufacturer shall list on each of the two copies of the certificate given to the manufacturer the number of each marking device authorized by the certificate. The manufacturer shall retain one copy of the certificate for the manufacturer's records and return the remaining copy with the marking devices to the Program employee whose name and address are given on the certificate as the recipient.

    (4) In order that all such marking devices bear identifying numbers, within one year after June 24, 1985, an establishment shall either replace each such marking device that does not bear an identifying number, or, under the direction of the inspector- in-charge, mark such marking device with a permanent identifying number.

(Recordkeeping requirements approved by the Office of Management and Budget under control number 0583-0015) [50 FR 21423, May 24, 1985]

§381.132 Labeling approval.

(a) No final labeling shall be used on any product unless the sketch labeling of such final labeling has been submitted for approval to the Food Labeling Division, Regulatory Programs, Food Safety and Inspection Service, and approved by such division, accompanied by FSIS Form, Application for Approval of Labels, Marking, and Devices, except for generically approved labeling authorized for use in §381.133(b) (2)-(9). The management of the official establishment or establishment certified under a foreign inspection system, in accordance with subpart T of this part, must maintain a copy of all labeling used, along with the product formulation and processing procedure, in

9 CFR §381.133

accordance with subpart Q of this part. Such records shall be made available to any duly authorized representative of the Secretary upon request.

(b) The Food Labeling Division shall permit submission for approval of only sketch labeling, as defined in §381.132(d), for all products, except as provided in §381.133(b) (2)-(9) and except for temporary use of final labeling as prescribed in paragraph (f) of this section.

(c) All labeling required to be submitted for approval as set forth in §381.132(b) shall be submitted in duplicate to the Food Labeling Division, Regulatory Programs, Food Safety and Inspection Service, U.S. Department of Agriculture, Washington, DC 20250. A parent company for a corporation may submit only one labeling application (in duplicate) for a product produced in other establishments that are owned by the corporation.

(d) "Sketch" labeling is a printer's proof or equivalent which clearly shows all labeling features, size, location, and indication of final color, as specified in subpart N of this part. FSIS will accept sketches that are hand drawn, computer generated or other reasonable facsimiles that clearly reflect and project the final version of the labeling. Indication of final color may be met by: submission of a color sketch, submission of a sketch which indicates by descriptive language the final colors, or submission with the sketch of previously approved final labeling that indicates the final colors.

(e) Inserts, tags, liners, pasters, and like devices containing printed or graphic matter and for use on, or to be placed within, containers and coverings of product shall be submitted for approval in the same manner as provided for labeling in §381.132(a), except that such devices which contain no reference to product and bear no misleading feature shall be used without submission for approval as prescribed in §381.133(b)(9).

(f)

  (1) Consistent with the requirements of this section, temporary approval for the use of a final label or other final labeling that may otherwise be deemed deficient in some particular may be granted by the Food Labeling Division. Temporary approvals may be granted for a period not to exceed 180 calendar days under the following conditions:

  (i) The proposed labeling would not misrepresent the product;

  (ii) The use of the labeling would not present any potential health, safety, or dietary problems to the consumer;

  (iii) Denial of the request would create undue economic hardship; and

  (iv) An unfair competitive advantage would not result from the granting of the temporary approval.

  (2) Extensions of temporary approvals may also be granted by the Food Labeling Division, provided that the applicant demonstrates that new circumstances, meeting the above criteria, have developed since the original temporary approval was granted.

[48 FR 11420, Mar. 18, 1983; 60 FR 67456, Dec. 29, 1995]

§381.133 Generically approved labeling.

(a) (1) An official establishment or an establishment certified under a foreign inspection system, in accordance with subpart T of this part, is authorized to use generically approved labeling, as defined in paragraph (b) of this section, without such labeling being submitted for approval to the Food Safety and Inspection Service in Washington or the field, provided the labeling is in accord with this section and shows all mandatory features in a prominent manner as required in subpart N of this part, and is not otherwise false or misleading in any particular.

(2) The Food Safety and Inspection Service shall select samples of generically approved labeling from the records maintained by official establishments and establishments certified under foreign inspection systems, in accordance with subpart T of this part, as required in §381.132, to determine compliance with labeling requirements. Any finding of false or misleading labeling shall institute the proceedings prescribed in §381.233.

(b) Generically approved labeling is labeling which complies with the following:

  (1) Labeling for a product which has a product standard as specified in subpart 381 of this subchapter or the Standards and Labeling Policy Book and which does not contain any special claims, such as quality claims, nutrient content claims, health claims, negative claims, geographical origin claims, or guarantees, or which is not a domestic product labeled in a foreign language;

  (2) Labeling for single-ingredient products (such as chicken legs or turkey breasts) which does not contain any special claims, such as quality claims, nutrient content claims, health claims, negative claims, geographical

9 CFR §381.133

origin claims, or guarantees, or which is not a domestic product labeled with a foreign language;

(3) Labeling for containers of products sold under contract specifications to Federal Government agencies, when such product is not offered for sale to the general public, provided that the contract specifications include specific requirements with respect to labeling, and are made available to the inspector-in-charge;

(4) Labeling for shipping containers which contain fully labeled immediate containers, provided such labeling complies with §381.127;

(5) Labeling for products not intended for human food, provided they comply with Secs. 381.152(c) and 381.193, and labeling for poultry heads and feet for export for processing as human food if they comply with §381.190(b);

(6) Poultry inspection legends, which comply with subpart M of this part;

(7) Inserts, tags, liners, pasters, and like devices containing printed or graphic matter and for use on, or to be placed within containers, and coverings of products, provided such devices contain no reference to product and bear no misleading feature;

(8) Labeling for consumer test products not intended for sale; and

(9) Labeling which was previously approved by the Food Labeling Division as sketch labeling, and the final labeling was prepared without modification or with the following modifications:

(i) All features of the labeling are proportionately enlarged or reduced, provided that all minimum size requirements specified in applicable regulations are met and the labeling is legible;

(ii) The substitution of any unit of measurement with its abbreviation or the substitution of any abbreviation with its unit of measurement, e.g., "lb." for "pound," or "oz." for "ounce," or of the word "pound" for "lb." or "ounce" for "oz.";

(iii) A master or stock label has been approved from which the name and address of the distributor are omitted and such name and address are applied before being used (in such case, the words "prepared for" or similar statement must be shown together with the blank space reserved for the insertion of the name and address when such labels are offered for approval);

(iv) Wrappers or other covers bearing pictorial designs, emblematic designs or illustrations, e.g., floral arrangements, illustrations of animals, fireworks, etc. are used with approved labeling (the use of such designs will not make necessary the application of labeling not otherwise required);

(v) A change in the language or the arrangement of directions pertaining to the opening of containers or the serving of the product;

(vi) The addition, deletion, or amendment of a dated or undated coupon, a cents-off statement, cooking instructions, packer product code information, or UPC product code information;

(vii) Any change in the name or address of the packer, manufacturer or distributor that appears in the signature line;

(viii) Any change in the net weight, provided that the size of the net weight statement complies with §381.121;

(ix) The addition, deletion, or amendment of recipe suggestions for the product;

(x) Any change in punctuation;

(xi) Newly assigned or revised establishment numbers for a particular establishment for which use of the labeling has been approved by the Food Labeling Division, Regulatory Programs;

(xii) The addition or deletion of open dating information;

(xiii) A change in the type of packaging material on which the label is printed;

(xiv) Brand name changes, provided that there are no design changes, the brand name does not use a term that connotes quality or other product characteristics, the brand name has no geographic significance, and the brand name does not affect the name of the product;

(xv) The deletion of the word "new" on new product labeling;

(xvi) The addition, deletion, or amendment of special handling statements, provided that the change is consistent with §381.125(a);

(xvii) The addition of safe handling instructions as required by §381.125(b);

(xviii) Changes reflecting a change in the quantity of an ingredient shown in the formula without a change in the order of predominance shown on the label, provided that the change in quantity of ingredients complies with any minimum or maximum limits for the use of such ingredients prescribed in §381.147 and subpart P of this part;

9 CFR §381.134

(xix) Changes in the color of the labeling, provided that sufficient contrast and legibility remain;

(xx) A change in the product vignette, provided that the change does not affect mandatory labeling information or misrepresent the content of the package;

(xxi) The addition, deletion, or substitution of the official USDA poultry grade shield; (xxii) A change in the establishment number by a corporation or parent company for an establishment under its ownership;

(xxiii) Changes in nutrition labeling that only involve quantitative adjustments to the nutrition labeling information, except for services sizes, provided the nutrition labeling information maintains its accuracy and consistency;

(xxiv) Deletion of any claim, and the deletion of non-mandatory features or non-mandatory information;

(xxv) The addition or deletion of a direct translation of the English language into a foreign language for products marked "for export only"; and

(xxvi) The use of the descriptive term "fresh" in accordance with §381.129(b)(6)(i) of this subchapter.

(xxvii) The use of the descriptive term "frozen" as required by §381.129(b)(6)(ii) of this subchapter.

[48 FR 11420, Mar. 18, 1983; as amended at 59 FR 14540, Mar. 28, 1994; 59 FR 40215, Aug. 8, 1994; 60 FR 44413, Aug. 25, 1995; 60 FR 67456, Dec. 29, 1995; 61 FR 66201, December 17, 1996]

§381.134 Requirement of formulas.

Copies of each label submitted for approval, shall when the Administrator requires in any specific case, be accompanied by a statement showing, by their common or usual names, the kinds and percentages of the ingredients comprising the poultry product and by a statement indicating the method or preparation of the product with respect to which the label is to be used. Approximate percentages may be given in cases where the percentages of ingredients may vary from time to time, if the limits of variation are stated.

[37 FR 9706, May 16, 1972, as amended at 39 FR 4569, Feb. 5, 1974; 59 FR 45196, Sept. 1, 1994; 60 FR 67456, Dec. 29, 1995]

§381.135 Irradiated poultry product.

(a) The labeling of packages of poultry product irradiated in conformance with §381.147(f)(4) of this part must bear the following logo along with a statement such as, "Treated with radiation" or "Treated by irradiation," in addition to all other labeling requirements of this subpart. The logo must be placed prominently and conspicuously in conjunction with the required statement and be colored green. The statement must appear as a qualifier contiguous to the product name and in letters of the same style, color, and type as the product name. Letters used for the qualifying statement shall be no less than one-third the size of the largest letter in the product name. Any labeling bearing the logo and any wording of explanation with respect to this logo must be approved as required by subparts M and N of this part.

9 CFR §381.139

(b) The product label must bear the handling statement "Keep Refrigerated" or "Keep Frozen," as appropriate, in conformance with §381.125 of this Subpart.
(c) Optional labeling statements about the purpose for radiation processing may be included on the product label in addition to the above stated requirements. Such statements must not be false or misleading.
[57 FR 43597, Sept. 21, 1992]

§381.136 Affixing of official identification.

(a) No official inspection legend or any abbreviation or other simulation thereof may be affixed to or placed on or caused to be affixed to or placed on any poultry product or container thereof, except by an inspector or under the supervision of an inspector or other person authorized by the Administrator, and no container bearing any such legend shall be filled except under such supervision.
(b) No official inspection legend shall be used on any poultry product or other article which does not qualify for such mark under the regulations.

§381.137 Evidence of labeling and devices approval.

No inspector shall authorize the use of any device bearing any official inspection legend unless he or she has on file evidence that such device has been approved in accordance with the provisions of this subpart.
[37 FR 9706, May 16, 1972, as amended at 60 FR 67458, Dec. 29, 1995]

§381.138 Unauthorized use or disposition of approved labeling or devices.

(a) Labeling and devices approved for use pursuant to §381.115 shall be used only for the purpose for which approved, and shall not be disposed of from the official establishment for which approved except with written approval of the Administrator. Any unauthorized use or disposition of approved labeling or devices bearing official inspection marks is prohibited and may result in cancellation of the approval.
(b) Labeling and containers bearing any official inspection marks, with or without the official establishment number, may be transported from one official establishment to any other official establishment, only if such shipments are made with the prior authorization of the inspector in charge at point of origin, who will notify the inspector in charge at destination concerning the date of shipment, quantity, and type of labeling material involved. Approved labeling and containers may be moved without restriction under this part between official establishments operated by the same person if such labeling and containers are approved for use at all such establishments. No such material shall be used at the establishment to which it is shipped unless such use conforms with the requirements of this subpart.

§381.139 Removal of official identifications.

(a) Every person who receives any poultry product in containers which bear any official inspection legend shall remove or deface such legend or destroy the containers upon removal of such articles from the containers.
(b) No person shall alter, detach, deface, or destroy any official identifications prescribed in Subpart M that were applied pursuant to the regulations, unless he is authorized to do so by an inspector or this section; and no person shall fail to use any such official identification when required by this part.

§381.140 Relabeling poultry products.

9 CFR §381.140

When it is claimed by the operator of an official establishment that some of its labeled poultry product, which has been transported to a location other than an official establishment, is in need of relabeling because the labeling has become mutilated or damaged, or for some other reason needs relabeling, the requests for relabeling the poultry product shall be sent to the Administrator and accompanied with a statement of the reasons therefor and the quantity of labeling required. Labeling material intended for relabeling inspected and passed product shall not be transported from an official establishment until permission has been received from the Administrator. The relabeling of inspected and passed product with official labels shall be done under the supervision of an inspector pursuant to the regulations in Part 362 of this chapter. The establishment shall reimburse the Inspection Service for any cost involved in supervising the relabeling of such product as provided in said regulations.

§381.141 [Removed and Reserved]
[47 FR 54287, Dec. 2, 1982; 60 FR 67458, Dec. 29, 1995]

§381.143 [Reserved]

§381.144 Packaging materials.

(a) Edible products may not be packaged in a container which is composed in whole or in part of any poisonous or deleterious substances which may render the contents adulterated or injurious to health. All packaging materials must be safe for the intended use within the meaning of section 409 of the Federal Food, Drug, and Cosmetic Act, as amended (FFDCA).

(b) Packaging materials entering the official establishment must be accompanied or covered by a guaranty, or statement of assurance, from the packaging supplier under whose brand name and firm name the material is marketed to the official establishment. The guaranty shall state that the material's intended use complies with the FFDCA and all applicable food additive regulations. The guaranty must identify the material, e.g., by the distinguishing brand name or code designation appearing on the packaging material shipping container; must specify the applicable conditions of use, including temperature limits and other pertinent limits specified under the FFDCA and food additive regulations; and must be signed by an authorized official of the supplying firm. The guaranty may be limited to a specific shipment of an article, in which case it may be part of or attached to the invoice covering such shipment, or it may be general and continuing, in which case, in its application to any article or other shipment of an article, it shall be considered to have been given at the date such article was shipped by the person who gives the guaranty. Guaranties consistent with the Food and Drug Administration's regulations regarding such guaranties (21 CFR 7.12 and 7.13) will be acceptable. The management of the establishment must maintain a file containing guaranties for all food contact packaging materials in the establishment. The file shall be made available to Program inspectors or other Department officials upon request. While in the official establishment, the identity of all packaging materials must be traceable to the applicable guaranty.

(c) The guaranty by the packaging supplier will be accepted by Program inspectors to establish that the use of material complies with the FFDCA and all applicable food additive regulations.

(d) The Department will monitor the use of packaging materials in official establishments to assure that the requirements of paragraph (a) of this section are met, and may question the basis for any guaranty described under paragraph (b) of this section. Official establishments and packaging suppliers providing written guaranties to those official establishments will be permitted an opportunity to provide information to designated Department officials as needed to verify the basis for any such guaranty. The required information will include, but is not limited to, manufacturing firm's name, trade name or code designation for the material, complete chemical composition, and use. Selection of a material for review does not in itself affect a material's acceptability. Materials may continue to be used during the review period. However, if information requested from the supplier is not provided within the time indicated in the request-a minimum of 30 days-any applicable guaranty shall cease to be effective and approval to continue using the specified packaging material in official establishments may be denied. The Administrator may extend this time where reasonable grounds for extension are shown, as, for example, where data must be obtained from suppliers.

(e) The Administrator may disapprove for use in official establishments packaging materials whose use cannot be confirmed as complying with the FFDCA and applicable food additive regulations. Before approval to use a packaging material is finally denied by the Administrator, the affected official establishment and the supplier of the

9 CFR §381.402

material shall be given notice and the opportunity to present their views to the Administrator. If the official establishment and the supplier do not accept the Administrator's determination, a hearing in accordance with applicable rules of practice will be held to resolve such dispute. Approval to use the materials pending the outcome of the presentation of views or hearing shall be denied if the Administrator determines that such use may present an imminent hazard to public health.

(f) Periodically, the Administrator will issue to inspectors a listing, by distinguishing brand name or code designation, of packaging materials that have been reviewed and that fail to meet the requirements of paragraph (a) of this section. Listed materials will not be permitted for use in official establishments. If a subsequent review of any material indicates that it meets the requirements of paragraph (a), the material will be deleted from the listing.

(g) Nothing in this section shall affect the authority of Program inspectors to refuse a specific material if he/she determines the material may render products adulterated or injurious to health.

[49 FR 2236, Jan. 19, 1984]

Subpart Y-Nutrition Labeling
Source: 58 FR 675, Jan. 6, 1993, unless otherwise noted.

§381.400 Nutrition labeling of poultry products.

(a) Nutrition labeling shall be provided for all poultry products intended for human consumption and offered for sale, except single-ingredient, raw products, in accordance with the requirements of §381.409, except as exempted under §381.500 of this subpart.

(b) Nutrition labeling may be provided for single-ingredient, raw poultry products in accordance with the requirements of §§381.409 and 381.445. Significant participation in voluntary nutrition labeling shall be measured by the Agency in accordance with §§381.443 and 381.444 of this subpart.

[58 FR 675, Jan. 6, 1993; 60 FR 197, Jan. 3, 1995]

§381.401 [Reserved]

§381.402 Location of nutrition information.

(a) Nutrition information on a label of a packaged poultry product shall appear on the label's principal display panel or on the information panel, except as provided in paragraphs (b) and (c) of this section.

(b) Nutrition information for gift packs may be shown at a location other than on the product label, provided that the labels for these products bear no nutrition claim. In lieu of on the product label, nutrition information may be provided by alternate means such as product label inserts.

(c) Poultry products in packages that have a total surface area available to bear labeling greater than 40 square inches but whose principal display panel and information panel do not provide sufficient space to accommodate all required information may use any alternate panel that can be readily seen by consumers for the nutrition information. In determining the sufficiency of available space for the nutrition information, the space needed for vignettes, designs, and other nonmandatory label information on the principal display panel may be considered.

[58 FR 675, Jan. 6, 1993, as amended at 59 FR 40215, Aug. 8, 1994]

§§381.403-381.407 [Reserved]

§381.408 Labeling of poultry products with number of servings.

The label of any package of a poultry product that bears a representation as to the number of servings contained in such package shall meet the requirements of §381.121(c)(7).

9 CFR §381.409

§381.409 Nutrition label content.

(a) All nutrient and food component quantities shall be declared in relation to a serving as defined in this section.

(b)

(1) The term "serving" or "serving size" means an amount of food customarily consumed per eating occasion by persons 4 years of age or older, which is expressed in a common household measure that is appropriate to the product. When the product is specially formulated or processed for use by infants or by toddlers, a serving or serving size means an amount of food customarily consumed per eating occasion by infants up to 12 months of age or by children 1 through 3 years of age, respectively.

(2) Except as provided in paragraphs (b)(8), (b)(12), and (b)(14) of this section and for products that are intended for weight control and are available only through a weight-control or weight-maintenance program, the serving size declared on a product label shall be determined from the "Reference Amounts Customarily Consumed Per Eating Occasion--General Food Supply" (Reference Amount(s)) that appear in §381.412(b) using the procedures described in this paragraph (b). For products that are both intended for weight control and available only through a weight-control program, a manufacturer may determine the serving size that is consistent with the meal plan of the program. Such products must bear a statement, "for sale only through the _____ program" (fill in the blank with the name of the appropriate weight- control program, e.g., Smith's Weight Control), on the principal display panel. However, the Reference Amounts in §381.412(b) shall be used for purposes of evaluating whether weight-control products that are available only through a weight-control program qualify for nutrition claims.

(3) The declaration of nutrient and food component content shall be on the basis of the product "as packaged" for all products, except that single-ingredient, raw products may be declared on the basis of the product "as consumed" as set forth in §381.445(a)(1). In addition to the required declaration on the basis of "as packaged" for products other than single ingredient, raw products, the declaration may also be made on the basis of "as consumed," provided that preparation and cooking instructions are clearly stated. (4) For products in discrete units (e.g., chicken wings, and individually packaged products within a multi-serving package), and for products which consist of two or more foods packaged and presented to be consumed together where the ingredient represented as the main ingredient is in discrete units (e.g., chicken wings and barbecue sauce), the serving size shall be declared as follows:

(i) If a unit weighs 50 percent or less of the Reference Amount, the serving size shall be the number of whole units that most closely approximates the Reference Amount for the product category.

(ii) If a unit weighs more than 50 percent but less than 67 percent of the Reference Amount, the manufacturer may declare one unit or two units as the serving size.

(iii) If a unit weighs 67 percent or more but less than 200 percent of the Reference Amount, the serving size shall be one unit.

(iv) If a unit weighs 200 percent or more of the Reference Amount, the manufacturer may declare one unit as the serving size if the whole unit can reasonably be consumed at a single eating occasion.

(v) For products that have Reference Amounts of 100 grams (or milliliter) or larger and are individual units within a multi-serving package, if a unit contains more than 150 percent but less than 200 percent of the Reference Amount, the manufacturer may decide whether to declare the individual unit as 1 or 2 servings.

(vi) For products which consist of two or more foods packaged and presented to be consumed together where the ingredient represented as the main ingredient is in discrete units (e.g., chicken wings and barbecue sauce), the serving size may be the number of discrete units represented as the main ingredient plus proportioned minor ingredients used to make the Reference Amount for the combined product as determined in §381.412(c).

(vii) For packages containing several individual single-serving containers, each of which is labeled with all required information including nutrition labeling as specified in this section (i.e., are labeled appropriately for individual sale as single-serving containers), the serving size shall be 1 unit.

(5) For products in large discrete units that are usually divided for consumption (e.g., pizza, pan of poultry lasagna), for unprepared products where the entire contents of the package is used to prepare large discrete units that are usually divided for consumption (e.g., pizza kit), and for products which consist of two or more foods packaged and presented to be consumed together where the ingredient represented as the main ingredient is a large discrete unit usually divided for consumption, the serving size shall be the fractional slice of the ready-to-eat product (e.g., 1/8 quiche, 1/4 pizza) that most closely approximates the Reference Amount for the

9 CFR §381.409

product category. The serving size may be the fraction of the package used to make the Reference Amount for the unprepared product determined in §381.412(d) or the fraction of the large discrete unit represented as the main ingredient plus proportioned minor ingredients used to make the Reference Amount of the combined product determined in §381.412(c). In expressing the fractional slice, manufacturers shall use 1/2, 1/3, 1/4, 1/5, 1/6, or smaller fractions that can be generated by further division by 2 or 3.

(6) For nondiscrete bulk products (e.g., whole turkey, turkey breast, ground poultry), and for products which consist of two or more foods packaged and presented to be consumed together where the ingredient represented as the main ingredient is a bulk product (e.g., turkey breast and gravy), the serving size shall be the amount in household measure that most closely approximates the Reference Amount for the product category and may be the amount of the bulk product represented as the main ingredient plus proportioned minor ingredients used to make the Reference Amount for the combined product determined in §381.412(c).

(7) For labeling purposes, the term "common household measure" or "common household unit" means cup, tablespoon, teaspoon, piece, slice, fraction (e.g., 1/4 pizza), ounce (oz), or other common household equipment used to package food products (e.g., jar or tray). In expressing serving size in household measures, except as specified in paragraphs (b)(7)(iv), (v), and (vi) of this section, the following rules shall be used:

(i) Cups, tablespoons, or teaspoons shall be used wherever possible and appropriate. Cups shall be expressed in 1/4- or 1/3-cup increments, tablespoons in whole number of tablespoons for quantities less than 1/4 cup but greater than or equal to 2 tablespoons (tbsp), 1, 1 1/3, 1 1/2, or 1 2/3 tbsp for quantities less than 2 tbsp but greater than or equal to 1 tbsp, and teaspoons in whole number of teaspoons for quantities less than 1 tbsp but greater than or equal to 1 teaspoon (tsp), and in 1/4-tsp increments for quantities less than 1 tsp.

(ii) If cups, tablespoons or teaspoons are not applicable, units such as piece, slice, tray, jar, and fraction shall be used.

(iii) If cups, tablespoons and teaspoons, or units such as piece, slice, tray, jar, or fraction are not applicable, ounces may be used. Ounce measurements shall be expressed in 0.5-ounce increments most closely approximating the Reference Amount with rounding indicated by the use of the term "about" (e.g., about 2.5 ounces).

(iv) A description of the individual container or package shall be used for single-serving containers and meal-type products and for individually packaged products within multi-serving containers (e.g., can, box, package, meal, or dinner). A description of the individual unit shall be used for other products in discrete units (e.g., wing, slice, link, or patty).

(v) For unprepared products where the entire contents of the package is used to prepare large discrete units that are usually divided for consumption (e.g., pizza kit), the fraction or portion of the package may be used.

(vi) For products that consist of two or more distinct ingredients or components packaged and presented to be consumed together (e.g., chicken wings with a glaze packet), the nutrition information may be declared for each component or as a composite. The serving size may be provided in accordance with the provisions of paragraphs (b)(4), (b)(5), and (b)(6) of this section.

(vii) For nutrition labeling purposes, a teaspoon means 5 milliliters (mL), a tablespoon means 15 mL, a cup means 240 mL, and 1 oz in weight means 28 grams (g).

(viii) When a serving size, determined from the Reference Amount in §381.412(b) and the procedures described in this section, falls exactly half way between two serving sizes (e.g., 2.5 tbsp), manufacturers shall round the serving size up to the next incremental size.

(8) A product that is packaged and sold individually and that contains less than 200 percent of the applicable Reference Amount shall be considered to be a single-serving container, and the entire content of the product shall be labeled as one serving, except for products that have Reference Amounts of 100 g (or mL) or larger, manufacturers may decide whether a package that contains more than 150 percent but less than 200 percent of the Reference Amount is 1 or 2 servings. Packages sold individually that contain 200 percent or more of the applicable Reference Amount may be labeled as a single-serving if the entire content of the package can reasonably be consumed at a single- eating occasion.

(9) A label statement regarding a serving shall be the serving size expressed in common household measures as set forth in paragraphs (b)(2) through (b)(8) of this section and shall be followed by the equivalent metric quantity in parenthesis (fluids in milliliters and all other foods in grams), except for single-serving containers.

9 CFR §381.409

(i) For a single-serving container, the parenthetical metric quantity, which will be presented as part of the net weight statement on the principal display panel, is not required except where nutrition information is required on a drained weight basis according to paragraph (b)(11) of this section. However, if a manufacturer voluntarily provides the metric quantity on products that can be sold as single servings, then the numerical value provided as part of the serving size declaration must be identical to the metric quantity declaration provided as part of the net quantity of contents statement.

(ii) The gram or milliliter quantity equivalent to the household measure should be rounded to the nearest whole number except for quantities that are less than 5 g (mL). The gram (mL) quantity between 2 and 5 g (mL) should be rounded to the nearest 0.5 g (mL) and the g (mL) quantity less than 2 g (mL) should be expressed in 0.1-g (mL) increments.

(iii) In addition, serving size may be declared in ounce, in parenthesis, following the metric measure separated by a slash where other common household measures are used as the primary unit for serving size, e.g., 1 slice (28 g/1 oz) for sliced chicken roll. The ounce quantity equivalent to the metric quantity should be expressed in 0.1-oz increments.

(iv) If a manufacturer elects to use abbreviations for units, the following abbreviations shall be used: tbsp for tablespoon, tsp for teaspoon, g for gram, mL for milliliter, and oz for ounce.

(10) Determination of the number of servings per container shall be based on the serving size of the product determined by following the procedures described in this section.

(i) The number of servings shall be rounded to the nearest whole number except for the number of servings between 2 and 5 servings and random weight products. The number of servings between 2 and 5 servings shall be rounded to the nearest 0.5 serving. Rounding should be indicated by the use of the term "about" (e.g., about 2 servings; about 3.5 servings).

(ii) When the serving size is required to be expressed on a drained solids basis and the number of servings varies because of a natural variation in unit size, the manufacturer may state the typical number of servings per container (e.g., usually 5 servings).

(iii) For random weight products, a manufacturer may declare "varied" for the number of servings per container provided the nutrition information is based on the Reference Amount expressed in ounces. The manufacturer may provide the typical number of servings in parenthesis following the "varied" statement (e.g., varied (approximately 8 servings per pound)).

(iv) For packages containing several individual single-serving containers, each of which is labeled with all required information including nutrition labeling as specified in this section (i.e., are labeled appropriately for individual sale as single-serving containers), the number of servings shall be the number of individual packages within the total package.

(v) For packages containing several individually packaged multi- serving units, the number of servings shall be determined by multiplying the number of individual multi-serving units in the total package by the number of servings in each individual unit.

(11) The declaration of nutrient and food component content shall be on the basis of product as packaged or purchased with the exception of products that are packed or canned in water, brine, or oil but whose liquid packing medium is not customarily consumed. Declaration of the nutrient and food component content of products that are packed in liquid which is not customarily consumed shall be based on the drained solids.

(12) Serving size for meal-type products as defined in §381.413(l) shall be the entire content (edible portion only) of the package.

(13) Another column of figures may be used to declare the nutrient and food component information in the same format as required by §381.409(e),

(i) Per 100 grams, 100 milliliters, or 1 ounce of the product as packaged or purchased.

(ii) Per one unit if the serving size of a product in discrete units in a multi-serving container is more than one unit.

(14) If a product consists of assortments of poultry products (e.g., variety packs) in the same package, nutrient content shall be expressed on the entire package contents or on each individual product.

(15) If a product is commonly combined with other ingredients or is cooked or otherwise prepared before eating, and directions for such combination or preparations are provided, another column of figures may be used to declare the nutrient contents on the basis of the product as consumed for the product alone (e.g., a cream soup mix may be labeled with one set of Daily Values for the dry mix (per serving), and another set for the serving of

9 CFR §381.409

the final soup when prepared (e.g., per serving of cream soup mix and 1 cup of vitamin D fortified whole milk)): Provided, that the type and quantity of the other ingredients to be added to the product by the user and the specific method of cooking and other preparation shall be specified prominently on the label.

(c) The declaration of nutrition information on the label or in labeling of a poultry product shall contain information about the level of the following nutrients, except for those nutrients whose inclusion, and the declaration of amounts, is voluntary as set forth in this paragraph. No nutrients or food components other than those listed in this paragraph as either mandatory or voluntary may be included within the nutrition label. Except as provided for in paragraph (f) or (g) of this section, nutrient information shall be presented using the nutrient names specified and in the following order in the formats specified in paragraph (d) or (e) of this section.

(1) "Calories, total," "Total calories," or "Calories": A statement of the caloric content per serving, expressed to the nearest 5-calorie increment up to and including 50 calories, and 10-calorie increment above 50 calories, except that amounts less than 5 calories may be expressed as zero. Energy content per serving may also be expressed in kilojoule units, added in parenthesis immediately following the statement of the caloric content.

(i) Caloric content may be calculated by the following methods. Where either specific or general food factors are used, the factors shall be applied to the actual amount (i.e., before rounding) of food components (e.g., fat, carbohydrate, protein, or ingredients with specific food factors) present per serving.

(A) Using specific Atwater factors (i.e., the Atwater method) given in Table 13, page 25, "Energy Value of Foods--Basis and Derivation," by A. L. Merrill and B. K. Watt, United States Department of Agriculture (USDA), Agriculture Handbook No. 74 (Slightly revised February 1973), which is incorporated by reference. Table 13 of the "Energy Value of Foods--Basis and Derivation," Agriculture Handbook No. 74 is incorporated as it exists on the date of approval. This incorporation by reference was approved by the Director of the Federal Register in accordance with 5 U.S.C. 552(a) and 1 CFR part 51. It is available for inspection at the Office of the Federal Register, suite 700, 800 North Capitol Street, NW., Washington, DC, or at the office of the FSIS Docket Clerk, Room 3171, South Building, 14th and Independence Avenue, SW., Washington, DC. Copies of the incorporation by reference are available from the Product Assessment Division, Regulatory Programs, Food Safety and Inspection Service, U.S. Department of Agriculture, Room 329, West End Court Building, Washington, DC 20250- 3700;

(B) Using the general factors of 4, 4, and 9 calories per gram for protein, total carbohydrate, and total fat, respectively, as described in USDA's Agriculture Handbook No. 74 (Slightly revised February 1973), pages 9-11, which is incorporated by reference. Pages 9-11, Agriculture Handbook No. 74 is incorporated as it exists on the date of approval. This incorporation by reference was approved by the Director of the Federal Register in accordance with 5 U.S.C. 552(a) and 1 CFR part 51. (The availability of this incorporation by reference is given in paragraph (c)(1)(i)(A) of this section.);

(C) Using the general factors of 4, 4, and 9 calories per gram for protein, total carbohydrate less the amount of insoluble dietary fiber, and total fat, respectively, as described in USDA's Agriculture Handbook No. 74 (Slightly revised February 1973), pages 9-11, which is incorporated by reference in accordance with 5 U.S.C. 552(a) and 1 CFR part 51. (The availability of this incorporation by reference is given in paragraph (c)(1)(i)(A) of this section.); or

(D) Using data for specific food factors for particular foods or ingredients approved by the Food and Drug Administration (FDA) and provided in parts 172 or 184 of 21 CFR, or by other means, as appropriate.

(ii) "Calories from fat": A statement of the caloric content derived from total fat as defined in paragraph (c)(2) of this section per serving, expressed to the nearest 5-calorie increment, up to and including 50 calories, and the nearest 10-calorie increment above 50 calories, except that label declaration of "calories from fat" is not required on products that contain less than 0.5 gram of fat per serving and amounts less than 5 calories may be expressed as zero. This statement shall be declared as provided in paragraph (d)(5) of this section.

(iii) "Calories from saturated fat" or "Calories from saturated" (VOLUNTARY): A statement of the caloric content derived from saturated fat as defined in paragraph (c)(2)(i) of this section per serving may be declared voluntarily, expressed to the nearest 5- calorie increment, up to and including 50 calories, and the nearest 10-calorie increment above 50 calories, except that amounts less than 5 calories may be expressed as zero. This statement shall be indented under the statement of calories from fat as provided in paragraph (d)(5) of this section.

9 CFR §381.409

(2) "Fat, total" or "Total fat": A statement of the number of grams of total fat per serving defined as total lipid fatty acids and expressed as triglycerides. Amounts shall be expressed to the nearest 0.5 (1/2)-gram increment below 5 grams and to the nearest gram increment above 5 grams. If the serving contains less than 0.5 gram, the content shall be expressed as zero.

(i) "Saturated fat" or "Saturated": A statement of the number of grams of saturated fat per serving defined as the sum of all fatty acids containing no double bonds, except that label declaration of saturated fat content information is not required for products that contain less than 0.5 gram of total fat per serving if no claims are made about fat or cholesterol content, and if "calories from saturated fat" is not declared. Saturated fat content shall be indented and expressed as grams per serving to the nearest 0.5 (1/2)-gram increment below 5 grams and to the nearest gram increment above 5 grams. If the serving contains less than 0.5 gram, the content shall be expressed as zero.

(A) "Stearic Acid" (VOLUNTARY): A statement of the number of grams of stearic acid per serving may be declared voluntarily, except that when a claim is made about stearic acid, label declaration shall be required. Stearic acid content shall be indented under saturated fat and expressed to the nearest 0.5 (1/2)-gram increment below 5 grams and the nearest gram increment above 5 grams. If the serving contains less than 0.5 gram, the content shall be expressed as zero.

(B) [Reserved]

(ii) "Polyunsaturated fat" or "Polyunsaturated" (VOLUNTARY): A statement of the number of grams of polyunsaturated fat per serving defined as cis,cis-methylene-interrupted polyunsaturated fatty acids may be declared voluntarily, except that when monounsaturated fat is declared, or when a claim about fatty acids or cholesterol is made on the label or in labeling of a product other than one that meets the criteria in §381.462(b)(1) for a claim for "fat free," label declaration of polyunsaturated fat is required. Polyunsaturated fat content shall be indented and expressed as grams per serving to the nearest 0.5 (1/2)-gram increment below 5 grams and to the nearest gram increment above 5 grams. If the serving contains less than 0.5 gram, the content shall be expressed as zero.

(iii) "Monounsaturated fat" or "Monounsaturated" (VOLUNTARY): A statement of the number of grams of monounsaturated fat per serving defined as cis-monounsaturated fatty acids may be declared voluntarily, except that when polyunsaturated fat is declared, or when a claim about fatty acids or cholesterol is made on the label or in labeling of a product other than one that meets the criteria in §381.462(b)(1) for a claim for "fat free," label declaration of monounsaturated fat is required. Monounsaturated fat content shall be indented and expressed as grams per serving to the nearest 0.5 (1/2)-gram increment below 5 grams and to the nearest gram increment above 5 grams. If the serving contains less than 0.5 gram, the content shall be expressed as zero.

(3) "Cholesterol": A statement of the cholesterol content per serving expressed in milligrams to the nearest 5-milligram increment, except that label declaration of cholesterol information is not required for products that contain less than 2 milligrams of cholesterol per serving and make no claim about fat, fatty acids, or cholesterol content, or such products may state the cholesterol content as zero. If the product contains 2 to 5 milligrams of cholesterol per serving, the content may be stated as "less than 5 milligrams."

(4) "Sodium": A statement of the number of milligrams of sodium per serving expressed as zero when the serving contains less than 5 milligrams of sodium, to the nearest 5-milligram increment when the serving contains 5 to 140 milligrams of sodium, and to the nearest 10- milligram increment when the serving contains greater than 140 milligrams.

(5) "Potassium" (VOLUNTARY): A statement of the number of milligrams of potassium per serving may be declared voluntarily, except that when a claim is made about potassium content, label declaration shall be required. Potassium content shall be expressed as zero when the serving contains less than 5 milligrams of potassium, to the nearest 5-milligram increment when the serving contains 5 to 140 milligrams of potassium, and to the nearest 10-milligram increment when the serving contains greater than 140 milligrams.

(6) "Carbohydrate, total" or "Total carbohydrate": A statement of the number of grams of total carbohydrate per serving expressed to the nearest gram, except that if a serving contains less than 1 gram, the statement "Contains less than 1 gram" or "less than 1 gram" may be used as an alternative, or, if the serving contains less than 0.5 gram, the content may be expressed as zero. Total carbohydrate content shall be calculated by subtraction of the sum of the crude protein, total fat, moisture, and ash from the total weight of the product. This calculation method is described in USDA's Agriculture Handbook No. 74 (Slightly revised February 1973), pages 2 and 3, which is incorporated by reference. Pages 2 and 3, Agriculture Handbook No. 74 is incorporated as it exists on

9 CFR §381.409

the date of approval. This incorporation by reference was approved by the Director of the Federal Register in accordance with 5 U.S.C. 552(a) and 1 CFR part 51. (The availability of this incorporation by reference is given in paragraph (c)(1)(i)(A) of this section.).

(i) "Dietary fiber": A statement of the number of grams of total dietary fiber per serving, indented and expressed to the nearest gram, except that if a serving contains less than 1 gram, declaration of dietary fiber is not required, or, alternatively, the statement "Contains less than 1 gram" or "less than 1 gram" may be used, and if the serving contains less than 0.5 gram, the content may be expressed as zero.

(A) "Soluble fiber" (VOLUNTARY): A statement of the number of grams of soluble dietary fiber per serving may be declared voluntarily except when a claim is made on the label or in labeling about soluble fiber, label declaration shall be required. Soluble fiber content shall be indented under dietary fiber and expressed to the nearest gram, except that if a serving contains less than 1 gram, the statement "Contains less than 1 gram" or "less than 1 gram" may be used as an alternative, and if the serving contains less than 0.5 gram, the content may be expressed as zero.

(B) "Insoluble fiber" (VOLUNTARY): A statement of the number of grams of insoluble dietary fiber per serving may be declared voluntarily except when a claim is made on the label or in labeling about insoluble fiber, label declaration shall be required. Insoluble fiber content shall be indented under dietary fiber and expressed to the nearest gram, except that if a serving contains less than 1 gram, the statement "Contains less than 1 gram" or "less than 1 gram" may be used as an alternative, and if the serving contains less than 0.5 gram, the content may be expressed as zero.

(ii) "Sugars": A statement of the number of grams of sugars per serving, except that label declaration of sugars content is not required for products that contain less than 1 gram of sugars per serving if no claims are made about sweeteners, sugars, or sugar alcohol content. Sugars shall be defined as the sum of all free mono- and disaccharides (such as glucose, fructose, lactose, and sucrose). Sugars content shall be indented and expressed to the nearest gram, except that if a serving contains less than 1 gram, the statement "Contains less than 1 gram" or "less than 1 gram" may be used as an alternative, and if the serving contains less than 0.5 gram, the content may be expressed as zero.

(iii) "Sugar alcohol" (VOLUNTARY): A statement of the number of grams of sugar alcohols per serving may be declared voluntarily on the label, except that when a claim is made on the label or in labeling about sugar alcohol or sugars when sugar alcohols are present in the product, sugar alcohol content shall be declared. For nutrition labeling purposes, sugar alcohols are defined as the sum of saccharide derivatives in which a hydroxyl group replaces a ketone or aldehyde group and whose use in the food is listed by FDA (e.g., mannitol or xylitol) or is generally recognized as safe (e.g., sorbitol). In lieu of the term "sugar alcohol," the name of the specific sugar alcohol (e.g., "xylitol") present in the product may be used in the nutrition label, provided that only one sugar alcohol is present in the product. Sugar alcohol content shall be indented and expressed to the nearest gram, except that if a serving contains less than 1 gram, the statement "Contains less than 1 gram" or "less than 1 gram" may be used as an alternative, and if the serving contains less than 0.5 gram, the content may be expressed as zero.

(iv) "Other carbohydrate" (VOLUNTARY): A statement of the number of grams of other carbohydrate per serving may be declared voluntarily. Other carbohydrate shall be defined as the difference between total carbohydrate and the sum of dietary fiber, sugars, and sugar alcohol, except that if sugar alcohol is not declared (even if present), it shall be defined as the difference between total carbohydrate and the sum of dietary fiber and sugars. Other carbohydrate content shall be indented and expressed to the nearest gram, except that if a serving contains less than 1 gram, the statement "Contains less than 1 gram" or "less than 1 gram" may be used as an alternative, and if the serving contains less than 0.5 gram, the content may be expressed as zero.

(7) "Protein": A statement of the number of grams of protein per serving expressed to the nearest gram, except that if a serving contains less than 1 gram, the statement "Contains less than 1 gram" or "less than 1 gram" may be used as an alternative, and if the serving contains less than 0.5 gram, the content may be expressed as zero. When the protein in products represented or purported to be for adults and children 4 or more years of age has a protein quality value that is a protein digestibility-corrected amino acid score of less than 20 expressed as a percent, or when the protein in a product represented or purported to be for children greater than 1 but less than 4 years of age has a protein quality value that is a protein digestibility- corrected amino acid score of less than

9 CFR §381.409

40 expressed as a percent, either of the following shall be placed adjacent to the declaration of protein content by weight: The statement "not a significant source of protein," or a listing aligned under the column headed "Percent Daily Value" of the corrected amount of protein per serving, as determined in paragraph (c)(7)(ii) of this section, calculated as a percentage of the Daily Reference Value (DRV) or Reference Daily Intake (RDI), as appropriate, for protein and expressed as percent of Daily Value. When the protein quality in a product as measured by the Protein Efficiency Ratio (PER) is less than 40 percent of the reference standard (casein) for a product represented or purported to be for infants, the statement "not a significant source of protein" shall be placed adjacent to the declaration of protein content. Protein content may be calculated on the basis of the factor of 6.25 times the nitrogen content of the food as determined by appropriate methods of analysis in accordance with §381.409(h), except when the procedure for a specific food requires another factor.

(i) A statement of the corrected amount of protein per serving, as determined in paragraph (c)(7)(ii) of this section, calculated as a percentage of the RDI or DRV for protein, as appropriate, and expressed as percent of Daily Value, may be placed on the label, except that such a statement shall be given if a protein claim is made for the product, or if the product is represented or purported to be for infants or children under 4 years of age. When such a declaration is provided, it shall be placed on the label adjacent to the statement of grams of protein and aligned under the column headed "Percent Daily Value," and expressed to the nearest whole percent. However, the percentage of the RDI for protein shall not be declared if the product is represented or purported to be for infants and the protein quality value is less than 40 percent of the reference standard.

(ii) The corrected amount of protein (grams) per serving for products represented or purported to be for adults and children 1 or more years of age is equal to the actual amount of protein (grams) per serving multiplied by the amino acid score corrected for protein digestibility. If the corrected score is above 1.00, then it shall be set at 1.00. The protein digestibility-corrected amino acid score shall be determined by methods given in sections 5.4.1, 7.2.1, and 8 in "Protein Quality Evaluation, Report of the Joint FAO/WHO Expert Consultation on Protein Quality Evaluation," Rome, 1990, which is incorporated by reference. Sections 5.4.1, 7.2.1, and 8 of the "Report of the Joint FAO/WHO Expert Consultation on Protein Quality Evaluation," as published by the Food and Agriculture Organization of the United Nations/World Health Organization, is incorporated as it exists on the date of approval. This incorporation by reference was approved by the Director of the Federal Register in accordance with 5 U.S.C. 552(a) and 1 CFR part 51. It is available for inspection at the Office of the Federal Register, suite 700, 800 North Capitol Street, NW., Washington, DC, or at the office of the FSIS Docket Clerk, Room 3171, South Building, 14th and Independence Avenue, SW., Washington, DC. Copies of the incorporation by reference are available from the Product Assessment Division, Regulatory Programs, Food Safety and Inspection Service, U.S. Department of Agriculture, Room 329, West End Court Building, Washington, DC 20250-3700. For products represented or purported to be for infants, the corrected amount of protein (grams) per serving is equal to the actual amount of protein (grams) per serving multiplied by the relative protein quality value. The relative protein quality value shall be determined by dividing the subject product's protein PER value by the PER value for casein. If the relative protein value is above 1.00, it shall be set at 1.00. (iii) For the purpose of labeling with a percent of the DRV or RDI, a value of 50 grams of protein shall be the DRV for adults and children 4 or more years of age, and the RDI for protein for children less than 4 years of age, infants, pregnant women, and lactating women shall be 16 grams, 14 grams, 60 grams, and 65 grams, respectively.

(8) Vitamins and minerals: A statement of the amount per serving of the vitamins and minerals as described in this paragraph, calculated as a percent of the RDI and expressed as percent of Daily Value. (i) For purposes of declaration of percent of Daily Value as provided for in paragraphs (d) through (g) of this section, products represented or purported to be for use by infants, children less than 4 years of age, pregnant women, or lactating women shall use the RDI's that are specified for the intended group. For products represented or purported to be for use by both infants and children under 4 years of age, the percent of Daily Value shall be presented by separate declarations according to paragraph (e) of this section based on the RDI values for infants from birth to 12 months of age and for children under 4 years of age. Similarly, the percent of Daily Value based on both the RDI values for pregnant women and for lactating women shall be declared separately on products represented or purported to be for use by both pregnant and lactating women. When such dual declaration is used on any label, it shall be included in all labeling, and equal prominence shall be given to both values in all such labeling. All other products shall use the RDI for adults and children 4 or more years of age.

(ii) The declaration of vitamins and minerals as a percent of the RDI shall include vitamin A, vitamin C,

9 CFR §381.409

calcium, and iron, in that order, and shall include any of the other vitamins and minerals listed in paragraph (c)(8)(iv) of this section when they are added, or when a claim is made about them. Other vitamins and minerals need not be declared if neither the nutrient nor the component is otherwise referred to on the label or in labeling or advertising and the vitamins and minerals are:

(A) Required or permitted in a standardized food (e.g., thiamin, riboflavin, and niacin in enriched flour) and that standardized food is included as an ingredient (i.e., component) in another product; or

(B) Included in a product solely for technological purposes and declared only in the ingredients statement. The declaration may also include any of the other vitamins and minerals listed in paragraph (c)(8)(iv) of this section when they are naturally occurring in the food. The additional vitamins and minerals shall be listed in the order established in paragraph (c)(8)(iv) of this section.

(iii) The percentages for vitamins and minerals shall be expressed to the nearest 2-percent increment up to and including the 10-percent level, the nearest 5-percent increment above 10 percent and up to and including the 50-percent level, and the nearest 10-percent increment above the 50-percent level. Amounts of vitamins and minerals present at less than 2 percent of the RDI are not required to be declared in nutrition labeling but may be declared by a zero or by the use of an asterisk (or other symbol) that refers to another asterisk (or symbol) that is placed at the bottom of the table and that is followed by the statement "Contains less than 2 percent of the Daily Value of this (these) nutrient (nutrients)." Alternatively, if vitamin A, vitamin C, calcium, or iron is present in amounts less than 2 percent of the RDI, label declaration of the nutrient(s) is not required if the statement "Not a significant source of _____ (listing the vitamins or minerals omitted)" is placed at the bottom of the table of nutrient values.

(iv) The following RDI's and nomenclature are established for the following vitamins and minerals which are essential in human nutrition:

Vitamin A, 5,000 International Units
Vitamin C, 60 milligrams
Calcium, 1.0 gram
Iron, 18 milligrams
Vitamin D, 400 International Units
Vitamin E, 30 International Units
Thiamin, 1.5 milligrams
Riboflavin, 1.7 milligrams
Niacin, 20 milligrams
Vitamin B_6, 2.0 milligrams
Folate, 0.4 milligram
Vitamin B_{12}, 6 micrograms
Biotin, 0.3 milligram
Pantothenic acid, 10 milligrams
Phosphorus, 1.0 gram
Iodine, 150 micrograms
Magnesium, 400 milligrams
Zinc, 15 milligrams
Copper, 2.0 milligrams

(v) The following synonyms may be added in parenthesis immediately following the name of the nutrient or dietary component:

Vitamin C--Ascorbic acid
Thiamin—Vitamin B_1
Riboflavin—Vitamin B_2
Folate—Folacin

9 CFR §381.409

Calories—Energy

(vi) A statement of the percent of vitamin A that is present as beta-carotene may be declared voluntarily. When the vitamins and minerals are listed in a single column, the statement shall be indented under the information on vitamin A. When vitamins and minerals are arrayed horizontally, the statement of percent shall be presented in parenthesis following the declaration of vitamin A and the percent of Daily Value of vitamin A in the product (e.g., "Percent Daily Value: Vitamin A 50 (90 percent as beta-carotene)"). When declared, the percentages shall be expressed in the same increments as are provided for vitamins and minerals in paragraph (c)(8)(iii) of this section. (9) For the purpose of labeling with a percent of the DRV, the following DRV's are established for the following food components based on the reference caloric intake of 2,000 calories:

Food component	Unit of measurement	DRV
Fat.............................	gram (g)	65
Saturated fatty acids......	do	20
Cholesterol....................	milligrams (mg)	300
Total carbohydrate........	grams (g)	300
Fiber...........................	do	25
Sodium........................	milligrams (mg)	2,400
Potassium.....................	do	3,500
Protein.........................	grams (g)	50

(d)
(1) Nutrient information specified in paragraph (c) of this section shall be presented on products in the following format, except on products on which dual columns of nutrition information are declared as provided for in paragraph (e) of this section, on those products on which the simplified format is permitted to be used as provided for in paragraph (f) of this section, on products for infants and children less than 4 years of age as provided for in §381.500(c), and on products in packages that have a total surface area available to bear labeling of 40 or less square inches as provided for in paragraph (g) of this section.
(i) The nutrition information shall be set off in a box by use of hairlines and shall be all black or one color type, printed on a white or other neutral contrasting background whenever practical.
(ii) All information within the nutrition label shall utilize: (A) A single easy-to-read type style,
(B) Upper and lower case letters,
(C) At least one point leading (i.e., space between two lines of text) except that at least four points leading shall be utilized for the information required by paragraphs (d)(7) and (d)(8) of this section, and
(D) Letters should never touch.
(iii) Information required in paragraphs (d)(3), (d)(5), (d)(7), and (d)(8) of this section shall be in type size no smaller than 8 point. Except for the heading "Nutrition Facts," the information required in paragraphs (d)(4), (d)(6), and (d)(9) of this section and all other information contained within the nutrition label shall be in type size no smaller than 6 point. When provided, the information described in paragraph (d)(10) of this section shall also be in type no smaller than 6 point.
(iv) The headings required by paragraphs (d)(2), (d)(4), and (d)(6) of this section (i.e., "Nutrition Facts," "Amount Per Serving," and "% Daily Value*"), the names of all nutrients that are not indented according to requirements of paragraph (c) of this section (i.e., Calories, Total fat, Cholesterol, Sodium, Potassium, Total carbohydrate, and Protein), and the percentage amounts required by paragraph (d)(7)(ii) of this section shall be

9 CFR §381.409

highlighted by bold or extra bold type or other highlighting (reverse printing is not permitted as a form of highlighting) that prominently distinguishes it from other information. No other information shall be highlighted.

(v) A hairline rule that is centered between the lines of text shall separate "Amount Per Serving" from the calorie statements required in paragraph (d)(5) of this section and shall separate each nutrient and its corresponding percent of Daily Value required in paragraphs (d)(7)(i) and (d)(7)(ii) of this section from the nutrient and percent of Daily Value above and below it.

(2) The information shall be presented under the identifying heading of "Nutrition Facts" which shall be set in a type size larger than all other print size in the nutrition label and, except for labels presented according to the format provided for in paragraph (d)(11) of this section, unless impractical, shall be set the full width of the information provided under paragraph (d)(7) of this section.

(3) Information on serving size shall immediately follow the heading. Such information shall include:

(i) "Serving Size": A statement of the serving size as specified in paragraph (b)(9) of this section.

(ii) "Servings Per Container": The number of servings per container, except that this statement is not required on single-serving containers as defined in paragraph (b)(8) of this section.

(4) A subheading "Amount Per Serving" shall be separated from serving size information by a bar.

(5) Information on calories shall immediately follow the heading "Amount Per Serving" and shall be declared in one line, leaving sufficient space between the declaration of "Calories" and "Calories from fat" to allow clear differentiation, or, if "Calories from saturated fat" is declared, in a column with total "Calories" at the top, followed by "Calories from fat" (indented), and "Calories from saturated fat" (indented).

(6) The column heading "% Daily Value," followed by an asterisk (e.g., "% Daily Value*"), shall be separated from information on calories by a bar. The position of this column heading shall allow for a list of nutrient names and amounts as described in paragraph (d)(7) of this section to be to the left of, and below, this column heading. The column heading "Percent Daily Value," "Percent DV," or "% DV" may be substituted for "% Daily Value."

(7) Except as provided for in paragraph (g) of this section, and except as permitted by §381.500(d)(2), nutrient information for both mandatory and any voluntary nutrients listed in paragraph (c) of this section that are to be declared in the nutrition label, except vitamins and minerals, shall be declared as follows:

(i) The name of each nutrient, as specified in paragraph (c) of this section, shall be given in a column and followed immediately by the quantitative amount by weight for that nutrient appended with a "g" for grams or "mg" for milligrams.

(ii) A listing of the percent of the DRV as established in paragraphs (c)(7)(iii) and (c)(9) of this section shall be given in a column aligned under the heading "% Daily Value" established in paragraph (d)(6) of this section with the percent expressed to the nearest whole percent for each nutrient declared in the column described in paragraph (d)(7)(i) of this section for which a DRV has been established, except that the percent for protein may be omitted as provided in paragraph (c)(7) of this section. The percent shall be calculated by dividing either the amount declared on the label for each nutrient or the actual amount of each nutrient (i.e., before rounding) by the DRV for the nutrient, except that the percent for protein shall be calculated as specified in paragraph (c)(7)(ii) of this section. The numerical value shall be followed by the symbol for percent (i.e., %). (8) Nutrient information for vitamins and minerals shall be separated from information on other nutrients by a bar and shall be arrayed horizontally (e.g., Vitamin A 4%, Vitamin C 2%, Calcium 15%, Iron 4%) or may be listed in two columns, except that when more than four vitamins and minerals are declared, they may be declared vertically with percentages listed under the column headed "% Daily Value."

(9) A footnote, preceded by an asterisk, shall be placed beneath the list of vitamins and minerals and shall be separated from that list by a hairline.

(i) The footnote shall state: Percent Daily Values are based on a 2,000 calorie diet. Your daily values may be higher or lower depending on your calorie needs.

	Calories:	2,000	2,500
Total fat	Less than	65 g	80 g

9 CFR §381.409

Saturated fat	Less than	20 g	25 g
Cholesterol	Less than	300 mg	300 mg
Sodium	Less than	2,400 mg	2,400 mg
Total carbohydrate		300 g	375 g
Dietary fiber		25 g	30 g

(ii) If the percent of Daily Value is given for protein in the Percent of Daily Value column as provided in paragraph (d)(7)(ii) of this section, protein shall be listed under dietary fiber, and a value of 50 g shall be inserted on the same line in the column headed "2,000" and value of 65 g in the column headed "2,500."

(iii) If potassium is declared in the column described in paragraph (d)(7)(i) of this section, potassium shall be listed under sodium and the DRV established in paragraph (c)(9) of this section shall be inserted on the same line in the numeric columns.

(iv) The abbreviations established in paragraph (g)(2) of this section may be used within the footnote.

(10) Caloric conversion information on a per-gram basis for fat, carbohydrate, and protein may be presented beneath the information required in paragraph (d)(9), separated from that information by a hairline. This information may be presented horizontally (i.e., "Calories per gram: Fat 9, Carbohydrate 4, Protein 4") or vertically in columns.

(11) (i) If the space beneath the information on vitamins and minerals is not adequate to accommodate the information required in paragraph (d)(9) of this section, the information required in paragraph (d)(9) may be moved to the right of the column required in paragraph (d)(7)(ii) of this section and set off by a line that distinguishes it and sets it apart from the percent of Daily Value information. The caloric conversion information provided for in paragraph (d)(10) of this section may be presented beneath either side or along the full length of the nutrition label.

(ii) If the space beneath the mandatory declaration of iron is not adequate to accommodate any remaining vitamins and minerals to be declared or the information required in paragraph (d)(9) of this section, the remaining information may be moved to the right and set off by a line that distinguishes it and sets it apart from the percent of Daily Value information given to the left. The caloric conversion information provided for in paragraph (d)(10) of this section may be presented beneath either side or along the full length of the nutrition label.

(iii) If there is not sufficient continuous vertical space (i.e., approximately 3 inches) to accommodate the required components of the nutrition label up to and including the mandatory declaration of iron, the nutrition label may be presented in a tabular display in which the footnote required by paragraph (d)(9) of the section is given to the far right of the label, and additional vitamins and minerals beyond the four that are required (i.e., vitamin A, vitamin C, calcium, and iron) are arrayed horizontally following declarations of the required vitamins and minerals.

9 CFR §381.409

(12) The following sample label illustrates the provisions of paragraph (d) of this section:

Nutrition Facts

Serving Size 1 cup (228g)
Servings Per Container 2

Amount Per Serving

Calories 260   Calories from Fat 120

	% Daily Value*
Total Fat 13g	20%
Saturated Fat 5g	25%
Cholesterol 30mg	10%
Sodium 660mg	28%
Total Carbohydrate 31g	10%
Dietary Fiber 0g	0%
Sugars 5g	
Protein 5g	

Vitamin A 4%   •   Vitamin C 2%
Calcium 15%   •   Iron 4%

* Percent Daily Values are based on a 2,000 calorie diet. Your daily values may be higher or lower depending on your calorie needs:

	Calories:	2,000	2,500
Total Fat	Less than	65g	80g
Sat Fat	Less than	20g	25g
Cholesterol	Less than	300mg	300mg
Sodium	Less than	2,400mg	2,400mg
Total Carbohydrate		300g	375g
Dietary Fiber		25g	30g

Calories per gram:
Fat 9   •   Carbohydrate 4   •   Protein 4

(13) (i) Nutrition labeling on the outer label of packages of poultry products that contain two or more products in the same packages (e.g., variety packs) or of packages that are used interchangeably for the same type of food (e.g., poultry salad containers) may use an aggregate display.

(ii) Aggregate displays shall comply with format requirements of paragraph (d) of this section to the maximum extent possible, except that the identity of each food shall be specified to the right of the "Nutrition Facts" title, and both the quantitative amount by weight (i.e., g/mg amounts) and the percent Daily Value for each nutrient shall be listed in separate columns under the name of each food.

(14) When nutrition labeling appears in a second language, the nutrition information may be presented in a separate nutrition label for each language or in one nutrition label with the information in the second language following that in English. Numeric characters that are identical in both languages need not be repeated (e.g., "Protein/ Proteinas 2 g"). All required information must be included in both languages.

(e) Nutrition information may be presented for two or more forms of the same product (e.g., both "raw" and "cooked") or for common combinations of foods as provided for in paragraph (b) of this section, or for different units (e.g., per 100 grams) as provided for in paragraph (b) of this section, or for two or more groups for which RDI's are established (e.g., both infants and children less than 4 years of age) as provided for in paragraph (c)(8)(i) of this section. When such dual labeling is provided, equal prominence shall be given to both sets of values. Information shall be presented in a format consistent with paragraph (d) of this section, except that:

(1) Following the subheading of "Amount Per Serving," there shall be two or more column headings accurately describing the forms of the same product (e.g., "raw" and "roasted"), the combinations of foods, the units, or the RDI groups that are being declared. The column representing the product as packaged and according to the label serving size based on the Reference Amount in §381.412(b) shall be to the left of the numeric columns.

(2) When the dual labeling is presented for two or more forms of the same product, for combinations of foods, or for different units, total calories and calories from fat (and calories from saturated fat, when declared) shall be listed in a column and indented as specified in paragraph (d)(5) of this section with quantitative amounts declared in columns aligned under the column headings set forth in paragraph (e)(1) of this section.

(3) Quantitative information by weight required in paragraph (d)(7)(i) of this section shall be specified for the form of the product as packaged and according to the label serving size based on the Reference Amount in §381.412(b).

(i) Quantitative information by weight may be included for other forms of the product represented by the additional column(s) either immediately adjacent to the required quantitative information by weight for the product as packaged and according to the label serving size based on the Reference Amount in §381.412(b) or as a footnote.

(A) If such additional quantitative information is given immediately adjacent to the required quantitative information, it shall be declared for all nutrients listed and placed immediately following and differentiated from the required quantitative information (e.g., separated by a comma). Such information shall not be put in a separate column.

(B) If such additional quantitative information is given in a footnote, it shall be declared in the same order as the nutrients are listed in the nutrition label. The additional quantitative information may state the total nutrient content of the product identified in the second column or the nutrient amounts added to the product as packaged for only those nutrients that are present in different amounts than the amounts declared in the required quantitative information. The footnote shall clearly identify which amounts are declared. Any subcomponents declared shall be listed parenthetically after principal components (e.g., 1/2 cup skim milk contributes an additional 40 calories, 65 mg sodium, 6 g total carbohydrate (6 g sugars), and 4 g protein). (ii) Total fat and its quantitative amount by weight shall be followed by an asterisk (or other symbol) (e.g., "Total fat (2 g)*") referring to another asterisk (or symbol) at the bottom of the nutrition label identifying the form(s) of the product for which quantitative information is presented.

(4) Information required in paragraphs (d)(7)(ii) and (d)(8) of this section shall be presented under the subheading "% DAILY VALUE" and in columns directly under the column headings set forth in paragraph (e)(1) of this section.

9 CFR §381.409

(5) The following sample label illustrates the provisions of paragraph (e) of this section:

Nutrition Facts

Serving Size 1/12 package (44g, about 1/4 cup dry mix)
Servings Per Container 12

Amount Per Serving	Mix	Baked
Calories	190	280
Calories from Fat	45	140
	% Daily Value**	
Total Fat 5g*	8%	24%
Saturated Fat 2g	10%	13%
Cholesterol 0mg	0%	23%
Sodium 300mg	13%	13%
Total Carbohydrate 34g	11%	11%
Dietary Fiber 0g	0%	0%
Sugars 18g		
Protein 2g		
Vitamin A	0%	0%
Vitamin C	0%	0%
Calcium	6%	8%
Iron	2%	4%

*Amount in Mix
**Percent Daily Values are based on a 2,000 calorie diet. Your daily values may be higher or lower depending on your calorie needs:

	Calories:	2,000	2,500
Total Fat	Less than	65g	80g
Sat Fat	Less than	20g	25g
Cholesterol	Less than	300mg	300mg
Sodium	Less than	2,400mg	2,400mg
Total Carbohydrate		300g	375g
Dietary Fiber		25g	30g

Calories per gram:
Fat 9 • Carbohydrate 4 • Protein 4

(f)

(1) Nutrition information may be presented in a simplified format as set forth herein when any required nutrients, other than the core nutrients (i.e., calories, total fat, sodium, total carbohydrate, and protein), are present in insignificant amounts. An insignificant amount shall be defined as that amount that may be rounded to zero in nutrition labeling, except that for total carbohydrate, dietary fiber, sugars and protein, it shall be an amount less than 1 gram.

9 CFR §381.408

(2) The simplified format shall include information on the following nutrients:

(i) Total calories, total fat, total carbohydrate, sodium, and protein;

(ii) Any of the following that are present in more than insignificant amounts: Calories from fat, saturated fat, cholesterol, dietary fiber, sugars, vitamin A, vitamin C, calcium, and iron; and

(iii) Any vitamins and minerals listed in paragraph (c)(8)(iv) of this section when they are added in fortified or fabricated foods.

(3) Other nutrients that are naturally present in the product in more than insignificant amounts may be voluntarily declared as part of the simplified format.

(4) Any required nutrient, other than a core nutrient, that is present in an insignificant amount may be omitted from the tabular listing, provided that the following statement is included at the bottom of the nutrition label, "Not a significant source of _____." The blank shall be filled in with the appropriate nutrient or food component. Alternatively, amounts of vitamins and minerals present in insignificant amounts may be declared by the use of an asterisk (or symbol) that is placed at the bottom of the table of nutrient values and that is followed by the statement "Contains less than 2 percent of the Daily Value of this (these) nutrient (nutrients)."

(5) Except as provided for in paragraph (g) of this section and in §381.500(c) and (d), nutrient information declared in the simplified format shall be presented in the same manner as specified in paragraphs (d) or (e) of this section, except that the footnote required in paragraph (d)(9) of this section is not required. When the footnote is omitted, an asterisk shall be placed at the bottom of the label followed by the statement "Percent Daily Values are based on a 2,000 calorie diet" and, if the term "Daily Value" is not spelled out in the heading, a statement that "DV" represents "Daily Value."

(g) Foods in packages that have a total surface area available to bear labeling of 40 or less square inches may modify the requirements of paragraphs (c) through (f) of this section and §381.402(a) by one or more of the following means:

(1) (i) Presenting the required nutrition information in a tabular or linear (i.e., string) fashion, rather than in vertical columns if the product has a total surface area available to bear labeling of less than 12 square inches, or if the product has a total surface area available to bear labeling of 40 or less square inches and the package shape or size cannot accommodate a standard vertical column or tabular display on any label panel. Nutrition information may be given in a linear fashion only if the package shape or size will not accommodate a tabular display.

(ii) When nutrition information is given in a linear display, the nutrition information shall be set off in a box by the use of a hairline. The percent Daily Value is separated from the quantitative amount declaration by the use of parenthesis, and all nutrients, both principal components and subcomponents, are treated similarly. Bolding is required only on the title "Nutrition Facts" and is allowed for nutrient names for "Calories," "Total fat," "Cholesterol," "Sodium," "Total carbohydrate," and "Protein."

(2) Using any of the following abbreviations:

Serving size—Serv size
Servings per container—Servings
Calories from fat—Fat cal
Calories from saturated fat—Sat fat cal
Saturated fat—Sat fat
Monounsaturated fat—Monounsat fat
Polyunsaturated fat—Polyunsat fat
Cholesterol—Cholest
Total carbohydrate—Total carb
Dietary fiber—Fiber
Soluble fiber—Sol fiber
Insoluble fiber—Insol fiber
Sugar alcohol—Sugar alc
Other carbohydrate—Other carb

(3) Omitting the footnote required in paragraph (d)(9) of this section and placing another asterisk at the bottom

9 CFR §381.408

of the label followed by the statement "Percent Daily Values are based on a 2,000 calorie diet" and, if the term "Daily Value" is not spelled out in the heading, a statement that "DV" represents "Daily Value."

(4) Presenting the required information on any other label panel.

(h) Compliance with this section shall be determined as follows:

(1) A production lot is a set of food production consumer units that are from one production shift. Alternatively, a collection of consumer units of the same size, type, and style produced under conditions as nearly uniform as possible, designated by a common container code or marking, constitutes a production lot.

(2) The sample for nutrient analysis shall consist of a composite of a minimum of six consumer units, each from a production lot. Alternatively, the sample for nutrient analysis shall consist of a composite of a minimum of six consumer units, each randomly chosen to be representative of a production lot. In each case, the units may be individually analyzed and the results of the analyses averaged, or the units would be composited and the composite analyzed. In both cases, the results, whether an average or a single result from a composite, will be considered by the Agency to be the nutrient content of a composite. All analyses shall be performed by appropriate methods and procedures used by the Department for each nutrient in accordance with the "Chemistry Laboratory Guidebook," or, if no USDA method is available and appropriate for the nutrient, by appropriate methods for the nutrient in accordance with the 1990 edition of the "Official Methods of Analysis" of the AOAC International, formerly Association of Official Analytical Chemists, 15th ed., which is incorporated by reference, unless a particular method of analysis is specified in §381.409(c), or, if no USDA, AOAC, or specified method is available and appropriate, by other reliable and appropriate analytical procedures as so determined by the Agency. The "Official Methods of Analysis" is incorporated as it exists on the date of approval. This incorporation by reference was approved by the Director of the Federal Register in accordance with 5 U.S.C. 552(a) and 1 CFR part 51. Copies may be purchased from the AOAC International, 2200 Wilson Blvd., Suite 400, Arlington, VA 22201. It is also available for inspection at the Office of the Federal Register Information Center, suite 700, 800 North Capitol Street, NW., Washington, DC.

(3) Two classes of nutrients are defined for purposes of compliance:

(i) Class I. Added nutrients in fortified or fabricated foods; and

(ii) Class II. Naturally occurring (indigenous) nutrients. If any ingredient which contains a naturally occurring (indigenous) nutrient is added to a food, the total amount of such nutrient in the final food product is subject to Class II requirements unless the same nutrient is also added, which would make the total amount of such nutrient subject to Class I requirements.

(4) A product with a label declaration of a vitamin, mineral, protein, total carbohydrate, dietary fiber, other carbohydrate, polyunsaturated or monounsaturated fat, or potassium shall be deemed to be misbranded under section 4(h) of the Poultry Products Inspection Act (21 U.S.C. 453(h)(4)) unless it meets the following requirements:

(i) Class I vitamin, mineral, protein, dietary fiber, or potassium. The nutrient content of the composite is at least equal to the value for that nutrient declared on the label.

(ii) Class II vitamin, mineral, protein, total carbohydrate, dietary fiber, other carbohydrate, polyunsaturated or monounsaturated fat, or potassium. The nutrient content of the composite is at least equal to 80 percent of the value for that nutrient declared on the label; Provided, That no regulatory action will be based on a determination of a nutrient value which falls below this level by an amount less than the variability generally recognized for the analytical method used in that product at the level involved, and inherent nutrient variation in a product.

(5) A product with a label declaration of calories, sugars, total fat, saturated fat, cholesterol, or sodium shall be deemed to be misbranded under section 4(h) of the Poultry Products Inspection Act (21 U.S.C. 453(h)(4)) if the nutrient content of the composite is greater than 20 percent in excess of the value for that nutrient declared on the label; Provided, That no regulatory action will be based on a determination of a nutrient value which falls above this level by an amount less than the variability generally recognized for the analytical method used in that product at the level involved, and inherent nutrient variation in a product.

(6) The amount of a vitamin, mineral, protein, total carbohydrate, dietary fiber, other carbohydrate, polyunsaturated or monounsaturated fat, or potassium may vary over labeled amounts within good manufacturing practice. The amount of calories, sugars, total fat, saturated fat, cholesterol, or sodium may vary

9 CFR §381.409

under labeled amounts within good manufacturing practice.

(7) Compliance will be based on the metric measure specified in the label statement of serving size.

(8) The management of the establishment must maintain records to support the validity of nutrient declarations contained on product labels. Such records shall be made available to the inspector or any duly authorized representative of the Agency upon request.

(9) The compliance provisions set forth in paragraph (h)(1) through (8) of this section shall not apply to single-ingredient, raw poultry products, including those that have been previously frozen, when nutrition labeling is based on the most current representative data base values contained in USDA's National Nutrient Data Bank or its published form, the Agriculture Handbook No. 8 series.

(Paperwork requirements were approved by the Office of Management and Budget under control number 0583-0088.)

[58 FR 675, Jan. 6, 1993; 58 FR 43788, Aug. 18, 1993, as amended at 58 FR 47628, Sept. 10, 1993; 59 FR 45196, Sept. 1, 1994; 60 FR 197, Jan. 3, 1995]

§381.410 [Reserved]

§381.411 [Reserved]

§381.412 Reference amounts customarily consumed per eating occasion.

(a) The general principles followed in arriving at the reference amounts customarily consumed per eating occasion (Reference Amount(s)), as set forth in paragraph (b) of this section, are:

(1) The Reference Amounts are calculated for persons 4 years of age or older to reflect the amount of food customarily consumed per eating occasion by persons in this population group. These Reference Amounts are based on data set forth in appropriate national food consumption surveys.

(2) The Reference Amounts are calculated for an infant or child under 4 years of age to reflect the amount of food customarily consumed per eating occasion by infants up to 12 months of age or by children 1 through 3 years of age, respectively. These Reference Amounts are based on data set forth in appropriate national food consumption surveys. Such Reference Amounts are to be used only when the product is specially formulated or processed for use by an infant or by a child under 4 years of age.

(3) An appropriate national food consumption survey includes a large sample size representative of the demographic and socioeconomic characteristics of the relevant population group and must be based on consumption data under actual conditions of use.

(4) To determine the amount of food customarily consumed per eating occasion, the mean, median, and mode of the consumed amount per eating occasion were considered.

(5) When survey data were insufficient, FSIS took various other sources of information on serving sizes of food into consideration. These other sources of information included:

(i) Serving sizes used in dietary guidance recommendations or recommended by other authoritative systems or organizations;

(ii) Serving sizes recommended in comments;

(iii) Serving sizes used by manufacturers and grocers; and (iv) Serving sizes used by other countries.

(6) Because they reflect the amount customarily consumed, the Reference Amount and, in turn, the serving size declared on the product label are based on only the edible portion of food, and not bone, seed, shell, or other inedible components.

(7) The Reference Amount is based on the major intended use of the product (e.g., a mixed dish measurable with a cup as a main dish and not as a side dish).

(8) The Reference Amounts for products that are consumed as an ingredient of other products, but that may also be consumed in the form in which they are purchased (e.g., ground poultry), are based on use in the form purchased.

(9) FSIS sought to ensure that foods that have similar dietary usage, product characteristics, and customarily consumed amounts have a uniform Reference Amount.

(b) The following Product Categories and Reference Amounts shall be used as the basis for determining serving sizes for specific products:

9 CFR §381.412

TABLE 1 Reference Amounts Customarily Consumed per Eating Occasion — Infant and Toddler Foods[1, 2, 3]

Product category	Reference amount
Infant & Toddler Foods:	
Dinner Dry Mix	15 g
Dinner, ready-to-serve, strained type...................	60 g
Dinner, soups, ready-to-serve junior type.............	110 g
Dinner, stew or soup ready-to-serve toddlers.......	170 g
Plain poultry and poultry sticks, ready-to-serve...	55 g

{1} These values represent the amount of food customarily consumed per eating occasion and were primarily derived from the 1977-1978 and the 1987-1988 Nationwide Food Consumption Surveys conducted by the U.S. Department of Agriculture.

{2} Unless otherwise noted in the Reference Amount column, the Reference Amounts are for the ready-to-serve or almost ready-to-serve form of the product (i.e., heat and serve). If not listed separately, the Reference Amount for the unprepared form (e.g., dehydrated cereal) is the amount required to make one Reference Amount of the prepared form.

{3} Manufacturers are required to convert the Reference Amount to the label serving size in a household measure most appropriate to their specific product using the procedures established by the regulation.

TABLE 2 Reference Amounts Customarily Consumed per Eating Occasion — General Food Supply [1, 2, 3, 4, 5]

Product category	Reference Amount Ready-to-serve	Reference Amount Ready-to-cook
Egg mixtures (western style omelet, souffle, egg foo young with poultry).	110 g	n/a
Salad and potato toppers; e.g., poultry bacon bits.	7 g	n/a
Bacon; e.g., poultry breakfast = bacon. strips.	15 g	26 g 18 g = breakfast strips
Dried; e.g., poultry jerky, dried poultry, poultry sausage products with a moisture/protein ratio of less than 2:1.	30 g	n/a
Snacks; e.g., poultry snack food sticks.	30 g	n/a
Luncheon products, poultry bologna, poultry Canadian style bacon, poultry crumbles, poultry luncheon loaf, potted poultry products, poultry taco filings.	55 g	n/a
Linked poultry sausage products, poultry franks, poultry Polish sausage, smoked or pickled poultry meat, poultry smoked sausage.	55 g	n/a 69 g = uncooked sausage
Entrees without sauce, poultry cuts, ready to cook poultry cuts, including marinated, tenderized, injected cuts of poultry, poultry corn dogs, poultry croquettes, poultry fritters, cured poultry ham products, adult pureed poultry.	85 g	114 g

9 CFR §381.412

Product category	Reference Amount Ready-to-serve	Reference Amount Ready-to-cook
Canned poultry, canned chicken, canned [4] turkey.	55 g	n/a
Entrees with sauce, turkey and gravy.	140 g	n/a
Mixed dishes NOT measurable with a cup; [5] e.g., poultry burrito, poultry enchiladas, poultry pizza, poultry quiche, all types of poultry sandwiches, cracker and poultry lunch-type packages, poultry gyro, poultry stromboli, poultry frank on a bun, poultry burger on a bun, poultry taco, chicken cordon bleu, poultry calzone, stuffed vegetables with poultry, poultry kabobs.	140 g (plus 55 g for products toppings)	n/a
Mixed dishes, measurables with a cup; e.g., poultry casserole, macaroni and cheese with poultry, poultry pot pie, poultry spaghetti with sauce, poultry chili, poultry chili with beans, poultry hash, creamed dried poultry, poultry ravioli in sauce, poultry a la king, poultry stew, poultry goulash, poultry lasagna, poultry filled pasta.	1 cup	n/a
Salads-pasta or potato, potato salad with poultry, macaroni and poultry salad.	140 g	n/a
Salads-all other, poultry salads, chicken salad, turkey salad.	100 g	n/a
Soups-all varieties.	245 g	n/a
Major main entree type sauce; e.g., spaghetti sauce with poultry.	125 g	n/a
Minor main entree sauce; e.g., pizza sauce with poultry, gravy.	1/4 cup	n/a
Seasoning mixes dry, freeze dry, dehydrated, concentrated soup mixes, bases, extracts, dried broths and stock/juice, freeze dry trail mix products with poultry.		
As reconstituted: Amount to make one Reference Amount of the final dish; e.g.-		
Gravy...	1/4 cup	n/a
Major main entree type sauce........................	125 g	n/a
Soup..	245 g	n/a
Entree measurable with a cup.........................	1 cup	n/a

{1} These values represent the amount of food customarily consumed per eating occasion and were primarily derived from the 1977-78 and the 1987-88 Nationwide Food Consumption Surveys conducted by the U.S. Department of Agriculture.
{2} Manufacturers are required to convert the Reference Amounts to the label serving size in a household measure most appropriate to their specific product using the procedures established by regulation.
{3} Examples listed under Product Category are not all inclusive or exclusive. Examples are provided to assist manufacturers in identifying appropriate product Reference Amount.
{4} If packed or canned in liquid, the Reference Amount is for the drained solids, except for products in which both the solids and liquids are customarily consumed.
{5} Pizza sauce is part of the pizza and is not considered to be a sauce topping.

(c) For products that have no Reference Amount listed in paragraph (b) of this section for the unprepared or the prepared form of the product and that consist of two or more foods packaged and presented to be consumed together (e.g., poultry lunch meat with cheese and crackers), the Reference Amount for the combined product shall be

9 CFR §381.412

determined using the following rules:

(1) For bulk products, the Reference Amount for the combined product shall be the Reference Amount, as established in paragraph (b) of this section, for the ingredient that is represented as the main ingredient plus proportioned amounts of all minor ingredients.

(2) For products where the ingredient represented as the main ingredient is one or more discrete units, the Reference Amount for the combined product shall be either the number of small discrete units or the fraction of the large discrete unit that is represented as the main ingredient that is closest to the Reference Amount for that ingredient as established in paragraph (b) of this section plus proportioned amounts of all minor ingredients. (3) If the Reference Amounts are in compatible units, they shall be summed (e.g., ingredients in equal volumes such as tablespoons). If the Reference Amounts are in incompatible units, the weights of the appropriate volumes should be used (e.g., grams of one ingredient plus gram weight of tablespoons of a second ingredient).

(d) If a product requires further preparation, e.g., cooking or the addition of water or other ingredients, and if paragraph (b) of this section provides a Reference Amount for the product in the prepared form, then the Reference Amount for the unprepared product shall be determined using the following rules:

(1) Except as provided for in paragraph (d)(2) of this section, the Reference Amount for the unprepared product shall be the amount of the unprepared product required to make the Reference Amount for the prepared product as established in paragraph (b) of this section.

(2) For products where the entire contents of the package is used to prepare one large discrete unit usually divided for consumption, the Reference Amount for the unprepared product shall be the amount of the unprepared product required to make the fraction of the large discrete unit closest to the Reference Amount for the prepared product as established in paragraph (b) of this section.

(e) The Reference Amount for an imitation or substitute product or altered product as defined in §381.413(d), such as a "low calorie" version, shall be the same as for the product for which it is offered as a substitute.

(f) The Reference Amounts set forth in paragraphs (b) through (e) of this section shall be used in determining whether a product meets the criteria for nutritional claims. If the serving size declared on the product label differs from the Reference Amount, and the product meets the criteria for the claim only on the basis of the Reference Amount, the claim shall be followed by a statement that sets forth the basis on which the claim is made. That statement shall include the Reference Amount as it appears in paragraph (b) of this section followed, in parenthesis, by the amount in common household measure if the Reference Amount is expressed in measures other than common household measures.

(g) The Administrator, on his or her own initiative or on behalf of any interested person who has submitted a labeling application, may issue a proposal to establish or amend a Product Category or Reference Amount identified in paragraph (b) of this section.

(1) Labeling applications and supporting documentation to be filed under this section shall be submitted in quadruplicate, except that the supporting documentation may be submitted on a computer disc copy. If any part of the material submitted is in a foreign language, it shall be accompanied by an accurate and complete English translation. The labeling application shall state the applicant's post office address.

(2) Pertinent information will be considered as part of an application on the basis of specific reference to such information submitted to and retained in the files of the Food Safety and Inspection Service. However, any reference to unpublished information furnished by a person other than the applicant will not be considered unless use of such information is authorized (with the understanding that such information may in whole or part be subject to release to the public) in a written statement signed by the person who submitted it. Any reference to published information should be accompanied by reprints or photostatic copies of such references.

(3) The availability for public disclosure of labeling applications, along with supporting documentation, submitted to the Agency under this section will be governed by the rules specified in subchapter D, title 9.

(4) Data accompanying the labeling application, such as food consumption data, shall be submitted on separate sheets, suitably identified. If such data has already been submitted with an earlier labeling application from the applicant, the present labeling application must provide the data.

(5) The labeling application must be signed by the applicant or by his or her attorney or agent, or (if a corporation) by an authorized official.

(6) The labeling application shall include a statement signed by the person responsible for the labeling

9 CFR §381.412

application, that to the best of his or her knowledge, it is a representative and balanced submission that includes unfavorable information, as well as favorable information, known to him or her pertinent to the evaluation of the labeling application.

(7) Labeling applications for a new Reference Amount and/or Product Category shall be accompanied by the following data which shall be submitted in the following form to the Director, Food Labeling Division, Regulatory Programs, Food Safety and Inspection Service, Washington, DC 20250:

(Date)

The undersigned, ------ submits this labeling application pursuant to 9 CFR 381.412 with respect to Reference Amount and/or Product Category.

Attached hereto, in quadruplicate, or on a computer disc copy, and constituting a part of this labeling application, are the following:

(i) A statement of the objective of the labeling application; (ii) A description of the product;
(iii) A complete sample product label including nutrition label, using the format established by regulation;
(iv) A description of the form in which the product will be marketed;
(v) The intended dietary uses of the product with the major use identified (e.g., turkey as a luncheon meat);
(vi) If the intended use is primarily as an ingredient in other foods, list of foods or food categories in which the product will be used as an ingredient with information on the prioritization of the use;
(vii) The population group for which the product will be offered for use (e.g., infants, children under 4 years of age);
(viii) The names of the most closely-related products (or in the case of foods for special dietary use and imitation or substitute foods, the names of the products for which they are offered as substitutes);
(ix) The suggested Reference Amount (the amount of edible portion of food as consumed, excluding bone, skin or other inedible components) for the population group for which the product is intended with full description of the methodology and procedures that were used to determine the suggested Reference Amount. In determining the Reference Amount, general principles and factors in paragraph (a) of this section should be followed.
(x) The suggested Reference Amount shall be expressed in metric units. Reference Amounts for foods shall be expressed in grams except when common household units such as cups, tablespoons, and teaspoons are more appropriate or are more likely to promote uniformity in serving sizes declared on product labels. For example, common household measures would be more appropriate if products within the same category differ substantially in density such as mixed dishes measurable with a cup.
(A) In expressing the Reference Amount in grams, the following general rules shall be followed:
(1) For quantities greater than 10 grams, the quantity shall be expressed in nearest 5 grams increment.
(2) For quantities less than 10 grams, exact gram weights shall be used.
(B) [Reserved]
(xi) A labeling application for a new subcategory of food with its own Reference Amount shall include the following additional information:
(A) Data that demonstrate that the new subcategory of food will be consumed in amounts that differ enough from the Reference Amount for the parent category to warrant a separate Reference Amount. Data must include sample size, and the mean, standard deviation, median, and modal consumed amount per eating occasion for the product identified in the labeling application and for other products in the category. All data must be derived from the same survey data.
(B) Documentation supporting the difference in dietary usage and product characteristics that affect the consumption size that distinguishes the product identified in the labeling application from the rest of the products in the category.
(xii) In conducting research to collect or process food consumption data in support of the labeling application, the following general guidelines should be followed.
(A) Sampled population selected should be representative of the demographic and socioeconomic characteristics of the target population group for which the food is intended.
(B) Sample size (i.e., number of eaters) should be large enough to give reliable estimates for customarily consumed amounts.
(C) The study protocol should identify potential biases and describe how potential biases are controlled for or, if not possible to control, how they affect interpretation of results.
(D) The methodology used to collect or process data including study design, sampling procedures, materials used (e.g., questionnaire, interviewer's manual), procedures used to collect or process data, methods or procedures used to control for unbiased estimates, and procedures used to correct for nonresponse, should be fully documented.
(xiii) A statement concerning the feasibility of convening associations, corporations, consumers, and other interested parties to engage in negotiated rulemaking to develop a proposed rule.

9 CFR §381.412

Yours very truly,

Applicant _____

By _____
(Indicate authority)

(8) Upon receipt of the labeling application and supporting documentation, the applicant shall be notified, in writing, of the date on which the labeling application was received. Such notice shall inform the applicant that the labeling application is undergoing Agency review and that the applicant shall subsequently be notified of the Agency's decision to consider for further review or deny the labeling application.

(9) Upon review of the labeling application and supporting documentation, the Agency shall notify the applicant, in writing, that the labeling application is either being considered for further review or that it has been summarily denied by the Administrator.

(10) If the labeling application is summarily denied by the Administrator, the written notification shall state the reasons therefor, including why the Agency has determined that the proposed Reference Amount and/or Product Category is false or misleading. The notification letter shall inform the applicant that the applicant may submit a written statement by way of answer to the notification, and that the applicant shall have the right to request a hearing with respect to the merits or validity of the Administrator's decision to deny the use of the proposed Reference Amount and/or Product Category.

(i) If the applicant fails to accept the determination of the Administrator and files an answer and requests a hearing, and the Administrator, after review of the answer, determines the initial determination to be correct, the Administrator shall file with the Hearing Clerk of the Department the notification, answer, and the request for a hearing, which shall constitute the complaint and answer in the proceeding, which shall thereafter be conducted in accordance with the Department's Uniform Rules of Practice.

(ii) The hearing shall be conducted before an administrative law judge with the opportunity for appeal to the Department's Judicial Officer, who shall make the final determination for the Secretary. Any such determination by the Secretary shall be conclusive unless, within 30 days after receipt of notice of such final determination, the applicant appeals to the United States Court of Appeals for the circuit in which the applicant has its principal place of business or to the United States Court of Appeals for the District of Columbia Circuit.

(11) If the labeling application is not summarily denied by the Administrator, the Administrator shall publish in the Federal Register a proposed rule to amend the regulations to authorize the use of the Reference Amount and/or Product Category. The proposal shall also summarize the labeling application, including where the supporting documentation can be reviewed. The Administrator's proposed rule shall seek comment from consumers, the industry, consumer and industry groups, and other interested persons on the labeling application and the use of the proposed Reference Amount and/or Product Category. After public comment has been received and reviewed by the Agency, the Administrator shall make a determination on whether the proposed Reference Amount and/or Product Category shall be approved for use on the labeling of poultry products.

(i) If the Reference Amount and/or Product Category is denied by the Administrator, the Agency shall notify the applicant, in writing, of the basis for the denial, including the reason why the Reference Amount and/or Product Category on the labeling was determined by the Agency to be false or misleading. The notification letter shall also inform the applicant that the applicant may submit a written statement by way of answer to the notification, and that the applicant shall have the right to request a hearing with respect to the merits or validity of the Administrator's decision to deny the use of the proposed Reference Amount and/or Product Category.

(A) If the applicant fails to accept the determination of the Administrator and files an answer and requests a hearing, and the Administrator, after review of the answer, determines the initial determination to be correct, the Administrator shall file with the Hearing Clerk of the Department the notification, answer, and the request for a hearing, which shall constitute the complaint and answer in the proceeding, which shall thereafter be conducted in accordance with the Department's Uniform Rules of Practice.

(B) The hearing shall be conducted before an administrative law judge with the opportunity for appeal to the Department's Judicial Officer, who shall make the final determination for the Secretary. Any such determination by the Secretary shall be conclusive unless, within 30 days after receipt of notice of such final determination, the applicant appeals to the United States Court of Appeals for the circuit in which the applicant has its principal place of business or to the United States Court of Appeals for the District of Columbia.

(ii) If the Reference Amount and/or Product Category is approved, the Agency shall notify the applicant, in writing, and shall also publish in the Federal Register a final rule amending the regulations to authorize the use of the Reference Amount and/or Product Category.

(Paperwork requirements were approved by the Office of Management and Budget under control number 0583-0088.)

[58 FR 675, Jan. 6, 1993; 58 FR 43789, Aug. 18, 1993, as amended at 58 FR 47628, Sept. 10, 1993; 59 FR 45198, Sept. 1, 1994; 60 FR 207, Jan. 3, 1995]

§381.413 Nutrient content claims; general principles.

(a) This section applies to poultry products that are intended for human consumption and that are offered for sale.

(b) A claim which, expressly or by implication, characterizes the level of a nutrient (nutrient content claim) of the type required in nutrition labeling pursuant to §381.409, may not be made on a label or in labeling of that product unless the claim is made in accordance with the applicable provisions in this subpart.

(1) An expressed nutrient content claim is any direct statement about the level (or range) of a nutrient in the product, e.g., "low sodium" or "contains 100 calories."

(2) An implied nutrient content claim is any claim that:

(i) Describes the product or an ingredient therein in a manner that suggests that a nutrient is absent or present in a certain amount (e.g., "high in oat bran"); or

(ii) Suggests that the product, because of its nutrient content, may be useful in maintaining healthy dietary practices and is made in association with an explicit claim or statement about a nutrient (e.g., "healthy, contains 3 grams (g) of fat").

(3) Except for claims regarding vitamins and minerals described in paragraph (q)(3) of this section, no nutrient content claims may be made on products intended specifically for use by infants and children less than 2 years of age unless the claim is specifically provided for in subpart Y of this part.

(4) Reasonable variations in the spelling of the terms defined in applicable provisions in this subpart and their synonyms are permitted provided these variations are not misleading (e.g., "hi" or "lo").

(c) Information that is required or permitted by §381.409 to be declared in nutrition labeling, and that appearsas part of the nutrition label, is not a nutrient content claim and is not subject to the requirements of this section. If such information is declared elsewhere on the label or in labeling, it is a nutrient content claim and is subject to the requirements for nutrient content claims.

(d) A "substitute" product is one that may be used interchangeably with another product that it resembles, i.e., that it is organoleptically, physically, and functionally (including shelf life) similar to, and that it is not nutritionally inferior to unless it is labeled as an "imitation."

(1) If there is a difference in performance characteristics that materially limits the use of the product, the product may still be considered a substitute if the label includes a disclaimer adjacent to the most prominent claim as defined in paragraph (j)(2)(iii) of this section, informing the consumer of such difference (e.g., "not recommended for frying").

(2) This disclaimer shall be in easily legible print or type and in a size no less than that required by §381.121(c) for the net quantity of contents statement, except where the size of the claim is less than two times the required size of the net quantity of contents statement, in which case the disclaimer statement shall be no less than one-half the size of the claim but no smaller than 1/16-inch minimum height, except as permitted by §381.500(d)(2).

(e)

(1) Because the use of a "free" or "low" claim before the name of a product implies that the product differs from other products of the same type by virtue of its having a lower amount of the nutrient, only products that have been specially processed, altered, formulated, or reformulated so as to lower the amount of the nutrient in the product, remove the nutrient from the product, or not include the nutrient in the product, may bear such a claim (e.g., "low sodium chicken noodle soup").

9 CFR §381.413

(2) Any claim for the absence of a nutrient in a product, or that a product is low in a nutrient when the product has not been specially processed, altered, formulated, or reformulated to qualify for that claim shall indicate that the product inherently meets the criteria and shall clearly refer to all products of that type and not merely to the particular brand to which the labeling attaches (e.g., "chicken breast meat, a low sodium food").

(f) A nutrient content claim shall be in type size and style no larger than two times that of the statement of identity and shall not be unduly prominent in type style compared to the statement of identity.

(g) Labeling information required in Secs. 381.413, 381.454, 381.456, 381.460, 381.461, 381.462, and 381.480, whose type size is not otherwise specified, is required to be in letters and/or numbers no less than 1/16 inch in height, except as permitted by §381.500(d)(2).

(h) [Reserved]

(i) Except as provided in §381.409 or in paragraph (q)(3) of this section, the label or labeling of a product may contain a statement about the amount or percentage of a nutrient if:

(1) The use of the statement on the product implicitly characterizes the level of the nutrient in the product and is consistent with a definition for a claim, as provided in subpart Y of this part, for the nutrient that the label addresses. Such a claim might be, "less than 10 g of fat per serving;"

(2) The use of the statement on the product implicitly characterizes the level of the nutrient in the product and is not consistent with such a definition, but the label carries a disclaimer adjacent to the statement that the product is not "low" in or a "good source" of the nutrient, such as "only 200 milligrams (mg) sodium per serving, not a low sodium product." The disclaimer must be in easily legible print or type and in a size no less than required by §381.121(c) for the net quantity of contents, except where the size of the claim is less than two times the required size of the net quantity of contents statement, in which case the disclaimer statement shall be no less than one-half the size of the claim but no smaller than 1/16-inch minimum height, except as permitted by §381.500(d)(2);

(3) The statement does not in any way implicitly characterize the level of the nutrient in the product and it is not false or misleading in any respect (e.g., "100 calories" or "5 grams of fat"), in which case no disclaimer is required.

(4) "Percent fat free" claims are not authorized by this paragraph. Such claims shall comply with §381.462(b)(6).

(j) A product may bear a statement that compares the level of a nutrient in the product with the level of a nutrient in a reference product. These statements shall be known as "relative claims" and include "light," "reduced," "less" (or "fewer"), and "more" claims.

(1) To bear a relative claim about the level of a nutrient, the amount of that nutrient in the product must be compared to an amount of nutrient in an appropriate reference product as specified in this paragraph (j).

(i) (A) For "less" (or "fewer") and "more" claims, the reference product may be a dissimilar product within a product category that can generally be substituted for one another in the diet or a similar product.

(B) For "light," "reduced," and "added" claims, the reference product shall be a similar product, and

(ii) (A) For "light" claims, the reference product shall be representative of the type of product that includes the product that bears the claim. The nutrient value for the reference product shall be representative of a broad base of products of that type; e.g., a value in a representative, valid data base; an average value determined from the top three national (or regional) brands, a market basket norm; or, where its nutrient value is representative of the product type, a market leader. Firms using such a reference nutrient value as a basis for a claim, are required to provide specific information upon which the nutrient value was derived, on request, to consumers and appropriate regulatory officials.

(B) For relative claims other than "light," including "less" and "more" claims, the reference product may be the same as that provided for "light" in paragraph (j)(1)(ii)(A) of this section or it may be the manufacturer's regular product, or that of another manufacturer, that has been offered for sale to the public on a regular basis for a substantial period of time in the same geographic area by the same business entity or by one entitled to use its trade name, provided the name of the competitor is not used on the labeling of the product. The nutrient values used to determine the claim when comparing a single manufacturer's product to the labeled product shall be either the values declared in nutrition labeling or the actual nutrient values, provided that the resulting labeling is internally consistent (i.e., that the values stated in the nutrition information, the nutrient values in the

9 CFR §381.413

accompanying information, and the declaration of the percentage of nutrient by which the product has been modified are consistent and will not cause consumer confusion when compared), and that the actual modification is at least equal to the percentage specified in the definition of the claim.

(2) For products bearing relative claims:

(i) The label or labeling must state the identity of the reference product and the percent (or fraction) of the amount of the nutrient in the reference product by which the nutrient has been modified, (e.g., "50 percent less fat than 'reference product'" or "1/3 fewer calories than 'reference product'"); and

(ii) This information shall be immediately adjacent to the most prominent claim in easily legible boldface print or type, in distinct contrast to other printed or graphic matter, that is no less than that required by §381.121(c) for net quantity of contents, except where the size of the claim is less than two times the required size of the net quantity of contents statement, in which case the referral statement shall be no less than one-half the size of the claim, but no smaller than 1/16-inch minimum height, except as permitted by §381.500(d)(2).

(iii) The determination of which use of the claim is in the most prominent location on the label or labeling will be made based on the following factors, considered in order:

(A) A claim on the principal display panel adjacent to the statement of identity;

(B) A claim elsewhere on the principal display panel;

(C) A claim on the information panel; or

(D) A claim elsewhere on the label or labeling.

(iv) The label or labeling must also bear:

(A) Clear and concise quantitative information comparing the amount of the subject nutrient in the product per labeled serving size with that in the reference product; and

(B) This statement shall appear adjacent to the most prominent claim or to the nutrition information.

(3) A relative claim for decreased levels of a nutrient may not be made on the label or in labeling of a product if the nutrient content of the reference product meets the requirement for a "low" claim for that nutrient.

(k) The term "modified" may be used in the statement of identity of a product that bears a relative claim that complies with the requirements of this part, followed immediately by the name of the nutrient whose content has been altered (e.g., "modified fat 'product'"). This statement of identity must be immediately followed by the comparative statement such as "contains 35 percent less fat than 'reference product'." The label or labeling must also bear the information required by paragraph (j)(2) of this section in the manner prescribed.

(l) For purposes of making a claim, a "meal-type product" shall be defined as a product that:

(1) Makes a significant contribution to the diet by weighing at least 6 ounces, but no more than 12 ounces per serving (container), and

(2) Contains ingredients from two or more of the following four food groups:

(i) Bread, cereal, rice and pasta group,

(ii) Fruits and vegetables group,

(iii) Milk, yogurt, and cheese group, and

(iv) Meat, poultry, fish, dry beans, eggs, and nuts group, and

(3) Is represented as, or is in a form commonly understood to be a breakfast, lunch, dinner, meal, main dish, entree, or pizza. Such representations may be made either by statements, photographs, or vignettes.

(m) [Reserved]

(n) Nutrition labeling in accordance with §381.409 shall be provided for any food for which a nutrient content claim is made.

(o) Compliance with requirements for nutrient content claims shall be in accordance with §381.409(h).

(p)

(1) Unless otherwise specified, the reference amount customarily consumed set forth in §381.412(b) through (e) shall be used in determining whether a product meets the criteria for a nutrient content claim. If the serving size declared on the product label differs from the reference amount customarily consumed, and the amount of the nutrient contained in the labeled serving does not meet the maximum or minimum amount criterion in the definition for the descriptor for that nutrient, the claim shall be followed by the criteria for the claim as required by §381.412(f) (e.g., "very low sodium, 35 mg or less per 55 grams").

(2) The criteria for the claim shall be immediately adjacent to the most prominent claim in easily legible print or type and in a size that is no less than that required by §381.121(c) for net quantity of contents, except where the size of the claim is less than two times the required size of the net quantity of contents statement, in which case

9 CFR §381.414

the criteria statement shall be no less than one-half the size of the claim but no smaller than 1/16-inch minimum height, except as permitted by §381.500(d)(2).

(q) The following exemptions apply:

(1) Nutrient content claims that have not been defined by regulation and that appear as part of a brand name that was in use prior to November 27, 1991, may continue to be used as part of that brand name, provided they are not false or misleading under section 4(h) of the Act (21 U.S.C. 453(h)(4)).

(2) [Reserved]

(3) A statement that describes the percentage of a vitamin or mineral in the food, including foods intended specifically for use by infants and children less than 2 years of age, in relation to a Reference Daily Intake (RDI) as defined in §381.409 may be made on the label or in the labeling of a food without a regulation authorizing such a claim for a specific vitamin or mineral.

(4) The requirements of this section do not apply to infant formulas and medical foods, as described in 21 CFR 101.13(q)(4).

(5) [Reserved]

(6) Nutrient content claims that were part of the name of a product that was subject to a standard of identity as of November 27, 1991, are not subject to the requirements of paragraph (b) of this section whether or not they meet the definition of the descriptive term.

(7) Implied nutrient content claims may be used as part of a brand name, provided that the use of the claim has been authorized by FSIS. Labeling applications requesting approval of such a claim may be submitted pursuant to §381.469.

[58 FR 675, Jan. 6, 1993; 58 FR 43789, Aug. 18, 1993, as amended at 58 FR 47628, Sept. 10, 1993; 59 FR 40215, Aug. 8, 1994; 59 FR 45198, Sept. 1, 1994; 60 FR 208, Jan. 3, 1995]

§381.414-381.442 [Reserved]

§381.443 Significant participation for voluntary nutrition labeling.

(a) In evaluating significant participation for voluntary nutrition labeling, FSIS will consider only the major cuts of single-ingredient, raw poultry products, as identified in §381.444, including those that have been previously frozen.

(b) FSIS will judge a food retailer to be participating at a significant level if the retailer provides nutrition labeling information for at least 90 percent of the major cuts of single- ingredient, raw poultry products, listed in §381.444, that it sells, and if the nutrition label is consistent in content and format with the mandatory program, or nutrition information is displayed at point-of-purchase in an appropriate manner.

(c) To determine whether there is significant participation by retailers under the voluntary nutrition labeling guidelines, FSIS will select a representative sample of companies allocated by type and size.

(d) FSIS will find that significant participation by food retailers exists if at least 60 percent of all companies that are evaluated are participating in accordance with the guidelines.

(e) FSIS will evaluate significant participation of the voluntary program every 2 years beginning in May 1995.

(1) If significant participation is found, the voluntary nutrition labeling guidelines shall remain in effect.

(2) If significant participation is not found, FSIS shall initiate rulemaking to require nutrition labeling on those products under the voluntary program.

§381.444 Identification of major cuts of poultry products.

The major cuts of single-ingredient, raw poultry products are: Whole chicken (without neck and giblets), chicken breast, chicken wing, chicken drumstick, chicken thigh, whole turkey (without necks and giblets; separate nutrient panels for white and dark meat permitted as an option), turkey breast, turkey wing, turkey drumstick, and turkey thigh.

9 CFR §381.445

§381.445 Guidelines for voluntary nutrition labeling of single- ingredient, raw products.

(a) Nutrition information on the cuts of single-ingredient, raw poultry products, including those that have been previously frozen, shall be provided in the following manner:

(1) If a retailer or manufacturer chooses to provide nutrition information on the label of these products, these products shall be subject to all requirements of the mandatory nutrition labeling program, except that nutrition labeling may be declared on the basis of either "as consumed" or "as packaged." In addition, the declaration of the number of servings per container need not be included in nutrition labeling of single-ingredient, raw poultry products, including those that have been previously frozen.

(2) A retailer may choose to provide nutrition information at the point-of-purchase, such as by posting a sign, or by making the information readily available in brochures, notebooks, or leaflet form in close proximity to the food. The nutrition labeling information may also be supplemented by a video, live demonstration, or other media. If a nutrition claim is made on point-of-purchase materials all of the requirements of the mandatory nutrition labeling program apply. However, if only nutrition information--and not a nutrition claim--is supplied on point-of-purchase materials:

(i) The requirements of the mandatory nutrition labeling program apply, but the nutrition information may be supplied on an "as packaged" or "as consumed," basis;

(ii) The listing of percent of Daily Value for the nutrients (except vitamins and minerals specified in §381.409(c)(8)) and footnote required by §381.409(d)(9) may be omitted; and

(iii) The point-of-purchase materials are not subject to any of the format requirements.

(b) [Reserved]

(c) The declaration of nutrition information may be presented in a simplified format as specified in §381.409(f) for the mandatory nutrition labeling program.

(d) The nutrition label data should be based on either raw or cooked edible portions of poultry cuts with skin. If data are based on cooked portions, the methods used to cook the products must be specified and should be those which do not add nutrients from other ingredients such as flour, breading, and salt. Additional nutritional data may be presented on an optional basis for the raw or cooked edible portions of the skinless poultry meat.

(e) Nutrient data that are the most current representative data base values contained in USDA's National Nutrient Data Bank or its published form, the Agriculture Handbook No. 8 series, may be used for nutrition labeling of single-ingredient, raw poultry products, including those that have been previously frozen. These data may be composite data that reflect different classes of turkey or other variables affecting nutrient content. Alternatively, data that reflect specific classes or other variables may be used, except that if data are used on labels attached to a product which is labeled as to class of poultry or other variables, the data must represent the product in the package when such data are contained in the representative data base. When data are used on labels attached to a product, the data must represent the edible poultry tissues present in the package.

(f) If the nutrition information is in accordance with paragraph (e) of this section, a nutrition label or labeling will not be subject to the Agency compliance review under §381.409(h), unless a nutrition claim is made on the basis of the representative data base values.

(g) Retailers may use data bases that they believe reflect the nutrient content of single-ingredient, raw poultry products, including those that have been previously frozen; however, such labeling shall be subject to the compliance procedures of paragraph (e) of this section and the requirements specified in this Subpart for the mandatory nutrition labeling program.

[58 FR 675, Jan. 6, 1993, as amended at 58 FR 47628, Sept. 10, 1993; 60 FR 209, Jan. 3, 1995]

§§381.446-381.453 [Reserved]

§381.454 Nutrient content claims for "good source," "high," and "more."

(a) *General requirements.* Except as provided in paragraph (e) of this section, a claim about the level of a nutrient in a product in relation to the Reference Daily Intake (RDI) or Daily Reference Value (DRV), established for that nutrient (excluding total carbohydrate) in §381.409(c), may only be made on the label or in labeling of the product if:

(1) The claim uses one of the terms defined in this section in accordance with the definition for that term;

9 CFR §381.454

(2) The claim is made in accordance with the general requirements for nutrient content claims in §381.413; and

(3) The product for which the claim is made is labeled in accordance with §381.409.

(b) *"High" claims.*

(1) The terms "high," "rich in," or "excellent source of" may be used on the label or in labeling of products, except meal-type products as defined in §381.413(l), provided that the product contains 20 percent or more of the RDI or the DRV per reference amount customarily consumed.

(2) The terms defined in paragraph (b)(1) of this section may be used on the label or in labeling of a meal-type product as defined in §381.413(l), provided that:

(i) The product contains a food that meets the definition of "high" in paragraph (b)(1) of this section; and

(ii) The label or labeling clearly identifies the food that is the subject of the claim (e.g., "the serving of broccoli in this meal is high in vitamin C").

(c) *"Good Source" claims.*

(1) The terms "good source," "contains," or "provides" may be used on the label or in labeling of products, except meal-type products as described in §381.413(l), provided that the product contains 10 to 19 percent of the RDI or the DRV per reference amount customarily consumed.

(2) The terms defined in paragraph (c)(1) of this section may be used on the label or in labeling of a meal-type product as defined in §381.413(l), provided that:

(i) The product contains a food that meets the definition of "good source" in paragraph (c)(1) of this section; and

(ii) The label or labeling clearly identifies the food that is the subject of the claim (e.g., "the serving of sweet potatoes in this meal is a good source of fiber").

(d) *Fiber claims.*

(1) If a nutrient content claim is made with respect to the level of dietary fiber, i.e., that the product is high in fiber, a good source of fiber, or that the product contains "more" fiber, and the product is not "low" in total fat as defined in §381.462(b)(2) or, in the case of a meal-type product, is not "low" in total fat as defined in §381.462(b)(3), then the labeling shall disclose the level of total fat per labeled serving size (e.g., "contains 12 grams (g) of fat per serving"); and

(2) The disclosure shall appear in immediate proximity to such claim and be in a type size no less than one-half the size of the claim.

(e) *"More" claims.*

(1) A relative claim using the terms "more" and "added" may be used on the label or in labeling to describe the level of protein, vitamins, minerals, dietary fiber, or potassium in a product, except meal-type products as defined in §381.413(l), provided that:

(i) The product contains at least 10 percent more of the RDI or the DRV for protein, vitamins, minerals, dietary fiber, or potassium (expressed as a percent of the Daily Value) per reference amount customarily consumed than an appropriate reference product as described in §381.413(j)(1); and

(ii) As required in §381.413(j)(2) for relative claims:

(A) The identity of the reference product and the percent (or fraction) that the nutrient is greater relative to the RDI or DRV are declared in immediate proximity to the most prominent such claim (e.g., "contains 10 percent more of the Daily Value for fiber than 'reference product'"); and

(B) Quantitative information comparing the level of the nutrient in the product per labeled serving size with that of the reference product that it replaces is declared adjacent to the most prominent claim or to the nutrition information (e.g., "fiber content of 'reference product' is 1 g per serving; 'this product' contains 4 g per serving"). (2) A relative claim using the terms "more" and "added" may be used on the label or in labeling to describe the level of protein, vitamins, minerals, dietary fiber, or potassium in meal-type products as defined in §381.413(l), provided that:

(i) The product contains at least 10 percent more of the RDI or the DRV for protein, vitamins, minerals, dietary fiber, or potassium (expressed as a percent of the Daily Value) per 100 g of product than an appropriate reference product as described in §381.413(j)(1); and

(ii) As required in §381.413(j)(2) for relative claims:

(A) The identity of the reference product and the percent (or fraction) that the nutrient is greater relative to the RDI or DRV are declared in immediate proximity to the most prominent such claim (e.g., "contains 10 percent

9 CFR §381.455

more of the Daily Value for fiber per 3 ounces (oz) than does 'reference product'"), and

(B) Quantitative information comparing the level of the nutrient in the meal-type product per specified weight with that of the reference product that it replaces is declared adjacent to the most prominent claim or to the nutrition information (e.g., "fiber content of 'reference product' is 2 g per 3 oz; 'this product' contains 5 g per 3 oz").

[59 FR 40215, Aug. 8, 1994, as amended at 60 FR 210, Jan. 3, 1995]

§381.455 [Reserved]

§381.456 Nutrient content claims for "light" or "lite."

(a) *General requirements*. A claim using the terms "light" or "lite" to describe a product may only be made on the label or in labeling of the product if:

(1) The claim uses one of the terms defined in this section in accordance with the definition for that term;

(2) The claim is made in accordance with the general requirements for nutrient content claims in §381.413; and

(3) The product for which the claim is made is labeled in accordance with §381.409.

(b) *"Light" claims*. The terms "light" or "lite" may be used on the label or in labeling of products, except meal-type products as defined in §381.413(l), without further qualification, provided that:

(1) If the product derives 50 percent or more of its calories from fat, its fat content is reduced by 50 percent or more per reference amount customarily consumed compared to an appropriate reference product as described in §381.413(j)(1); or

(2) If the product derives less than 50 percent of its calories from fat:

(i) The number of calories is reduced by at least one-third (33 1/3 percent) per reference amount customarily consumed compared to an appropriate reference product as described in §381.413(j)(1); or

(ii) Its fat content is reduced by 50 percent or more per reference amount customarily consumed compared to the appropriate reference product as described in §381.413(j)(1); and

(3) As required in §381.413(j)(2) for relative claims:

(i) The identity of the reference product and the percent (or fraction) that the calories and the fat were reduced are declared in immediate proximity to the most prominent such claim (e.g., "1/3 fewer calories and 50 percent less fat than the market leader"); and

(ii) Quantitative information comparing the level of calories and fat content in the product per labeled serving size with that of the reference product that it replaces is declared adjacent to the most prominent claim or to the nutrition information (e.g., "lite 'this product'--200 calories, 4 grams (g) fat; regular 'reference product'-- 300 calories, 8 g fat per serving"); and

(iii) If the labeled product contains less than 40 calories or less than 3 g fat per reference amount customarily consumed, the percentage reduction for that nutrient need not be declared.

(4) A "light" claim may not be made on a product for which the reference product meets the definition of "low fat" and "low calorie."

(c)

(1) (i) A product for which the reference product contains 40 calories or less and 3 g fat or less per reference amount customarily consumed may use the terms "light" or "lite" without further qualification if it is reduced by 50 percent or more in sodium content compared to the reference product; and

(ii) As required in §381.413(j)(2) for relative claims: (A) The identity of the reference product and the percent (or fraction) that the sodium was reduced are declared in immediate proximity to the most prominent such claim (e.g., "50 percent less sodium than the market leader"); and

(B) Quantitative information comparing the level of sodium per labeled serving size with that of the reference product it replaces is declared adjacent to the most prominent claim or to the nutrition information (e.g., "lite 'this product'--500 milligrams (mg) sodium per serving; regular 'reference product'--1,000 mg sodium per serving").

(2) (i) A product for which the reference product contains more than 40 calories or more than 3 g fat per reference amount customarily consumed may use the terms "light in sodium" or "lite in sodium" if it is reduced by 50 percent or more in sodium content compared to the reference product, provided that "light" or "lite" is presented in immediate proximity with "in sodium" and the entire term is presented in uniform type size, style,

9 CFR §381.456

color, and prominence; and

(ii) As required in §381.413(j)(2) for relative claims:

(A) The identity of the reference product and the percent (or fraction) that the sodium was reduced are declared in immediate proximity to the most prominent such claim (e.g., "50 percent less sodium than the market leader"); and

(B) Quantitative information comparing the level of sodium per labeled serving size with that of the reference product it replaces is declared adjacent to the most prominent claim or to the nutrition information (e.g., or "lite 'this product'--170 mg sodium per serving; regular 'reference product'--350 mg per serving").

(3) Except for meal-type products as defined in §381.413(l), a "light in sodium" claim may not be made on a product for which the reference product meets the definition of "low in sodium."

(d)

(1) The terms "light" or "lite" may be used on the label or in labeling of a meal-type product as defined in §381.413(l), provided that:

(i) The product meets the definition of:

(A) "Low in calories" as defined in §381.460(b)(3); or (B) "Low in fat" as defined in §381.462(b)(3); and

(ii)(A) A statement appears on the principal display panel that explains whether "light" is used to mean "low fat," "low calories," or both (e.g., "Light Delight, a low fat meal"); and

(B) The accompanying statement is no less than one-half the type size of the "light" or "lite" claim.

(2) (i) The terms "light in sodium" or "lite in sodium" may be used on the label or in labeling of a meal-type product as defined in §381.413(l), provided that the product meets the definition of "low in sodium" as defined in §381.461(b)(5)(i); and

(ii) "Light" or "lite" and "in sodium" are presented in uniform type size, style, color, and prominence.

(3) The terms "light" or "lite" may be used in the brand name of a product to describe the sodium content, provided that:

(i) The product is reduced by 50 percent or more in sodium content compared to the reference product;

(ii) A statement specifically stating that the product is "light in sodium" or "lite in sodium" appears:

(A) Contiguous to the brand name; and

(B) In uniform type size, style, color, and prominence as the product name; and

(iii) As required in §381.413(j)(2) for relative claims:

(A) The identity of the reference product and the percent (or fraction) that the sodium was reduced are declared in immediate proximity to the most prominent such claim; and

(B) Quantitative information comparing the level of sodium per labeled serving size with that of the reference product it replaces is declared adjacent to the most prominent claim or to the nutrition information.

(e) Except as provided in paragraphs (b) through (d) of this section, the terms "light" or "lite" may not be used to refer to a product that is not reduced in fat by 50 percent, or, if applicable, in calories by 1/3 or, when properly qualified, in sodium by 50 percent unless:

(1) It describes some physical or organoleptic attribute of the product such as texture or color and the information (e.g., "light in color" or "light in texture") so stated, clearly conveys the nature of the product; and

(2) The attribute (e.g., "color" or "texture") is in the same style, color, and at least one-half the type size as the word "light" and in immediate proximity thereto.

(f) If a manufacturer can demonstrate that the word "light" has been associated, through common use, with a particular product to reflect a physical or organoleptic attribute to the point where it has become part of the statement of identity, such use of the term "light" shall not be considered a nutrient content claim subject to the requirements in this part.

(g) The term "lightly salted" may be used on a product to which has been added 50 percent less sodium than is normally added to the reference product as described in §381.413(j)(1)(i)(B) and (j)(1)(ii)(B), provided that if the product is not "low in sodium" as defined in §381.461(b)(4), the statement "not a low sodium food," shall appear adjacent to the nutrition information and the information required to accompany a relative claim shall appear on the

9 CFR §381.460

label or labeling as specified in §381.413(j)(2).
[58 FR 675, Jan. 6, 1993; 58 FR 43789, Aug. 18, 1993, as amended at 59 FR 40215, Aug. 8, 1994; 60 FR 210, Jan.3, 1995]

§§381.457-381.459 [Reserved]

§381.460 Nutrient content claims for calorie content.

(a) *General requirements.* A claim about the calorie or sugar content of a product may only be made on the label or in labeling of the product if:
    (1) The claim uses one of the terms defined in this section in accordance with the definition for that term;
    (2) The claim is made in accordance with the general requirements for nutrient content claims in §381.413; and
    (3) The product for which the claim is made is labeled in accordance with §381.409.

(b) *Calorie content claims.*
    (1) The terms "calorie free," "free of calories," "no calories," "zero calories," "without calories," "trivial source of calories," "negligible source of calories," or "dietarily insignificant source of calories" may be used on the label or in labeling of products, provided that:
    (i) The product contains less than 5 calories per reference amount customarily consumed and per labeled serving size; and
    (ii) If the product meets this condition without the benefit of special processing, alteration, formulation, or reformulation to lower the caloric content, it is labeled to clearly refer to all products of its type and not merely to the particular brand to which the label attaches.
    (2) The terms "low calorie," "few calories," "contains a small amount of calories," "low source of calories," or "low in calories" may be used on the label or in labeling of products, except meal-type products as defined in §381.413(l), provided that:
    (i) (A) The product has a reference amount customarily consumed greater than 30 grams (g) or greater than 2 tablespoons (tbsp) and does not provide more than 40 calories per reference amount customarily consumed; or
    (B) The product has a reference amount customarily consumed of 30 g or less or 2 tbsp or less and does not provide more than 40 calories per reference amount customarily consumed and per 50 g (for dehydrated products that must be reconstituted before typical consumption with water or a diluent containing an insignificant amount, as defined in §381.409(f)(1), of all nutrients per reference amount customarily consumed, the per-50-g criterion refers to the "as prepared" form).
    (ii) If the product meets these conditions without the benefit of special processing, alteration, formulation, or reformulation to lower the caloric content, it is labeled to clearly refer to all products of its type and not merely to the particular brand to which the label attaches.
    (3) The terms defined in paragraph (b)(2) of this section may be used on the label or in labeling of a meal-type product as defined in §381.413(l), provided that:
    (i) The product contains 120 calories or less per 100 g of product; and
    (ii) If the product meets this condition without the benefit of special processing, alteration, formulation, or reformulation to lower the calorie content, it is labeled to clearly refer to all products of its type and not merely to the particular brand to which it attaches.
    (4) The terms "reduced calorie," "reduced in calories," "calorie reduced," "fewer calories," "lower calorie," or "lower in calories" may be used on the label or in labeling of products, except meal-type products as defined in §381.413(l), provided that:
    (i) The product contains at least 25 percent fewer calories per reference amount customarily consumed than an appropriate reference product as described in §381.413(j)(1); and
    (ii) As required in §381.413(j)(2) for relative claims:
    (A) The identity of the reference product and the percent (or fraction) that the calories differ between the two products are declared in immediate proximity to the most prominent such claim (e.g., lower calorie 'product'--"33 1/3 percent fewer calories than our regular 'product'"); and
    (B) Quantitative information comparing the level of calories in the product per labeled serving size with

9 CFR §381.460

that of the reference product that it replaces is declared adjacent to the most prominent claim or to the nutrition information (e.g., "calorie content has been reduced from 150 to 100 calories per serving").

(iii) Claims described in paragraph (b)(4) of this section may not be made on the label or in labeling of products if the reference product meets the definition for "low calorie."

(5) The terms defined in paragraph (b)(4) of this section may be used on the label or in labeling of a meal-type product as defined in §381.413(l), provided that:

(i) The product contains at least 25 percent fewer calories per 100 g of product than an appropriate reference product as described in §381.413(j)(1); and

(ii) As required in §381.413(j)(2) for relative claims:

(A) The identity of the reference product and the percent (or fraction) that the calories differ between the two products are declared in immediate proximity to the most prominent such claim (e.g., "calorie reduced 'product', 25% less calories per ounce (oz) (or 3 oz) than our regular 'product'"); and

(B) Quantitative information comparing the level of calories in the product per specified weight with that of the reference product that it replaces is declared adjacent to the most prominent claim or to the nutrition information (e.g., "calorie content has been reduced from 110 calories per 3 oz to 80 calories per 3 oz").

(iii) Claims described in paragraph (b)(5) of this section may not be made on the label or in labeling of products if the reference product meets the definition for "low calorie."

(c) *Sugar content claims.*

(1) Terms such as "sugar free," "free of sugar," "no sugar," "zero sugar," "without sugar," "sugarless," "trivial source of sugar," "negligible source of sugar," or "dietarily insignificant source of sugar" may reasonably be expected to be regarded by consumers as terms that represent that the product contains no sugars or sweeteners, e.g., "sugar free," or "no sugar," as indicating a product which is low in calories or significantly reduced in calories. Consequently, except as provided in paragraph (c)(2) of this section, a product may not be labeled with such terms unless:

(i) The product contains less than 0.5 g of sugars, as defined in §381.409(c)(6)(ii), per reference amount customarily consumed and per labeled serving size or, in the case of a meal-type product, less than 0.5 g of sugars per labeled serving size;

(ii) The product contains no ingredient that is a sugar or that is generally understood by consumers to contain sugars unless the listing of the ingredient in the ingredients statement is followed by an asterisk that refers to the statement below the list of ingredients, which states: "Adds a trivial amount of sugar," "adds a negligible amount of sugar," or "adds a dietarily insignificant amount of sugar;" and

(iii) (A) It is labeled "low calorie" or "reduced calorie" or bears a relative claim of special dietary usefulness labeled in compliance with paragraphs (b)(2), (b)(3), (b)(4), or (b)(5) of this section; or

(B) Such term is immediately accompanied, each time it is used, by either the statement "not a reduced calorie product," "not a low calorie product," or "not for weight control."

(2) The terms "no added sugar," "without added sugar," or "no sugar added" may be used only if:

(i) No amount of sugars, as defined in §381.409(c)(6)(ii), or any other ingredient that contains sugars that functionally substitute for added sugars is added during processing or packaging;

(ii) The product does not contain an ingredient containing added sugars such as jam, jelly, or concentrated fruit juice;

(iii) The sugars content has not been increased above the amount present in the ingredients by some means such as the use of enzymes, except where the intended functional effect of the process is not to increase the sugars content of a product, and a functionally insignificant increase in sugars results;

(iv) The product that it resembles and for which it substitutes normally contains added sugars; and

(v) The product bears a statement that the product is not "low calorie" or "calorie reduced" (unless the product meets the requirements for a "low" or "reduced calorie" product) and that directs consumers' attention to the nutrition panel for further information on sugar and calorie content.

(3) Paragraph (c)(1) of this section shall not apply to a factual statement that a product, including products intended specifically for infants and children less than 2 years of age, is unsweetened or contains no added sweeteners in the case of a product that contains apparent substantial inherent sugar content, e.g., juices.

(4) The terms "reduced sugar," "reduced in sugar," "sugar reduced," "less sugar," "lower sugar," or "lower

in sugar" may be used on the label or in labeling of products, except meal-type products as defined in §381.413(l), provided that:

(i) The product contains at least 25 percent less sugars per reference amount customarily consumed than an appropriate reference product as described in §381.413(j)(1); and

(ii) As required in §381.413(j)(2) for relative claims:

(A) The identity of the reference product and the percent (or fraction) that the sugars differ between the two products are declared in immediate proximity to the most prominent such claim (e.g., "this product contains 25 percent less sugar than our regular product"); and

(B) Quantitative information comparing the level of the sugar in the product per labeled serving size with that of the reference product that it replaces is declared adjacent to the most prominent claim or to the nutrition information (e.g., "sugar content has been lowered from 8 g to 6 g per serving").

(5) The terms defined in paragraph (c)(4) of this section may be used on the label or in labeling of a meal-type product as defined in §381.413(l), provided that:

(i) The product contains at least 25 percent less sugars per 100 g of product than an appropriate reference product as described in §381.413(j)(1); and

(ii) As required in §381.413(j)(2) for relative claims:

(A) The identity of the reference product and the percent (or fraction) that the sugars differ between the two products are declared in immediate proximity to the most prominent such claim (e.g., "reduced sugar 'product'--25% less sugar than our regular 'product'"); and

(B) Quantitative information comparing the level of the nutrient in the product per specified weight with that of the reference product that it replaces is declared adjacent to the most prominent claim or to the nutrition information (e.g., "sugar content has been reduced from 17 g per 3 oz to 13 g per 3 oz").

[59 FR 40216, Aug. 8, 1994; 60 FR 211, Jan. 3, 1995]

§381.461 Nutrient content claims for the sodium content.

(a) *General requirements.* A claim about the level of sodium in a product may only be made on the label or in labeling of the product if:

(1) The claim uses one of the terms defined in this section in accordance with the definition for that term;

(2) The claim is made in accordance with the general requirements for nutrient content claims in §381.413; and

(3) The product for which the claim is made is labeled in accordance with §381.409.

(b) *Sodium content claims.*

(1) The terms "sodium free," "free of sodium," "no sodium," "zero sodium," "without sodium," "trivial source of sodium," "negligible source of sodium," or "dietarily insignificant source of sodium" may be used on the label or in labeling of products, provided that:

(i) The product contains less than 5 milligrams (mg) of sodium per reference amount customarily consumed and per labeled serving size or, in the case of a meal-type product, less than 5 mg of sodium per labeled serving size;

(ii) The product contains no ingredient that is sodium chloride or is generally understood by consumers to contain sodium unless the listing of the ingredient in the ingredients statement is followed by an asterisk that refers to the statement below the list of ingredients, which states: "Adds a trivial amount of sodium," "adds a negligible amount of sodium" or "adds a dietarily insignificant amount of sodium;" and

(iii) If the product meets these conditions without the benefit of special processing, alteration, formulation, or reformulation to lower the sodium content, it is labeled to clearly refer to all products of its type and not merely to the particular brand to which the label attaches.

(2) The terms "very low sodium" or "very low in sodium" may be used on the label or in labeling of products, except meal-type products as defined in §381.413(l), provided that:

(i) (A) The product has a reference amount customarily consumed greater than 30 grams (g) or greater than 2 tablespoons (tbsp) and contains 35 mg or less sodium per reference amount customarily consumed; or

(B) The product has a reference amount customarily consumed of 30 g or less or 2 tbsp or less and contains 35 mg or less sodium per reference amount customarily consumed and per 50 g (for dehydrated products

9 CFR §381.461

that must be reconstituted before typical consumption with water or a diluent containing an insignificant amount, as defined in §381.409(f)(1), of all nutrients per reference amount customarily consumed, the per-50-g criterion refers to the "as prepared" form); and

(ii) If the product meets these conditions without the benefit of special processing, alteration, formulation, or reformulation to lower the sodium content, it is labeled to clearly refer to all products of its type and not merely to the particular brand to which the label attaches.

(3) The terms defined in paragraph (b)(2) of this section may be used on the label or in labeling of a meal-type product as defined in §381.413(l), provided that:

(i) The product contains 35 mg or less of sodium per 100 g of product; and

(ii) If the product meets this condition without the benefit of special processing, alteration, formulation, or reformulation to lower the sodium content, it is labeled to clearly refer to all products of its type and not merely to the particular brand to which the label attaches.

(4) The terms "low sodium," "low in sodium," "little sodium," "contains a small amount of sodium," or "low source of sodium" may be used on the label and in labeling of products, except meal-type products as defined in §381.413(l), provided that:

(i) (A) The product has a reference amount customarily consumed greater than 30 g or greater than 2 tbsp and contains 140 mg or less sodium per reference amount customarily consumed; or

(B) The product has a reference amount customarily consumed of 30 g or less or 2 tbsp or less and contains 140 mg or less sodium per reference amount customarily consumed and per 50 g (for dehydrated products that must be reconstituted before typical consumption with water or a diluent containing an insignificant amount, as defined in §381.409(f)(1), of all nutrients per reference amount customarily consumed, the per-50-g criterion refers to the "as prepared" form); and

(ii) If the product meets these conditions without the benefit of special processing, alteration, formulation, or reformulation to lower the sodium content, it is labeled to clearly refer to all products of its type and not merely to the particular brand to which the label attaches.

(5) The terms defined in paragraph (b)(4) of this section may be used on the label or in labeling of a meal-type product as defined in §381.413(l), provided that:

(i) The product contains 140 mg or less sodium per 100 g of product; and

(ii) If the product meets these conditions without the benefit of special processing, alteration, formulation, or reformulation to lower the sodium content, it is labeled to clearly refer to all products of its type and not merely to the particular brand to which the label attaches.

(6) The terms "reduced sodium," "reduced in sodium," "sodium reduced," "less sodium," "lower sodium," or "lower in sodium" may be used on the label or in labeling of products, except meal-type products as defined in §381.413(l), provided that:

(i) The product contains at least 25 percent less sodium per reference amount customarily consumed than an appropriate reference product as described in §381.413(j)(l); and

(ii) As required in §381.413(j)(2) for relative claims:

(A) The identity of the reference product and the percent (or fraction) that the sodium differs between the two products are declared in immediate proximity to the most prominent such claim (e.g., "reduced sodium 'product', 50 percent less sodium than regular 'product'"); and

(B) Quantitative information comparing the level of sodium in the product per labeled serving size with that of the reference product that it replaces is declared adjacent to the most prominent claim or to the nutrition information (e.g., "sodium content has been lowered from 300 to 150 mg per serving").

(iii) Claims described in paragraph (b)(6) of this section may not be made on the label or in labeling of a product if the nutrient content of the reference product meets the definition for "low sodium."

(7) The terms defined in paragraph (b)(6) of this section may be used on the label or in labeling of a meal-type product as defined in §381.413(l), provided that:

(i) The product contains at least 25 percent less sodium per 100 g of product than an appropriate reference product as described in §381.413(j)(l); and

(ii) As required in §381.413(j)(2) for relative claims:

(A) The identity of the reference product and the percent (or fraction) that the sodium differs between the

9 CFR §381.462

two products are declared in immediate proximity to the most prominent such claim (e.g., "reduced sodium 'product'--30% less sodium per 3 oz than our 'regular product'"); and

(B) Quantitative information comparing the level of sodium in the product per specified weight with that of the reference product that it replaces is declared adjacent to the most prominent claim or to the nutrition information (e.g., "sodium content has been reduced from 220 mg per 3 oz to 150 mg per 3 oz").

(iii) Claims described in paragraph (b)(7) of this section may not be made on the label or in labeling of products if the nutrient content of the reference product meets the definition for "low sodium."

(c) The term "salt" is not synonymous with "sodium." Salt refers to sodium chloride. However, references to salt content such as "unsalted," "no salt," "no salt added" are potentially misleading.

(1) The term "salt free" may be used on the label or in labeling of products only if the product is "sodium free" as defined in paragraph (b)(1) of this section.

(2) The terms "unsalted," "without added salt," and "no salt added" may be used on the label or in labeling of products only if:

(i) No salt is added during processing;

(ii) The product that it resembles and for which it substitutes is normally processed with salt; and

(iii) If the product is not sodium free, the statement "not a sodium free product" or "not for control of sodium in the diet" appears adjacent to the nutrition information of the product bearing the claim.

(3) Paragraph (c)(2) of this section shall not apply to a factual statement that a product intended specifically for infants and children less than 2 years of age is unsalted, provided such statement refers to the taste of the product and is not false or otherwise misleading.

[58 FR 40216, Aug. 8, 1994, as amended at 60 FR 213, Jan. 3, 1995; 60 FR 5762, Jan. 30, 1995]

§381.462 Nutrient content claims for fat, fatty acids, and cholesterol content.

(a) *General requirements.* A claim about the level of fat, fatty acid, and cholesterol in a product may only be made on the label or in labeling of products if:

(1) The claim uses one of the terms defined in this section in accordance with the definition for that term;

(2) The claim is made in accordance with the general requirements for nutrient content claims in §381.413; and

(3) The product for which the claim is made is labeled in accordance with §381.409.

(b) *Fat content claims.*

(1) The terms "fat free," "free of fat," "no fat," "zero fat," "without fat," "nonfat," "trivial source of fat," "negligible source of fat," or "dietarily insignificant source of fat" may be used on the label or in labeling of products, provided that:

(i) The product contains less than 0.5 gram (g) of fat per reference amount customarily consumed and per labeled serving size or, in the case of a meal-type product, less than 0.5 g of fat per labeled serving size;

(ii) The product contains no added ingredient that is a fat or is generally understood by consumers to contain fat unless the listing of the ingredient in the ingredients statement is followed by an asterisk that refers to the statement below the list of ingredients, which states: "Adds a trivial amount of fat," "adds a negligible amount of fat," or "adds a dietarily insignificant amount of fat"; and (iii) If the product meets these conditions without the benefit of special processing, alteration, formulation, or reformulation to lower the fat content, it is labeled to clearly refer to all products of its type and not merely to the particular brand to which the label attaches.

(2) The terms "low fat," "low in fat," "contains a small amount of fat," "low source of fat," or "little fat" may be used on the label and in labeling of products, except meal-type products as defined in §381.413(l), provided that:

(i) (A) The product has a reference amount customarily consumed greater than 30 g or greater than 2 tablespoons (tbsp) and contains 3 g or less of fat per reference amount customarily consumed; or

(B) The product has a reference amount customarily consumed of 30 g or less or 2 tbsp or less and contains 3 g or less of fat per reference amount customarily consumed and per 50 g (for dehydrated products that must be reconstituted before typical consumption with water or a diluent containing an insignificant

9 CFR §381.462

amount, as defined in §381.409(f)(1), of all nutrients per reference amount customarily consumed, the per-50-g criterion refers to the "as prepared" form).

(ii) If the product meets these conditions without the benefit of special processing, alteration, formulation, or reformulation to lower the fat content, it is labeled to clearly refer to all products of its type and not merely to the particular brand to which the label attaches.

(3) The terms defined in paragraph (b)(2) of this section may be used on the label or in labeling of a meal-type product as defined in §381.413(l), provided that:

(i) The product contains 3 g or less of total fat per 100 g of product and not more than 30 percent of calories from fat; and

(ii) If the product meets these conditions without the benefit of special processing, alteration, formulation, or reformulation to lower the fat content, it is labeled to clearly refer to all products of its type and not merely to the particular brand to which the label attaches.

(4) The terms "reduced fat," "reduced in fat," "fat reduced," "less fat," "lower fat," or "lower in fat" may be used on the label or in labeling of products, except meal-type products as defined in §381.413(l), provided that:

(i) The product contains at least 25 percent less fat per reference amount customarily consumed than an appropriate reference product as described in §381.413(j)(1); and

(ii) As required in §381.413(j)(2) for relative claims:

(A) The identity of the reference product and the percent (or fraction) that the fat differs between the two products are declared in immediate proximity to the most prominent such claim (e.g., "reduced fat--50 percent less fat than our regular 'product'"); and

(B) Quantitative information comparing the level of fat in the product per labeled serving size with that of the reference product that it replaces is declared adjacent to the most prominent claim or to the nutrition information (e.g., "fat content has been reduced from 8 g to 4 g per serving").

(iii) Claims described in paragraph (b)(4) of this section may not be made on the label or in labeling of a product if the nutrient content of the reference product meets the definition for "low fat."

(5) The terms defined in paragraph (b)(4) of this section may be used on the label or in labeling of a meal-type product as defined in §381.413(l), provided that:

(i) The product contains at least 25 percent less fat per 100 g of product than an appropriate reference product as described in §381.413(j)(1); and

(ii) As required in §381.413(j)(2) for relative claims:

(A) The identity of the reference product and the percent (or fraction) that the fat differs between the two products are declared in immediate proximity to the most prominent such claim (e.g., "reduced fat 'product', 33 percent less fat per 3 oz than our regular 'product'"); and

(B) Quantitative information comparing the level of fat in the product per specified weight with that of the reference product that it replaces is declared adjacent to the most prominent such claim or to the nutrition information (e.g., "fat content has been reduced from 8 g per 3 oz to 5 g per 3 oz").

(iii) Claims described in paragraph (b)(5) of this section may not be made on the label or in labeling of a product if the nutrient content of the reference product meets the definition for "low fat."

(6) The term "_____ percent fat free" may be used on the label or in labeling of products, provided that:

(i) The product meets the criteria for "low fat" in paragraph (b)(2) or (b)(3) of this section;

(ii) The percent declared and the words "fat free" are in uniform type size; and

(iii) A "100 percent fat free" claim may be made only on products that meet the criteria for "fat free" in paragraph (b)(1) of this section, that contain less than 0.5 g of fat per 100 g, and that contain no added fat.

(iv) A synonym for "_____ percent fat free" is "_____ percent lean."

(c) *Fatty acid content claims.*

(1) The terms "saturated fat free," "free of saturated fat," "no saturated fat," "zero saturated fat," "without saturated fat," "trivial source of saturated fat," "negligible source of saturated fat," or "dietarily insignificant source of saturated fat" may be used on the label or in labeling of products, provided that:

(i) The product contains less than 0.5 g of saturated fat and less than 0.5 g trans fatty acids per reference

9 CFR §381.462

amount customarily consumed and per labeled serving size or, in the case of a meal-type product, less than 0.5 g of saturated fat and less than 0.5 g trans fatty acids per labeled serving size;

(ii) The product contains no ingredient that is generally understood by consumers to contain saturated fat unless the listing of the ingredient in the ingredients statement is followed by an asterisk that refers to the statement below the list of ingredients, which states: "Adds a trivial amount of saturated fat," "adds a negligible amount of saturated fat," or "adds a dietarily insignificant amount of saturated fat;" and

(iii) If the product meets these conditions without the benefit of special processing, alteration, formulation, or reformulation to lower saturated fat content, it is labeled to clearly refer to all products of its type and not merely to the particular brand to which the label attaches.

(2) The terms "low in saturated fat," "low saturated fat," "contains a small amount of saturated fat," "low source of saturated fat," or "a little saturated fat" may be used on the label or in labeling of products, except meal-type products as defined in §381.413(l), provided that:

(i) The product contains 1 g or less of saturated fat per reference amount customarily consumed and not more than 15 percent of calories from saturated fat; and

(ii) If the product meets these conditions without benefit of special processing, alteration, formulation, or reformulation to lower saturated fat content, it is labeled to clearly refer to all products of its type and not merely to the particular brand to which the label attaches.

(3) The terms defined in paragraph (c)(2) of this section may be used on the label or in labeling of a meal-type product as defined in §381.413(l), provided that:

(i) The product contains 1 g or less of saturated fat per 100 g and less than 10 percent calories from saturated fat; and

(ii) If the product meets these conditions without the benefit of special processing, alteration, formulation, or reformulation to lower saturated fat content, it is labeled to clearly refer to all products of its type and not merely to the particular brand to which the label attaches.

(4) The terms "reduced saturated fat," "reduced in saturated fat," "saturated fat reduced," "less saturated fat," "lower saturated fat," or "lower in saturated fat" may be used on the label or in labeling of products, except meal-type products as defined in §381.413(l), provided that:

(i) The product contains at least 25 percent less saturated fat per reference amount customarily consumed than an appropriate reference product as described in §381.413(j)(1); and

(ii) As required in §381.413(j)(2) for relative claims:

(A) The identity of the reference product and the percent (or fraction) that the saturated fat differs between the two products are declared in immediate proximity to the most prominent such claim (e.g., "reduced saturated fat 'product', contains 50 percent less saturated fat than the national average for 'product'"); and

(B) Quantitative information comparing the level of saturated fat in the product per labeled serving size with that of the reference product that it replaces is declared adjacent to the most prominent claim or to the nutrition information (e.g., "saturated fat reduced from 3 g to 1.5 g per serving").

(iii) Claims described in paragraph (c)(4) of this section may not be made on the label or in labeling of a product if the nutrient content of the reference product meets the definition for "low saturated fat."

(5) The terms defined in paragraph (c)(4) of this section may be used on the label or in labeling of a meal-type product as defined in §381.413(l), provided that:

(i) The product contains at least 25 percent less saturated fat per 100 g of product than an appropriate reference product as described in §381.413(j)(1); and

(ii) As required in §381.413(j)(2) for relative claims:

(A) The identity of the reference product and the percent (or fraction) that the saturated fat differs between the two products are declared in immediate proximity to the most prominent such claim (e.g., "reduced saturated fat 'product', 50 percent less saturated fat than our regular 'product'"); and

(B) Quantitative information comparing the level of saturated fat in the product per specified weight with that of the reference product that it replaces is declared adjacent to the most prominent claim or to the nutrition information (e.g., "saturated fat content has been reduced from 2.5 g per 3 oz to 1.5 g per 3 oz").

(iii) Claims described in paragraph (c)(5) of this section may not be made on the label or in labeling of a product if the nutrient content of the reference product meets the definition for "low saturated fat."

(d) *Cholesterol content claims.*

(1) The terms "cholesterol free," "free of cholesterol," "zero cholesterol," "without cholesterol," "no

9 CFR §381.462

cholesterol," "trivial source of cholesterol," "negligible source of cholesterol," or "dietarily insignificant source of cholesterol" may be used on the label or in labeling of products, provided that:

(i) The product contains less than 2 milligrams (mg) of cholesterol per reference amount customarily consumed and per labeled serving size or, in the case of a meal-type product as defined in §381.413(l), less than 2 mg of cholesterol per labeled serving size;

(ii) The product contains no ingredient that is generally understood by consumers to contain cholesterol, unless the listing of the ingredient in the ingredients statement is followed by an asterisk that refers to the statement below the list of ingredients, which states: "Adds a trivial amount of cholesterol," "adds a negligible amount of cholesterol," or "adds a dietarily insignificant amount of cholesterol";

(iii) The product contains 2 g or less of saturated fat per reference amount customarily consumed or, in the case of a meal-type product as defined in §381.413(l), 2 g or less of saturated fat per labeled serving size; and

(iv) If the product meets these conditions without the benefit of special processing, alteration, formulation, or reformulation to lower cholesterol content, it is labeled to clearly refer to all products of its type and not merely to the particular brand to which it attaches; or

(v) If the product meets these conditions only as a result of special processing, alteration, formulation, or reformulation, the amount of cholesterol is reduced by 25 percent or more from the reference product it replaces as described in §381.413(j)(1) and for which it substitutes as described in §381.413(d) that has a significant (e.g., 5 percent or more of a national or regional market) market share. As required in §381.413(j)(2) for relative claims:

(A) The identity of the reference product and the percent (or fraction) that the cholesterol was reduced are declared in immediate proximity to the most prominent such claim (e.g., "cholesterol free 'product', contains 100 percent less cholesterol than 'reference product'"); and

(B) Quantitative information comparing the level of cholesterol in the product per labeled serving size with that of the reference product that it replaces is declared adjacent to the most prominent claim or to the nutrition information (e.g., "contains no cholesterol compared with 30 mg in one serving of 'reference product'").

(2) The terms "low in cholesterol," "low cholesterol," "contains a small amount of cholesterol," "low source of cholesterol," or "little cholesterol" may be used on the label or in labeling of products, except meal-type products as defined in §381.413(l), provided that:

(i) (A) If the product has a reference amount customarily consumed greater than 30 g or greater than 2 tbsp:

(1) The product contains 20 mg or less of cholesterol per reference amount customarily consumed; and

(2) The product contains 2 g or less of saturated fat per reference amount customarily consumed; or

(B) If the product has a reference amount customarily consumed of 30 g or less or 2 tbsp or less:

(1) The product contains 20 mg or less of cholesterol per reference amount customarily consumed and per 50 g (for dehydrated products that must be reconstituted before typical consumption with water or a diluent containing an insignificant amount, as defined in §381.409(f)(1), of all nutrients per reference amount customarily consumed, the per-50-g criterion refers to the "as prepared" form); and

(2) The product contains 2 g or less of saturated fat per reference amount customarily consumed.

(ii) If the product meets these conditions without the benefit of special processing, alteration, formulation, or reformulation to lower cholesterol content, it is labeled to clearly refer to all products of its type and not merely to the particular brand to which the label attaches; or

(iii) If the product contains 20 mg or less of cholesterol only as a result of special processing, alteration, formulation, or reformulation, the amount of cholesterol is reduced by 25 percent or more from the reference product it replaces as described in §381.413(j)(1) and for which it substitutes as described in §381.413(d) that has a significant (e.g., 5 percent or more of a national or regional market) market share. As required in §381.413(j)(2) for relative claims:

(A) The identity of the reference product and the percent (or fraction) that the cholesterol has been reduced are declared in immediate proximity to the most prominent such claim (e.g., "low cholesterol 'product', contains 85 percent less cholesterol than our regular 'product'"); and

9 CFR §381.462

(B) Quantitative information comparing the level of cholesterol in the product per labeled serving size with that of the reference product that it replaces is declared adjacent to the most prominent claim or to the nutrition information (e.g., "cholesterol lowered from 30 mg to 5 mg per serving").

(3) The terms defined in paragraph (d)(2) of this section may be used on the label or in labeling of a meal-type product as defined in §381.413(l), provided that:

(i) The product contains 20 mg or less of cholesterol per 100 g of product;

(ii) The product contains 2 g or less of saturated fat per 100 g of product; and

(iii) If the product meets these conditions without the benefit of special processing, alteration, formulation, or reformulation to lower cholesterol content, it is labeled to clearly refer to all products of its type and not merely to the particular brand to which the label attaches.

(4) The terms "reduced cholesterol," "reduced in cholesterol," "cholesterol reduced," "less cholesterol," "lower cholesterol," or "lower in cholesterol" may be used on the label or in labeling of products or products that substitute for those products as specified in §381.413(d), excluding meal-type products as defined in §381.413(l), provided that:

(i) The product has been specifically formulated, altered, or processed to reduce its cholesterol by 25 percent or more from the reference product it replaces as described in §381.413(j)(1) and for which it substitutes as described in §381.413(d) that has a significant (e.g., 5 percent or more of a national or regional market) market share;

(ii) The product contains 2 g or less of saturated fat per reference amount customarily consumed; and

(iii) As required in §381.413(j)(2) for relative claims:

(A) The identity of the reference product and the percent (or fraction) that the cholesterol has been reduced are declared in immediate proximity to the most prominent such claim (e.g., "25 percent less cholesterol than 'reference product'"); and

(B) Quantitative information comparing the level of cholesterol in the product per labeled serving size with that of the reference product that it replaces is declared adjacent to the most prominent claim or to the nutrition information (e.g., "cholesterol lowered from 55 mg to 30 mg per serving").

(iv) Claims described in paragraph (d)(4) of this section may not be made on the label or in labeling of a product if the nutrient content of the reference product meets the definition for "low cholesterol."

(5) The terms defined in paragraph (d)(4) of this section may be used on the label or in labeling of a meal-type product as defined in §381.413(l), provided that:

(i) The product has been specifically formulated, altered, or processed to reduce its cholesterol by 25 percent or more from the reference product it replaces as described in §381.413(j)(1) and for which it substitutes as described in §381.413(d) that has a significant (e.g., 5 percent or more of a national or regional market) market share;

(ii) The product contains 2 g or less of saturated fat per 100 g of product; and

(iii) As required in §381.413(j)(2) for relative claims:

(A) The identity of the reference product and the percent (or fraction) that the cholesterol has been reduced are declared in immediate proximity to the most prominent such claim (e.g., "25% less cholesterol than 'reference product'"); and

(B) Quantitative information comparing the level of cholesterol in the product per specified weight with that of the reference product that it replaces is declared adjacent to the most prominent claim or to the nutrition information (e.g., "cholesterol content has been reduced from 35 mg per 3 oz to 25 mg per 3 oz).

(iv) Claims described in paragraph (d)(5) of this section may not be made on the label or in labeling of a product if the nutrient content of the reference product meets the definition for "low cholesterol."

(e) *"Lean" and "Extra Lean" claims.*

(1) The term "lean" may be used on the label or in labeling of a product, provided that the product contains less than 10 g of fat, 4.5 g or less of saturated fat, and less than 95 mg of cholesterol per 100 g of product and per reference amount customarily consumed for individual foods, and per 100 g of product and per labeled serving size for meal-type products as defined in §381.413(l).

(2) The term "extra lean" may be used on the label or in labeling of a product, provided that the product contains less than 5 g of fat, less than 2 g of saturated fat, and less than 95 mg of cholesterol per 100 g of product and per reference amount customarily consumed for individual foods, and per 100 g of product and per labeled serving size for meal-type products as defined in §381.413(l).

9 CFR §381.469

[58 FR 675, Jan. 6, 1993; 58 FR 43789, Aug. 18, 1993, as amended at 58 FR 47628, Sept. 10, 1993; 59 FR 40216, Aug. 8, 1994; 60 FR 214, Jan. 3, 1995]

§381.463 Nutrient content claims for "healthy."

(a) The term "healthy," or any other derivative of the term "health," may be used on the labeling of any poultry product, provided that the product is labeled in accordance with §381.409 and §381.413.

(b)

(1) The product shall meet the requirements for "low fat" and "low saturated fat," as defined in §381.462, except that single-ingredient, raw products may meet the total fat and saturated fat criteria for "extra lean" in §381.462.

(2) The product shall not contain more than 60 milligrams (mg) of cholesterol per reference amount customarily consumed, per labeled serving size, and, only for foods with reference amounts customarily consumed of 30 grams (g) or less or 2 tablespoons (tbsp) or less, per 50 g, and, for dehydrated products that must be reconstituted with water or a diluent containing an insignificant amount, as defined in §381.409(f)(1), of all nutrients, the per-50-g criterion refers to the prepared form, except that:

(i) A meal-type product, as defined in §381.413(l), and including meal-type products that weigh more than 12 ounces (oz) per serving (container), shall not contain more than 90 mg of cholesterol per labeled serving size; and

(ii) Single-ingredient, raw products may meet the cholesterol criterion for "extra lean" in §381.462.

(3) The product shall not contain more than 360 mg of sodium, except that it shall not contain more than 480 mg of sodium effective through January 1, 2000, per reference amount customarily consumed, per labeled serving size, and, only for foods with reference amounts customarily consumed of 30 g or less or 2 tbsp or less, per 50 g, and, for dehydrated products that must be reconstituted with water or a diluent containing an insignificant amount, as defined in §381.409(f)(1), of all nutrients, the per-50-g criterion refers to the prepared form, except that:

(i) A meal-type product, as defined in §381.413(l), and including meal-type products that weigh more than 12 oz per serving (container), shall not contain more than 480 mg of sodium, except that it shall not contain more than 600 mg of sodium effective through January 1, 2000, per labeled serving size; and

(ii) The requirements of this paragraph (b)(3) do not apply to single-ingredient, raw products.

(4) The product shall contain 10 percent or more of the Reference Daily Intake or Daily Reference Value as defined in §381.409 for vitamin A, vitamin C, iron, calcium, protein, or fiber per reference amount customarily consumed prior to any nutrient addition, except that:

(i) A meal-type product, as defined in §381.413(l), and including meal-type products that weigh at least 6 oz but less than 10 oz per serving (container), shall meet the level for two of the nutrients per labeled serving size; and

(ii) A meal-type product, as defined in §381.413(l), and including meal-type products that weigh 10 oz or more per serving (container), shall meet the level for three of the nutrients per labeled serving size.

[59 FR 24228, May 10, 1994; 60 FR 217, Jan. 3, 1995; 63 *FR* 7281, Feb. 13, 1998]

§§381.464-381.468 [Reserved]

§381.469 Labeling applications for nutrient content claims.

(a) This section pertains to labeling applications for claims, express or implied, that characterize the level of any nutrient required to be on the label or in labeling of product by this subpart.

(b) Labeling applications included in this section are:

(1) Labeling applications for a new (heretofore unauthorized) nutrient content claim,

(2) Labeling applications for a synonymous term (i.e., one that is consistent with a term defined by regulation) for characterizing the level of a nutrient, and

9 CFR §381.469

(3) Labeling applications for the use of an implied claim in a brand name.

(c) Labeling applications and supporting documentation to be filed under this section shall be submitted in quadruplicate, except that the supporting documentation may be submitted on a computer disc copy. If any part of the material submitted is in a foreign language, it shall be accompanied by an accurate and complete English translation. The labeling application shall state the applicant's post office address.

(d) Pertinent information will be considered as part of an application on the basis of specific reference to such information submitted to and retained in the files of the Food Safety and Inspection Service. However, any reference to unpublished information furnished by a person other than the applicant will not be considered unless use of such information is authorized (with the understanding that such information may in whole or part be subject to release to the public) in a written statement signed by the person who submitted it. Any reference to published information should be accompanied by reprints or photostatic copies of such references.

(e) If nonclinical laboratory studies accompany a labeling application, the applicant shall include, with respect to each nonclinical study included with the application, either a statement that the study has been, or will be, conducted in compliance with the good laboratory practice regulations as set forth in Part 58 of Chapter 1, Title 21, or, if any such study was not conducted in compliance with such regulations, a brief statement of the reason for the noncompliance.

(f) If clinical investigations accompany a labeling application, the applicant shall include, with respect to each clinical investigation included with the application, either a statement that the investigation was conducted in compliance with the requirements for institutional review set forth in Part 56 of Chapter 1, Title 21, or was not subject to such requirements in accordance with §56.194 or §56.105, and that it was conducted in compliance with the requirements for informed consents set forth in Part 50 of Chapter 1, Title 21.

(g) The availability for public disclosure of labeling applications, along with supporting documentation, submitted to the Agency under this section will be governed by the rules specified in Subchapter D, Title 9.

(h) The data specified under this section to accompany a labeling application shall be submitted on separate sheets, suitably identified. If such data has already been submitted with an earlier labeling application from the applicant, the present labeling application must provide the data.

(i) The labeling application must be signed by the applicant or by his or her attorney or agent, or (if a corporation) by an authorized official.

(j) The labeling application shall include a statement signed by the person responsible for the labeling application, that to the best of his or her knowledge, it is a representative and balanced submission that includes unfavorable information, as well as favorable information, known to him or her pertinent to the evaluation of the labeling application.

(k)

(1) Labeling applications for a new nutrient content claim shall be accompanied by the following data which shall be submitted in the following form to the Director, Food Labeling Division, Regulatory Programs, Food Safety and Inspection Service, Washington, DC 20250:

(Date)

The undersigned, ------ submits this labeling application pursuant to 9 CFR 381.469 with respect to (statement of the claim and its proposed use).

Attached hereto, in quadruplicate, or on a computer disc copy, and constituting a part of this labeling application, are the following:

(i) A statement identifying the nutrient content claim and the nutrient that the term is intended to characterize with respect to the level of such nutrient. The statement shall address why the use of the term as proposed will not be misleading. The statement shall provide examples of the nutrient content claim as it will be used on labels or labeling, as well as the types of products on which the claim will be used. The statement shall also specify the level at which the nutrient must be present or what other conditions concerning the product must be met for the appropriate use of the term in labels or labeling, as well as any factors that would make the use of the term inappropriate.

(ii) A detailed explanation supported by any necessary data of why use of the food component characterized by the claim is of importance in human nutrition by virtue of its presence or absence at the levels that such claim would describe. This explanation shall also state what nutritional benefit to the public will derive from use of the claim as proposed and why such benefit is not available through the use of existing terms defined by regulation. If the claim is

9 CFR §381.469

intended for a specific group within the population, the analysis shall specifically address nutritional needs of such group, and scientific data sufficient for such purpose, and data and information to the extent necessary to demonstrate that consumers can be expected to understand the meaning of the term under the proposed conditions of use.

(iii) Analytical data that demonstrates the amount of the nutrient that is present in the products for which the claim is intended. The assays should be performed on representative samples in accordance with 381.409(h). If no USDA or AOAC methods are available, the applicant shall submit the assay method used, and data establishing the validity of the method for assaying the nutrient in the particular food. The validation data shall include a statistical analysis of the analytical and product variability.

(iv) A detailed analysis of the potential effect of the use of the proposed claim on food consumption, and any corresponding changes in nutrient intake. The analysis shall specifically address the intake of nutrients that have beneficial and negative consequences in the total diet. If the claim is intended for a specific group within the population, the above analysis shall specifically address the dietary practices of such group, and shall include data sufficient to demonstrate that the dietary analysis is representative of such group.

Yours very truly,

Applicant _____

By _____
(Indicate authority)

(2) Upon receipt of the labeling application and supporting documentation, the applicant shall be notified, in writing, of the date on which the labeling application was received. Such notice shall inform the applicant that the labeling application is undergoing Agency review and that the applicant shall subsequently be notified of the Agency's decision to consider for further review or deny the labeling application.

(3) Upon review of the labeling application and supporting documentation, the Agency shall notify the applicant, in writing, that the labeling application is either being considered for further review or that it has been summarily denied by the Administrator.

(4) If the labeling application is summarily denied by the Administrator, the written notification shall state the reasons therefor, including why the Agency has determined that the proposed nutrient content claim is false or misleading. The notification letter shall inform the applicant that the applicant may submit a written statement by way of answer to the notification, and that the applicant shall have the right to request a hearing with respect to the merits or validity of the Administrator's decision to deny the use of the proposed nutrient content claim.

(i) If the applicant fails to accept the determination of the Administrator and files an answer and requests a hearing, and the Administrator, after review of the answer, determines the initial determination to be correct, the Administrator shall file with the Hearing Clerk of the Department the notification, answer, and the request for a hearing, which shall constitute the complaint and answer in the proceeding, which shall thereafter be conducted in accordance with the Department's Uniform Rules of Practice.

(ii) The hearing shall be conducted before an administrative law judge with the opportunity for appeal to the Department's Judicial Officer, who shall make the final determination for the Secretary. Any such determination by the Secretary shall be conclusive unless, within 30 days after receipt of notice of such final determination, the applicant appeals to the United States Court of Appeals for the circuit in which the applicant has its principal place of business or to the United States Court of Appeals for the District of Columbia Circuit.

(5) If the labeling application is not summarily denied by the Administrator, the Administrator shall publish in the Federal Register a proposed rule to amend the regulations to authorize the use of the nutrient content claim. The proposal shall also summarize the labeling application, including where the supporting documentation can be reviewed. The Administrator's proposed rule shall seek comment from consumers, the industry, consumer and industry groups, and other interested persons on the labeling application and the use of the proposed nutrient content claim. After public comment has been received and reviewed by the Agency, the Administrator shall make a determination on whether the proposed nutrient content claim

shall be approved for use on the labeling of poultry products.

(i) If the claim is denied by the Administrator, the Agency shall notify the applicant, in writing, of the basis for the denial, including the reason why the claim on the labeling was determined by the Agency to be false or misleading. The notification letter shall also inform the applicant that the applicant may submit a written statement by way of answer to the notification, and that the applicant shall have the right to request a hearing with respect to the merits or validity of the Administrator's decision to deny the use of the proposed nutrient content claim.

(A) If the applicant fails to accept the determination of the Administrator and files an answer and requests a hearing, and the Administrator, after review of the answer, determines the initial determination to be correct, the Administrator shall file with the Hearing Clerk of the Department the notification, answer, and the request for a hearing, which shall constitute the complaint and answer in the proceeding, which shall thereafter be conducted in accordance with the Department's Uniform Rules of Practice.

(B) The hearing shall be conducted before an administrative law judge with the opportunity for appeal to the Department's Judicial Officer, who shall make the final determination for the Secretary. Any such determination by the Secretary shall be conclusive unless, within 30 days after receipt of the notice of such final determination, the applicant appeals to the United States Court of Appeals for the circuit in which the applicant has its principal place of business or to the United States Court of Appeals for the District of Columbia Circuit.

(ii) If the claim is approved, the Agency shall notify the applicant, in writing, and shall also publish in the Federal Register a final rule amending the regulations to authorize the use of the claim.

(l)

(1) Labeling applications for a synonymous term shall be accompanied by the following data which shall be submitted in the following form to the Director, Food Labeling Division, Regulatory Programs, Food Safety and Inspection Service, Washington, DC 20250:

(Date)

The undersigned, ------ submits this labeling application pursuant to 9 CFR 381.469 with respect to (statement of the synonymous term and its proposed use in a nutrient content claim that is consistent with an existing term that has been defined under subpart Y of part 381).

Attached hereto, in quadruplicate, or on a computer disc copy, and constituting a part of this labeling application, are the following:

(i) A statement identifying the synonymous term, the existing term defined by a regulation with which the synonymous term is claimed to be consistent, and the nutrient that the term is intended to characterize the level of. The statement shall address why the use of the synonymous term as proposed will not be misleading. The statement shall provide examples of the nutrient content claim as it will be used on labels or labeling, as well as the types of products on which the claim will be used. The statement shall also specify whether any limitations not applicable to the use of the defined term are intended to apply to the use of the synonymous term.

(ii) A detailed explanation supported by any necessary data of why use of the proposed term is requested, including whether the existing defined term is inadequate for the purpose of effectively characterizing the level of a nutrient. This explanation shall also state what nutritional benefit to the public will derive from use of the claim as proposed, and why such benefit is not available through use of existing terms defined by regulation. If the claim is intended for a specific group within the population, the analysis shall specifically address nutritional needs of such group, scientific data sufficient for such purpose, and data and information to the extent necessary to demonstrate that consumers can be expected to understand the meaning of the term under the proposed conditions of use.

Yours very truly,

Applicant _____

By _____
(Indicate authority)

9 CFR §381.469

(2) Upon receipt of the labeling application and supporting documentation, the applicant shall be notified, in writing, of the date on which the labeling application was received. Such notice shall inform the applicant that the labeling application is undergoing Agency review and that the applicant shall subsequently be notified of the Agency's decision to consider for further review or deny the labeling application.

(3) Upon review of the labeling application and supporting documentation, the Agency shall notify the applicant, in writing, that the labeling application is either being considered for further review or that it has been summarily denied by the Administrator.

(4) If the labeling application is summarily denied by the Administrator, the written notification shall state the reasons therefor, including why the Agency has determined that the proposed synonymous term is false or misleading. The notification letter shall inform the applicant that the applicant may submit a written statement by way of answer to the notification, and that the applicant shall have the right to request a hearing with respect to the merits or validity of the Administrator's decision to deny the use of the proposed synonymous term.

(i) If the applicant fails to accept the determination of the Administrator and files an answer and requests a hearing, and the Administrator, after review of the answer, determines the initial determination to be correct, the Administrator shall file with the Hearing Clerk of the Department the notification, answer, and the request for a hearing, which shall constitute the complaint and answer in the proceeding, which shall thereafter be conducted in accordance with the Department's Uniform Rules of Practice.

(ii) The hearing shall be conducted before an administrative law judge with the opportunity for appeal to the Department's Judicial Officer, who shall make the final determination for the Secretary. Any such determination by the Secretary shall be conclusive unless, within 30 days after receipt of notice of such final determination, the applicant appeals to the United States Court of Appeals for the circuit in which the applicant has its principal place of business or to the United States Court of Appeals for the District of Columbia Circuit.

(5) If the claim is approved, the Agency shall notify the applicant, in writing, and shall publish in the Federal Register a notice informing the public that the synonymous term has been approved for use.

(m)

(1) Labeling applications for the use of an implied nutrient content claim in a brand name shall be accompanied by the following data which shall be submitted in the following form to the Director, Food Labeling Division, Regulatory Programs, Food Safety and Inspection Service, Washington, DC 20250:

(Date)

The undersigned, ------ submits this labeling application pursuant to 9 CFR 381.469 with respect to (statement of the implied nutrient content claim and its proposed use in a brand name).

Attached hereto, in quadruplicate, or on a computer disc copy, and constituting a part of this labeling application, are the following:

(i) A statement identifying the implied nutrient content claim, the nutrient the claim is intended to characterize, the corresponding term for characterizing the level of such nutrient as defined by a regulation, and the brand name of which the implied claim is intended to be a part. The statement shall address why the use of the brand-name as proposed will not be misleading. The statement shall provide examples of the types of products on which the brand name will appear. It shall also include data showing that the actual level of the nutrient in the food would qualify the label of the product to bear the corresponding term defined by regulation. Assay methods used to determine the level of a nutrient shall meet the requirements stated under labeling application format in paragraph (k)(1)(iii) of this section.

(ii) A detailed explanation supported by any necessary data of why use of the proposed brand name is requested. This explanation shall also state what nutritional benefit to the public will derive from use of the brand name as proposed. If the branded product is intended for a specific group within the population, the analysis shall specifically address nutritional needs of such group and scientific data sufficient for such purpose.

Yours very truly,

9 CFR §381.469

Applicant _____

By _____

(2) Upon receipt of the labeling application and supporting documentation, the applicant shall be notified, in writing, of the date on which the labeling application was received. Such notice shall inform the applicant that the labeling application is undergoing Agency review and that the applicant shall subsequently be notified of the Agency's decision to consider for further review or deny the labeling application.

(3) Upon review of the labeling application and supporting documentation, the Agency shall notify the applicant, in writing, that the labeling application is either being considered for further review or that it has been summarily denied by the Administrator.

(4) If the labeling application is summarily denied by the Administrator, the written notification shall state the reasons therefor, including why the Agency has determined that the proposed implied nutrient content claim is false or misleading. The notification letter shall inform the applicant that the applicant may submit a written statement by way of answer to the notification, and that the applicant shall have the right to request a hearing with respect to the merits or validity of the Administrator's decision to deny the use of the proposed implied nutrient content claim.

(i) If the applicant fails to accept the determination of the Administrator and files an answer and requests a hearing, and the Administrator, after review of the answer, determines the initial determination to be correct, the Administrator shall file with the Hearing Clerk of the Department the notification, answer, and the request for a hearing, which shall constitute the complaint and answer in the proceeding, which shall thereafter be conducted in accordance with the Department's Uniform Rules of Practice.

(ii) The hearing shall be conducted before an administrative law judge with the opportunity for appeal to the Department's Judicial Officer, who shall make the final determination for the Secretary. Any such determination by the Secretary shall be conclusive unless, within 30 days after receipt of notice of such final determination, the applicant appeals to the United States Court of Appeals for the circuit in which the applicant has its principal place of business or to the United States Court of Appeals for the District of Columbia Circuit.

(5) If the labeling application is not summarily denied by the Administrator, the Administrator shall publish a notice of the labeling application in the Federal Register seeking a comment on the use of the implied nutrient content claim. The notice shall also summarize the labeling application, including where the supporting documentation can be reviewed. The Administrator's notice shall seek comment from consumers, the industry, consumer and industry groups, and other interested persons on the labeling application and the use of the implied nutrient content claim. After public comment has been received and reviewed by the Agency, the Administrator shall make a determination on whether the implied nutrient content claim shall be approved for use on the labeling of poultry products.

(i) If the claim is denied by the Administrator, the Agency shall notify the applicant, in writing, of the basis for the denial, including the reason why the claim on the labeling was determined by the Agency to be false or misleading. The notification letter shall also inform the applicant that the applicant may submit a written statement by way of answer to the notification, and that the applicant shall have the right to request a hearing with respect to the merits or validity of the Administrator's decision to deny the use of the proposed implied nutrient content claim.

(A) If the applicant fails to accept the determination of the Administrator and files an answer and requests a hearing, and the Administrator, after review of the answer, determines the initial determination to be correct, the Administrator shall file with the Hearing Clerk of the Department the notification, answer, and the request for a hearing, which shall constitute the complaint and answer in the proceeding, which shall thereafter be conducted in accordance with the Department's Uniform Rules of Practice.

(B) The hearing shall be conducted before an administrative law judge with the opportunity for appeal to the Department's Judicial Officer, who shall make the final determination for the Secretary. Any such determination by the Secretary shall be conclusive unless, within 30 days after receipt of the notice of such final determination, the applicant appeals to the United States Court of Appeals for the circuit in which the applicant has its principal place of business or to the United States Court of Appeals for the District of

9 CFR §381.470

Columbia Circuit.

(ii) If the claim is approved, the Agency shall notify the applicant, in writing, and shall also publish in the Federal Register a notice informing the public that the implied nutrient content claim has been approved for use.

(Paperwork requirements were approved by the Office of Management and Budget under control number 0583-0088.)

[58 FR 675, Jan. 6, 1993, as amended at 59 FR 45196, Sept. 1, 1994; 60 FR 217, Jan. 3, 1995]

§381.470-381.479 [Reserved]

§381.480 Label statements relating to usefulness in reducing or maintaining body weight.

(a) *General requirements.* Any product that purports to be or is represented for special dietary use because of usefulness in reducing body weight shall bear:

(1) Nutrition labeling in conformity with §381.409 of this subpart, unless exempt under that section, and

(2) A conspicuous statement of the basis upon which the product claims to be of special dietary usefulness.

(b) *Nonnutritive ingredients.*

(1) Any product subject to paragraph (a) of this section that achieves its special dietary usefulness by use of a nonnutritive ingredient (i.e., one not utilized in normal metabolism) shall bear on its label a statement that it contains a nonnutritive ingredient and the percentage by weight of the nonnutritive ingredient.

(2) A special dietary product may contain a nonnutritive sweetener or other ingredient only if the ingredient is safe for use in the product under the applicable law and regulations of this chapter. Any product that achieves its special dietary usefulness in reducing or maintaining body weight through the use of a nonnutritive sweetener shall bear on its label the statement required by paragraph (b)(1) of this section, but need not state the percentage by weight of the nonnutritive sweetener. If a nutritive sweetener(s) as well as nonnutritive sweetener(s) is added, the statement shall indicate the presence of both types of sweetener; e.g., "Sweetened with nutritive sweetener(s) and nonnutritive sweetener(s)."

(c) *"Low calorie" foods.* A product purporting to be "low calorie" must comply with the criteria set forth for such foods in §381.460.

(d) *"Reduced calorie" foods and other comparative claims.* A product purporting to be "reduced calorie" or otherwise containing fewer calories than a reference food must comply with the criteria set forth for such foods in §381.460(b) (4) and (5).

(e) *"Label terms suggesting usefulness as low calorie or reduced calorie foods".*

(1) Except as provided in paragraphs (e)(2) and (e)(3) of this section, a product may be labeled with terms such as "diet," "dietetic," "artificially sweetened," or "sweetened with nonnutritive sweetener" only if the claim is not false or misleading, and the product is labeled "low calorie" or "reduced calorie" or bears another comparative calorie claim in compliance with the applicable provisions in this subpart.

(2) Paragraph (e)(1) of this section shall not apply to any use of such terms that is specifically authorized by regulation governing a particular food, or, unless otherwise restricted by regulation, to any use of the term "diet" that clearly shows that the product is offered solely for a dietary use other than regulating body weight, e.g., "for low sodium diets."

(3) Paragraph (e)(1) of this section shall not apply to any use of such terms on a formulated meal replacement or other product that is represented to be of special dietary use as a whole meal, pending the issuance of a regulation governing the use of such terms on foods.

(f) *"Sugar free" and "no added sugar".* Criteria for the use of the terms "sugar free" and "no added sugar" are provided for in §381.460(c).

[58 FR 675, Jan. 6, 1993; 58 FR 43789, Aug. 18, 1993, as amended at 58 FR 47628, Sept. 10, 1993; 60 FR 217, Jan. 3, 1995]

§§381.481-381.499 [Reserved]

9 CFR §381.500

§381.500 Exemption from nutrition labeling.

(a) The following poultry products are exempt from nutrition labeling:
    (1) Food products produced by small businesses, provided that the labels for these products bear no nutrition claims or nutrition information,
    (i) A food product, for purposes of the small business exemption, is defined as a formulation, not including distinct flavors which do not significantly alter the nutritional profile, sold in any size package in commerce.
    (ii) For purposes of this paragraph, a small business is any single-plant facility or multi-plant company/firm that employs 500 or fewer people and produces no more than the following amounts of pounds of the product qualifying the firm for exemption from this subpart:
    (A) During the first year of implementation of nutrition labeling, from July 1994 to July 1995, 250,000 pounds or less,
    (B) During the second year of implementation of nutrition labeling, from July 1995 to July 1996, 175,000 pounds or less, and
    (C) During the third year of implementation and subsequent years thereafter, 100,000 pounds or less.
    (iii) For purposes of this paragraph, calculation of the amount of pounds shall be based on the most recent 2-year average of business activity. Where firms have been in business less than 2 years or where products have been produced for less than 2 years, reasonable estimates must indicate that the annual pounds produced will not exceed the amounts specified.
    (2) Products intended for further processing, provided that the labels for these products bear no nutrition claims or nutrition information,
    (3) Products that are not for sale to consumers, provided that the labels for these products bear no nutrition claims or nutrition information,
    (4) Products in small packages that are individually wrapped packages of less than 1/2 ounce net weight, provided that the labels for these products bear no nutrition claims or nutrition information,
    (5) Products custom slaughtered or prepared,
    (6) Products intended for export, and
    (7) The following products prepared and served or sold at retail provided that the labels or the labeling of these products bear no nutrition claims or nutrition information:
    (i) Ready-to-eat products that are packaged or portioned at a retail store or similar retail-type establishment; and (ii) Multi-ingredient products (e.g. sausage) processed at a retail store or similar retail-type establishment.

(b) Restaurant menus generally do not constitute labeling or fall within the scope of these regulations.

(c)
    (1) Foods represented to be specifically for infants and children less than 2 years of age shall bear nutrition labeling as provided in paragraph (c)(2) of this section, except such labeling shall not include calories from fat, calories from saturated fat, saturated fat, stearic acid, polyunsaturated fat, monounsaturated fat, and cholesterol.
    (2) Foods represented or purported to be specifically for infants and children less than 4 years of age shall bear nutrition labeling except that:
    (i) Such labeling shall not include declarations of percent of Daily Value for total fat, saturated fat, cholesterol, sodium, potassium, total carbohydrate, and dietary fiber;
    (ii) Nutrient names and quantitative amounts by weight shall be presented in two separate columns;
    (iii) The heading "Percent Daily Value" required in §381.409(d)(6) shall be placed immediately below the quantitative information by weight for protein;
    (iv) The percent of the Daily Value for protein, vitamins, and minerals shall be listed immediately below the heading "Percent Daily Value"; and
    (v) Such labeling shall not include the footnote specified in §381.409(d)(9).

(d)
    (1) Products in packages that have a total surface area available to bear labeling of less than 12 square inches are exempt from nutrition labeling, provided that the labeling for these products bear no nutrition claims or other nutrition information. The manufacturer, packer, or distributor shall provide, on the label of

packages that qualify for and use this exemption, an address or telephone number that a consumer can use to obtain the required nutrition information (e.g., "For nutrition information call 1-800-123-4567").

(2) When such products bear nutrition labeling, either voluntarily or because nutrition claims or other nutrition information is provided, all required information shall be in a type size no smaller than 6 point or all upper case type of 1/16-inch minimum height, except that individual serving-size packages of poultry products that have a total area available to bear labeling of 3 square inches or less may provide all required information in a type size no smaller than 1/32-inch minimum height.

[58 FR 675, Jan. 6, 1993, as amended at 58 FR 47628, Sept. 10, 1993; 59 FR 45198, Sept. 1, 1994; 60 FR 217, Jan.3, 1995]

APPENDIX

- REFERENCES TO KEY FDA & FSIS NUTRITION LABELING RULES

- FORTHCOMING ACTIONS PLANNED BY FDA & FSIS

- BIBLIOGRAPHY & ADDITIONAL READING

TABLE A.1
Key FDA & FSIS Nutrition Labeling Rules Published Since NLEA

Date	FR reference	Subject of rule
November 27, 1991	56 FR 60421 and 60478	Proposals to define nutrient content claims for food labeling
January 6, 1993	58 FR 2066 et al.	FDA final regulations implementing various NLEA requirements, and FSIS rules modeled after FDA program
June 18, 1993	58 FR 33715	FDA proposed rule governing nutrition labeling of dietary supplements (pre-DSHEA)
October 14, 1993	58 FR 53254 et al.	Folic acid health claim and fortification proposals
January 4, 1994	59 FR 354	Original final rule on nutrition labeling of dietary supplements
January 3, 1995	60 FR 173	FSIS final rule codifying all nutrition labeling requirements, rather than incorporating FDA rules by reference
December 21, 1995	60 FR 66206	Nutrient content claim synonyms, shortened health claims, and labeling exemptions proposed
December 28, 1995	60 FR 67176	DSHEA-inspired requirements for dietary supplements proposed
December 28, 1995	60 FR 67184	"High potency" and "antioxidant" definitions proposed for nutrient content claims
December 28, 1995	60 FR 67194	FDA proposes statement of identity and labeling requirements for dietary supplements
December 29, 1995	60 FR 67474	FSIS proposal for "expressed nutrient content claim" rule for products using standardized name
January 4, 1996	61 FR 296	Health claim for oat products proposed
February 2, 1996	61 FR 3885	FDA proposes nutrient content and health claims recordkeeping requirements
February 12, 1996	61 FR 5349	FDA proposes more flexible use of "healthy"
March 5, 1996	61 FR 8752	New health claim rule for folate/neural tube defects
March 14, 1996	61 FR 10480	Correction issued for proposed dietary supplement labels
August 2, 1996	61 FR 40320	FDA rule revoking restaurant menu exemption for claims
August 7, 1996	61 FR 40963	FDA low-volume product exemption; model notice
August 16, 1996	61 FR 42741	Final rule and nutrient data for voluntary labeling of raw fruits, vegetables, and fish
August 23, 1996	61 FR 43433	Final rule authorizing health claim relating reduced tooth decay and sugar alcohols

TABLE A.1
Key FDA & FSIS Nutrition Labeling Rules Published Since NLEA

Date	FR reference	Subject of rule
September 27, 1996	61 FR 50771, 50774	Proposal of two sets of procedures for notifying FDA of certain activities related to dietary supplements
November 20, 1996	61 FR 58991	Revocation of milk standards conflicting with labeling rules
November 29, 1996	61 FR 60661	Advance notice concerning FDA regulation of medical foods
December 13, 1996	61 FR 65490	Proposal to amend RDI list in FSIS rules
January 23, 1997	62 FR 3584	FDA final action authorizing health claim linking soluble fiber from oat products to reduced coronary heart disease
March 4, 1997	62 FR 9826	Proposal to develop a national system for determining label information for package net contents
March 17, 1997	62 FR 12579	Proposal by FDA to establish a time frame for reviewing petitions for adoption of new health claims
March 31, 1997	62 FR 15343	Clarification of January 1997 oats/CHD health-claim rule
April 1, 1997	62 FR 15390	Partial stay of FDA provisions in §101.65(d) pertaining to implied claims for "healthy" and sodium content
May 22, 1997	62 FR 28230	FDA regulations specifying 9-month time frame for agency response to health claim petitions
May 22, 1997	62 FR 28234	Proposal to revise oats/CHD health-claim rules in §101.81 to include soluble fiber from psyllium seed husk
June 4, 1997	62 FR 30678	FDA proposal to restrict use of ephedrine alkaloids in dietary supplement products
June 9, 1997	62 FR 31338	Final nutrient content claims rule authorizing use of the term "plus" as a synonym for "added"
September 23, 1997	62 FR 49881	Uniform compliance date set for January 1, 2000
September 23, 1997	62 FR 49826 62 FR 49859 62 FR 49868 62 FR 49883 62 FR 49886	Five dietary supplement final rules: -Statement of identity, nutrition facts and ingredient labeling, compliance policy guide revocation -Requirements for nutrient content claims, health claims, and statements of nutritional support -Definition of "high potency" and "antioxidant" for use in nutrient content claims for supplements and other foods -Notification procedures for informing FDA about statements on dietary supplements -Premarket notification for a new dietary ingredient
February 13, 1998	63 FR 7279	Interim final FSIS rule delaying until January 1, 2000 the provisions pertaining to implied claims for "healthy" and lower sodium content

TABLE A.1
Key FDA & FSIS Nutrition Labeling Rules Published Since NLEA

Date	FR reference	Subject of rule
February 18, 1998	63 FR 8103	Final rule revising FDA health claim on soluble fiber and CHD to include psyllium seed husk
March 25, 1998	63 FR 14349	Revision of FDA's definition of "healthy" to accommodate frozen or canned single-ingredient fruits and vegetables and certain enriched cereal-grain products
April 29, 1998	63 FR 23633	Comments by FDA on final report by Dietary Supplement Label Commission
April 29, 1998	63 FR 23624	FDA proposal concerning regulation of structure/function statements on dietary supplements

TABLE A.2
Nutrition Labeling Actions Planned by FDA & FSIS

Schedule for planned action	FR reference of related action	Subject of rule
N/A	March 4, 1997; 62 FR 9826 (proposal)	Developing a national system for determining label information for package net contents
June 1998	December 29, 1995; 60 FR 67474	Completion of FSIS proposal for "expressed nutrient content claim" rule for products using standardized name
October 1998	December 13, 1996; 61 FR 65490	FSIS adoption of proposal for revised list of RDIs for meat and poultry
October 1998	April 22, 1998; 63 FR 19852 (withdrawal of 1994 proposal)	Re-proposal of FSIS rules for health claims on meat and poultry products
N/A	April 29, 1998; 63 FR 23624	Regulation of structure/function statements on dietary supplements

BIBLIOGRAPHY & ADDITIONAL READING

- *A Food Labeling Guide*, U.S. Dept. Of Health and Human Services, Food and Drug Administration, Center for Food Safety and Applied Nutrition, September 1994

- *Food and Drug Administration Nutrition Labeling Information Study — December 1996, Raw Fruit/Vegetables and Raw Fish*, U.S. Dept. Of Health and Human Services, Food and Drug Administration, May 1997 (see *FR* notice dated May 9, 1997; 62 *FR* 25635)

- *Food Labeling Questions and Answers: For Guidance to Facilitate the Process of Developing or Revising Labels for Foods Other than Dietary Supplements*, U.S. Dept. Of Health and Human Services, Food and Drug Administration, August 1993

- *Food Labeling Questions and Answers Volume II: A Guide for Restaurants and Other Retail Establishments*, U.S. Dept. Of Health and Human Services, Food and Drug Administration, August 1995

- *Report of the Commission on Dietary Supplement Labels*, U.S. Commission on Dietary Supplement Labels, November 1997

Note: Many of the FDA and FSIS labeling-related documents are available on the agencies' web sites at http://www.fda.gov and http://www.usda.gov/fsis, respectively.

INDEX

Note that this *Index* is arranged as follows:

- The first section contains citations for the **FDA** labeling rules — page **I.1**

- The second contains entries for the FSIS **meat product** labeling regulations in 9 *CFR* Part 317 — page **I.21**

- The third section provides references to the FSIS **poultry** product labeling rules in 9 *CFR* Part 381 — page **I.25**

		Regs in Chapter 9	Review Chap #
FDA RULES—			
Abbreviations	Exempt foods	101.105(n)	
	Small packages	101.9(j)(13)(ii)(B)	3
Added (nutrients)	Nutrient content claims	101.54(e)	5
Added flavor	Ingredients statement	101.22(h)	
Additive	Incidental	101.100(a)(3); 101.22(h)(2)	
Aggregate display	Nutrition information	101.9(d)(13)(ii)	3
	Sample label	101.9(d)(13)(ii)	3
Alternate display panel		101.9(j)(17)	3
Amount per serving	Subheading requirement	101.9(d)(4)	3
Annual gross sales	Labeling exemptions	101.9(j)(1)	3
Antioxidants	Nutrient content claims	101.54(f)	8
Antioxidant vitamins and cancer	Health claims	101.71(c)	6
Artificial color(ing)	Definition	101.22(a)(4)	
	Insufficient space	101.22(d)	
	Statement required	101.22(c)	
Artificial flavor(ing)	Definition	101.22	
	Insufficient space	101.22(d)	
	Smoke flavoring	101.22(h)(6)	
	Statement required	101.22(c)	
Assortments	Exemptions	101.100(a)(1)	3
Banana clusters (prior to retail sale)	Exemptions	101.100(i)	
Beta-carotene	Declaration of nutrient information	101.9(c)(8)(vi)	
Bulk foods	Exemptions	101.100(a)(2)	3
	Exceptions	101.9(j)(16)	3
Butter	Ingredient declarations	101.67(c)	
	Nutrient content claims	101.67(a)	
Butter, cheese, or ice cream	Coloring declaration	101.22(k)(3)	
Calcium	High levels	101.54(c)	5
Calcium and osteoporosis	Criteria for claim	101.72(c)(2)(i)	6
	Health claims	101.72	6
	Model health claim	101.72(e)	6
	Nature of food making claim	101.72(c)(ii)	6
	Relationship between calcium and osteoporosis	101.72(a), (b)	6
Caloric conversions	Nutrition information	101.9(d)(10)	3
Calorie content claims	Calorie free	101.60(b)(1)	5
	Low calorie	101.60(b)(2)	5
	Reduced calorie	101.60(b)(4)	5
Calorie free	Definition	101.60(b)(1)	5
Calories	Nutrient content claims	101.60(b)	5

FDA RULES—		Regs in Chapter 9	Review Chap #
Calories from fat	Declaration of nutrient information	101.9(c)(1)(ii)	4
Calories from saturated fat	Voluntary declaration	101.9(c)(1)(iii)	4
Calories, total	Declaration of nutrient information	101.9(c)(1)(i)	4
Cancer	Health claims–fiber-containing grain products, fruits, and vegetables	101.76	6
	Health claims–fruits and vegetables	101.78	6
	Health claims–dietary lipids	101.73	6
Carbohydrate, other	Voluntary declaration	101.9(c)(6)(iv)	4
Carbohydrate, total	Declaration of nutrient information	101.9(c)(6)	4
Caries, dental	Health claims-sugar alcohols	101.80	6
Characterizing ingredient	Flavoring	101.22(i)	
Chemical preservative	Definition	101.22(a)(5)	
	Ingredient statement	101.22(j)	
	Insufficient space	101.22(d)	
	Statement required	101.22(c)	
Cholesterol	Declaration of nutrient information	101.9(c)(3)	4
	Health claims–coronary heart disease	101.75	6
Cholesterol content claims	Cholesterol free	101.62(d)(1)	5
	Low cholesterol	101.62(d)(2)	5
	Nutrient content claims	101.62(d)	5
	Reduced cholesterol	101.62(d)(4)	5
Cholesterol free	Definition	101.62(d)(1)	5
Claims	Health (disease prevention)	101.14	6
	Misbranding	101.9(k)(1)	
	Nutrient	101.13	5
Class I	Nutrient information	101.9(g)(3)(i)	4
Class II	Nutrient information	101.9(g)(3)(ii)	4
Coloring	Butter, cheese, or ice cream	101.22(k)(3)	
	Declaration if not subject to certification	101.22(k)(2)	
	Declaration if not subject to certification	101.22(k)(2)	
	Ingredient statement	101.22(k)	
Compliance	Alternative means	101.9(g)(9)	4
	Uniform compliance date	N/A	1
	Determination	101.9(g)	4
Coronary heart disease	Health claims–fiber-containing fruits, vegetables, and grain products	101.77	6
	Health claims–saturated fat and cholesterol	101.75	6
D-erythroascorbic acid	In fabricated food	101.33	
Daily reference value (DRV)		101.9(c)(9)	4

		Regs in Chapter 9	Review Chap #
FDA RULES—			
Declaration of net quantity of contents	Exemptions from labeling rules	101.105	
Definitions	Artificial color(ing)	101.22(a)(4)	
	Artificial flavor(ing)	101.22	
	Calorie free	101.60(b)(1)	5
	Chemical preservative	101.22(a)(5)	
	Cholesterol free	101.62(d)(1)	5
	Fat free	101.62(b)(1)	5
	Fresh	101.95(a)	7
	Freshly frozen, fresh frozen, or frozen fresh	101.95(b)	7
	Health claims	101.14(a)(1)	6
	Information panel	101.2(a)	3
	Insignificant amounts	101.9(f)(1)	4
	Lot	101.9(g)(1)	4
	Low cholesterol	101.62(d)(2)	5
	Low fat	101.62(b)(2)	5
	Low in saturated fat	101.62(c)(2)	5
	Low sodium	101.61(b)(4)	5
	Low-calorie, low in calories	101.60(b)(2)	5
	Main dish	101.13(m)	5
	Meal product	101.13(l)	5
	Natural flavor(ing)	101.22(a)(3)	
	No saturated fat	101.62(c)(1)	5
	Reduced calorie	101.60(b)(4)	5
	Reduced cholesterol	101.62(d)(4)	5
	Reduced fat	101.62(b)(4)	5
	Reduced saturated fat	101.62(c)(4)	5
	Reduced sodium	101.61(b)(6)	5
	Spice	101.22(a)(2)	
	Very low in sodium	101.61(b)(2)	5
Dental caries and health claim		101.80	6
Descriptive claims		Part 101, Subpart F	7
Diabetes	Foods for special dietary use	105.67	
Dietary fiber (see also Fiber)			
Dietary fiber	Declaration of nutrient information	101.9(c)(6)(i)	4
Dietary fiber and cancer	Health claims	101.71(a)	6
Dietary fiber and cardiovascular disease	Health claims	101.71(b)	6
Dietary lipids and cancer	Criteria for claim	101.73(c)(2)(i)	6
	Health claims	101.73	6
	Model claim	101.73(e)	6
	Nature of food making claim	101.73(c)(2)(ii)	6

		Regs in Chapter 9	Review Chap #
FDA Rules—			
Dietary supplements	Relationship between fat and cancer	101.73(a), (b)	6
	Declarations	101.36(b)	4, 8
	Definition for health claims	101.14(a)(3)	8
	Ephedrine alkaloids		8
	Allowed structure/function statements and prohibited drug/disease claims		8
	Exceptions	101.9(j)(6)	3, 8
	Exemptions	101.36(h)	8
	Format	101.36(b)-(e)	8
	General	101.36	8
	Percent daily value	101.36(b)	4, 8
	Protein products	101.17(d)	
	Sample labels	101.36(e)	8
	With RDI or DRV	101.36(b)(2)	4, 8
	Without RDI or DRV	101.36(b)(3)	8
	Proprietary blends	101.36(b)(4)	8
	Structure/function statements	101.14(a)(6)	8
Disease prevention (see Health claims)			
Disqualifying nutrient levels	Health claims	101.14(a)(5)	6
Distributor	Name used on labeling	101.5(c)	
DRV (see Daily reference value)			
Dual labeling	Nutrition information	101.9(e)(2)	3
Eggs	Exceptions	101.9(j)(14)	3
English language	Nutrition information	101.9(d)(14); 101.15(c)	
Enriched	Nutrient content claims	101.54(e)	5
Examples of graphic enhancements on labeling		Part 101, Appendix B	3
Excellent source of	Nutrient content claims	101.54(b)	5
Exceptions	Bulk foods	101.9(j)(16)	3
	Dietary supplements	101.9(j)(6)	3, 8
	Eggs	101.9(j)(14)	3
	Fish or game meat	101.9(j)(11), (12)	3
	Foods for infants and children	101.9(j)(5)	3
	Foods shipped in bulk form	101.9(j)(9)	3
	Infant formula	101.9(j)(7)	3
	Information panel requirements	101.2(a)(1)-(3)	3
	Insignificant nutrient amounts	101.9(j)(4)	3, 4
	Insufficient space for information	101.2(d); 101.9(j)(13), (j)(17); 101.15(b)	3
	Medical foods	101.9(j)(8)	3
	Multiunit retail packages	101.9(j)(15)	3
	Raw fruit, vegetables, and fish	101.9(j)(10); 101.42; 101.45	3

		Regs in Chapter 9	Review Chap #
FDA RULES—			
	Small packages	101.9(j)(13)	3
Exempt foods	Declaring net quantity of contents	101.105	
	Examples of net contents declaration	101.105(m)	
	Format of net contents declaration	101.105(h)	
	Location of net contents declaration	101.105(e)	
	Reasonable variation of net quantity of contents	101.105(q)	
	Weights and measures	101.105(b)	
Exemptions	Assortments	101.100(a)(1)	
	Banana clusters prior to retail sale	101.100(i)	
	Bulk food	101.100(a)(2)	
	Dietary supplements	101.36(f)	3, 8
	Fish fillets (wrapped) prior to retail sale	101.100(h)	
	Food repackaged in retail establishment	101.100(b)	3
	Food repacked at another establishment	101.100(d)(1)-(3)	3
	Food repacked at another establishment–conditions affecting exemption	101.100(e)	
	Foods served for immediate consumption	101.9(j)(2)(ii)	3
	Incidental additives	101.100(a)(3)	
	Low-volume products	101.9(j)(18)	3
	Markers without label declaration	101.100(g)	
	Nutrient content claims	101.13(q)	5
	Open containers	101.100(c)	
	Processed cheese	101.100(f)	
	Restaurant foods	101.9(j)(2)(i)	3
	Small businesses	101.9(j)(1), (j)(18)	3
	Soft drinks	101.2(c)(4)	
	Specific foods exempted	101.100	
	Sulfiting agents	101.100(a)(4)	
	Temporary–authorized food labeling experiments	101.108	
	Type size requirements	101.2(c)(1)-(3)	3
Exemptions and special requirements		101.9(j)	
	Petitions requesting	101.103	
Exemptions from labeling requirements		Part 101, Subpart G	
Exemptions or special requirements for ingredients	Petitions requesting	101.103	
Experiments (food labeling)	Temporary exemptions	101.108	
Extra	Nutrient content claims	101.54	5
Extra lean	Use of the term	101.62(e)	5

	Regs in Chapter 9	Review Chap #

FDA Rules—			
Fabricated foods	With D-erythroascorbic acid	101.33	
Fat (see Cholesterol, Saturated fat, etc)			
Fat content claims	Cholesterol claims	101.62(d)	5
	Fat free	101.62(b)(1)	5
	Fatty acid claims	101.62(c)	5
	Lean or extra lean	101.62(e)	5
	Low fat	101.62(b)(2)	5
	Low in saturated fat	101.62(c)(2)	5
	No saturated fat	101.62(c)(1)	5
	Reduced fat	101.62(b)(4)	5
	Reduced saturated fat	101.62(c)(4)	5
	Substitutes		5
Fat free	Definition	101.62(b)(1)	5
Fat free claims	Percentage fat free	101.13(i)(4); 101.62(b)(6)	5
Fat, monounsaturated	Voluntary declaration	101.9(c)(2)(iii)	4
Fat, polyunsaturated	Voluntary declaration	101.9(c)(2)(ii)	4
Fat, saturated	Declaration of nutrient information	101.9(c)(2)(i)	4
Fat, total	Declaration of nutrient information	101.9(c)(2)	4
Fatty acid content	Nutrient content claims	101.62(c)	4
FDA Modernization Act	Nutrient content and health claims		5, 6
Fiber	Health claims	101.76; 101.77	6
	Nutrient content claims	101.54(d)	5
Fiber, dietary	Declaration of nutrient information	101.9(c)(6)(i)	4
Fiber, insoluble	Voluntary declaration	101.9(c)(6)(i)(B)	4
Fiber, soluble	Voluntary declaration	101.9(c)(6)(i)(A)	4
Fiber-containing grain products, fruits, and vegetables/cancer	Criteria for claim	101.76(c)(2)(i)	
	Health claims	101.76	6
	Model claim	101.76(e)	6
	Nature of food making claim	101.76(c)(2)(ii)	6
	Relationship between fat, fiber, and cancer	101.76(a), (b)	6
Fiber-containing grain products, fruits, and vegetables/heart disease	Criteria for claim	101.77(c)(2)(i)	6
	Health claims	101.77	6
	Model claim	101.77(e)	6
	Nature of food making claim	101.77(c)(2)(ii)	6
	Relationship between fat, fiber, and coronary heart disease	101.77(a), (b)	6
Fish fillets (wrapped) prior to retail sale	Exemptions	101.100(h)	

FDA RULES—		Regs in Chapter 9	Review Chap #
Fish or game meat	Exceptions	101.9(j)(11), (12)	3
Fish, raw		101.9(j)(10); 101.42; 101.45	3
Flavor enhancers	Protein hydrolysates	101.22(h)(7)	
Flavorings	Ingredients statement	101.22(h)	
	Juice (minor amounts)	101.30(c)	
	Shipments to processor	101.22(g)	
	Supplier certifications	101.22(i)(4)	
Folate and neural tube defects	Criteria for claim	101.79(c)(2)(i)	6
	Health claims	101.79	6
	Model claim with examples	101.79(d)	6
	Nature of food making claim	101.79(c)(2)(ii)	6
	Relationship between folate and defects	101.79(a), (b)	6
Font size requirements (See Type size)			
Food labeling experiments	Proposal to FDA	101.108(b)	
	Temporary exemptions	101.108	
Food repackaged in retail establishment	Exemptions	101.100(b)	3
	Exemptions	101.100(d)(1)-(3)	
Food repacked at another establishment–conditions affecting exemption	Exemptions	101.100(e)	3
Foods exempted from labeling		101.100	3
Foods for infants and children	Exceptions	101.9(j)(5)	3
Foods for special dietary use		Part 105	
	Criteria	101.9(a)(4)	
Foods for use by children	Warnings and notices	101.17(a)(2)	
Foods labeled for use in multiple forms	Nutrition information	101.9(e)	3
Foods not in package form		101.9(a)(2)	3
Foods served for immediate consumption	Exemptions	101.9(j)(2)(ii)	3
Foods shipped in bulk form	Exceptions	101.9(j)(9)	3
Foods sold in bulk form	Exceptions	101.9(j)(16)	3
Foods with added flavor	Ingredients statement	101.22(h)	
Foods with separately packaged ingredients	Nutrition information	101.9(h)	3
Footnotes	Percent daily values	101.9(d)(9)	3
Foreign language	Nutrition information	101.9(d)(14); 101.15(c)	
Format	Nutrition information	101.9(d)	3
	Prominence	101.15	
	Sample label	101.9(d)(12), Part 101 Appendix B	2, 3

FDA Rules—		Regs in Chapter 9	Review Chap #
Fortified	Nutrient content claims	101.54(e)	
Fresh	Definition	101.95(a)	7
Fresh, freshly frozen, fresh frozen, or frozen fresh	Labeling requirements	101.95	7
	Provisions for use of the term	101.95(c)	7
Freshly frozen, fresh frozen, or frozen fresh	Definition	101.95(b)	7
Fruit juice	Percentage declarations	101.30	
Fruit, raw		101.9(j)(10); 101.42; 101.45	3
Fruits	Health claims	101.76; 101.77; 101.78	6
Fruits and vegetables/cancer	Criteria for claim	101.78(c)(2)(i)	6
	Health claims	101.78	6
	Model claim	101.78(e)	6
	Nature of food making claim	101.78(c)(2)(ii)	6
	Relationship between fat and cancer	101.78(a), (b)	6
Fruits, vegetables	Chemical preservatives	101.22(f)	
Fruits, vegetables, and grain products that contain fiber/cancer	Health claims	101.76	6
Fruits, vegetables, and grain products that contain fiber/coronary heart disease	Health claims	101.77	6
Game meat	Healthy claims	101.65(d)(3)	7
Game meat or fish		101.9(j)(11), (12); 101.65(d)(3)	3, 7
General definition	Nutrient content claims on products with a standard of identity	130.10	5, 7
Glass containers	Warnings and notices	101.17(a)(3)	
Good source	Nutrient content claims	101.54(c)	5
Grains or grain products	Health claims	101.76; 101.77	
Guidelines	Voluntary–raw fruit, vegetables, and fish	101.45	3
Health claims	Antioxidant vitamins and cancer	101.71(c)	6
	Applicability of rules	101.14(g)	6
	Calcium and osteoporosis	101.72	6
	Claims not authorized	101.71	6
	Definition	101.14(a)(1)	6
	Dental caries and sugar alcohols	101.80	6
	Dietary fiber and cancer	101.71(a)	6
	Dietary fiber and cardiovascular disease	101.71(b)	6
	Dietary lipids and cancer	101.73	6
	Dietary supplement–definition	101.14(a)(3)	6, 8
	Disease or health-related condition	101.14(a)(6)	6

		Regs in Chapter 9	Review Chap #
FDA RULES—			
	Disqualifying nutrient levels	101.14(a)(5)	6
	Eligibility	101.14(b)	6
	Fat and cholesterol	101.75	6
	FDA adoption	101.14(d)(2)	6
	Fiber-containing grain products, fruits, and vegetables/cancer	101.76	6
	Fiber-containing grain products, fruits, and vegetables/heart disease	101.77	6
	Folate and neural tube defects	101.79	6
	Fruits and vegetables/cancer	101.78	6
	Fruits/cancer	101.76 and 101.78	6
	Fruits/heart disease	101.77	6
	General requirements	101.14	6
	Grains/cancer	101.76	6
	Grains/heart disease	101.77	6
	Implied claims	101.14(a)(1)	6
	Inappropriate substance levels	101.14(e)(4)	6
	Infant formulas	101.14(f)(1)	6
	Jelly bean rule	101.14(e)(6)	6
	Medical foods	101.14(f)(2)	6
	Nutritive value–definition	101.14(a)(3)	6
	Oats/heart disease	101.81	6
	Petitioning FDA to issue regulations	101.70(a)	6
	Petitions–agency action	101.70(j)	
	Petitions–format	101.70(f)	
	Petitions–public disclosure	101.70(e)	
	Petitions–time frame	101.70(j)	6
	Prohibitions	101.14(e); 101.71	
	Psyllium/heart disease	101.81	6
	RACC	101.12(g)	3
	Saturated fat and cholesterol/coronary heart disease	101.75	6
	Sodium and hypertension	101.74	6
	Substance–definition	101.14(a)(2)	
	Third-party references	101.14(a)(1)	
	Unauthorized claims	101.71	6
	Validity	101.14(c)	
	Vegetables/cancer	101.76;101.78	6
	Vegetables/heart disease	101.77	6
	Zinc and immune function in the elderly	101.71(d)	6
Healthy	Criteria for use of the term	101.65(d)(2)	7

		Regs in Chapter 9	Review Chap #
FDA Rules—			
Healthy claims	Enriched cereal grain products: exemption	101.65(d)(2)(iv)	7
	Fruits and vegetables: exemption for raw, canned, or frozen	101.65(d)(2)(iv)	7
	Implied nutrient content claims	101.65(d)(2)	7
	On main dish products	101.65(d)(4)	7
	On raw, single ingredient game meat	101.65(d)(3)	7
	On raw, single ingredient seafood	101.65(d)(3)	7
	10% nutrient contribution	101.65(d)	7
Heart disease	Health claims–fiber-containing fruits, vegetables, and grain products	101.77	6
	Health claims–saturated fat and cholesterol	101.75	6
	Health claims–oats and soluble fiber	101.81	6
	Health claims–psyllium and soluble fiber	101.81	6
High	Nutrient content claims	101.54(b)	5
High blood pressure	Health claims–sodium	101.74	6
High potency	Nutrient content claims	101.54(f)	8
Hypertension	Health claims–sodium	101.74	6
Hypoallergenic foods	Label statements	105.62(a)-(c)	
Identity labeling	Common or usual names	101.3(b)	
	Foods with optional forms	101.3(c)	
	Imitations	101.3(e)	3
	Package form food	101.3	
	PDP	101.3(a), (d)	3
Imitations	Labeling rules	101.3(e); 101.13(d)	
Implied claims	Criteria for use	101.65(a)	7
	Exceptions	101.65(b)(1)-(6)	7
	General claims accompanying explicit claims	101.65(d)	7
	Health (disease prevention)	101.14(a)(1)	6
	Healthy–use of the term	101.65(d)	7
	Nutrient content	101.13(b); 101.65	5
	Specific implied claims	101.65(c)	7
	Statements that are not claims	101.65(b)(1)-(6)	7
Incidental additive	Exemptions	101.100(a)(3)	
	Ingredients statement	101.22(h)(2)	
Infant foods	Label statements	105.65(a)	
Infant formula	Exceptions	101.9(j)(7)	
	Health claims	101.14(f)(1)	6
	Labeling exemptions	107.30	3
Inferior products		101.3(e)(2); 104.20	

I.10   INDEX OF FDA RULES

		Regs in Chapter 9	Review Chap #
FDA Rules—			
Information panel	Definition	101.2(a)	3
	Exceptions–package with alternate PDP	101.2(a)(2); 101.9(j)(17)	3
	Exceptions–PDP on container top	101.2(a)(3)	3
	Exceptions–too small	101.2(a)(1); 101.9(j)(13), (17)	3
	Ingredients statement	101.4(a)	
	Package form food	101.2	
	Percentage juice declarations	101.30(e), (g)	
Information requirements	Insufficient space	101.2(d); 101.9(j)(13), (j)(17)	
		101.2(d)(1)	
Ingredient declarations	Added water	101.4(c)	
	Chemical preservative	101.22(j)	
	Coloring	101.22(k)	
	Exceptions to specific ingredient names	101.4(b)(1)-(22)	
	Exemptions	101.100; 101.4(a)(1)	3
	Ingredients containing other ingredients	101.4(b)(2)	
	Label designations for standardized foods	130.11	
	Labeling rules	101.4	3
	Nondairy characterization	101.4(d)	
	Percentages	101.4(e)	
	Specific names	101.4(b)	
Insignificant amounts	Definition	101.9(f)(1)	3, 4
	Labeling exemptions	101.9(j)(4)	3, 4
	Simplified information format	101.9(f)(2)	3
Insufficient space for information	General provisions	101.2(d); 101.9(j)(13), (17)	3
	Loss of exemption	101.15(b)	
	Nutrition information	101.9(d)(11)	3
	Petition for acceptable alternative	101.2(f); 102.19(b)	
Jelly bean rule	Health claims	101.14(e)(6)	6
	"Healthy" implied claims	101.65(d)	7
	10% nutrient contribution	101.14(e)(6)	6
Juice (fruit or vegetable)	As flavoring only	101.30(c)	
	Brix values	101.30(h)(1)	
	Expressed	101.30(j)	
	Labeling restrictions	101.30(l)	
	Modified properties	101.30(k)	
	Percentage declarations	101.30	
Kosher and kosher-style foods	Use of term	101.29	
Label format	Sample label	101.9(d)(12)	2, 3

FDA Rules—		Regs in Chapter 9	Review Chap #
Labeling exemptions	Infant formula	107.30	
Lean	Use of the term	101.62(e)	5
Light or lite	Nutrient content claims	101.13(j); 101.56(b)	5
Linear display	Sample label	101.9(j)(13)(ii)(A)	3
Lipids (see Dietary lipids)			
Lot	Definition for food labeling	101.9(g)(1)	4
Low calorie foods	Label terms	105.66(e)	
	Special dietary use	105.66(c)	
Low cholesterol	Definition	101.62(d)(2)	5
Low fat	Definition	101.62(b)(2)	5
Low in saturated fat	Definition	101.62(c)(2)	5
Low sodium	Definition	101.61(b)(4)	5
Low volume	Exemption	101.9(j)(18)	3
Low-calorie, low in calories	Definition	101.60(b)(2)	5
Main dish	Definition	101.13(m)	5
Manufacturer	Name and place of business of manufacturer, packer, or distributor	101.5(a)	3
Margarine	Food standards; labeling	166.40	
Markers without label declaration	Exemptions	101.100(g)	
Meal product	Definition	101.13(l)	5
Medical foods	Exceptions	101.9(j)(8)	3
	Health claims	101.14(f)(2)	6
Metric quantity	Requirements	101.9(b)(7), (g)(7)	
Minerals (see Vitamins and minerals)			
Misbranding	Characterizing ingredient	101.22(i)	
	Chemical preservative	101.22(j)	
	Claims	101.9(k)(1)	5, 6, 7
	Coloring	101.22(k)	
	Criteria	101.9(g)(4), (g)(5); 101.18	
	Dietary supplements	101.9(k); 101.36(h)	8
	False or misleading representations	101.18(a)	
	Geographical origin	101.18(c)	
	Health claims	101.9(k)(1)	6
	Imitations	101.3(e)	
	Ingredient information	101.18(b)	
Model health claims	Suggested language for each claim	101.72(e)–101.81(e)	Table 6.3
Modernization Act of 1997	Nutrient content and health claims		5, 6
Monier-Williams procedure	Sulfites	Part 101, Appendix A	
Monosodium glutamate	Ingredients statement	101.22(h)(5)	
Monounsaturated fat	Voluntary declaration	101.9(c)(2)(iii)	4
More	Nutrient content claims	101.54(e)	5
Multiple forms of same food	Nutrition information	101.9(e)	3

		Regs in Chapter 9	Review Chap #
FDA RULES—			
Multiple ingredients packaged separately	Nutrition information	101.9(h)	3
Multiunit retail packages	Exceptions	101.9(j)(15)	3
Name of business	Label declaration required	101.5(b)	3
Natural flavor(ing)	Definition	101.22(a)(3)	
Net quantity of contents	For exempt foods	101.105(a)	
	Multiunit retail package–exempt foods	101.105(s)	3
	Reasonable variation for exempt foods	101.105(q)	
	Weights and measures for exempt foods	101.105(b)	
No saturated fat	Definition	101.62(c)(1)	5
Nonconformity with standard of identity	Named instead by nutrient content claim and a standardized term	101.13(d); 130.10(a)	5
Number of servings	Determining number per container	101.9(b)(8)	3
	Labeling with net quantity of servings	101.8(a)	3
Nutrient content claims	Added	101.54(e)	5
	Added, fortified, or enriched	101.13(j)	5
	Amount of a nutrient	101.13(i)	5
	Antioxidants	101.54(g)	8
	Brand names–exceptions	101.13(q)	
	Butter	101.67	
	Calories or sugar	101.60	5
	Cholesterol content	101.62(d)	5
	Claims of "free" or "low"	101.13(e)	5
	Enriched	101.54(e)	5
	Excellent source of	101.54(b)	5
	Exceptions to location	101.13(h)	
	Exemptions	101.13(q)	5
	Expressed claims	101.13(b)(1)	5
	Extra	101.54	5
	Fat content	101.62(b)	5
	Fat, fatty acid, and cholesterol	101.62	5
	Fat substitutes		5
	Fatty acid content	101.62(c)	5
	Fiber	101.54(d)	5
	Foods named by a nutrient content claim and a standardized term	130.10	5
	Fortification	101.54(e)(2)(ii); 104.20	
	Fortified	101.54(e)	5
	General definition for expressed nutrient content claims on products with a standard of identity	130.10	5, 7
	General principles	101.13	5

INDEX OF FDA RULES   I.13

FDA Rules—

		Regs in Chapter 9	Review Chap #
	Good source, high, and more	101.54	5
	Healthy	101.65(d)	7
	High	101.54(b)	5
	High potency	101.54(f)	8
	Imitations	101.13(d)	5
	Implied claims	101.13(b)(2); 101.65	5
	Less (or fewer)	101.13(j)	5
	Light or lite	101.13(j); 101.56(b)	5
	Location–referral statement	101.13(g)	5
	Low calorie	101.60(b)(2)	5
	Main dish product	101.13(m)	5
	Meal product	101.13(l)	5
	Modified	101.13(k)	5
	More	101.13(j); 101.54(e)	5
	No special processing	101.60(b)(2)(ii); 101.61(b)(2)(ii), 101.61(b)(4)(ii)	5
	Percentage fat free	101.13(i)(4); 101.62(b)(6)	5
	Percentage of a nutrient	101.13(i)	5
	Petitions	101.69(a)	
	Petitions–implied claim in brand name	101.69(o)(1)	
	Petitions–new	101.69(m)(1)	5
	Petitions–public disclosure	101.69(g)	
	Petitions–synonymous term	101.69(n)(1)	
	Plus	101.54	5
	RACC	101.12(g)	3
	Reduced	101.13(j)	5
	Reference foods	101.13(j)(1)	5
	Relative claims	101.13(j)	5
	Required information	101.13(c)	5
	Rich in	101.54(b)	5
	Sodium content	101.61	5
	Substitutes	101.13(d)	5
	Synonyms	101.13	5
	Type size	101.13(f)	3, 5
Nutrients	Class I and II	101.9(g)(3)	4
Nutrition facts	Heading required	101.9(d)(2)	3
	Sample label	101.9(d)(12)	2, 3
Nutrition information	Abbreviations	101.9(d)(9)(iv)	3
	Aggregate display	101.9(d)(13)(ii)	3
	As prepared	101.9(e)(1)	3
	As purchased	101.9(e)(1)	3
	Caloric conversions	101.9(d)(10)	3

		Regs in Chapter 9	Review Chap #
FDA RULES—			
	Class I and II nutrients	101.9(g)(4)	4
	Declaration–basis for	101.9(b)(9)	4
	Dietary supplements	101.36(b)	8
	Dual labeling	101.9(e)(2)	3
	Establishing compliance	101.9(g)(4)	4
	Exemptions	101.9(j)	3
	Footnotes	101.9(d)(9)	3
	Format	101.9(d)	3
	Format–simplified	101.9(f)	3
	Insignificant amounts	101.9(f)	3, 4
	Insufficient space for information	101.9(d)(11)	3
	Insufficient vertical space	101.9(d)(11)(iii)	3
	Location	101.2; 101.9(i)	3
	Multiple ingredients packaged separately	101.9(h)	3
	On basis of food as consumed	101.9(h)(4)	3
	Raw fruit, vegetables, and fish	101.42-101.45	3
	Required nutrients	101.9(c)	4
	Sample label	101.9(d)(12)	2, 3
	Second language	101.9(d)(14)	
	Serving size	101.9(d)(3)(i)	3
	Servings per container	101.9(d)(3)(ii)	3
	Solicitation of requests for information	101.9(a)(3)	
	Special requirements	101.9(j)	
	Variety packs	101.9(b)(4), (d)(13), (h)(3)	3
	Vitamins and minerals	101.9(d)(8)	4
	Voluntary declarations	101.9(c)	4
Nutrition labeling of food	General provisions	101.9	1, 3
Nutritional inferiority		101.3(e)(2); 104.20	5
Oats	Health claims–heart disease	101.81	6
Open containers	Exemptions	101.100(c)	
Osteoporosis	Health claims–calcium	101.72	6
Other carbohydrate	Voluntary declaration	101.9(c)(6)(iv)	4
Oysters (canned)–declaration of quantity of contents	Food standards	161.30	
Package form food		101.9(a)(1)	3
Packer	Name used on labeling	101.5(c)	
Percent daily value	Based on DRV or RDI	101.9(c)(8)-(9)	4
	Dietary supplements	101.36(b)(4)	4, 8
	Subheading required	101.9(d)(6)	3
Percentage fat free	Claims	101.13(i)(4); 101.62(b)(6)	5
Percentage juice declaration	Beverages containing juice	101.30(b)	

		Regs in Chapter 9	Review Chap #
FDA Rules—			
Pesticides	Fruits and vegetables	101.22(f)	
Petitions	Health claims	101.70	6
	Health claims–agency action	101.70(j)	
	Health claims–format	101.70(f)	
	Health claims–public disclosure	101.70(e)	
	Nutrient content claims	101.69	5
	Nutrient content claims–format	101.69(i)	
	Nutrient content claims–implied claim in brand name	101.69(o)(1)	
	Nutrient content claims–new	101.69(m)(1)	
	Nutrient content claims–public disclosure	101.69(g)	
	Nutrient content claims–synonymous term	101.69(n)(1)	
	Petitioning FDA to issue health claim rules	101.70(a)	
	RACC	101.12(h)	
	Requesting exemptions and special requirements	101.103	
	Time frame for health claims	101.70(j)	6
Plus	Nutrient content claims	101.54	5
Polyunsaturated fat	Voluntary declaration	101.9(c)(2)(ii)	4
Potassium	Voluntary declaration	101.9(c)(5); 101.9(d)(9)(iii)	4
Pressurized containers	Warnings and notices	101.17	
Principal display panel (PDP)	Alternate panels	101.1	3
	Area	101.1	3
	Area–cylindrical package	101.1(b)	
	Area–other package shapes	101.1(c)	
	Area–rectangular package	101.1(a)	
	Information requirements	101.2(d); 101.3(a)	
	Insufficient space for information	101.2(f); 101.9(j)(17); 102.19(b)	3
	Package form food	101.1	3
	Package lids as PDPs	101.2(d)(2)	3
	Percentage juice declarations	101.30(f)	
Processed cheese	Exemptions	101.100(f)	
Prominence of required statements		101.15	
Protein	Declaration of nutrient information	101.9(c)(7)	4
	DRV or RDI	101.9(c)(7)(iii)	4
Protein hydrolysates	Flavor enhancers	101.22(h)(7); 102.22	
Protein products	Warnings and notices	101.17(d)	
Protein, corrected amount	Declaration of nutrient information	101.9(c)(7)(ii)	4

FDA RULES—		Regs in Chapter 9	Review Chap #
Psyllium	Health claims–heart disease	101.81	6
Puerto Rico	Labeling exemptions	101.15(c)	
RACC		101.12(b)	3
	Baking powder		3
	Candies		3
	Salt and substitutes	101.12(b)	3
	Promoted use differs from RACC	101.9(b)(11)	3
Raw fruit, vegetables, and fish	General provisions	101.9(j)(10); 101.42; 101.45	3
	Identification of the 20 most frequently consumed products	101.44	3
	Substantial compliance with voluntary guidelines	101.43	3
	Voluntary guidelines	101.42; 101.45	3
RDI (see Reference daily intake)			
Reduced calorie	Definition	101.60(b)(4)	5
Reduced calorie foods	Label terms	105.66(e)	5
	Special dietary use	105.66(d)	
Reduced cholesterol	Definition	101.62(d)(4)	5
Reduced fat	Definition	101.62(b)(4)	5
Reduced saturated fat	Definition	101.62(c)(4)	5
Reduced sodium	Definition	101.61(b)(6)	5
Reference amounts customarily consumed (RACC)	Aerated foods	101.12(e)	3
	Basis for nutrient and health claims	101.12(g)	3, 5, 6
	FDA calculations methods	101.12(a)	
	Imitations or substitutes	101.12(d)	
	Multiple-food products with no RACC	101.12(f)	3
	Petitions to establish or modify	101.12(h)	
	Unprepared foods	101.12(c)	
	Values	101.12(b)	
Reference daily intake (RDI)		101.9(c)(8)(iv)	4
Reference foods	Nutrient content claims	101.13(j)(1)	5
Relative claims	Nutrient content	101.13(j)	5
Representations	Flavoring	101.22(i)	
Required information	Prominence	101.15(a)	
Required nutrients	Label declaration	101.9(c)	4
Restaurant foods		101.10	3
	Labeling exemptions	101.9(j)(2)(i)	3
Rich in	Nutrient content claims	101.54(b)	5
Saccharin	Exceptions	101.11(c)	
	Retail establishment notice	101.11(a)	

FDA RULES—		Regs in Chapter 9	Review Chap #
Salt	Ingredients statement	101.22(h)(4)	
	RACC	101.12(b)	3
Salt free	Nutrient content claims	101.61(c)(1)	5
Sample labels	Aggregate display	101.9(d)(13)(ii)	
	Dietary supplements	101.36(c)(9)	8
	Dual labeling	101.9(e)(5)	3
	Enhancements used by FDA	Part 101, Appendix B	3
	Exceptions	101.9(j)(13)(ii)(A)	3
	Insufficient vertical space	101.9(d)(11)(iii)	3
	Linear or tabular display	101.9(j)(13)(ii)(A)	3
	Nutrition information	101.9(d)(12)	2, 3
Saturated fat	Declaration of nutrient information	101.9(c)(2)(i)	4
Saturated fat and cholesterol/coronary heart disease	Criteria for claim	101.75(c)(2)(i)	6
	Health claims	101.75	6
	Model claim	101.75(e)	6
	Nature of food making claim	101.75(c)(2)(ii)	6
	Relationship between fat and heart disease	101.75(a), (b)	6
Seafood	Healthy claims	101.65(d)(3)	7
Second language	Nutrition information	101.9(d)(14); 101.15(c)	
Self-pressurized containers	Warnings and notices	101.17	
Separately packaged ingredients	Nutrition information	101.9(h)	
Serving size	Common household measure	101.9(b)(5)	3
	General	101.8; 101.9(b)	3
	Label declaration	101.9(b)(7)	3
	Meals, main dishes	101.9(b)(3)	3
	Metric quantity	101.9(b)(7)	3
	Nutrition information	101.9(d)(3)(i)	3
	RACC (Reference amount customarily consumed)	101.9(b)(2); 101.12(b)	3
	Single serving containers	101.9(b)(2)(i)(I); 101.9(b)(6)	3
	Special cases–discrete units	101.9(b)(2)(i)	3
	Special cases–large discrete units	101.9(b)(2)(ii)	3
	Special cases–nondiscrete bulk products	101.9(b)(2)(iii)	3
	Variety packs	101.9(b)(4)	3
	Voluntary product standard	101.8(b)	
Servings per container	Label declaration	101.9(d)(3)(ii)	3
Simplified format	Nutrition information	101.9(f)	3

FDA RULES—		Regs in Chapter 9	Review Chap #
Single serving container	Criteria	101.9(b)(2)(i)(I); 101.9(b)(6)	
Small businesses	Exemptions	101.9(j)(1)(i), (j)(18)	3
Small packages	Exceptions	101.9(j)(13)	3
Smoke flavoring	Artificial flavors	101.22(h)(6)	
Sodium	Declaration of nutrient information	101.9(c)(4)	4
Sodium (reducing intake)	Foods for special dietary use	105.67	
Sodium and hypertension	Criteria for claim	101.74(c)(2)(i)	6
	Health claims	101.74	6
	Model claim	101.74(e)	6
	Nature of food making claim	101.74(c)(2)(ii)	6
	Relationship between sodium and hypertension	101.74(a), (b)	6
Sodium content claims	Low sodium	101.61(b)(4)	5
	Nutrient content claims	101.61(b)	5
	Reduced sodium	101.61(b)(6)	5
	Specific terms on labeling	101.61(b), (c)	5
	Unsalted	101.61(c)(3)	5, 7
	Very low sodium	101.61(b)(2)	5
Soft drinks	Bottled manufactured before 10/31/75	101.2(c)(4)(i)	3
	Labeling exemptions	101.2(c)(4)(ii)	
Solicitation of requests for information		101.9(a)(3)	
Soluble fiber (see also Dietary fiber)			
Soluble fiber	Voluntary declaration	101.9(c)(6)(i)(A)	4
Special dietary use	Criteria	101.9(a)(4)	3
Special requirements		101.9(j)	3
Specific food labeling requirements		Part 101, Subpart B	
	D-erythroascorbic acid	101.33	
	Juice (fruit or vegetable)	101.30	
	Kosher and kosher-style foods	101.29	
	Spices, flavorings, colorings, chemical preservatives	101.22	
Specific nutrition labeling requirements and guidelines		Part 101, Subpart C	3, 4
Specific requirements for claims that are neither nutrient content nor health claims		Part 101, Subpart F	7
Specific requirements for health claims		Part 101, Subpart E	6
Specific requirements for nutrient content claims		Part 101, Subpart D	5
Spices		101.22	
	Definition	101.22(a)(2)	
Standard of identity	General definition for expressed nutrient content claims	130.10	5, 7

		Regs in Chapter 9	Review Chap #
FDA RULES—			
Structure/function statements	Dietary supplements	101.14(a)(6)	8
	Examples of allowed statements		8
Substitutes		101.12(d); 101.13(d)	5
Sugar	Declaration of nutrient information	101.9(c)(6)(ii)	4
	Nutrient content claims	101.60(c)	5
Sugar alcohol	Voluntary declaration	101.9(c)(6)(iii)	3
	Health claims-tooth decay	101.80	6
Sugarless claims	Criteria	101.60(c)	5
Sulfiting agents	Exemptions	101.100(a)(4)	
	Monier-Williams procedure	Part 101, Appendix A	
Tabular display	Sample label	101.9(j)(13)(ii)(A)	3
Temporary exemptions	Authorized food labeling experiments	101.108	
	Authorized food labeling experiments	101.108	
Third-party references (see Health claims)			
Tooth decay	Health claims-sugar alcohols	101.80	6
Type size requirements	Exemptions	101.2(c)(1)-(3)	3
	Principal display and information panels	101.2(c)	3
	Sample label	Part 101 Appendix B	3
Variety packs	Nutrition information	101.9(b)(4), (d)(13), (h)(3)	3
Vegetable juice	Percentage declarations	101.30	
Vegetables	Health claims	101.76; 101.77; 101.78	6
Vegetables, raw		101.9(j)(10); 101.42; 101.45	3
Very low in sodium	Definition	101.61(b)(2)	5
Vitamins and minerals	Declaration of nutrient information	101.9(c)(8)	4, 8
	DRVs	101.9(c)(9)	4
	Percent daily value or % RDI	101.9(c)(8)(i), (ii)	4
	RDIs	101.9(c)(8)(iv)	4
	Synonyms	101.9(c)(8)(v)	4
Voluntary declarations	Nutrition information	101.9(c)	4
Voluntary labeling	Raw fruit, vegetables, and fish	101.9(j)(10); 101.42-45, Part 101 - Appx C and D	3
	Substantial compliance	101.43	3
Warnings and notices	Food labeling statements	101.17	
Zinc and immune function in the elderly	Health claims	101.71(d)	6

[End of entries for FDA regulations.]

		Regs in Chap 10, 11	Review Chap#
FSIS MEAT PRODUCT RULES—			
Calorie content	Nutrient content claims	317.360	5
Calorie free	Nutrient content claims	317.360(b)(1)	5
Calories declaration	Nutrition labeling	317.309(c)(1)	4
Carbohydrate content	Nutrition labeling	317.309(c)(6)	4
Cholesterol content	Nutrition labeling	317.309(c)(3)	4
Cholesterol free	Nutrient content claims	317.362(d)	5
Daily reference value (DRV)	Nutrition labeling	317.309(c)(9)	4
Definitions			
	Label	301.2(oo)	1, 3
	Labeling	301.2(pp)	1, 3
Dietary fiber	Nutrition labeling	317.309(c)(6)(i)	4
Exceptions, exemptions	Advance labeling approval	317.3	
	Generically approved labeling	317.5	
	Labeling formats for meat products	317.309(b)	3
	Meat product exemptions	317.300, 317.400	3
Fat declaration	Nutrition labeling	317.309(c)(2)	4
Fat free	Nutrient content claims	317.362(b)(1)	5
Fat, fatty acids, and cholesterol	Nutrient content claims	317.362	5
Fiber	Nutrient content claims	317.354(d)	5
Generically approved labeling		317.5	
Good source	Nutrient content claims	317.354(c)	5
Good source, high, and more	Nutrient content claims	317.354	5
Healthy claims	Nutrient content claims	317.363	5, 7
High in	Nutrient content claims	317.354(b)	5
Information panel	General requirements	317.2(d)	
	Nutrition labeling formats	317.309	3
Ingredients statement		317.2(c), (f)	
Labeling	Approval, approved	317.4-6	3
	Cured products	317.17(a,c)	
Labeling applications	Nutrition labels	317.369	3
Labels	Abbreviations	317.3(a)	
	Definition	317.2(a)	
	Obsolete	317.14	
	Requirements	317.1	
	Required features	317.2(b-m)	
	Standard requirements	Part 317, Subpart A	
Lean or extra lean	Nutrient content claims	317.362(e)	5
Light or lite	Nutrient content claims	317.356	5
Low fat	Nutrient content claims	317.362(b)(2)	5
Low in calories	Nutrient content claims	317.360(b)(3)	5
More	Nutrient content claims	317.354(e)	5
Name of product	On PDP	317.2(c), (e)	

INDEX OF FSIS RULES

Regs in Chap 10, 11   Review Chap#

FSIS Meat Product Rules—			
Number of servings		317.308	3
Nutrient content claims	Calorie content	317.360	5
	Calorie free	317.360(b)(1)	5
	Cholesterol free	317.362(d)	5
	Fat free	317.362(b)(1)	5
	Fat, fatty acids, and cholesterol	317.362	5
	Fiber	317.354(d)	5
	General principles	317.313	5
	Good source	317.354(c)	5
	Good source, high, and more	317.354	5
	Healthy	317.363	7
	High in	317.354(b)	5
	Labeling applications	317.369	5
	Lean or extra lean	317.362(e)	5
	Light or lite	317.356	5
	Low fat	317.362(b)(2)	5
	Low in calories	317.360(b)(3)	5
	More	317.354(e)	5
	Sodium	317.361	5
	Sodium free	317.361(b)(1)	5
	Statements relating to usefulness in reducing or maintaining body weight	317.380	
	Substitute product	317.313(d)	5
	Sugar free	317.360(c)(1)	5
Nutrition label content		317.309	3
Nutrition labeling	Alternate panel	317.302(c)	3
	Applicability	317.300, 317.400	3
	Caloric conversion	317.309(d)(10)	3
	Calories declaration	317.309(c)(1)	4
	Carbohydrate content	317.309(c)(6)	4
	Cholesterol content	317.309(c)(3)	4
	Daily reference value (DRV)	317.309(c)(9)	4
	Dietary fiber	317.309(c)(6)(i)	4
	Exemptions	317.400	3
	Fat declaration	317.309(c)(2)	4
	Format	317.309(d)	3
	Format–multiple products in one package	317.309(d)(13)	3
	Format–simplified	317.309(f)(2)	3
	Format–small packages	317.309(g)	3
	Format–vitamins and minerals	317.309(d)(8)	3
	General requirements	317.300	3
	Gift packs	317.302(b)	3

FSIS MEAT PRODUCT RULES—		Regs in Chap 10, 11	Review Chap#
	Identification of major cuts	317.344	3
	Label content	317.309	3
	Location of	317.302	3
	Meat products, general	317.300	
	Minerals and vitamins	317.309(c)(8)	4
	Multiple forms of same product	317.309(e)	3
	Multiple products in one package	317.309(d)(13)	3
	Number of servings	317.308	3
	PDP	317.302, 317.309	3
	Potassium content–voluntary	317.309(c)(5)	4
	Principal display panel	317.302, 317.309	3
	Protein content	317.309(c)(7)	4
	Reference amounts customarily consumed (RACC)	317.312(a)	3
	Reference daily intake (RDI)	317.309(c)(8)(iv)	4
	Sample labels	317.309(d)(12), (e)(5)	2, 3
	Second language	317.309(d)(14)	
	Serving size	317.309(b)	3
	Simplified format	317.309(f)(2)	3
	Single-ingredient, raw products	317.343, 317.345	3
	Small packages	317.309(g)	3
	Sodium content	317.309(c)(4)	4
	Standards	Part 317, Subpart B	
	Sugar content	317.309(c)(6)(ii)	4
	Vitamins and minerals	317.309(c)(8)	4
	Voluntary labeling – single- ingredient, raw products	317.343, 317.345	3
PDP	General requirements	317.2(d)	
	Nutrition labeling formats	317.309	3
Potassium content–voluntary	Nutrition labeling	317.309(c)(5)	4
Principal display panel	General requirements, nutrition labeling	317.2(d), 317.309	3
Protein content	Nutrition labeling	317.309(c)(7)	4
Reducing body weight	Nutrient content claims	317.380	5
Reference amounts customarily consumed (RACC)	Nutrition labeling	317.312(a)	3
Reference daily intake (RDI)	Nutrition labeling	317.309(c)(8)(iv)	4
Serving size	Nutrition labeling	317.309(b)(1)	3
Sodium content	Nutrient content claims	317.361	4
	Nutrition labeling	317.309(c)(4)	4
Sodium free	Nutrient content claims	317.361(b)(1)	5

FSIS MEAT PRODUCT RULES—		Regs in Chap 10, 11	Review Chap#
Statements relating to usefulness in reducing or maintaining body weight		317.380	
Sugar content	Nutrition labeling	317.309(c)(6)(ii)	4
Sugar free	Nutrient content claims	317.360(c)(1)	5
Vitamins and minerals	Nutrition labeling	317.309(c)(8)	4
Voluntary labeling – single-ingredient, raw products	Nutrition labeling	317.343, 317.345	3

[End of entries for meat product rules.]

		Regs in Chap 10, 11	Review Chap#
FSIS POULTRY PRODUCT RULES—			
Calorie content	Nutrient content claims	381.460	5
Calorie free	Nutrient content claims	381.460(b)(1)	5
Calories declaration	Nutrition labeling	381.409(c)(1)	4
Carbohydrate content	Nutrition labeling	381.409(c)(6)	4
Cholesterol content	Nutrition labeling	381.409(c)(3)	4
Cholesterol free	Nutrient content claims	381.462(d)	5
Daily reference value (DRV)	Nutrition labeling	381.409(c)(9)	4
Definitions			
	Label	381.1(b)(29)	1, 3
	Labeling	381.1(b)(30)	1, 3
Dietary fiber	Nutrition labeling	381.409(c)(6)(i)	4
Dietary food claims, labeling		381.124	
Fat declaration	Nutrition labeling	381.409(c)(2)	4
Exceptions, exemptions	Advance labeling approval	381.131	
	General labeling approval	381.133(b)	
	Labeling for interchange between two types of poultry	381.118(f)	
	Labeling formats	381.409(b)	3
	Poultry product exemptions	381.500	3
Fat free	Nutrient content claims	381.462(b)(1)	5
Fat, fatty acids, and cholesterol content	Nutrient content claims	381.462	5
Fiber	Nutrient content claims	381.454(d)	5
Fresh	Labeling	381.129(b)(6), 381.133	3
Frozen	Labeling	381.129(b)(6), 381.133	3
Generic labeling approval		381.133(b)	
Good source	Nutrient content claims	381.454(c)	5
Good source, high, and more	Nutrient content claims	381.454	5
Healthy claims	Nutrient content claims	381.463	7
High in	Nutrient content claims	381.454(b)	5
Information panel	General requirements	381.116	
	Nutrition labeling formats	381.409	3
Ingredients statement		381.118	
Label content		381.409	
	Area of the principal display panel	381.116(b)(3)	3
	Dietary food claims	381.124	
	Exemptions	381.121(c)(9)	3
	False or misleading labels	381.129 - 381.130	
	Information panel	381.116(c)(1), 381.118(e)	3
	Principal display panel	381.116, 381.121(c)(1)	3
	Type size	381.121(c)(3)	3
Labeling applications	Nutrient content claims	381.469	5

		Regs in Chap 10, 11	Review Chap#
FSIS POULTRY PRODUCT RULES—			
Lean or extra lean	Nutrient content claims	381.462(e)	5
Light, or lite	Nutrient content claims	381.456	5
Low fat	Nutrient content claims	381.462(b)(2)	5
Low in calories	Nutrient content claims	381.460(b)(3)	5
Minerals and vitamins	Nutrition labeling	381.409(c)(8)	4
More	Nutrient content claims	381.454(e)	5
Name of product		381.117	
Number of servings		381.408	3
Nutrient content claims	Calorie content	381.460	5
	Calorie free	381.460(b)(1)	5
	Cholesterol free	381.462(d)	5
	Fat free	381.462(b)(1)	5
	Fat, fatty acids, and cholesterol	381.462	5
	Fiber	381.454(d)	5
	General principles	381.413	5
	Good source	381.454(c)	5
	Good source, high, and more	381.454	5
	Healthy	381.463	7
	High in	381.454(b)	5
	Labeling applications	381.469	5
	Lean or extra lean	381.462(e)	5
	Light or lite	381.456	5
	Low fat	381.462(b)(2)	5
	Low in calories	381.460(b)(3)	5
	More	381.454(e)	5
	Sodium	381.461	5
	Sodium free	381.461(b)(1)	5
	Statements relating to usefulness in reducing or maintaining body weight	381.480	
	Substitute product	381.413(d)	5
	Sugar free	381.460(c)(1)	5
Nutrition label content		381.409	3
Nutrition labeling	Alternate panel	381.116, 381.402	3
	Applicability	381.400, 381.500	3
	Caloric conversion	381.409(d)(10)	4
	Calories declaration	381.409(c)(1)	4
	Carbohydrate content	381.409(c)(6)	4
	Cholesterol content	381.409(c)(3)	4
	Daily reference value (DRV)	381.409(c)(9)	4
	Dietary fiber	381.409(c)(6)(i)	4
	Exemptions	381.400, 381.500	3
	Fat declaration	381.409(c)(2)	4

FSIS POULTRY PRODUCT RULES—			
	Format	381.409(d)	3
	Format–multiple products in one package	381.409(d)(13)	3
	Format–simplified	381.409(f)(2)	3
	Format–small packages	381.409(g)	3
	Format–vitamins and minerals	381.409(d)(8)	3
	General requirements	381.400	3
	Gift packs	381.402(b)	3
	Identification of major cuts	381.444	3
	Label content	381.409	3
	Location of	381.402	3
	Minerals and vitamins	381.409(c)(8)	4
	Multiple forms of same product	381.409(e)	3
	Multiple products in one package	381.409(d)(13)	3
	Number of servings	381.408	3
	PDP	381.116, 381.409	3
	Potassium content–voluntary	381.409(c)(5)	4
	Poultry products, general	381.400	3
	Principal display panel	381.116, 381.409	3
	Protein content	381.409(c)(7)	4
	Reference amounts customarily consumed (RACC)	381.412(a)	3
	Reference daily intake (RDI)	381.409(c)(8)(iv)	4
	Sample labels	381.409(d)(12), (e)(5)	2, 3
	Second language	381.409(d)(14)	
	Serving size	381.409(b)	3
	Simplified format	381.409(f)(2)	3
	Single-ingredient, raw products	381.443, 381.445	3
	Small packages	381.409(g)	3
	Sodium content	381.409(c)(4)	4
	Sugar content	381.409(c)(6)(ii)	4
	Vitamins and minerals	381.409(c)(8)	4
	Voluntary labeling – single-ingredient, raw products	381.443, 381.445	3
Nutrition labeling, poultry products	Exemptions	381.500	3
	Gift packs	381.402(b)	3
	Identification of major cuts	381.444	3
	Label content	381.409	3
	Location	381.402	3
	Number of servings	381.408	3
	Nutrient content claims	381.413	5
	Reference amounts	381.412	3

FSIS POULTRY PRODUCT RULES—		Regs in Chap 10, 11	Review Chap#
	Standards–general	Part 381, Subpart Y	
	Voluntary–Raw, single-ingredient products	381.443, 381.445	3
PDP	General requirements	381.116	
	Nutrition labeling	381.409	3
Potassium content–voluntary	Nutrition labeling	381.409(c)(5)	4
Principal display panel	Area of	381.116(b)(3)	
	Information panel	381.116(c), 381.118(e)	
	Nutrition labeling information	381.409	3
	On a container of any other shape	381.116(b)(3)	3
	On a cylindrical package	381.116(b)(2)	
	On a rectangular package	381.116(b)(1)	
Raw, single-ingredient products	Voluntary labeling	381.445	3
Reducing or maintaining body weight	Nutrient content claims	381.480	5
Reference amounts customarily consumed (RACC)	Nutrition labeling	381.412(a)	3
Reference daily intake (RDI)	Nutrition labeling	381.409(c)(8)(iv)	4
Single-ingredient, raw products	Voluntary labeling	381.445	3
Sodium content	Nutrient content claims	381.461	5
	Nutrition labeling	381.409(c)(4)	4
Sodium free	Nutrient content claims	381.461(b)(1)	5
Sugar content	Nutrition labeling	381.409(c)(6)(ii)	4
Sugar free	Nutrient content claims	381.460(c)(1)	5
Two types of poultry	No label change required	381.118(f)	
Vitamins and minerals	Nutrition labeling	381.409(c)(8)	4
Voluntary labeling – single-ingredient, raw products	Nutrition labeling	381.443, 381.445	3

[End of entries for poultry product rules.]

ABOUT THE AUTHOR

Tracy Altman, Ph.D. (taltman@jetra.com) is the principal of a management consulting firm in Denver, Colorado, specializing in learning and knowledge transfer between industries and organizational units. Her academic background is in Chemical Engineering, Computer Science, and Policy Analysis. Current research interests involve measuring the impacts of regulatory policy on the private sector.